DESIGN AND DEVELOPMENT OF MEDICAL ELECTRONIC INSTRUMENTATION

DESIGN AND DEVELOPMENT OF MEDICAL ELECTRONIC INSTRUMENTATION

A Practical Perspective of the Design, Construction, and Test of Medical Devices

DAVID PRUTCHI
MICHAEL NORRIS

A JOHN WILEY & SONS, INC., PUBLICATION

Library of Congress Cataloging-in-Publication Data:

Prutchi, David.
    Design and development of medical electronic instrumentation: a practical perspective of
the design, construction, and test of material devices / David Prutchi, Michael Norris.
        p. cm.
    Includes bibliographical references and index.
    ISBN 0-471-67623-3 (cloth)
     1. Medical instruments and apparatus–Design and construction. I. Norris, Michael. II.
Title.

R856.P78 2004
681'.761–dc22

                                                                2004040853

Printed in the United States of America

10 9 8 7 6 5 4 3 2 1

In memory of Prof. Mircea Arcan,
who was a caring teacher, a true friend,
and a most compassionate human being.
—David

CONTENTS

PREFACE

The medical devices industry is booming. Growth in the industry has not stopped despite globally fluctuating economies. The main reason for this success is probably the self-sustaining nature of health care. In essence, the same technology that makes it possible for people to live longer engenders the need for more health-care technologies to enhance the quality of an extended lifetime. It comes as no surprise, then, that the demand for trained medical-device designers has increased tremendously over the past few years. Unfortunately, college courses and textbooks most often provide only a cursory view of the technology behind medical instrumentation. This book supplements the existing literature by providing background and examples of how medical instrumentation is actually designed and tested. Rather than delve into deep theoretical considerations, the book will walk you through the various practical aspects of implementing medical devices.

The projects presented in the book are truly unique. College-level books in the field of biomedical instrumentation present block-diagram views of equipment, and high-level hobby books restrict their scope to science-fair projects. In contrast, this book will help you discover the challenge and secrets of building practical electronic medical devices, giving you basic, tested blocks for the design and development of new instrumentation. The projects range from simple biopotential amplifiers all the way to a computer-controlled defibrillator. The circuits actually work, and the schematics are completely readable. The project descriptions are targeted to an audience that has an understanding of circuit design as well as experience in electronic prototype construction. You will understand all of the math if you are an electrical engineer who still remembers Laplace transforms, electromagnetic fields, and programming. However, the tested modular circuits and software are easy to combine into practical instrumentation even if you look at them as "black boxes" without digging into their theoretical basis. We will also assume that you have basic knowledge of physiology, especially how electrically excitable cells work, as well as how the aggregate activities of many excitable cells result in the various biopotential signals that can be detected from the body. For a primer (or a refresher), we recommend reading Chapters 6 and 7 of *Intermediate Physics for Medicine and Biology*, 3rd ed., by Russell K. Hobbie (1997).

Whether you are a student, hobbyist, or practicing engineer, this book will show you how easy it is to get involved in the booming biomedical industry by building sophisticated instruments at a small fraction of the comparable commercial cost.

The book addresses the practical aspects of amplifying, processing, simulating, and evoking these biopotentials. In addition, in two chapters we address the issue of safety in the development of electronic medical devices, bypassing the difficult math and providing lots of insider advice.

In Chapter 1 we present the development of amplifiers designed specifically for the detection of biopotential signals. A refresher on op-amp-based amplifiers is presented in the context of the amplification of biopotentials. Projects for this chapter include chloriding silver electrodes, high-impedance electrode buffer array, pasteless bioelectrode, single-ended electrocardiographic (ECG) amplifier array, body potential driver, differential biopotential amplifier, instrumentation-amplifier biopotential amplifier, and switched-capacitor surface array electromyographic amplifier.

In Chapter 2 we look at the frequency content of various biopotential signals and discuss the need for filtering and the basics of selecting and designing RC filters, active filters, notch filters, and specialized filters for biopotential signals. Projects include a dc-coupled biopotential amplifier with automatic offset cancellation, biopotential amplifier with dc rejection, ac-coupled biopotential amplifier front end, bootstrapped ac-coupled biopotential amplifier, biopotential amplifier with selectable RC bandpass filters, state-variable filter with tunable cutoff frequency, twin-T notch filter, gyrator notch filter, universal harmonic eliminator notch comb filter, basic switched-capacitor filters, slew-rate limiter, ECG amplifier with pacemaker spike detection, "scratch and rumble" filter for ECG, and an intracardiac electrogram evoked-potential amplifier.

In Chapter 3 we introduce safety considerations in the design of medical device prototypes. We include a survey of applicable standards and a discussion on mitigating the dangers of electrical shock. We also look at the way in which equipment should be tested for compliance with safety standards. Projects include the design of an isolated biopotential amplifier, transformer-coupled analog isolator module, carrier-based optically coupled analog isolator, linear optically coupled analog isolator with compensation, isolated eight-channel 12-bit analog-to-digital converter, isolated analog-signal multiplexer, ground bond integrity tester, microammeter for safety testing, and basic high-potential tester.

In Chapter 4 we discuss international regulations regarding electromagnetic compatibility and medical devices. This includes mechanisms of emission of and immunity against radiated and conducted electromagnetic disturbances as well as design practices for electromagnetic compatibility. Projects include a radio-frequency spectrum analyzer, near-field H-field and E-field probes, comb generator, conducted emissions probe, line impedance stabilization network, electrostatic discharge simulators, conducted-disturbance generator, magnetic field generator, and wideband transmitter for susceptibility testing.

In Chapter 5 we present the new breed of "smart" sensors that can be used to detect physiological signals with minimal design effort. We discuss analog-to-digital conversion of physiological signals as well as methods for high-resolution spectral analysis. Projects include a universal sensor interface, sensor signal conditioners, using the PC sound card as a data acquisition card, voltage-controlled oscillator for dc-correct signal acquisition through a sound card, as well as fast Fourier transform and high-resolution spectral estimation software.

In Chapter 6 we discuss the need for artificial signal sources in medical equipment design and testing. The chapter covers the basics of digital signal synthesis, arbitrary signal generation, and volume conductor experiments. Projects include a general-purpose signal generator, direct-digital-synthesis sine generator, two-channel digital arbitrary waveform generator, multichannel analog arbitrary signal source, cardiac simulator for pacemaker testing, and how to perform volume-conductor experiments with a voltage-to-current converter and physical models of the body.

In Chapter 7 we look at the principles and clinical applications of electrical stimulation of excitable tissues. Projects include the design of stimulation circuits for implantable pulse generators, fabrication of implantable stimulation electrodes, external neuromuscular stimulator, TENS device for pain relief, and transcutaneous/transcranial pulsed-magnetic neural stimulator.

In Chapter 8 we discuss the principles of cardiac pacing and defibrillation, providing a basic review of the electrophysiology of the heart, especially its conduction deficiencies and arrhythmias. Projects include a demonstration implantable pacemaker, external cardiac pacemaker, impedance plethysmograph, intracardiac impedance sensor, external defibrillator, intracardiac defibrillation shock box, and cardiac fibrillator.

The Epilogue is an engineer's perspective on bringing a medical device to market. The regulatory path, Food and Drug Administration (FDA) classification of medical devices, and process of submitting applications to the FDA are discussed and we look at the value of patents and how to recruit venture capital.

Finally, in Appendix A we provide addresses, Web sites, telephone numbers, and fax numbers for suppliers of components used in the projects described in the book. The contents of the book's ftp site, which contains software and information used for many of these projects, is given in Appendix B.

DAVID PRUTCHI
MICHAEL NORRIS

DISCLAIMER

The projects in this book are presented solely as examples of engineering building blocks used in the design of experimental electromedical devices. The construction of any and all experimental systems must be supervised by an engineer experienced and skilled with respect to such subject matter and materials, who will assume full responsibility for the safe and ethical use of such systems.

The authors do not suggest that the circuits and software presented herein can or should be used by the reader or anyone else to acquire or process signals from, or stimulate the living tissues of, human subjects or experimental animals. Neither do the authors suggest that they can or should be used in place of or as an adjunct to professional medical treatment or advice. Sole responsibility for the use of these circuits and/or software or of systems incorporating these circuits and/or software lies with the reader, who must apply for any and all approvals and certifications that the law may require for their use. Furthermore, safe operation of these circuits requires the use of isolated power supplies, and connection to external signal acquisition/processing/monitoring equipment should be done only through signal isolators with the proper isolation ratings.

The authors and publisher do not make any representations as to the completeness or accuracy of the information contained herein, and disclaim any liability for damage or injuries, whether caused by or arising from a lack of completeness, inaccuracy of information, misinterpretation of directions, misapplication of circuits and information, or otherwise. **The authors and publisher expressly disclaim any implied warranties of merchantability and of fitness of use for any particular purpose, even if a particular purpose is indicated in the book.**

References to manufacturers' products made in this book do not constitute an endorsement of these products but are included for the purpose of illustration and clarification. It is not the authors' intent that any technical information and interface data presented in this book supersede information provided by individual manufacturers. In the same way, various government and industry standards cited in the book are included solely for the purpose of reference and should not be used as a basis for design or testing.

Since some of the equipment and circuitry described in this book may relate to or be covered by U.S. or other patents, the authors disclaim any liability for the infringement of such patents by the making, using, or selling of such equipment or circuitry, and suggest that anyone interested in such projects seek proper legal counsel.

Finally, the authors and publisher are not responsible to the reader or third parties for any claim of special or consequential damages, in accordance with the foregoing disclaimer.

ABOUT THE AUTHORS

David Prutchi is Vice President of Engineering at Impulse Dynamics, where he is responsible for the development of implantable devices intended to treat congestive heart failure, obesity, and diabetes. His prior experience includes the development of Sulzer-Intermedics' next-generation cardiac pacemaker, as well as a number of other industrial and academic positions conducting biomedical R&D and developing medical electronic instrumentation. David Prutchi holds a Ph.D. in biomedical engineering from Tel-Aviv University and conducted postdoctoral research at Washington University, where he taught a graduate course in neuroelectric systems. Dr. Prutchi has over 40 technical publications and in excess of 60 patents in the field of active implantable medical devices.

Michael Norris is a Senior Electronics Engineer at Impulse Dynamics, where he has developed many cardiac stimulation devices, cardiac contractility sensors, and physiological signal acquisition systems. His 25 years of experience in electronics include the development of cardiac stimulation prototype devices at Sulzer-Intermedics as well as the design, construction, and deployment of telemetric power monitoring systems at Nabla Inc. in Houston, and instrumentation and controls at General Electric. Michael Norris has authored various technical publications and holds patents related to medical instrumentation.

1

BIOPOTENTIAL AMPLIFIERS

In general, signals resulting from physiological activity have very small amplitudes and must therefore be amplified before their processing and display can be accomplished. The specifications and lists of characteristics of biopotential amplifiers can be as long and confusing as those for any other amplifier. However, for most typical medical applications, the most relevant amplifier characterizing parameters are the seven described below.

1. Gain. The signals resulting from electrophysiological activity usually have amplitudes on the order of a few microvolts to a few millivolts. The voltage of such signals must be amplified to levels suitable for driving display and recording equipment. Thus, most biopotential amplifiers must have gains of 1000 or greater. Most often the gain of an amplifier is measured in decibels (dB). Linear gain can be translated into its decibel form through the use of

$$\text{Gain(dB)} = 20 \log_{10}(\text{linear gain})$$

2. Frequency response. The frequency bandwidth of a biopotential amplifier should be such as to amplify, without attenuation, all frequencies present in the electrophysiological signal of interest. The bandwidth of any amplifier, as shown in Figure 1.1, is the difference between the upper cutoff frequency f_2 and the lower cutoff frequency f_1. The gain at these cutoff frequencies is 0.707 of the gain in the midfrequency plateau. If the percentile gain is normalized to that of the midfrequency gain, the gain at the cutoff frequencies has decreased to 70.7%. The cutoff points are also referred to as the *half-power points*, due to the fact that at 70.7% of the signal the power will be $(0.707)^2 = 0.5$. These are also known as the -3-dB points, since the gain at the cutoff points is lower by 3 dB than the gain in the midfrequency plateau: $-3\,\text{dB} = 20 \log_{10}(0.707)$.

3. Common-mode rejection. The human body is a good conductor and thus will act as an antenna to pick up electromagnetic radiation present in the environment. As shown in Figure 1.2, one common type of electromagnetic radiation is the 50/60-Hz wave and its harmonics coming from the power line and radiated by power cords. In addition, other spectral components are added by fluorescent lighting, electrical machinery, computers,

Design and Development of Medical Electronic Instrumentation   By David Prutchi and Michael Norris
ISBN 0-471-67623-3   Copyright © 2005 John Wiley & Sons, Inc.

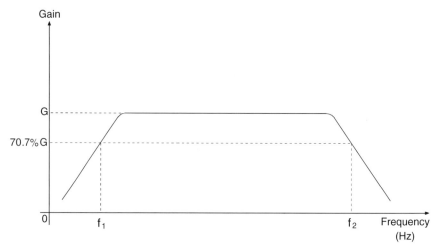

Figure 1.1   Frequency response of a biopotential amplifier.

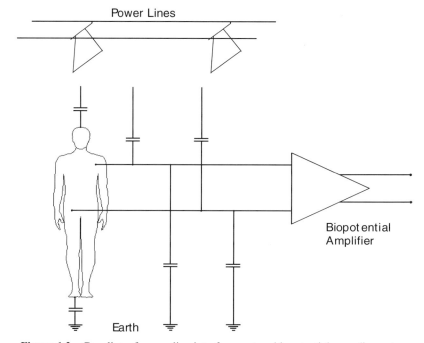

Figure 1.2   Coupling of power line interference to a biopotential recording setup.

and so on. The resulting interference on a single-ended bioelectrode is so large that it often obscures the underlying electrophysiological signals.

The *common-mode rejection ratio* (CMRR) of a biopotential amplifier is measurement of its capability to reject common-mode signals (e.g., power line interference), and it is defined as the ratio between the amplitude of the common-mode signal to the amplitude of an equivalent differential signal (the biopotential signal under investigation) that would produce the same output from the amplifier. *Common-mode rejection* is often expressed in decibels according to the relationship

$$\text{Common-mode rejection (CMR) (dB)} = 20 \log_{10}\text{CMRR}$$

4. Noise and drift. Noise and drift are additional unwanted signals that contaminate a biopotential signal under measurement. Both noise and drift are generated within the amplifier circuitry. The former generally refers to undesirable signals with spectral components above 0.1 Hz, while the latter generally refers to slow changes in the baseline at frequencies below 0.1 Hz.

The noise produced within amplifier circuitry is usually measured either in microvolts peak to peak (μV_{p-p}) or microvolts root mean square (RMS) (μV_{RMS}), and applies as if it were a differential input voltage. Drift is usually measured, as noise is measured, in microvolts and again, applies as if it were a differential input voltage. Because of its intrinsic low-frequency character, drift is most often described as peak-to-peak variation of the baseline.

5. Recovery. Certain conditions, such as high offset voltages at the electrodes caused by movement, stimulation currents, defibrillation pulses, and so on, cause transient interruptions of operation in a biopotential amplifier. This is due to saturation of the amplifier caused by high-amplitude input transient signals. The amplifier remains in saturation for a finite period of time and then drifts back to the original baseline. The time required for the return of normal operational conditions of the biopotential amplifier after the end of the saturating stimulus is known as *recovery time*.

6. Input impedance. The input impedance of a biopotential amplifier must be sufficiently high so as not to attenuate considerably the electrophysiological signal under measurement. Figure 1.3a presents the general case for the recording of biopotentials. Each electrode–tissue interface has a finite impedance that depends on many factors, such as the type of interface layer (e.g., fat, prepared or unprepared skin), area of electrode surface, or temperature of the electrolyte interface.

In Figure 1.3b, the electrode–tissue has been replaced by an equivalent resistance network. This is an oversimplification, especially because the electrode–tissue interface is not merely a resistive impedance but has very important reactive components. A more correct representation of the situation is presented in Figure 1.3c, where the final signal recorded as the output of a biopotential amplifier is the result of a series of transformations among the parameters of voltage, impedance, and current at each stage of the signal transfer. As shown in the figure, the electrophysiological activity is a current source that causes current flow i_e in the extracellular fluid and other conductive paths through the tissue. As these extracellular currents act against the small but nonzero resistance of the extracellular fluids R_e, they produce a potential V_e, which in turn induces a small current flow i_{in} in the circuit made up of the reactive impedance of the electrode surface X_{Ce} and the mostly resistive impedance of the amplifier Z_{in}. After amplification in the first stage, the currents from each of the bipolar contacts produce voltage drops across input resistors R_{in} in the summing amplifier, where their difference is computed and amplified to finally produce an output voltage V_{out}.

The skin between the potential source and the electrode can be modeled as a series impedance, split between the outer (epidermis) and the inner (dermis) layers. The outer layer of the epidermis—the stratum corneum—consists primarily of dead, dried-up cells which have a high resistance and capacitance. For a 1-cm^2 area, the impedance of the stratum corneum varies from 200 kΩ at 1 Hz down to 200 Ω at 1 MHz. Mechanical abrasion will reduce skin resistance to between 1 and 10 kΩ at 1 Hz.

7. Electrode polarization. Electrodes are usually made of metal and are in contact with an electrolyte, which may be electrode paste or simply perspiration under the electrode. Ion–electron exchange occurs between the electrode and the electrolyte, which results in voltage known as the *half-cell potential*. The front end of a biopotential amplifier must be able to deal with extremely weak signals in the presence of such dc polarization components. These dc potentials must be considered in the selection of a biopotential amplifier gain, since they can saturate the amplifier, preventing the detection of low-level ac components. International standards regulating the specific performance of biopotential recording systems

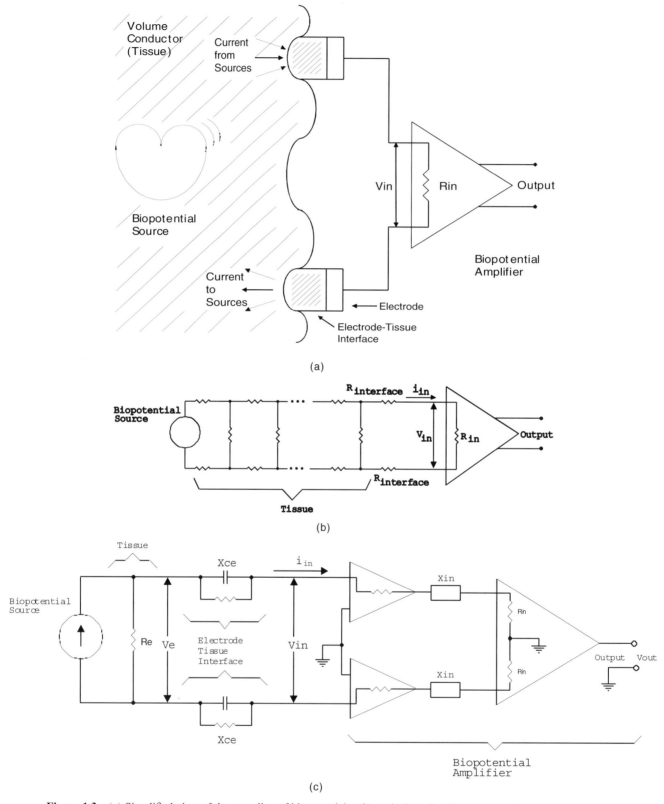

Figure 1.3 (*a*) Simplified view of the recording of biopotentials; (*b*) equivalent circuit; (*c*) generalized equivalent circuit.

usually specify the electrode offsets that are commonly present for the application covered by the standard. For example, the standards issued by the Association for the Advancement of Medical Instrumentation (AAMI) specify that electrocardiography (ECG) amplifiers must tolerate a dc component of up to $\pm300\,mV$ resulting from electrode–skin contact.

Commercial ECG electrodes have electrode offsets that are usually low enough, ensuring little danger of exceeding the maximum allowable dc input offset specifications of the standards. However, the design of a biopotential amplifier must consider that there are times when the dc offset may be much larger. For example, neonatal ECG monitoring applications often use sets of stainless-steel needle electrodes, whose offsets are much higher than those of commercial self-adhesive surface ECG electrodes. In addition, many physicians still prefer to use nondisposable suction cup electrodes (which have a rubber squeeze bulb attached to a silver-plated brass hemispherical cup). After the silver plating wears off, these brass cup electrodes can introduce very large offsets.

LOW-POLARIZATION SURFACE ELECTRODES

Silver (Ag) is a good choice for metallic skin-surface electrodes because silver forms a slightly soluble salt, silver chloride (AgCl), which quickly saturates and comes to equilibrium. A cup-shaped electrode provides enough volume to contain an electrolyte, including chlorine ions. In these electrodes, the skin never touches the electrode material directly. Rather, the interface is through an ionic solution.

One simple method to fabricate Ag/AgCl electrodes is to use electrolysis to chloride a silver base electrode (e.g., a small silver disk or silver wire). The silver substrate is immersed in a chlorine-ion-rich solution, and electrolysis is performed using a common 9-V battery connected via a series 10-$k\Omega$ potentiometer and a milliammeter. The positive terminal of the battery should be connected to the silver metal, and a plate of platinum or silver should be connected to the negative terminal and used as the opposite electrode in the solution. Our favorite electrolyte is prepared by mixing 1 part distilled water (the supermarket kind is okay), 1/2 part HCl 25%, and $FeCl_3$ at a rate of 0.5 g per milliliter of water.

If you want to make your own electrodes, use refined silver metal (99.9 to 99.99% Ag) to make the base electrode. Before chloriding, degrease and clean the silver using a concentrated aqueous ammonia solution (10 to 25%). Leave the electrodes immersed in the cleaning solution for several hours until all traces of tarnish are gone. Rinse thoroughly with deionized water (supermarket distilled water is okay) and blot-dry with clean filter paper. Don't touch the electrode surface with bare hands after cleaning. Suspend the electrodes in a suitably sized glass container so that they don't touch the sides or bottom. Pour the electrolyte into the container until the electrodes are covered, but be careful not to immerse the solder connections or leads that you will use to hook up to the electrode.

When the silver metal is immersed, the silver oxidation reaction with concomitant silver chloride precipitation occurs and the current jumps to its maximal value. As the thickness of the AgCl layer deposited increases, the reaction rate decreases and the current drops. This process continues, and the current approaches zero. Adjust the potentiometer to get an initial current density of about $2.5\,mA/cm^2$, making sure that no hydrogen bubbles evolve at the return electrode (large platinum or silver plate). You should remove the electrode from the solution once the current density drops to about $10\,\mu A/cm^2$. Coating should take no more than 15 to 20 minutes. Once done, remove the electrodes and rinse them thoroughly but carefully under running (tap) water.

An alternative to the electrolysis method is to immerse the silver electrode in a strong bleach solution. Yet another way of making a Ag/AgCl electrode is to coat by dipping the silver metal in molten silver chloride. To do so, heat AgCl in a small ceramic crucible with a gas flame until it melts to a dark brown liquid, then simply dip the electrode in the molten silver chloride.

> **Warning!**   The materials used to form Ag/AgCl electrodes are relatively dangerous. Do not breathe dust or mist and do not get in eyes, on skin, or on clothing. When working with these materials, safety goggles must be worn. Contact lenses are not protective devices. Appropriate eye and face protection must be worn instead of, or in conjunction with, contact lenses. Wear disposable protective clothing to prevent exposure. Protective clothing includes lab coat and apron, flame- and chemical-resistant coveralls, gloves, and boots to prevent skin contact. Follow good hygiene and housekeeping practices when working with these materials. Do not eat, drink, or smoke while working with them. Wash hands before eating, drinking, smoking, or applying cosmetics.

If you don't want to fabricate your own electrodes, you can buy all sorts of very stable Ag/AgCl electrodes from In Vivo Metric. They make them using a very fine grained homogeneous mixture of silver and silver chloride powder, which is then compressed and sintered into various configurations. Alternatively, Ag/AgCl electrodes are cheap enough that you may get a few pregelled disposable electrodes free just by asking at the nurse's station in the emergency department or cardiology service of your local hospital.

Recording gel is available at medical supply stores (also from In Vivo Metric). However, if you really want a home brew, heat some sodium alginate (pure seaweed, commonly used to thicken food) and water with low-sodium salt (e.g., Morton Lite Salt) into a thick soup that when cooled can be applied between the electrodes and skin. Note that there is no guarantee that this concoction will be hypoallergenic! A milder paste can be made by dissolving 0.9 g of pure NaCl in 100 mL of deionized water. Add 2 g of pharmaceutical-grade Karaya gum and agitate in a magnetic stirrer for 2 hours. Add 0.09 g of methyl paraben and 0.045 g of propyl paraben as preservatives and keep in a clean capped container.

SINGLE-ENDED BIOPOTENTIAL AMPLIFIERS

Most biopotential amplifiers are operational-amplifier-based circuits. As a refresher, the voltage present at the output of the operational amplifier is proportional to the differential voltage across its inputs. Thus, the noninverting input produces an in-phase output signal, while the inverting input produces an output signal that is 180° out of phase with the input.

In the circuit of Figure 1.4, an input signal V_{in} is presented through resistor R_{in} to the inverting input of an ideal operational amplifier. Resistor R_f provides feedback from the amplifier's output to its inverting input. The noninverting input is grounded, and due to the fact that in an ideal op-amp the setting conditions at one input will effectively set the same conditions at the other input, point A can be treated as it were also grounded. The power connections have been deleted for the sake of simplicity.

Ideal op-amps have an infinite input impedance, which implies that the input current i_{in} is zero. The inverting input will neither sink nor source any current. According to Kirchhoff's current law, the total current at junction A must sum to zero. Hence,

$$- i_{in} = i_f$$

But by Ohm's law, the currents are defined by

$$i_{in} = \frac{V_{in}}{R_{in}}$$

and

$$i_f = - \frac{V_{out}}{R_f}$$

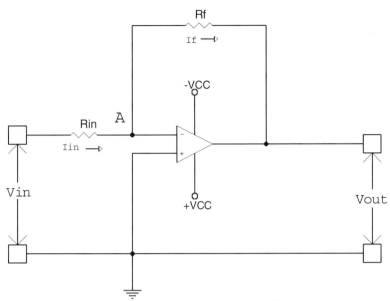

Figure 1.4 Inverting voltage amplifier.

Therefore, by substitution and by solving for V_{out},

$$V_{out} = \frac{R_f V_{in}}{R_{in}}$$

This equation can be rewritten as

$$V_{out} = -GV_{in}$$

where G represents the voltage gain constant R_f/R_{in}.

The circuit presented in Figure 1.5 is a noninverting voltage amplifier, also known as a *noninverting follower*, which can be analyzed in a similar manner. The setting of the noninverting input at input voltage V_{in} will force the same potential at point A. Thus,

$$i_{in} = \frac{V_{in}}{R_{in}}$$

and

$$i_f = \frac{V_{out} - V_{in}}{R_f}$$

But in the noninverting amplifier $i_{in} = i_{out}$, so by replacing and solving for V_{out}, we obtain

$$V_{out} = \left(1 + \frac{R_f}{R_{in}}\right)V_{in}$$

The voltage gain in this case is

$$G = 1 + \frac{R_f}{R_{in}}$$

A special case of this configuration is shown in Figure 1.6. Here $R_f = 0$, and R_{in} is unnecessary, which leads to a resistance ratio $R_f/R_{in} = 0$, which in turn results in unity gain. This configuration, termed a *unity-gain buffer* or *voltage follower*, is often used in biomedical instrumentation to couple a high-impedance signal source, through the (almost) infinite input impedance of the op-amp, to a low-impedance processing circuit connected to the very low impedance output of the op-amp.

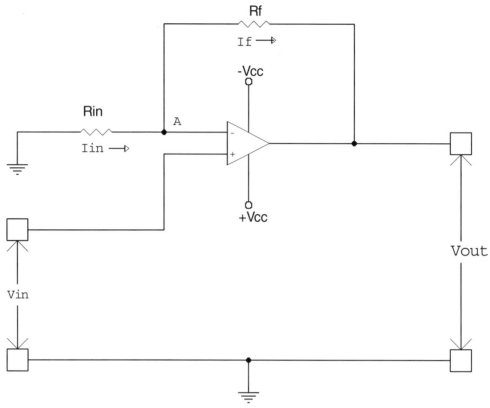

Figure 1.5   Noninverting op-amp voltage amplifier; also known as a noninverting follower.

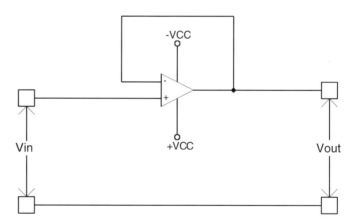

Figure 1.6   A unity-gain buffer is a special case of the noninverting voltage amplifier in which the resistance ratio is $R_f/R_{in} = 0$, which translates into unity gain. This configuration is often used in biomedical instrumentation to buffer a high-impedance signal source.

ULTRAHIGH-IMPEDANCE ELECTRODE BUFFER ARRAYS

A group of ultrahigh-impedance, low-power, low-noise op-amp voltage followers is commonly used as a buffer for signals collected from biopotential electrode arrays. These circuits are usually placed in close proximity to the subject or preparation to avoid contamination and degradation of biopotential signals. The circuit of Figure 1.7 comprises 32 unity-gain

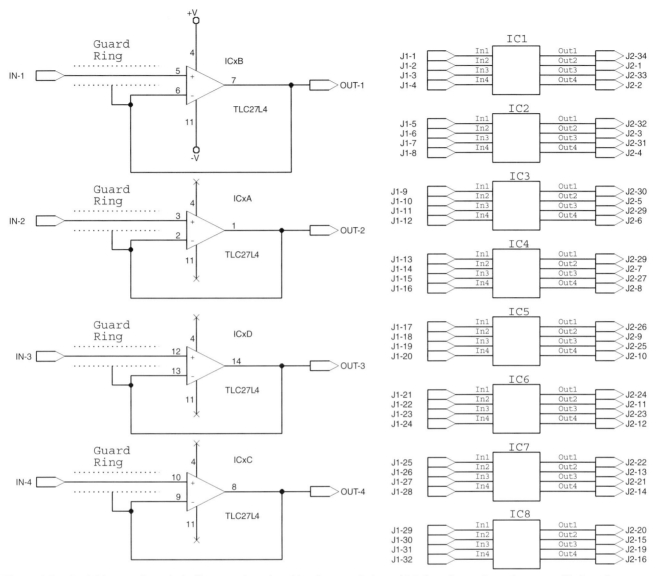

Figure 1.7 CMOS-input unity-gain buffers are often placed in close proximity to high-impedance electrodes to provide impedance conversion, making it possible to transmit the signal over relatively long distances without picking up noise, despite the fact that the contact impedance of the electrodes may range into the thousands of megohms.

buffers, which present an ultrahigh input impedance to an array of up to 32 electrodes. Each buffer in the array is implemented using a LinCMOS[1] precision op-amp operated as a unity-gain voltage follower. An output signal has the same amplitude as that of its corresponding input. The output impedance is very low, however (in the few kilohm range) and can source or sink a maximum of 25 mA. As a result of this impedance transformation, the signal at the buffer's output can be transmitted over long distances without picking up noise, despite the fact that the contact impedance of the electrodes may range into the thousands of megohms. Power for the circuit must be symmetrical ±3 to ±9 V dc with real or virtual ground.

In the circuit, input signals at J1 are buffered by eight TLC27L4 precision quad op-amp. The buffered output is available at J2. Despite its apparent simplicity, the circuit must be

---

[1]LinCMOS is a trademark of Texas Instruments Incorporated.

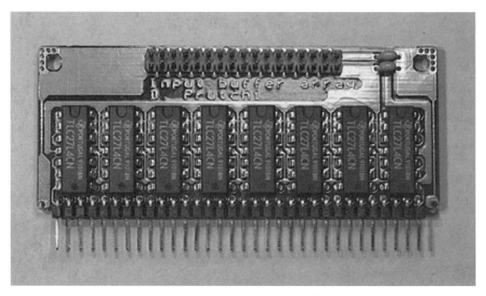

Figure 1.8   Printed circuit board for a high-input-impedance buffer array. The output of each channel is used to drive guard rings which form low-impedance isopotential barriers that shield all input paths from leakage currents.

laid out and constructed with care to take advantage of the op-amp's high input impedance. As shown in the PCB layout of Figure 1.8, the output of each channel is used to drive guard rings that form low-impedance isopotential barriers that shield all input paths from leakage currents.

The selection of op-amps from the TLC27 family has the additional advantage that electrostatic display (ESD) protection circuits that may degrade high input impedance are unnecessary because LinCMOS chips have internal safeguards against high-voltage static charges. Applications requiring ultrahigh input impedances (on the order of $10^{10}\,\Omega$) necessitate additional precautions to minimize stray leakage. These precautions include maintaining all surfaces of the printed circuit board (PCB), connectors, and components free of contaminants, such as smoke particles, dust, and humidity. Residue-free electronic-grade aerosols can be used effectively to dust off particles from surfaces. Humidity must be leached out from the relatively hygroscopic PCB material by drying the circuit board in a low-pressure oven at 40°C for 24 hours and storing in sealed containers with dry silica gel. If even higher input impedances are required, approaching the maximal input impedance of the TLC24L4, you may consider using Teflon[2] PCB material instead of the more common glass–epoxy type.

Typical applications for this circuit include active *medallions*, which are electrode connector blocks mounted in close proximity to the subject or preparation. The low input noise ($68\,\mathrm{nV}/\sqrt{\mathrm{Hz}}$) and high bandwidth (dc—10 kHz) make it suitable for a broad range of applications. For example, 32 standard Ag/AgCl electroencephalography (EEG) electrodes for a brain activity mapper could be connected to such a medallion placed on a headcap.

Figure 1.9 shows another application for the circuit as an active electrode array in electromyography (EMG). Here eight arrays were used to pick up muscle signals from 256 points. Connectors J1 in each of the circuits were made of L-shaped gold-plated pins that are used as electrodes to form an array with a spatial sampling period of 2.54 mm (given by the pitch of a standard connector with 0.1-in. pin center to center). The outputs of the op-amp buffers can then carry signals to the main biopotential signal amplifiers and signal processors

[2]Teflon is a trademark of the DuPont Corporation.

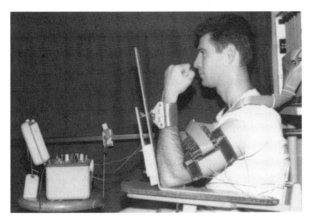

Figure 1.9 Eight high-input-impedance buffer arrays are used to detect muscle signals from 256 points for a high-resolution large-array surface electromyography system. Arrays of gold-plated pins soldered directly to array inputs are used as the electrodes.

using a long flat cable. Power could be supplied either locally, using a single 9-V battery and two 10-kΩ resistors, to create a virtual ground, or directly from a remotely placed symmetrical isolated power supply.

Low-impedance op-amp outputs are compatible with the inputs of most biopotential amplifiers. Wires from J2 can be connected to the inputs of instrumentation just as normal electrodes would. The isolated common post of the biopotential amplifiers should be connected to the ground electrode on the subject or preparation as well as to the ground point of the buffer array.

PASTELESS BIOPOTENTIAL ELECTRODES

Op-amp voltage followers are often used to buffer signals detected from biopotential sources with intrinsically high input impedance. One such application is detecting biopotential signals through capacitive bioelectrodes. One area in which these electrodes are particularly useful is in the measurement and analysis of biopotentials in humans subjected to conditions similar to those existing during flight. Knowledge regarding physiological reactions to flight maneuvers has resulted in the development of devices capable of predicting, detecting, and preventing certain conditions that might endanger the lives of crew members. For example, the detection of gravitationally induced loss of consciousness (loss of consciousness caused by extreme g-forces during sharp high-speed flight maneuvers in war planes) may save many pilots and their aircraft by allowing an onboard computer to take over the controls while the aviator regains consciousness [Whinnery et al., 1987]. G_{z+}-induced loss of consciousness (GLOC) detection is achieved through the analysis of various biosignals, the most important of which is the electroencephalogram (EEG).

Another new application is the use of the electrocardiography (ECG) signal to synchronize the inflation and deflation of pressure suits adaptively to gain an increase in the level of gravitational accelerations that an airman is capable of tolerating. Additional applications, such as the use of the processed electromyography (EMG) signal as a measure of muscle fatigue and pain as well as an analysis of eye blinks and eyeball movement through the detection of biopotentials around the eye as a measure of pilot alertness, constitute the promise of added safety in air operations.

One problem in making these techniques practical is that most electrodes used for the detection of bioelectric signals require skin preparation to decrease the electrical impedance of the skin–electrode interface. This preparation often involves shaving, scrubbing the skin, and applying an electrolyte paste: actions unacceptable as part of routine preflight procedures. In addition, the electrical interface characteristics deteriorate during long-term use of these electrodes as a result of skin reactions and electrolyte drying. Dry or *pasteless electrodes* can be used to get around the constraints of electrolyte–interface electrodes. Pasteless electrodes incorporate a bare or dielectric-coated metal plate, in direct contact with the skin, to form a very high impedance interface. By using an integral high-input-impedance amplifier, it is possible to record a signal through the capacitive or resistive interface.

Figure 1.10 presents the constitutive elements of a capacitive pasteless bioelectrode. In it, a highly dielectric material is used to form a capacitive interface between the skin and a conductive plate electrode. Ideally, this dielectric layer has infinite leakage resistance, but in reality this resistance is finite and decreases as the dielectric deteriorates. Signals presented to the buffer stage result from capacitive coupling of biopotentials to the network formed by series resistor R1 and the input impedance Z_{in} of the buffer amplifier. In addition, circuitry that is often used to protect the buffer stage from ESD further attenuates available signals. Shielding is usually provided in the enclosure of a bioelectrode assembly to protect it from interfering noise. The signal at the output of the buffer amplifier has low impedance and can be relayed to remotely placed processing apparatus without attenuation. External power must be supplied for operation of the active buffer circuitry.

A dielectric substance is used in capacitive biopotential electrodes to form a capacitor between the skin and the recording surface. Thin layers of aluminum anodization, pyre varnish, silicon dioxide, and other dielectrics have been used in these electrodes. For example, 17.5-μm (0.7-mil) film is easily prepared by anodic treatment, resulting in electrode plates that have a dc resistance greater than 1 GΩ and a capacitance of 5000 pF at

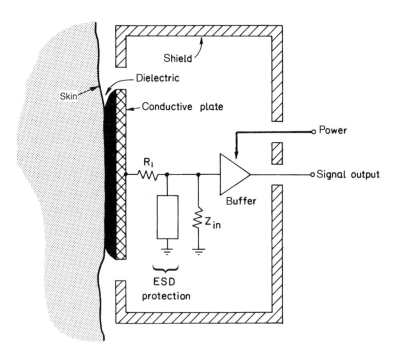

Figure 1.10   Block diagram of a typical capacitive active bioelectrode. A highly dielectric material is used to form a capacitive interface between the skin and a conductive plate electrode. Signals presented to the buffer stage result from capacitive coupling of biopotentials to the network formed by series resistor R1 and the input impedance Z_{in} of the buffer amplifier. (Reprinted from Prutchi and Sagi-Dolev [1993], with permission from the Aerospace Medical Association.)

30 Hz. Unfortunately, standard anodization breaks down in the presence of saline (e.g., from sweat), making the electrodes unreliable for long-term use.

A relatively new anodization process was used by Lisa Sagi-Dolev, the former head of R&D at the Israeli Airforce Aeromedical Center, and one of us [Prutchi and Sagi-Dolev, 1993] to manufacture pasteless EEG electrodes that could be embedded in flight helmets. The hard anodization Super coating process developed by the Sanford Process Corporation[3] is formed on the surface of an aluminum part and penetrates in a uniform manner, making it very stable and resistant. The main characteristics of this type of coating are hardness (strength types Rockwell 50c–70c), high resistance to erosion (exceeding military standard MIL-A-8625), high resistance to corrosion (complete stability after 1200 hours in a saltwater chamber), stable dielectric properties at high voltages (up to 1500 V with a coating thickness of 50 μm, and up to 4500 V with a coating thickness of 170 μm), and high uniformity.

Hard anodization Super has been authorized as a coating for aluminum kitchen utensils, and it proves to be very stable even under high temperatures and the presence of corrosive substances used while cooking. The coating does not wear off with the use of abrasive scrubbing pads and detergents. These properties indicate that no toxic substances are released in the presence of heat, alkaline or acid solutions, and organic solvents. This makes its use safe as a material in direct contact with skin, and resistant to sweat, body oils, and erosion due to skin friction.

Figure 1.11 is a circuit diagram of a prototype active pasteless bioelectrode. The biopotential source is coupled to buffer IC1A through resistor R1 and the capacitor formed by the biological tissues, aluminum oxide dielectric, and aluminum electrode plate. Operational amplifier IC1A is configured as a unity-gain buffer and is used to transform the extremely high impedance of the electrode interface into a low-impedance source that can carry the biopotential signal to processing equipment with low loss and free of

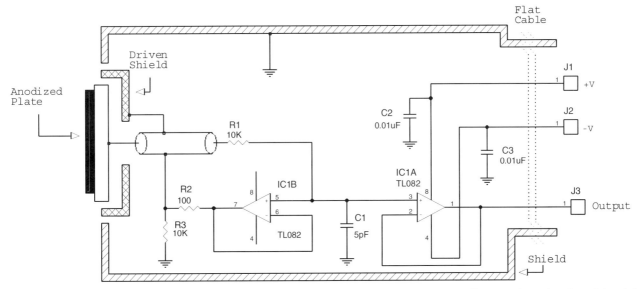

Figure 1.11 Schematic diagram of a capacitive active bioelectrode. Biopotentials are coupled to buffer IC1A through resistor R1 and the capacitor formed by the biological tissues, aluminum oxide dielectric, and aluminum electrode plate. Operational amplifier IC1A is configured as a unity-gain buffer. IC1B drives a shield that protects the input from current leakage and noise. Resistors R3 and R2 reduce the gain of the shield driver to just under unity to improve the stability of the guarding circuit. C1 limits the bandwidth of input signals buffered by IC1.

---

[3]Hard anodization Super is a process licensed by the Sanfor Process Corporation (United States) to Elgat Aerospace Finishing Services (Israel) and is described in Elgat Technical Publication 100, *Hard Anodizing: "Super" Design and Applications*.

contamination. IC1B, also a unity-gain buffer, is fed by the input signal, and its output drives a shield that protects the input from leaks and noise. Resistors R3 and R2 reduce the gain of the shield driver to just under unity in order to improve the stability of the guarding circuit. Capacitor C1 limits the bandwidth of input signals buffered by IC1A. The circuit is powered by a single supply of ±4 V dc. Miniature power supply decoupling capacitors are mounted in close proximity to the op-amp.

IC1A and IC1B are each one-half of a TLC277 precision dual op-amp's IC. Here again, the selection of op-amps from the TLC27 family has the additional advantage that ESD protection circuits which may degrade high input impedance are unnecessary because LinCMOS chips have internal safeguards against high-voltage static charges. Note that this circuit shows no obvious path for op-amp dc bias current. This is true if we assume that all elements are ideal or close to ideal. However, the imperfections in the electrode anodization, as well as in the dielectric separations and circuit board, provide sufficient paths for the very weak dc bias required by the TL082 op-amp.

The circuit is constructed on a miniature PCB in which ground planes, driven shield planes, and rings have been etched. The circuit is placed on top of a 1-cm^2 plate of thin aluminum coated with hard anodization Super used as the bioelectrode. A grounded conductive film layer shields the encapsulated bioelectrode and flexible printed circuit ribbon cable, which carries power for both the circuit and the signal output.

Figure 1.12 presents a prototype bioelectrode array designed to record frontal EEG signals measured differentially (between positions Fp1 and Fp2 of the International 10-20 System), as required for an experimental GLOC detection system. One of the bioelectrodes contains the same circuitry as that described above. The second, in addition to the buffer and shield drive circuits, also contains a high-accuracy monolithic instrumentation amplifier and filters. Such a configuration provides high-level filtered signals which may be carried to remotely placed processing stages with minimal signal contamination from noisy electronics in the helmet and elsewhere in the cockpit.

A miniaturized version of the circuit may be assembled on a single flexible printed circuit. Driven and ground shields, as well as the flat cables used to interconnect the electrodes and carry power and output lines, may be etched on the same printed circuit. As shown in Figure 1.13, the thin assembly may then be encapsulated and embedded at the appropriate position within the inner padding of a flight helmet. Nonactive reference for the instrumentation amplifier may be established by using conductive foam lining the headphone cavities (approximating positions A1 and A2 of the International 10-20 System) or as cushioning for the chin strap.

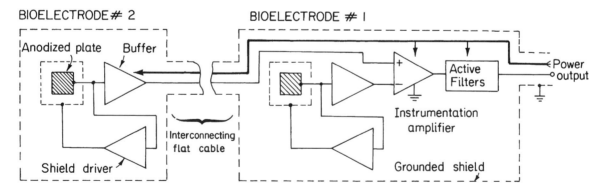

Figure 1.12 Block diagram of a capacitive bioelectrode array with integrated amplification and filter circuits designed to record frontal EEG signals. One of the bioelectrodes contains the same circuitry as Figure 1.11. The second also contains a high-accuracy monolithic instrumentation amplifier and filters. (Reprinted from Prutchi and Sagi-Dolev [1993], with permission from the Aerospace Medical Association.)

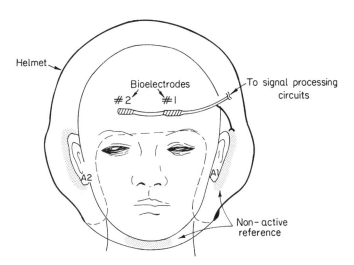

Figure 1.13 A miniaturized version of the capacitive bioelectrode array may be assembled on a single flexible printed circuit. This assembly can be encapsulated and embedded at the appropriate position within the inner padding of a flight helmet for differential measurement of the EEG between positions Fp1 and Fp2 of the International 10-20 System. Conductive foam is used to establish non-active reference either at positions A1 and A2 or at the chin of the subject. (Reprinted from Prutchi and Sagi-Dolev [1993], with permission from the Aerospace Medical Association.)

EEG and ECG signals recorded using the new pasteless bioelectrodes compare very well to recordings obtained through standard Ag/AgCl electrodes. Figure 1.14 presents a digitized tracing of a single-lead ECG signal detected with a capacitive pasteless bioelectrode as well as with a standard Ag/AgCl electrode. Figure 1.15 shows digitized EEG signals recorded from a frontal differential pair with a reference at A2 using a pasteless biopotential electrode array and with standard Ag/AgCl electrodes.

SINGLE-ENDED BIOPOTENTIAL AMPLIFIER ARRAYS

Single-ended op-amp amplifiers were in the past used as front-end stages for biopotential amplifiers. As we will see later, the advent of low-cost integrated instrumentation amplifiers has virtually eliminated the need to design single-ended biopotential amplifiers, and as such, the use of single-ended biopotential amplifiers is not recommended. Despite this, this section has strong educational value because it demonstrates the design principles of using single-ended amplifiers, which are common in the stages that follow the bioamplifier's front end. Figure 1.16 shows an array of 16 single-ended biopotential amplifiers. A number of these circuits may be stacked up to form very large arrays, which made them common for applications such as body potential mapping electrocardiography in the days when single op-amps were expensive.

Each biopotential amplification channel features high-impedance ESD-protected inputs, current limiting, and defibrillation protection. Individual shield drives are used to protect each input lead from external noise. Each channel provides a fixed gain of 1000 within a fixed (-3-dB) bandpass of 0.2 to 100 Hz. The chief advantage of the single-ended configuration is its simplicity, but this comes at the cost of lacking high immunity to common-mode signals. Because of this, single-ended biopotential amplifiers are usually found in equipment that incorporates other ways of suppressing common-mode signals. In this circuit, an onboard adjustable 50/60-Hz notch filter is connected at the output of each channel. The schematic diagram of Figure 1.17 shows how each channel

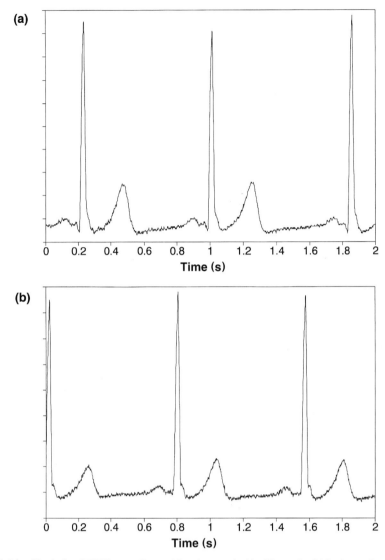

Figure 1.14 Single-lead ECG recordings: (a) using an Ag/AgCl standard bioelectrode; (b) using the capacitive active bioelectrode. (Reprinted from Prutchi and Sagi-Dolev [1993], with permission from the Aerospace Medical Association.)

is built around one-half of two TL064 quad op-amps. Eight copies of this circuit constitute the 16 identical biopotential amplification channels. Operation of a single channel is described in the following discussion.

A biopotential signal detected by a bioelectrode is coupled to the noninverting inputs of the first-stage amplifier and the shield driver amplifier. The input impedance is given mostly by the input impedance of the front-stage op-amps, yielding >100 MΩ paralleled with 100 pF. R1 limits the current that can flow through the input lead, while diodes D1 and D2 shunt to ground any signal that exceeds their zener voltage. This arrangement protects the inputs of the amplifiers from ESD and from the high voltages present during cardiac defibrillation. Furthermore, it protects the subject from currents that may leak back from the amplifiers or associated circuitry.

The shield driver is configured as a unity-gain buffer. The actual drive, however, determined by R2 and R3, is set to 99% of the signal magnitude at the inner wire to stabilize

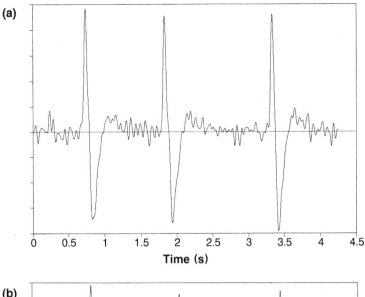

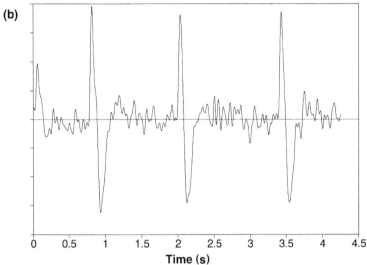

Figure 1.15   EEG measured differentially between positions Fp1 and Fp2 showing eyeblink EMG arti-facts: (a) using an Ag/AgCl standard bioelectrode; (b) using the capacitive active bioelectrode. (Reprinted from Prutchi and Sagi-Dolev [1993], with permission from the Aerospace Medical Association.)

the driver circuit while reducing the effective input cable capacitance by two orders of magnitude. The first amplification stage has a gain determined by

$$G_1 = 1 + \frac{R5}{R4} = 11$$

C2 and R5 form a low-pass filter with a (-3-dB) cutoff frequency of 160 Hz, which stabilizes the amplifier's operation. In addition, R1 and C1 (plus the capacitances of D1 and D2) also form a low-pass filter, which further prevents oscillatory behavior and rejects high-frequency noise.

The amplified signal is high-pass filtered by C3 and R13, with a (-3-dB) cutoff frequency of 0.16 Hz, before being amplified by the second stage. The gain of this stage is set by

$$G_2 = 1 + \frac{R8}{R7} = 101$$

Figure 1.16 Array of 16 single-ended biopotential amplifiers. A number of these circuits may be stacked up to form very large arrays, making them ideally suited for applications such as body potential mapping electrocardiography.

The last processing stage of each channel is an active notch filter, which can be tuned to the power line frequency by adjusting R12. Supply voltage to this circuit must be symmetrical and within the range of ± 5 V (minimum) to ± 18 V (absolute maximum). Two 9-V alkaline batteries can be used efficiently due to the circuit's very low power consumption. Capacitors C9–C12 are used to decouple the power supply and filter noise from the op-amp power lines.

To minimize electrical interference, the circuit should be built with a compact layout on an appropriate printed circuit board or small piece of stripboard. The construction of the circuit is straightforward, but care must be taken to keep wiring as short and clean as possible. Leads to the bioelectrodes should be low-loss coaxial cables, whose shields are connected to their respective shield drives at J1 (J1x-2 for left-side channels and J1y-1 for right-side channels). The circuit's ground should be connected to the subject's reference (*patient ground*) electrode. When connected to a test subject, the circuit must always be powered from batteries or through a properly rated isolation power supply. The same isolation requirements apply to the outputs of the amplifier channels.

It is important to note that the performance of a complete system is determined primarily by its input circuitry. Equivalent input noise is practically that of the first stage (approximately $10\,\mu V_{p\text{-}p}$ within the amplifier's -3-dB bandwidth of 0.2 to 100 Hz).

BODY POTENTIAL DRIVERS

Rejection of common-mode signals in the prior circuit example is limited to the single-ended performance of the input-stage op-amp and the 50/60-Hz rejection of the notch filter. Often, however, environmental noise (e.g., power line interference) is so large that common-mode potentials eclipse the weak biopotentials that can be picked up through single-ended amplifiers. Notch filters do not necessarily remove interfering signals in a substantial manner either. The first few harmonics of the power line constitute strong interfering signals in the recording of biopotentials. The range of these signals, however, is by no means confined to 100 or 200 Hz. High-frequency interference originating from fluorescent and other high-efficiency lamps commonly occurs with a maximal spectral density of approximately 1 kHz and with amplitudes of up to 50% of the 50/60-Hz harmonic.

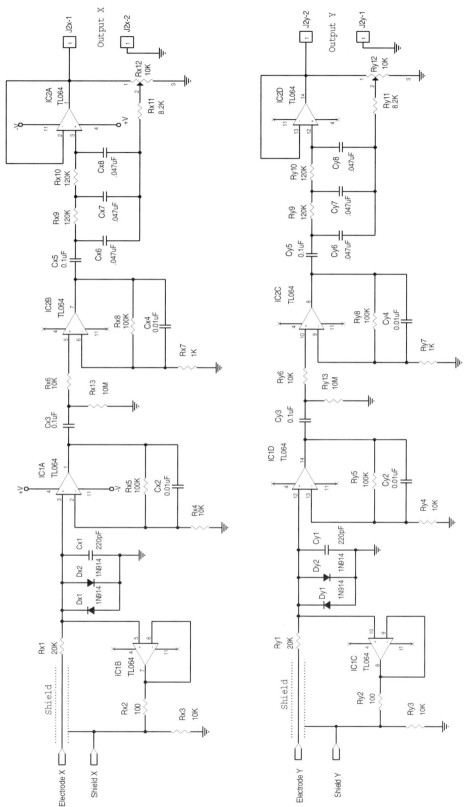

Figure 1.17 Each channel of the single-ended biopotential amplifier array is built around one-half of two TL064 quad op-amps. Eight copies of this circuit constitute the 16 identical biopotential amplification channels.

19

A way of improving the common-mode rejection problem is to use single-ended amplifiers concurrently with *body potential driver* (BPD) circuits to cancel out common-mode signals. Power line and other contaminating common-mode signals are capacitively coupled to the body, causing current to flow through it and into ground. The body, acting as a resistor through which a current flows, causes a voltage difference between any two points on it. The goal of a BPD is to detect and eliminate this voltage, effectively reducing common-mode signals between biopotential detection electrodes in the vicinity of its *sense electrode*.

A BPD is implemented by detecting the common-mode potential in the area of interest and then feeding into the body a 180° version of the same signal. A feedback loop is thus established which cancels out the common-mode potential. Circuits that have feedback are inherently unstable, and oscillatory behavior must be prevented to make a BPD useful. This, however, limits the BPD to a range well under its first resonance. The performance of the circuit within this range is dependent on the internal delay of the loop and varies according to the frequency of common-mode signal components.

The common-mode potential used for a BPD is often acquired from the outputs of the front stages of differential biopotential amplifiers. In electrocardiography, for example, a composite signal is often generated by summing the various differential leads. This signal is inverted and fed back to the subject's body through the right-leg electrode. This practice, commonly referred to as *right-leg driving*, is not optimal, especially at higher frequencies where the additional delay caused by the front stages and summing circuits degrades BPD performance.

Superior performance can be obtained by implementing a separate BPD circuit which uses an additional electrode (sense). Any modern operational amplifier operated in open-loop mode (with a feedback capacitor in the order of a few picofarads) can be used as the heart of the BPD [Levkov, 1982, 1988]. In the circuit of Figure 1.18, the common-mode signal is measured between the sense and common electrodes. This signal is applied through current-limiting resistor R2 to the inverting input of one-half of op-amp IC1. Operated in open-loop mode, a 180° out-of-phase signal is injected into the body through the drive electrode in order to cancel the common-mode voltage. D3 and D4 clip the BPD output so as not to exceed a safe current determined by resistor R3. In addition, this measure protects the circuit from defibrillation pulses. D1 and D2 are used to protect the input of the BPD from ESD and other transients. The low-pass filter formed by R2 and C5, as well as the presence of feedback capacitor C2, stabilize the circuit and prevent it from entering into oscillation.

The output of the BPD op-amp is rectified by the full-wave bridge formed by D5–D8 and then amplified by the differential amplifier built using the other half of IC1. The output of this op-amp is measured and displayed by the bar graph voltmeter formed by IC3 in conjunction with a 10-element LED display DISP1. The LM3914 bar graph driver IC has constant-current outputs, and thus series resistors are not required with the LEDs. The current is controlled by the value of resistors R8 and R9. Resistor values also set the range over which the input voltage produces a moving dot on the display. Power for the circuit is supplied by a single 9-V alkaline battery. The −9-V supply required by IC1 is generated using IC2, an integrated-circuit voltage converter. C3, D9, and C4 are required by IC2 to produce an inverted output of the power fed through pin 8.

An additional advantage of using the BPD is the possibility of monitoring the skin–electrode impedance of every electrode connected to the input of a single-ended biopotential amplifier system. To do so, a test voltage V_{test} fed into the inverting input of the BPD through J1-4 induces an additional component on each of the amplified output signals. Phased demodulation of one of these signals removes components corresponding to detected biopotentials, leaving only an amplified version of the detected test signal V_i. Assuming that an ideal BPD is used, the amplitude of this signal depends on the

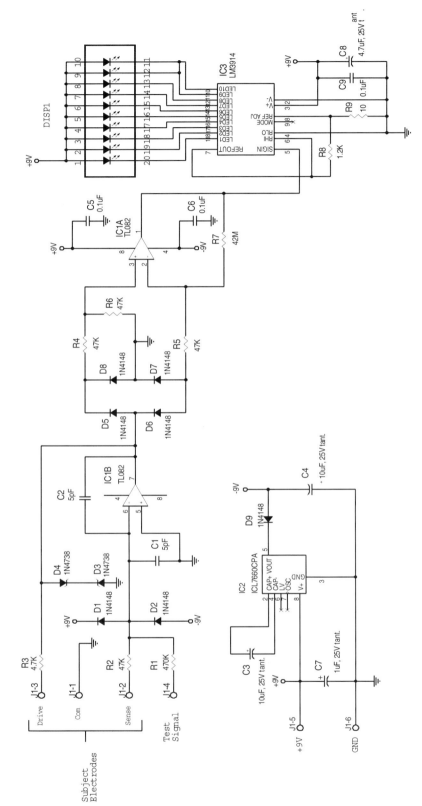

Figure 1.18 A body potential driver is implemented by detecting the common mode potential in the area of interest and then feeding the body a 180° version of the same signal. A feedback loop is thus established, which cancels out common-mode potentials.

21

skin–electrode impedance and is given by

$$V_i = G_i \frac{V_{\text{test}}(Z_i + R2)}{R1}$$

where G_i is the gain of each amplifier in the array.

For simplicity and convenience, the test signal can be generated by a computer and phased demodulation can be implemented in software. Impedance tests can be performed just prior to data collection as well as at selected times throughout an experiment, making it easy to locate faulty electrode–skin connections even in large amplifier arrays. Further theoretical and practical considerations regarding the construction of large single-ended biopotential amplifier arrays may be found in a paper by Van Rijn et al. [1990].

To use the BPD circuit in conjunction with biopotential amplifiers, connect the BPD reference terminal (J1-1) to the reference electrode (*subject ground*) of the biopotential amplifier system. Place the sense electrode (e.g., a standard Ag/AgCl ECG electrode) in contact with the body in the proximity of the biopotential amplifier's active electrode(s) and connect it to J1-2 of the BPD circuit using shielded cable (with the shield connected to J1-1). A similar electrode placed at a distant point on the body should be connected to the "drive" output (J1-3) of the BPD. Upon hooking up a 9-V alkaline battery to the appropriate power inputs (+ terminal to J1-5 and − terminal to J1-6), common-mode signals should be neutralized. The moving dot on the display shows the relative maximum amplitude of the BPD voltage. This can be used to assess the conditions of the recording environment.

In general, use of a separate sense electrode is not be recommended for any newly designed equipment. Whenever active common-mode suppression is required, the instrument should be designed such that the common-mode potential used for BPD is obtained from the outputs of the biopotential amplifier's front end. However, a stand-alone BPD such as the one shown in Figure 1.19 can be used to boost the performance of older

Figure 1.19   A body potential driver can be constructed as a stand-alone unit powered by a 9-V battery. This circuit can be used in conjunction with existing biopotential amplifiers to boost the common-mode rejection of older equipment. The LED display shows the relative maximum amplitude of the BPD voltage to assess the conditions of the recording environment.

equipment. For example, when the BPD is used in conjunction with an existing single-ended ECG channel, J1-1 should be connected to the right-leg cable, and the other two electrodes can be placed at convenient sites on the body.

DIFFERENTIAL AMPLIFIERS

When a differential voltage is applied to the input terminals of an op-amp as depicted in Figure 1.20, the transfer function of the inverting follower must be rewritten as

$$V_{\text{out}} = -\frac{R_f}{R_{\text{in}}}(V_1 - V_2)$$

Similarly, the transfer function of the noninverting follower must be modified to

$$V_{\text{out}} = \left(1 + \frac{R_f}{R_{\text{in}}}\right)(V_1 - V_2)$$

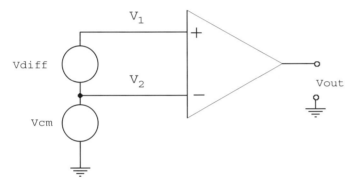

Figure 1.20   Differential and common-mode voltages applied to the input of an op-amp.

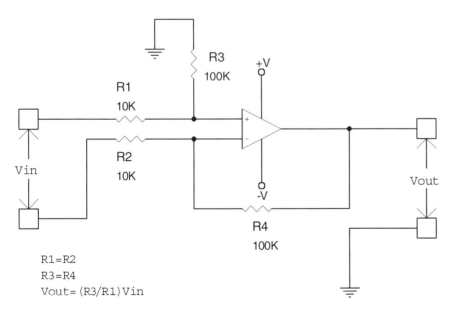

Figure 1.21   Differential amplifier implemented with an op-amp.

Figure 1.21 presents a differential amplifier based on a single op-amp. If R1 = R2 and R3 = R4, the gain of the stage is given by

$$G = \frac{R3}{R2} = \frac{R4}{R1}$$

In this case, the transfer function is

$$V_{out} = V_{in} \frac{R4}{R1}$$

or

$$V_{out} = V_{in} \frac{R3}{R2}$$

where $V_1 - V_2$ is the differential voltage V_{in}.

The balance of a differential amplifier is critical to preserve the property of an ideal op-amp by which its common-mode rejection ratio is infinite. If $V_1 = V_2$, an output voltage of zero should be obtained, disregarding any common-mode voltage V_{CM}. If the resistor equalities R1 = R2 and R3 = R4 are not preserved, the common-mode rejection deteriorates.

The main problem regarding use of a simple differential amplifier as a biopotential amplifier is its low input impedance. Especially in older equipment, where this configuration was used to amplify differential biopotentials, high-input-impedance JFET transistors or MOSFET-input op-amp unity-gain voltage followers were used to buffer each input of the differential amplifier. Despite the enhanced CMR of the differential amplifier configuration over that of a single-ended system, use of a BPD circuit can increase considerably the CMR of differential biopotential amplifiers. This is especially true regarding the rejection of interfering signals with high-frequency components.

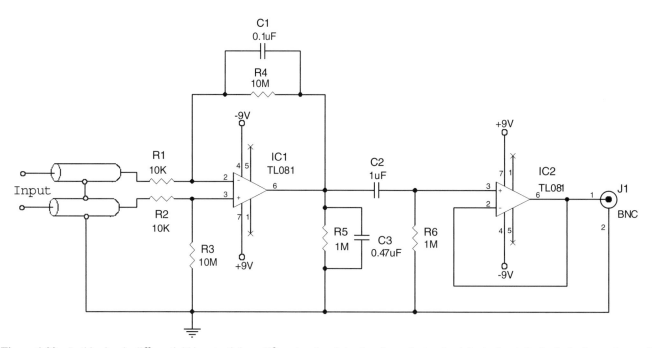

Figure 1.22   In this simple differential biopotential amplifier, signals originating from electrophysiological activity in the body are detected by measuring the potential differences between electrodes connected to the inputs. If the sensing bioelectrodes are placed in the proximity of the biopotential source, common-mode electrical interference affects both probes more or less equally and are rejected by the differential amplifier stage.

Simple Differential Biopotential Amplifier

Figure 1.22 presents the circuit diagram of a simple differential biopotential amplifier. Potential differences originating from electrophysiological activity in the body may be detected with this circuit by attaching bioelectrodes and measuring the potential differences between them. If the sensing bioelectrodes are placed in the proximity of a biopotential source, electrical interference induced from the power line or originating from other sources of biopotentials in more remote parts of the body will affect both probes more or less equally. The changes of signal detected simultaneously by both electrodes are rejected by the first stage of the preamplifier. This stage is made up by op-amp IC1 wired as a differential amplifier.

For low-frequency signals, the gain of the differential stage is given by

$$G = \frac{R3}{R2} = \frac{R4}{R1} = \frac{10\,M\Omega}{10\,k\Omega} = 1000$$

At high frequencies, however, C1 has low impedance, forcing the first stage to act as a low-pass filter. In addition to limiting the bandwidth of the amplifier, C1 and C3 damp oscillations and instabilities of the circuit. Note that a gain of 1000 requires that large dc offset voltages not be present on the biopotential signal. At this gain, the circuit will stop operating if the offset voltage exceeds a mere 10 mV. If higher offset voltages are expected, the gain of the amplifier formed around IC1 must be decreased. For example, to use this circuit as part of a surface ECG amplifier, the gain must be recalculated to cope with offset potentials of up to ±300 mV.

The output of IC1 is ac-coupled via C2 to IC2. The −3-dB cutoff for the high-pass filter formed by C2 and R6 is approximately 0.16 Hz. The filtered signal is then buffered by unity-gain voltage follower IC2. To minimize electrical interference, the circuit should be built with a compact layout on an appropriate printed circuit board or small piece of stripboard. The construction of the circuit is straightforward, but care must be taken to keep wiring as short and clean as possible. Leads to the electrodes are coaxial cables with their shields connected to ground at the circuit board.

This circuit is very useful to demonstrate how to measure the CMR and input impedance of a biopotential amplifier. First, test and calibrate the circuit. You will need a two-channel oscilloscope and a signal generator. Take the following steps:

1. Connect the oscilloscope and the signal generator to the biopotential amplifier as shown in Figure 1.23a.
2. Apply a 10-Hz signal of 1-mV amplitude as measured by channel 2 of the oscilloscope.
3. Verify that the output signal is an amplified version of the input signal.
4. Determine the theoretical gain of the equivalent circuit and confirm that the output signal has an amplitude of $G_{\text{noninverting}} \cdot 1$ mV.
5. Without changing the settings of the instruments, connect the oscilloscope and the signal generator to the biopotential amplifier as shown in Figure 1.23b.
6. Verify that the output signal is an amplified and inverted (opposite phase) version of the input signal. Determine the theoretical gain of the equivalent circuit and confirm that the output signal has an amplitude of $G_{\text{inverting}} \cdot 1$ mV.

To measure the CMR, do the following:

1. Connect the equipment as shown in Figure 1.24.
2. Adjust the signal generator to produce a 60-Hz 5-$V_{\text{p-p}}$ common-mode input signal $V_{\text{in CM}}$.
3. Measure the corresponding common-mode output voltage $V_{\text{out CM}}$.

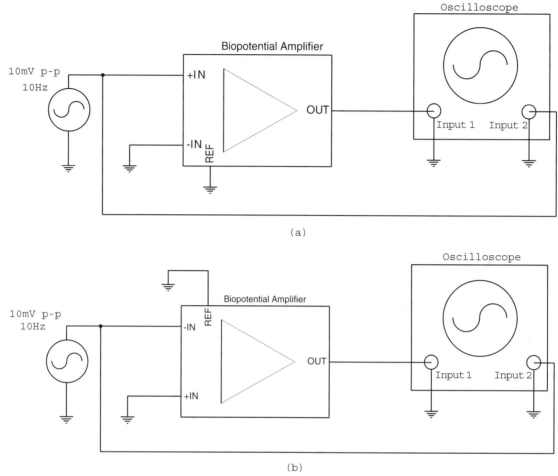

Figure 1.23   The gain of a differential biopotential amplifier can be measured by injecting a signal (e.g., 10-Hz sinusoidal) at an amplitude similar to that expected from the biopotential (e.g., 1 mV) to the inputs. (*a*) When the amplifer is configured as single-ended, the output signal should be an amplified version of the input signal. (*b*) When the circuit is reconfigured to use the inverting input, the output signal should be an amplified and inverted (opposite phase) version of the input signal.

4. Calculate the common-mode gain $G_{CM} = V_{out\ CM}/V_{in\ CM}$.
5. Considering that the differential gain of this biopotential amplifier is given by the ratio between the resistor pairs (i.e., $G_{differential} = 10\ M\Omega/10\ k\Omega = 1000$), calculate the common-mode rejection ratio, CMRR $= G_{differential}/G_{CM}$, and common-mode rejection, CMR(dB) $= 20\log_{10}(G_{differential}/G_{CM})$.

Next, measure the input impedance of the biopotential amplifier. You will need an ohmmeter (e.g., a digital multimeter or VOM) and a 10-MΩ multiturn potentiometer in addition to the oscilloscope and signal generator. Follow this procedure:

1. Connect the equipment as depicted in Figure 1.25.
2. Adjust the signal generator to produce a 100-Hz sinusoidal wave with an amplitude of 1 mV$_{p\text{-}p}$. This signal is measured by channel 2 of the oscilloscope.
3. Set the potentiometer to 0 Ω and measure the amplifier's peak-to-peak output voltage on channel 1. Record this value.
4. Carefully adjust the 10-MΩ potentiometer until the voltage measured on channel 1 reaches half the value recorded in the preceding step.

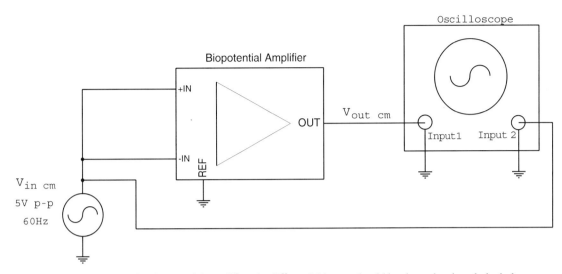

Figure 1.24   To measure the CMR of a biopotential amplifier, the differential inputs should be shorted and a relatively large common-mode signal (e.g., 60-Hz 5-V_{p-p} sinusoidal) injected between the shorted differential inputs and the biopotential amplifier's common reference input. The common-mode rejection is then calculated as CMR (dB) = $20 \log_{10}(G_{\text{differential}}/G_{\text{CM}})$.

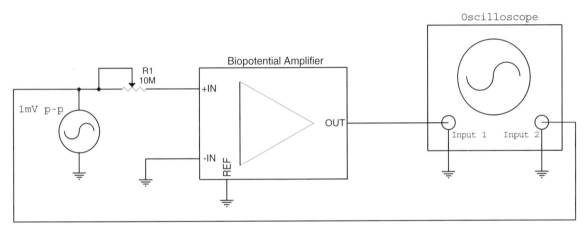

Figure 1.25   To measure the input impedance of the biopotential amplifier, inject an in-band test signal of known amplitude (e.g., 1 mV) to the biopotential amplifier's inputs through a 10-MΩ potentiometer. Adjust the potentiometer until the output voltage is half of the amplitude obtained with the potentiometer set to 0 Ω. The resistance of the potentiometer at the half-output point is equal to the input impedance of the biopotential amplifier at the test frequency. This measurement should be repeated for a number of in-band frequencies to compute the capacitive and resistive components of the input.

5. Without changing the setting of the potentiometer, measure its resistance with the ohmeter. This value is equal to the input impedance of the biopotential amplifier at the specified frequency.

6. Repeat the experiment for various frequencies from 0.1 Hz to 20 kHz. Use appropriate settings for the oscilloscope's time base. Compute the capacitive and resistive components of the input impedance based on the data obtained.

OP-AMP INSTRUMENTATION AMPLIFIERS

An alternative to the simple differential amplifier is the multiple op-amp configuration presented in Figure 1.26. This differential configuration, known as an *instrumentation*

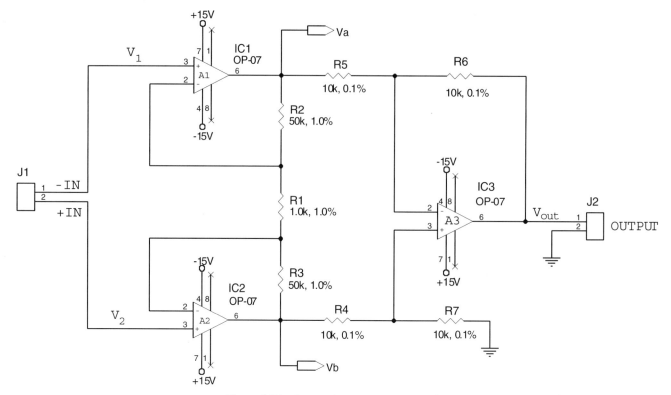

Figure 1.26 Op-amp instrumentation amplifier.

amplifier, has the advantage of preserving the high input impedance of the noninverting follower, yet offering gain. Input amplifiers A1 and A2 can be analyzed as noninverting amplifiers which produce output voltages of

$$V_A = \left(1 + \frac{R2}{R1}\right) V_1 - \frac{(R2)V_2}{R1}$$

and

$$V_B = \left(1 + \frac{R3}{R1}\right) V_2 - \frac{(R3)V_1}{R1}$$

Initially, if it is assumed that the gain of amplifier A3 is unity and that R2 = R3, then

$$V_{out} = V_B - V_A$$

Substituting into the two preceding equations gives the output voltage:

$$V_{out} = (V_2 - V_1)\left(1 + \frac{2R2}{R1}\right)$$

Whenever the gain of amplifier A3 is greater than unity, this equation must be multiplied by the gain of the differential stage. If R4 = R5 and R6 = R7, the overall voltage gain of the instrumentation amplifier is given by

$$V_{out} = (V_2 - V_1)\left(1 + \frac{2R2}{R1}\right)\frac{R6}{R5}$$

Biopotential amplifiers are seldom built these days using individual op-amps. Instead, an *integrated circuit instrumentation amplifier* (ICIA) combines in a single package most of the components required to make an instrumentation amplifier. ICIAs typically require one or two external resistors to set their gain. These resistors do not affect the high CMRR value or the high input impedance achieved in ICIAs through precise matching of their internal components.

Instrumentation Biopotential Amplifier

The circuit of Figure 1.27 is a typical ICIA-based biopotential amplifier. This high-input-impedance circuit combines a programmable-gain instrumentation amplifier and an ac-coupled (LPF: $-3\,$dB at $0.5\,$Hz; HPF: $-3\,$dB at $500\,$Hz) configurable bandpass filter to form a highly versatile, compact, stand-alone biopotential amplifier. Differential amplification of the biopotential signal is achieved with high CMR ($>90\,$dB at $60\,$Hz) through the use of a high-accuracy monolithic instrumentation amplifier IC. The low-noise ($<1\,\mu V_{p\text{-}p}$ between 0.5 and $100\,$Hz) front end can be programmed to have a gain of 10, 100, or 1000, while a fixed second stage and a configurable third stage further amplify the signal to an overall gain of up to 1 million.

Typical applications for this biopotential amplifier are as a front-end and main amplifier for standard and topographic EEG, evoked potential tests (BAER, MLAR, VER, SER), and for cognitive signals and long-latency studies. The heart of the circuit is IC1, Burr-Brown's INA102 programmable monolithic IC instrumentation amplifier. Biopotentials are dc-coupled to the instrumentation amplifier through current-limiting resistors R1 and R2. An INA102 gain of 1, 10, 100, or 1000 is selected by programming jumpers JP1 and JP2 as shown in Table 1.1. Since the amplifier is dc-coupled, care must be exercised in the selection of gain so that the amplifier is not saturated by dc offset voltages accompanying the biopotential signal. For example, to use this circuit as part of a surface ECG amplifier, the gain must be calculated to cope with offset potentials of up to $\pm300\,$mV.

The INA102 is ac-coupled ($-3\,$dB at $0.5\,$Hz) to a second amplification stage with a fixed gain of 100. Resistor R5 and capacitor C2 form a low-pass filter with -3-dB cutoff at $500\,$Hz. R6, R7, R8, C3, C4, and C5 are used to select the desired passband of two stages of filtering. R6–R8 and C3–C5, along with one-half of IC3, form a third-order ($-18\,$dB/octave) Butterworth low-pass active filter stage. The design of these filters is discussed in Chapter 2.

Finally, the ac output of the filter is presented to an inverting amplifier prior to output. The gain of this last stage is given by $G_{IC3B} = R10/R9$. As shown in Figure 1.28, components R6–R9 and C3–C5 can be soldered onto a DIP header which is inserted in a 14-pin DIP socket. A number of these DIP-header modules may be assembled to provide an assortment of desired passband and gain characteristics. If the listed resistors and capacitors are used, the low-pass -3-dB point is fixed at $22\,$Hz, with a third-stage gain of 10.

The supply voltage to the circuit must be symmetrical and within the range $\pm5\,$V (minimum) to $\pm16\,$V (absolute maximum). Rated specifications are obtained using a supply of $\pm15\,$V. Diodes D1 and D2 provide protection against incorrect supply voltage polarity, and capacitors C7–C14 are used to decouple and filter the power supply. Because of the very small quiescent maximal supply current used by this circuit, a pair of 9-V alkaline batteries constitute a suitable power supply for most applications. Preferably, leads to the bioelectrodes should be low-loss low-capacitance coaxial cables whose shields are connected to the subject ground terminal of the biopotential amplifier. Construction of this biopotential amplifier is simple and straightforward, but care must be taken to keep all wiring as short and as clean as possible.

In using this biopotential amplifier it is desirable to keep the gain of the first stage low (e.g., 10), and to reach the required overall gain by selecting a high gain for the third stage. In addition, optimal rejection of unwanted signal components is best achieved by careful

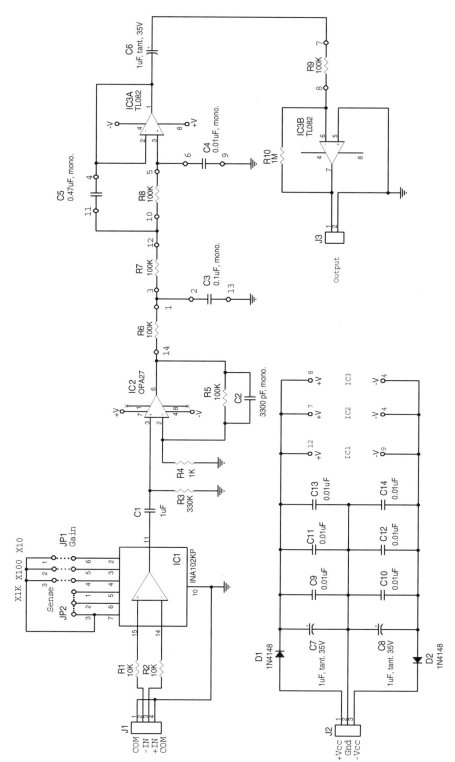

Figure 1.27 This high-input-impedance circuit combines a Burr-Brown's INA102 programmable-gain instrumentation amplifier and an ac-coupled (LPF: -3 dB at 0.5 Hz, HPF: -3 dB at 500 Hz) configurable bandpass filter to form a highly versatile, compact, stand-alone biopotential amplifier.

**TABLE 1.1   The Gain of the INA102 ICIA of
Figure 1.27 Is Jumper-Programmable According
to These Settings for JP1 and JP2**

Gain	JP1	JP2
1	None	2–3
10	1–6	2–3
100	2–5	2–3
1000	3–4	1–2

Figure 1.28 The components that select gain and bandpass filter characteristics (R6–R9 and C3–C5) for the ICIA-based biopotential amplifier can be soldered onto a DIP header which is inserted in a standard 14-pin DIP socket. A number of these DIP-header modules may be assembled to provide an assortment of desired passband and gain characteristics.

selection and preparation of electrode placement on the subject, and by keeping the band-pass characteristics of the biopotential amplifier as tight as possible. When connected to a test subject, the circuit must always be powered from batteries or through a properly rated isolation power supply. The same isolation requirements apply to the output of the amplifier.

To test and calibrate the unit, you will need a two-channel oscilloscope and a signal generator. Take the following steps:

1. Assemble the DIP header according to your requirements and install in the 14-pin socket.

2. After verifying the connections, power the biopotential amplifier circuit with a symmetrical power supply.

3. Short both inputs (J1-2 and J1-3) of the biopotential amplifier to ground (J1-1,4). Configure JP1 and JP2 for a gain of 10.

4. Connect the oscilloscope's input to the output of the biopotential amplifier circuit (J3-1) and the oscilloscope's ground to that of the biopotential amplifier's (J3-2).

5. The output signal should be stable and should present no oscillatory behavior or drift. At very high gains, the peak-to-peak input noise of the circuit can be measured.

6. Short the inverting (J1-2) input terminal of the biopotential amplifier to the subject ground terminal (J1-1,4), and connect these to the ground terminal of a signal

generator. Connect the output of the signal generator to the noninverting input (J1-2) of the biopotential amplifier.

7. Adjust the signal generator to produce a sinusoidal wave with an amplitude of $1\,\mathrm{mV_{p\text{-}p}}$ at a frequency within the passband of the filter configuration selected.

8. Check that changes in the configuration of JP1 and JP2 cause corresponding changes in the amplitude of the output signal.

9. Set the gain to 10, and using the second channel of the oscilloscope, check that there is no phase difference between the signal at the output of the ICIA and that at the noninverting input.

10. Without changing the settings of the instruments, short the noninverting (J1-3) input terminal of the biopotential amplifier to the subject ground terminal (J1-1,4), and connect these to the ground terminal of the signal generator. Connect the output of the signal generator to the inverting input (J1-2) of the biopotential amplifier.

11. Verify that the output signal is an amplified and inverted version (opposite phase) of the input signal. Verify that the gain remained constant.

While the signal generator is connected, monitor the output of the biopotential amplifier while increasing and decreasing the frequency of the signal generator. You can verify your choice of components used for the filter stages by observing that the decay in output amplitude indeed occurs at the expected frequencies. The procedure is as follows:

1. Set the gain of the biopotential amplifier front gain to unity.

2. Adjust the input sine wave to exactly 0.07 V and the frequency to the midpoint of the bandpass expected. Make this adjustment as accurately as possible.

3. Check that the output signal is of the amplitude expected. Readjust the signal generator if necessary.

4. Slowly increase the input frequency until the output amplitude decreases to 0.05 V (70.7% of the midrange gain). Measure the frequency at this point. This is the high-frequency cutoff point of the biopotential amplifier.

5. Repeat the preceding steps for gain factors of 10, 100, and 1000 using appropriate settings for the signal generator and the oscilloscope.

6. Reset the gain of the biopotential amplifier front end to unity.

7. Connect a 1-μF nonpolar capacitor in series between the signal generator and the input to the biopotential amplifier.

8. Slowly sweep the frequency of the input signal starting from dc and measure the frequency at the two points where the output signal is 0.05 V (70.7% of the midrange gain). This is the low-frequency cutoff point of the biopotential amplifier.

9. Plot the response of this last configuration on a semilogarithmic graph.

This amplifier is suitable for applications involving low-level low-frequency signals. Thus, you may want to measure the amplifier's equivalent noise level. To do this you will need a digital storage oscilloscope or chart recorder. Follow this procedure:

1. Short both inputs of the biopotential amplifier to the patient ground terminal.

2. Connect the oscilloscope to the output of the biopotential amplifier.

3. Set the oscilloscope for a 10-second total sweep and dc coupling.

4. Set the overall gain of the biopotential amplifier to 100,000.

5. Set the gain of the oscilloscope up to a point where the peak events of the wide fuzzy noise signal can be measured.

6. Typical peak-to-peak noise measurements are found by reading the maximum peak-to-peak voltage noise of the circuit's output for three observation periods of 10 seconds each, then dividing by the gain of the amplifier (i.e., 100,000).

You may want to compare the CMR of the ICIA to that of the differential amplifier described in the earlier project. Use a procedure similar to the one used earlier to measure the CMR of the instrumentation biopotential amplifier:

1. Connect the equipment, shorting the differential inputs as shown in Figure 1.24.

2. Adjust the signal generator to produce a 60-Hz, 5-V_{p-p} input signal $V_{in\ CM}$.

3. Measure the corresponding common-mode output voltage $V_{out\ CM}$.

4. Calculate the common-mode gain $G_{CM} = V_{out\ CM}/V_{in\ CM}$.

Using the measured gain of the biopotential amplifier, calculate the common-mode rejection:

$$CMR(dB) = 20\log_{10}\frac{G_{differential}}{G_{CM}}$$

Switched-Capacitor Instrumentation Biopotential Amplifier

Differential biopotential signal recordings can be done through circuits other than classical op-amp differential or instrumentation amplifiers. The simplified circuit in Figure 1.29 implements a simple but precise instrumentation amplifier that uses a switched-capacitor building block. It converts differential signals from a preamplified electrode pair to a single-ended output while rejecting common-mode signals in an effective manner. In this circuit, a solid-state dual-pole dual-throw (DPDT) switch block converts the differential input to a ground-referred single-ended signal which may then be amplified by a noninverting follower op-amp configuration.

During the time that the input switches are closed (Φ = odd), sampling capacitor C_S acquires the input signal $V_{in\ diff}$. When the input switches open (Φ = even), hold capacitor C_H receives the sampled charge. Switching C_S continuously between the input

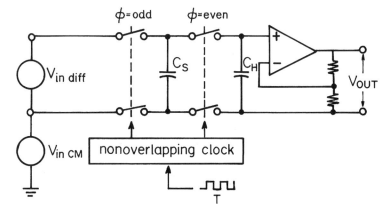

Figure 1.29 In this switched-capacitor instrumentation amplifier, a solid-state DPDT switch block converts the differential input to a ground-referred single-ended signal which is then amplified by a noninverting follower operational amplifier configuration.

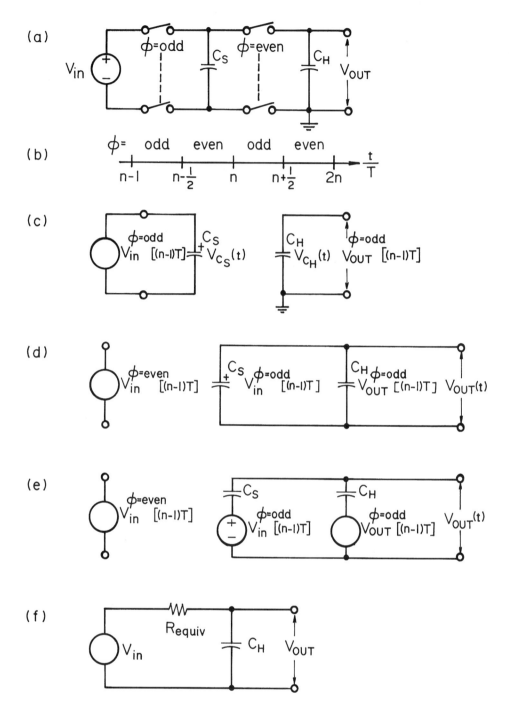

Figure 1.30 Switched-capacitor instrumentation amplifier timing and equivalent circuits: (*a*) basic equivalent circuit; (*b*) switching timing; (*c*) first odd-phase equivalent circuit; (*d*) first even-phase equivalent; (*e*) generalized even-phase equivalent circuit; (*f*) analog equivalent circuit.

voltage and C_H causes output voltage V_{out} to track the difference between the circuit's inputs $V_{in\ diff}$, rejecting common-mode voltages $V_{in\ CM}$. Assuming that the characteristics of the op-amp approach those of an ideal op-amp, the transfer characteristics of the circuit may be determined by following conventional network analysis methods. The equivalent circuit is presented in Figure 1.30*a*. For the purpose of analysis, input voltage V_{in} is assumed to be constant during the sampling period. It is further assumed that the switches are thrown back and forth continuously with a clock period T according to the timing diagram of Figure 1.30*b* and that their connections shift instantly with no overlap.

During the first odd-phase interval $(n-1) \leq t/T < (n - \frac{1}{2})$, the circuit is equivalent to Figure 1.30*c*, and C_S is charged instantaneously to $V_{in}^{\Phi = odd}(n-1)$:

$$V_{C_S}(t) = V_{in}^{\Phi = odd}[(n-1)T] = V_{in}^{\Phi = odd}(n-1)$$

While the odd-phase output voltage $V_{out}^{\Phi = odd}$ is equal to the voltage of V_{C_H},

$$V_{C_H}(t) = V_{out}^{\Phi = odd}[(n-1)T] = V_{out}^{\Phi = odd}(n-1)$$

During the even-phase interval $(n - \frac{1}{2}) \leq t/T < n$ that follows, the circuit is equivalent to that of Figure 1.30*d*, and charges are redistributed between C_S and C_H, which results in a new output voltage. The analysis of the charge transaction is simplified by assuming the alternative equivalent circuit of Figure 1.30*e* with uncharged capacitors. By applying the initial voltages of the capacitors represented by the sources as step functions edged at $t = (n-1)T$, the new output voltage is

$$V_{out}^{\Phi = even}\left(n - \frac{1}{2}\right) = \frac{C_S}{C_S + C_H} V_{in}^{\Phi = odd}(n-1) + \frac{C_H}{C_S + C_H} V_{out}^{\Phi = odd}(n-1)$$

During the odd-phase interval $n \leq t/T < (n + \frac{1}{2})$, capacitor C_H remains undisturbed, and thus the output voltage is represented by the expression

$$V_{out}^{\Phi = odd}(n) = V_{out}^{\Phi = even}\left(n - \frac{1}{2}\right)$$

A general expression representing the odd-phase output voltage may now be written

$$V_{out}^{\Phi = odd}(n) = \frac{C_S}{C_S + C_H} V_{in}^{\Phi = odd}(n-1) + \frac{C_H}{C_S + C_H} V_{out}^{\Phi = odd}(n-1)$$

Applying the *z*-transform to this equation, we obtain

$$V_{out}^{\Phi = odd}(z) = \frac{C_S}{C_S + C_H} V_{in}^{\Phi = odd}(z) + \frac{C_H z^{-1}}{C_S + C_H} V_{out}^{\Phi = odd}(z)$$

The odd-phase discrete-frequency-domain transfer function may then be solved directly:

$$H^{\Phi_{in} = odd; \Phi_{out} = odd}(z) = \frac{V_{out}^{\Phi = odd}(z)}{V_{in}^{\Phi = odd}(z)} = \frac{1}{1 + C_H/C_S} \frac{z^{-1}}{1 - \dfrac{C_H/C_S}{1 + C_H/C_S} z^{-1}}$$

By replacing z by $e^{j\omega T}$ and using Euler's formula, the time-domain representation of this equation can be written as

$$H^{\Phi_{in} = odd; \Phi_{out} = odd}(e^{j\omega T}) = \frac{V_{out}^{\Phi = odd} e^{j\omega T}}{V_{in}^{\Phi = odd} e^{j\omega T}}$$

$$= \frac{1}{(1 + C_H/C_S) \cos \omega T - C_H/C_S + j(1 + C_H/C_S) \sin \omega T}$$

where ω is the frequency of an applied sinusoidal signal and T is the clock period. From this equation it is possible to determine the magnitude response and phase shift of the switched-capacitor instrumentation block. In addition, a capacitance ratio C_H/C_S that results in an appropriate frequency response may be selected. Assuming that the switching frequency allows for a large oversampling of the desired passband, z can be approximated by the continuous Laplace term $(1 + sT)$; then

$$H^{\Phi_{in}=odd;\Phi_{out}=odd}(z) \xrightarrow{z=1+sT} H(s) = \frac{1}{sT(1 + C_H/C_S) + 1}$$

Comparing this to the continuous frequency-domain transfer function of a simple RC low-pass filter yields

$$H(s) = \frac{1}{(s/\omega_1) + 1}$$

where $\omega_1 = 1/RC$. Then a cutoff frequency of $f_{-3dB} = 1/2\pi RC$ may be obtained through a capacitance ratio of

$$\frac{C_H}{C_S} = \frac{1}{2\pi f_{-3dB} T}$$

The analogy to an RC low-pass filter is not a mere coincidence, because the DPDT switches together with C_S constitute the parallel switched-capacitor resistor realization of Figure 1.30f, whose value is given by

$$R_{equivalent} = TC_S$$

The capacitance of C_S must be computed to comply with a desired input impedance, which is dependent on the sampling frequency. It must be noted that because of the fact that common-mode signals are not sampled, their spectral content may well exceed the sampling frequency without encountering aliasing.

It may be seen from the analysis presented above that after charge reorganization has been achieved during an even-phase interval, the output voltage is held as long as the sampling capacitor does not bring a new sample in contact with the hold capacitor. This property may be used to reject stimulation artifacts (i.e., high-amplitude spikes caused by currents caused by a pulse generator intended to cause tissue stimulation) by extending the even-phase interval, making it slightly longer than the stimulation pulse to be rejected. This technique effectively isolates the stimulation artifact from the high-gain amplification and processing circuitry, following the instrumentation stage, allowing for the immediately consecutive detection of biopotentials.

Although switched-capacitor instrumentation stages are not very common in patient monitors, they are often used as the core of biopotential amplifiers in implantable devices (e.g., pacemakers). In addition, many modern analog signal processing applications rely on switched-capacitor sampled-data processing techniques implemented through the use of CMOS charge manipulation circuits. In these applications, CMOS application-specific ICs (ASICs) contain switches and capacitors that are used as an economical means of mass producing sophisticated signal processing functions, such as amplification, analog arithmetic, nonlinear functions, and filtering [Allen and Sanchez-Sinencio, 1984].

In the circuit of Figure 1.31, IC3, a monolithic charge-balanced dual switched-capacitor instrumentation building block (Linear Technology's LTC1043), implements all of the required charge manipulation functions. Within this integrated circuit, a nonoverlapping clock controls two DPDT CMOS switch sections. If the switched-capacitor stage IC3 would be connected directly to a differential biopotential source detected by small-area surface electrodes, and the sampling frequency is chosen to be 100 times that of the highest spectral component of the signal, the optimal value of the sampling capacitor would result in the picofarad range to present an input impedance in the gigaohm range.

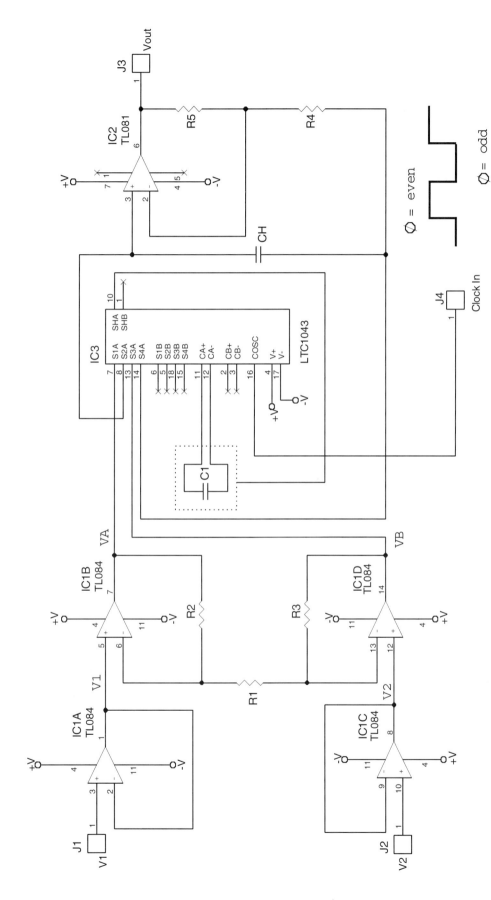

Figure 1.31 In the switched-capacitor-based biopotential instrumentation amplifier, the op-amps of IC1 form an ultrahigh-impedance differential-in/differential-out amplifier, while IC2 form a differential stage. a switched-capacitor block together with IC2 form a differential stage.

37

Using off-the-shelf components, however, the construction of such a circuit would introduce parasitic capacitances in the same order of magnitude as the sample-and-hold capacitances, resulting in errors that the internal charge-balancing circuitry within the integrated circuit cannot cancel. For this reason, this design includes two noninverting amplifiers IC1B and IC1D which present the switched-capacitor block with a signal level that is compatible with a larger-valued sampling capacitor, effectively eliminating the problems related to parasitic capacitances. The output of each of these amplifiers is given by

$$V_A = \left(1 + \frac{R2}{R1}\right)V_1 - \frac{(R2)V_2}{R1}$$

and

$$V_B = \left(1 + \frac{R3}{R1}\right)V_2 - \frac{(R3)V_1}{R1}$$

Thus, if R2 = R3, the theoretical differential voltage presented to the sampling capacitor is

$$V_A - V_B = (V_2 - V_1)\left(1 + 2\frac{R2}{R1}\right)$$

CMOS op-amps IC1A and IC1C are configured as unity-gain buffers and serve as ultrahigh impedance to low-impedance transformers so that the biopotential signal may be carried with negligible loss and contamination to the instrumentation stage. In critical applications, these could be mounted in close proximity to the electrodes used to detect the biopotentials. In addition, if the biopotential amplifier can be mounted close enough to the subject, IC1A and IC1C may be omitted.

In order not to reduce the high common-mode rejection that may be achieved through use of a switched-capacitor instrumentation block, the use of high-precision components is mandatory, so that the gain of the chain formed by IC1A and IC1B will closely match that of IC1C and IC1D. In addition, an adequate layout of the printed circuit board or breadboard, using guard rings and shielding the sampling capacitor from external parasitic capacitances, is necessary to preserve the common-mode rejection from being degraded. This also helps maintain the inherent ultrahigh impedance of the CMOS input buffers.

An additional high-performance CMOS operational amplifier IC2, configured as a noninverting follower, amplifies the single-ended output of the instrumentation stage. The ultrahigh input impedance of this amplifier ensures that the performance of the switched-capacitor stage is not affected by the output load. The dc gain of the noninverting follower is given by

$$G_{IC2} = 1 + \frac{R5}{R4}$$

which is multiplied by its own transfer function, the dc gain of the input amplifiers and buffers, their transfer function, and the transfer function of the switched-capacitor instrumentation block to yield the frequency-dependent gain of the complete system. However, the flat-response bandwidth of any modern operational amplifier is by far wider than that of biopotential signals, and by selecting a very high sampling frequency and the correct capacitance ratio, a virtually flat frequency response within the bandwidth of interest is achievable.

Figure 1.32 shows an array of these switched-capacitor instrumentation amplifiers used to detect myoelectric signals from muscle fibers stimulated by an electrical current. Artifacts induced by the high-voltage surface neuromuscular stimulation can be rejected by extending the even-phase switching interval during stimulation. To do so, an external clock drives the switched-capacitor timing logic. Just prior to stimulation, the clock is isolated from the amplifiers by a logic AND gate, and all switched-capacitor blocks are set unconditionally to even-phase mode. Shortly after stimulation ceases, switching at clock speed is restored.

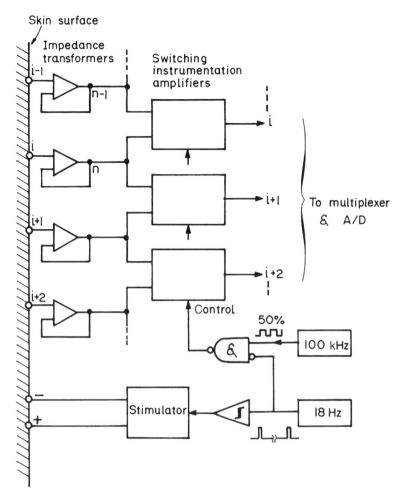

Figure 1.32   Switched-capacitor biopotential amplifier array setup for detecting propagating activity in electrically stimulated muscle. Stimulation artifacts are isolated from the high-gain amplification stages by extending the even-phase switching interval throughout a stimulation pulse.

Initial rejection of the stimulus artifact is effected through differential measurement of EMG signals. Rippling caused by the instantaneous change in switching period are minimized by sample-and-hold and low-pass filter properties of the switched-capacitor instrumentation amplifier. Further rejection of the induced discontinuity are easily rejected through a low-pass filter as well as through decreasing the slew rate of the output op-amp.

Figure 1.33 presents results from such an array recording used to record EMG signals from the biceps brachii muscle. Signals were differentially recorded using 32 surface electrodes 2.54 mm apart. The ultrahigh-impedance buffer array circuit presented earlier in this chapter was used as the electrode buffer. Switched capacitors were clocked at 100 kHz by a 50% duty-cycle oscillator. A separate oscillator triggered a high-voltage surface neuromuscular stimulator at a rate of 18 Hz. One-half millisecond before generation of a 1.06-ms compensated stimulation pulse, all switched-capacitor stages are forced to an even-phase state and remain that way for an additional 1 ms after the stimulation pulse ceases.

The 31 single-differential EMG signals clearly show a compound potential originating under channel 20 and propagating bidirectionally toward channels 1 and 31. The artifact resulting from stimulation currents and switching discontinuity in the biopotential amplifiers

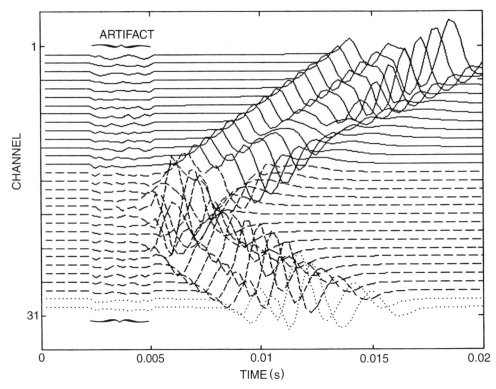

Figure 1.33 Bipolar array recording of composite propagating EMG activity evoked by a single surface stimulation pulse. The artifact resulting from stimulation currents and switching discontinuity is indicated between brackets. [Reprinted from *Med. Eng. Phys.*, 17, D. Prutchi, A High-Resolution Large Array (HRLA) Surface EMG System, pages 442–454, 1995, with permission from Elsevier.]

is present on all channels simultaneosly. Please note that the level of interference it causes (shown between brackets) is minimal compared to the amplitude of evoked potentials.

Recently, multichannel recordings such as these have gained wide popularity in research and diagnostic uses of electrophysiological activity. For example, 32-channel systems have been used in EEG and evoked potentials, and 64-channel systems have been developed for body potential mapping (BPM) ECG. In electromyography, surface electrode arrays have been used to detect propagating electrical activity in skeletal muscles, enabling noninvasive estimation of muscle fiber conduction velocity and innervation zone.

REFERENCES

Allen, P. E., and E. Sanchez-Sinencio, *Switched Capacitor Circuits*, Van Nostrand Reinhold, New York, 1984.

Levkov, Ch. L., Amplification of Biosignals by Body Potential Driving, *Medical and Biological Engineering and Computing*, 20, 248–250, 1982.

Levkov, Ch. L., Amplification of Biosignals by Body Potential Driving. Analysis of the Circuit Performance, *Medical and Biological Engineering and Computing*, 26, 389–396, 1988.

Prutchi, D., and A. Sagi-Dolev, New Technologies for In-Flight Pasteless Bioelectrodes, *Aviation, Space and Environmental Medicine*, 64, 552–556, 1993.

Van Rijn, A. C., A. Peper, and C. A. Grimbergen, High-Quality Recording of Bioelectric Events—Part 1: Interference Reduction, Theory and Practice, *Medical and Biological Engineering and Computing*, 28, 389–397, 1990.

Whinnery, J. E., D. H. Glaister, and R. R. Burton, G_{z+}-Induced Loss of Consciousness and Aircraft Recovery, *Aviation Space and Environmental Medicine*, 58, 600–603, 1987.

2

BANDPASS SELECTION FOR BIOPOTENTIAL AMPLIFIERS

As shown in Table 2.1, common biopotential signals span the range dc to 10 kHz. Under ideal conditions, a biopotential amplifier with wideband response would serve most applications. However, the presence of common-mode potentials, electrode polarization, and other interfering signals often obscure the biopotential signal under investigation. As such, the frequency response of a biopotential amplifier should be tuned to the specific spectral content expected from the application at hand.

Spectral analysis is the most common way of determining the bandwidth required to process physiological signals. For a first estimate, however, the rigors of spectral analysis can be avoided simply by evaluating the durations of high- and low-frequency components of the signal. Koide [1996] proposed a method for estimating the -3-dB bandpass based on acceptable distortion.

The duration of the highest-frequency component, t_{HF}, is estimated from a stereotypical signal to be the minimum rise or fall time of a signal variation. The duration of the lowest-frequency component, t_{LF}, on the other hand, is measured from the tilt of the baseline or of the lowest-frequency component of interest. Koide illustrated this with an example. Figure 2.1 shows a stereotypical intracellular potential measured from the pacemaker cells in a mammalian heart SA node. In this example, $t_{HF} = 75$ ms and $t_{LF} = 610$ ms. Using the formulas of Table 2.2, the amplification system must have a -3-dB bandpass of 0.0026 to 41.3 Hz to reproduce the signal with negligible distortion (1%). Acceptable distortion, usually considered to be 5% or less for physiological signals, would require a narrower -3-dB bandpass, of 0.013 to 18.7 Hz.

WIDEBAND BIOPOTENTIAL AMPLIFIER

The biopotential amplifier circuit described by the schematic diagrams of Figures 2.2 and 2.3 covers the complete frequency range of commonly recorded biopotentials with high CMR. In this circuit, a Burr-Brown INA110AG ICIA is dc-coupled to the electrodes via current-limiting resistors R22 and R23. Two Ohmic Instruments IS-1-3.3DP semiconductor

Design and Development of Medical Electronic Instrumentation   By David Prutchi and Michael Norris
ISBN 0-471-67623-3   Copyright © 2005 John Wiley & Sons, Inc.

TABLE 2.1   Frequency Ranges of Various Biopotential Signals

Application		Frequency Range
Action potentials detected with transmembrane pipette electrodes		dc–2 kHz
Electroneurogram (ENG): nerve bundle potentials detected with needle electrode		10 Hz–1 kHz
Electroretinogram (ERG): potentials generated by retina in response to a flash of light; detected with implanted electrodes		0.2–200 Hz
Electrooculogram (EOG): eye potentials used to measure eye position; detected with surface electrode pairs: left/right and above/below eyes		dc–100 Hz
Electrogastrogram (EGG): stomach potentials detected with surface electrodes placed on abdomen		0.01–0.55 Hz
Electroencephalogram (EEG): rhythmic brain potentials detected with surface electrodes placed on head	Delta waves	0.5–4 Hz
	Theta waves	4–7.5 Hz
	Alpha waves	7.5–13 Hz
	Low beta waves	13–15 Hz
	Beta waves	15–20 Hz
	High beta waves	20–38 Hz
	Gamma waves	38–42 Hz
Brain evoked potentials: brain potentials evoked by stimuli; detected with surface electrodes placed on head	Visual evoked potential (VEP)	1–300 Hz
	Auditory evoked potential (AEP)	100 Hz–3 kHz
	Somatosensory evoked potential (SSEP)	2 Hz–3 kHz
Electrocardiogram (ECG): heart potentials detected with surface electrodes placed on chest, back, and/or limbs	Heart rates (R-R intervals)	0.5–3.5 Hz
	R-R variability due to thermoregulation	0.01–0.04 Hz
	R-R variability due to baroreflex dynamics	0.04–0.15 Hz
	R-R variability due to respiration	0.15–0.4 Hz
	P,QRS,T complex	0.05–100 Hz
	Ventricular late potentials	40–200 Hz
	Bandwith requirement for clinical ECG/rate monitors	0.67–40 Hz
Clinical cardiac electrophysiology: analysis of cardiac potentials detected with catheter electrodes placed in contact with the myocardium	Intracardiac electrograms	10 Hz–1 kHz
	Monophasic action potentials (MAPs)	dc–2 kHz
Electromyogram (EMG): muscle potentials detected with surface electrodes or indwelling needle electrodes	Surface EMG	2–500 Hz
	Motor unit action potentials	5 Hz–10 kHz
	Single fiber electromyogram	500 Hz–10 kHz
Galvanic skin response (GSR): battery potentials produced by sweat on skin electrodes		dc–5 Hz

current limiters are used for redundant protection of the subject from leakage currents. C25, C27, and C26 are used to protect the amplifier from high-frequency currents, such as those used in electrosurgery and ablation procedures. R21 and R24 limit the impedance of each input to 10 MΩ referred to the circuit's isolated ground. These resistors provide enough bias to maintain high immunity to common-mode signals without the need of a patient ground electrode. Diodes D7–D10 are used to protect the inputs of IC5 from high-voltage transients such as those expected from defibrillation and electrostatic discharge.

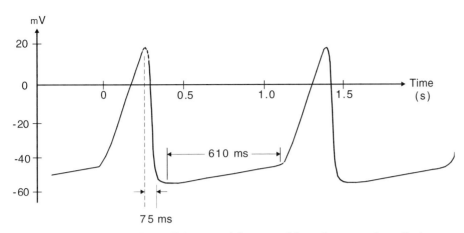

Figure 2.1 A stereotypical intracellular potential measured from the pacemaker cells in a mammalian heart SA node has a minimum rise time of $t_{HF} = 75$ ms and a tilt of $t_{LF} = 610$ ms. The -3-dB bandpass needed to reproduce this signal with 1% distortion is of 0.0026 to 41.3 Hz.

TABLE 2.2   Approximate -3-dB Frequencies Required for the Reproduction of Physiological Signals with Negligible and Acceptable Levels of Distortion

-3-dB Point	Negligible Distortion (1%)	Acceptable Distortion (5%)
High-pass (Hz)	$\leq \dfrac{0.0016}{t_{LF}(s)}$	$\leq \dfrac{0.008}{t_{LF}(s)}$
Low-pass (Hz)	$\leq \dfrac{3.1}{t_{HF}(s)}$	$\leq \dfrac{1.4}{t_{HF}(s)}$

IC5 is powered via a Burr-Brown ISO107 isolation amplifier IC3, which generates isolated ± 12 V when powered from ± 12 V. L1, C6, C7, and C8 form a pi filter to clean the isolated $+12$-V power line generated by the ISO107 from switching noise. An identical network is used to filter the negative isolated supply rail. A pi filter formed by L3, C15, C16, and C17 is used to decouple the positive input power rail of IC3 so that switching noise within IC3 does not find its way into the postisolation amplifier stages (IC2 and IC4). R10 and C21 form a low-pass filter with -3dB of approximately 7 kHz to eliminate any remaining trace of the carrier used to convey the signal across IC3's isolation barrier.

Op-amp IC2B is configured as a noninverting amplifier. The gain of this amplifier can be selected through potentiometers R16–R20 that are switched via SW3. These provide different levels of feedback to IC2B, depending on their setting. The output of this amplification stage is filtered via R9 and C20 with a -3dB low-pass cutoff of approximately 7 kHz to eliminate any residual switching noise that may have coupled through the input or power supply and amplified by IC2B. The amplified signal is buffered and further low-pass filtered via fixed-gain noninverting amplifier IC4 before being presented to the output connector J1.

Since the biopotential amplifier is dc-coupled and there are occasions when the signal of interest may have a relatively large input dc offset, an automatic zero-offset circuit has been implemented through IC2A. Whenever offset nulling is desired, momentary pushbutton switch SW1 should be pressed. Doing so presents a sample of the output signal to the integrating sample-and-hold (S&H) circuit formed by IC2A, R5, and C9. The hold capacitor should be a low-leakage type, and the path between the capacitor and the inverting

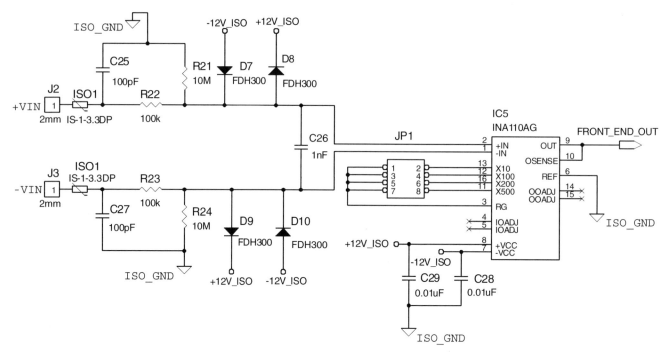

Figure 2.2 This wideband dc-coupled biopotential amplifier front end covers the complete frequency range of commonly recorded biopotentials. A Burr-Brown INA110AG ICIA is dc-coupled to the electrodes via current-limiting resistors R22 and R23 and IS-1-3.3DP fault-current limiters. Capacitors and diodes are used to protect the amplifier from high-frequency currents, such as those used in electrosurgery and ablation procedures as well as from high-voltage transients such as those that may be expected from defibrillation and electrostatic discharge.

input of the op-amp should be shielded against stray and leakage currents through a guard ring on the circuit board. The output of the integrator/S&H is summed with the output of IC2B via R4. The output of IC2A is also attenuated via R6 and R8 and summed with the output of the isolation amplifier (IC3) via R7. This trick allows offsets that would otherwise saturate amplifier IC2B to be canceled in a very effective way.

A typical application for a dc-coupled wideband biopotential amplifier is the measurement of transmembrane potentials as well as for the detection of cardiac monophasic action potentials (MAPs). Dc coupling is important for these applications because they are usually related to measuring the timing and amplitude of shifts in potentials that have a dc offset.

BIOPOTENTIAL AMPLIFIER WITH DC REJECTION

Unlike transmembrane and MAP signals, most biopotential recordings made with extracellular electrodes contain signals of physiological interest only at frequencies above dc. In fact, very low frequency components are usually the result of unwanted electrochemical processes at the electrodes, generating potentials that disturb or obscure the signal of interest. Dc potentials caused by electrode polarization and "injury" currents limit the gain that can be given to input stages so as to keep them away from saturation. In addition, the effect of changes in electrode contact, temperature, and hydration induce slow changes in the level of polarization, which shows in dc-coupled biopotential recordings as baseline wander.

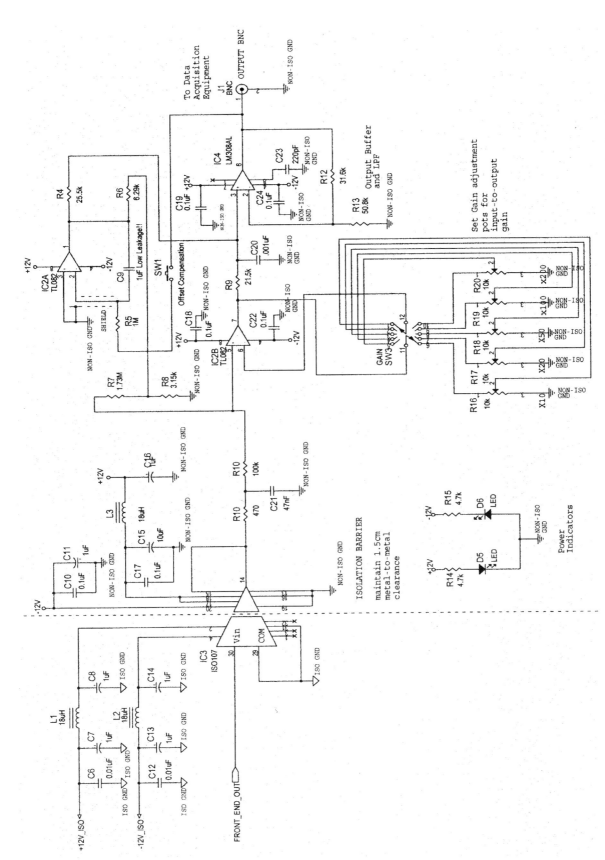

Figure 2.3 The output of the ISO107 isolation amplifier is fed to IC2B, which has its gain selectable through switch SW3. The circuit built around IC2A nulls dc offsets automatically when SW1 is closed. The features of this biopotential amplifier make it an ideal choice for recording cardiac monophasic action potentials (MAPs) using electrodes in direct contact with the heart.

45

In the biopotential amplifier of Figure 2.4, dc and very low frequency potentials are prevented from propagating beyond the front-end amplifier through a technique commonly referred to as *dc rejection*. In the circuit, signals picked up by electrodes attached to the patient's skin are dc-coupled and amplified by IC1, a Burr-Brown INA110 instrumentation amplifier IC. The gain of the front-end stage is programmable between unity and 500 by jumpers JP11–JP14. Potentiometer R17 is used to trim the input offset to IC1. Since IC1 is dc-coupled, care must be exercised in the selection of gain so that the amplifier is not saturated by dc offset voltages accompanying the biopotential signal. For example, to use this circuit as part of a surface ECG amplifier, the gain must be calculated to cope with offset potentials of up to ± 300 mV. In general, IC1's gain should be kept low so that dc-coupling does not result in its saturation.

To reject dc, IC4C together with R11 and C17 are used to offset IC1's reference to suppress a baseline composed of components in the range dc to 0.48 Hz. Once the dc component is removed, the dc-free biopotential signals are amplified via IC4A and IC4B. Notice that we used clipping diodes at the inputs and feedback paths of this specific implementation. Our application involved measuring the small electrical response of cardiac cells after the delivery of large stimuli. If you build this circuit, you may chose to leave D4–D7 and D9–D12 out of the circuit.

Galvanic isolation is provided by IC2, a Burr-Brown ISO107 isolation amplifier IC. In addition to providing a signal channel across the isolation barrier, the ISO107 has an internal dc–dc converter which powers the isolated side of the ISO107 circuitry as well as providing isolated power (± 15 V at ± 15 mA typical) for the rest of the circuitry of the isolated front end (i.e., IC1 and IC4). The output gain of IC2 is selected through jumpers JP4–JP6 to provide gains of 1, 10, or 100. IC3's output is then low-pass filtered by IC3.

AC-COUPLED INSTRUMENTATION BIOPOTENTIAL AMPLIFIER FRONT END

The circuit of Figure 2.5 embodies the classic implementation of a medium-impedance (10-MΩ) instrumentation biopotential amplifier based on the popular AD521 ICIA by Analog Devices. The gain of this circuit is adjustable between 10 and 1000 and maintains a CMR of at least 110 dB. This circuit offers superior dynamic performance with a minimal ac-coupled signal bandwidth (-3 dB) of 40 kHz and low noise ($1\,\mu V_{p\text{-}p}$ at $G = 10$, 0.1–100 Hz). This circuit is an example of a biopotential amplifier front end suitable for recording EMG or ECG signals or as a general-purpose high-impedance ac-coupled transducer amplifier.

The heart of the circuit is IC1, the monolithic IC instrumentation amplifier. Biopotentials are ac-coupled to the amplifier's inputs through C1 and C2. Although instrumentation amplifiers have differential inputs, bias currents would charge stray capacitances at the amplifier's input. As such, resistors R1 and R2 are required to provide a dc path to ground for the amplifier's input bias currents. These resistors limit the impedance of each input to 10 MΩ referred to ground. The high-pass filter, formed by the ac-coupling capacitor and the bias shunt resistor on each of the ICIA's inputs, has a -3-dB cutoff frequency of 0.12 Hz.

The gain of IC1 is determined by the ratio between R3 and R4. Using a 20-kΩ multiturn potentiometer, and given that the value of the range-setting resistor R3 is 100 kΩ, the differential gain of the amplifier can be trimmed between 5 and 1000. The output offset of the amplifier can be trimmed through R5, which, at any given gain, introduces an output offset equal and opposite to the input offset voltage multiplied by the gain. Thus, the total output offset can be reduced to zero by adjusting this potentiometer. The instrumentation amplifier provides a low-impedance output (0.1 Ω) with a permissible swing of ± 10 V and can source or sink up to 10 mA.

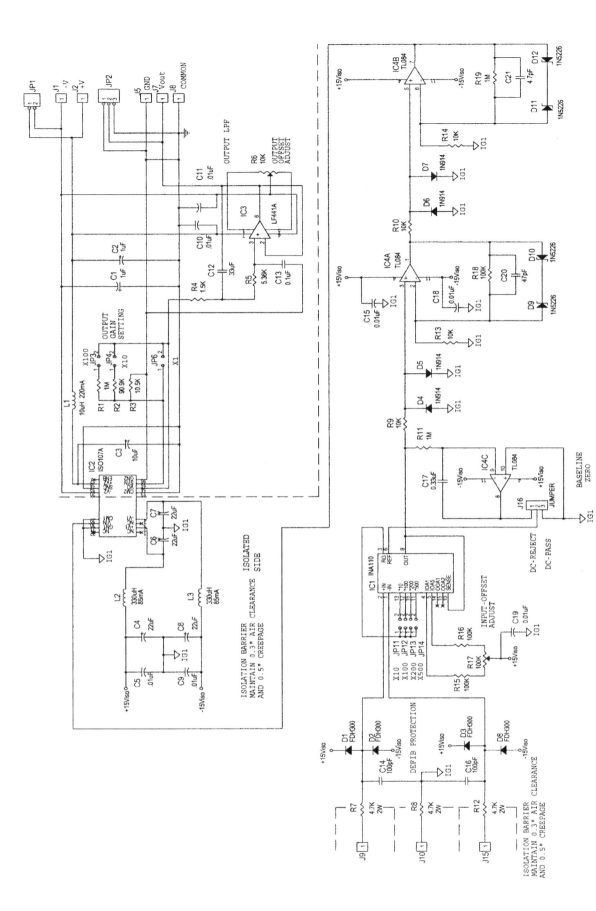

Figure 2.4 Dc and very low frequency potentials are prevented from propagating beyond the front-end amplifier through a technique commonly referred to as *dc rejection*. Here, IC4C, together with R11 and C17, are used to offset IC1's reference to suppress a baseline composed of components in the range dc to 0.48 Hz.

47

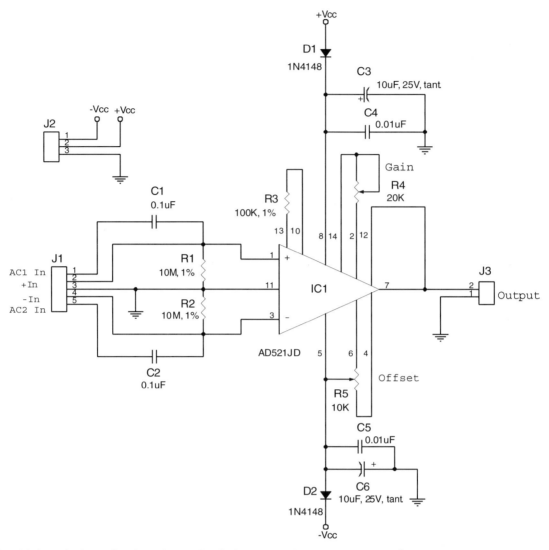

Figure 2.5   This is a classic medium-impedance (10-MΩ) instrumentation biopotential amplifier based on the popular Analog Devices' AD521 ICIA. The gain is adjustable between 10 and 1000 and maintains a CMR of at least 110 dB. The 40-kHz signal bandwidth makes this front end suitable for recording EMG or ECG signals or as a general-purpose high-impedance ac-coupled transducer amplifier.

Supply voltage to the AD521 must be symmetrical and within the range ±5 V (minimum) to ±18 V (absolute maximum). D1 and D2 provide protection against incorrect supply voltage polarity, and capacitors C3–C6 are used to decouple and filter the power supply. With a quiescent maximal supply current of 5 mA, a pair of 9-V alkaline batteries constitute a suitable power supply for most applications.

To minimize electrical interference the circuit should be built with a compact layout on an appropriate PCB or small piece of strip board. The construction of the circuit is straightforward, but care must be taken to keep wiring as short and clean as possible. Preferably, leads to the bioelectrodes should be low-loss low-capacitance coaxial cables, whose shields are connected to the subject ground terminal of the circuit. The construction of systems incorporating this circuit is simple, but care must be taken to keep all wiring as short and as clean as possible. When connected to a test subject, the circuit must always be powered from batteries or through a properly rated isolation power supply. The same isolation requirements apply to the output of the amplifier.

BOOTSTRAPPED AC-COUPLED BIOPOTENTIAL AMPLIFIER

Direct ac coupling of the instrumentation amplifier's inputs by way of *RC* high-pass filters across the inputs degrades the performance of the amplifier. This practice loads the input of the amplifier, which substantially lowers input impedance and degrades the CMRR of the differential amplifier. Although unity-gain input buffers can be used to present a high-input impedance to the biopotential source, any impedance mismatch in the ac coupling of these to an instrumentation amplifier stage degrades the CMR performance of the biopotential amplifier.

Suesserman has proposed an interesting modification of the standard biopotential instrumentation amplifier to yield an ac-coupled differential amplifier that retains all of the superior performance inherent in dc-coupled instrumentation amplifier designs. The circuit of Figure 2.6 is described by Suesserman [1994] in U.S. patent 5,300,896. If capacitors C3 and C4 were not present, the circuit of Figure 2.6 would be very similar to that of the ac-coupled instrumentation amplifier described earlier in the chapter. ICIA IC1 without C3 and C4 would be ac-coupled to the biopotential signal via capacitors C1 and C2. Just as in the earlier ac-coupled biopotential amplifier, resistors R1, R2, R3, and R4 are needed to provide a dc path to ground for the amplifier's input bias currents. In this circuit, these resistors would limit the ac impedance of each input to $2\,M\Omega$ (R1 + R2 and R3 + R4) referred to ground.

With C3 and C4 as part of the circuit, however, ac voltages from the outputs of the ICIA's differential input stage are fed to the inverting inputs of their respective amplifiers. This causes the ac voltage drop across R1 and R4 to be virtually zero. Ac current flow through resistors R1 and R4 is practically zero, while dc bias currents can flow freely to ground. This technique is known as *bootstrapping*, referring allegorically to the way in which the amplifier nulls its own ac input currents, as when one pulls his or her own bootstraps to put boots on.

Since bootstrapping capacitors C3 and C4 almost completely eliminate ac current flow through R1 and R4, the input current through ac-coupling capacitors C1 and C2 would also drop close to zero, which by Ohm's law translates into an almost infinite input impedance (since $R = V/i$; R tends to ∞ as i approaches 0). Suesserman described this biopotential amplifier as having an impressive 120-dB CMRR (at 100 Hz) with an input impedance of more than $75\,M\Omega$.

PASSIVE FILTERS

The simplest filters are those that comprise only passive components. These filters contain some combination of resistive (*R*), capacitive (*C*), and inductive (*L*) elements. The inductive and/or capacitive components are required because these elements present varying impedance to ac currents at different frequencies. As a refresher, you may remember that inductive reactance increases with frequency, whereas capacitive reactance decreases with frequency. Most passive filters used in the processing of biopotential signals are the resistive–capacitive or *RC* kind. This is because relatively large and heavy inductors would be required to implement filters at the low-frequency bands where biopotential signals reside, making inductive–capacitive (*LC*) filters impractical.

Despite their simplicity, *RC* filters are very common and effective in processing a wide variety of biopotential signals. Take, for example, the complete biopotential amplifier

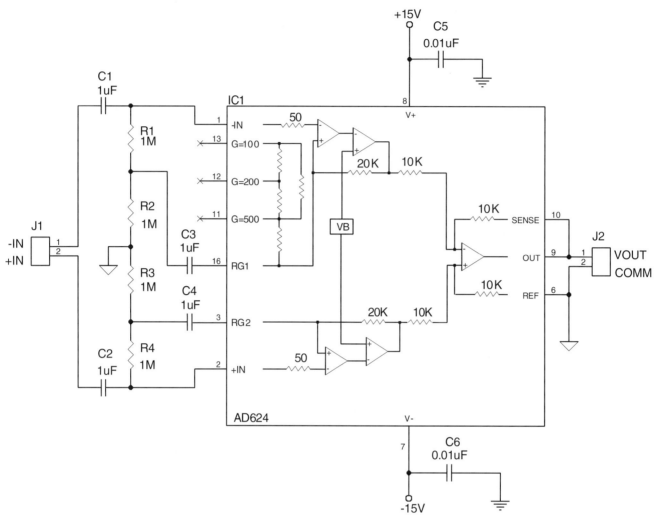

Figure 2.6   This bootstrapped design yields an ac-coupled differential amplifier that retains all of the superior performance inherent in dc-coupled instrumentation amplifiers. Ac voltages from the outputs of the ICIA's differential input stage are fed to the inverting inputs of their respective amplifiers via C3 and C4. This causes the ac voltage drop across R1 and R4 to be virtually zero. Ac current flow through resistors R1 and R4 is practically zero, while dc bias currents can flow freely to ground.

presented in the schematic circuits of Figures 2.7 through 2.12. In this design, biopotential signals are amplified by IC5, a Burr-Brown INA128U instrumentation amplifier. Its gain is fixed at 138 through resistor R7. The input of the amplifier is protected from high-voltage transients and electrosurgery currents by a network of resistors, capacitors, and diodes. Back-to-back zener diodes D2 and D4 clamp high-voltage transients induced into the electrodes by defibrillation currents to a level that can be handled by the rest of the protection network. C21 acts as a shunt for radio-frequency currents that may be induced into the electrodes and leads by sources of electromagnetic interference. This capacitor by itself has inherent filtering capability for high-frequency alternating current because capacitive reactance X_C (in ohms) is inversely proportional to frequency:

$$X_C = \frac{1}{2\pi f C}$$

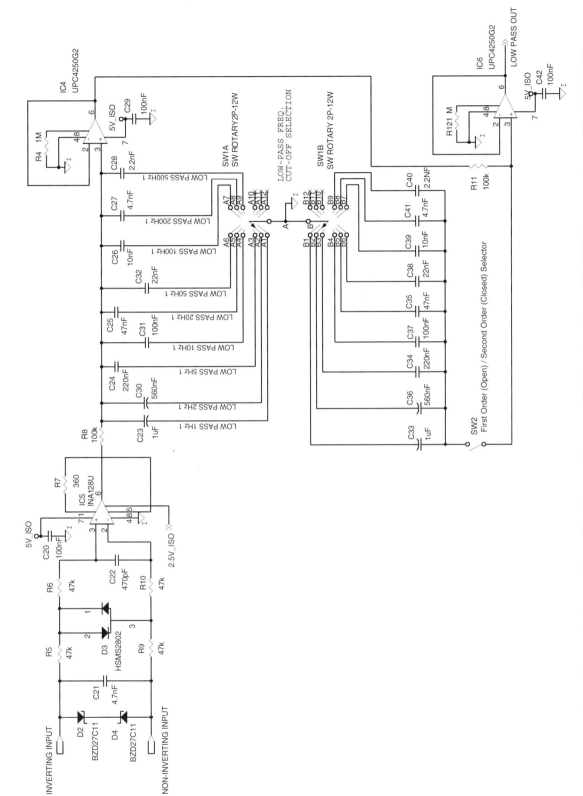

Figure 2.7 In this biopotential amplifier, biopotentials are amplified by a Burr-Brown INA128U instrumentation amplifier. R5, R6, R9, R10, and C22 implement a low-pass filter with a cutoff of approximately 3.6 kHz. The biopotential amplifier's main low-pass filters are implemented by two cascaded *RC* passive filters with selectable cutoff frequency. IC4 buffers the signals between the cascaded sections. The two *RC* sections are identical, therefore setting a pole at the same frequency. However, the effect of the second *RC* can be suppressed by disconnecting its capacitor through switch SW2.

51

As such, this capacitor is practically a short circuit for RF currents, while it leaves low-frequency signals pass unimpeded.

R5, R6, R9, R10, and C22 implement a low-pass filter. Because of the virtual-ground property governing the inputs of an op-amp, we can assume that the low-pass characteristics of this filter are given by a 94-kΩ resistor (either R5 and R6 in series, or R9 and R10 in series) and the 470-pF capacitor. The -3-dB cutoff frequency for an RC low-pass filter is

$$f_{-3dB} = \frac{1}{2\pi RC}$$

which provides the input network with a first-order low-pass cutoff of approximately 3.6 kHz.

The biopotential amplifier's main low-pass filters are implemented by two cascaded RC passive filters. An op-amp unity-gain follower (IC4) buffers the signals between the cascaded sections. Both RC sections are identical, therefore setting a pole at the same frequency. However, the effect of the second RC can be suppressed by disconnecting its capacitor through switch SW2. When SW2 is open, signals at the output of IC4 are fed to unity-gain buffer IC6 through R11. Since the input impedance of IC6 is practically infinite, R11 has no effect on the signal. However, when SW2 is closed, R11 and the capacitor selected by SW1B form a low-pass filter. The nominal cutoff frequencies that can be selected for the second-order filter were selected to be close to 1, 2, 5, 10, 20, 50, 100, 200, and 500 Hz. The exact -3-dB cutoff frequencies are shown in Table 2.3.

The high-pass filters are implemented in essentially the same way as the low-pass sections. In Figure 2.8, however, the RC elements are reversed. Each high-pass section has a capacitor (C50 and C53) which opposes current flow with an impedance that varies inversely with frequency, and a resistor of selectable value that shunts the load. Both RC sections are identical, therefore setting a pole at the same frequency. However, the effect of the second RC can be suppressed by shorting C53 through SW5. Op-amp IC13 buffers the signal between the stages. The nominal cutoff frequencies that can be selected for the second-order filter were selected to be close to 1, 2, 5, 10, 20, 50, 100, 200, and 500 Hz. The exact -3-dB cutoff frequencies are shown in Table 2.4.

The first follower (IC13) in the high-pass filter is implemented using a LTC1152 instead of a UPC4250 op-amp as in the case of the other followers because higher current output is required to drive the lowest resistor values associated with the highest -3-dB cutoff

TABLE 2.3 Low-Pass -3-dB Cutoff Frequencies for the Biopotential Amplifier of Figure 2.7 Selected through SW1[a]

SW1 Position	-3-dB Cutoff Frequency for Second-Order Low-Pass Filter (Hz) (SW2 Closed)	-3-dB Cutoff Frequency for First-Order Low-Pass Filter (Hz) (SW2 Open)
1	1.02	1.59
2	1.83	2.84
3	4.65	7.23
4	10.22	15.92
5	21.75	33.86
6	46.47	72.34
7	102.2	159.2
8	217.5	338.6
9	464.7	723.4

[a] SW2 selects between first- or second-order response.

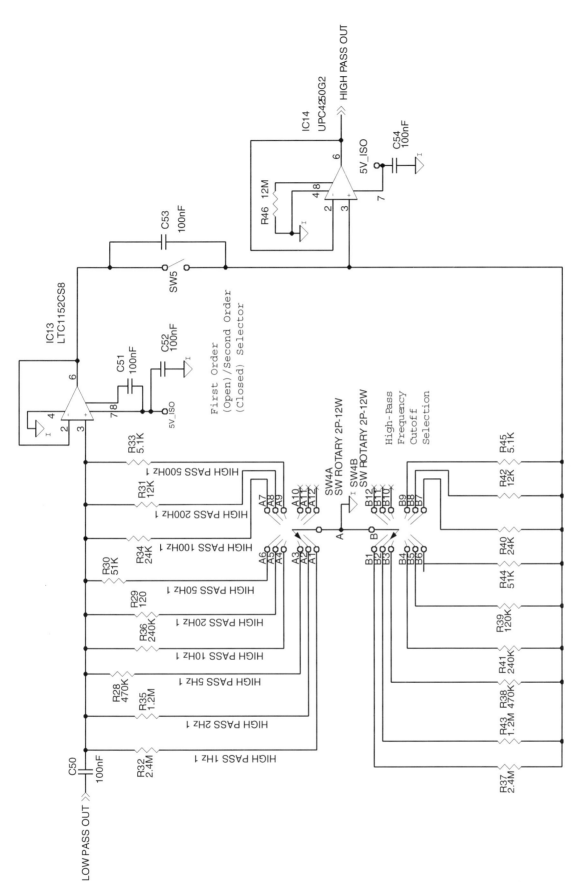

Figure 2.8 The high-pass filters for the amplifier of Figure 2.7 are implemented in essentially the same way as the low-pass sections. Each high-pass section has capacitors (C50 and C53) which oppose current flow with an impedance that varies inversely with frequency and a resistor of selectable value that shunts the load. The *RC* sections are identical, therefore setting a pole at the same frequency. However, the effect of the second *RC* can be suppressed by shorting C53 through SW5. Op-amp IC13 buffers the signal between the stages.

TABLE 2.4   High-Pass −3-dB Cutoff Frequencies for the Biopotential Amplifier of Figure 2.8 Selected through SW4[a]

SW4 Position	−3-dB Cutoff Frequency for Second-Order High-Pass Filter (Hz) (SW5 Open)	−3-dB Cutoff Frequency for First-Order High-Pass Filter (Hz) (SW5 Closed)
1	1.03	0.66
2	2.06	1.33
3	5.27	3.39
4	10.32	6.63
5	20.65	13.26
6	48.59	31.21
7	103.3	66.31
8	206.5	132.6
9	485.9	312.1

[a] SW5 selects between first- or second-order response.

frequencies when SW5 is closed. Figure 2.9 shows the notch filters that are used to filter power line interference. More detail will be presented on notch filters later in the chapter. Suffice it to say for now that one filter (built around IC15 and IC17) has a notch at around 50 Hz, while the other (built around IC16 and IC18) has a notch at 60 Hz. Trimmers R59 and R60 are used to fine-tune the notch frequency, while R57 and R58 select the notch depth.

The circuit of Figure 2.10 can be made to process the output signal coming out of the notch filters. This circuit performs full-wave rectification on the input signal. Zero-threshold rectification is achieved by placing the rectifier diodes (D6) within the feedback loop of op-amp IC11. Full-wave rectification results from adding an inverted half-wave-rectified signal at double amplitude to the original signal in the summing amplifier IC12. Full-wave rectification is an operation that is often used when the desired information can be extracted by analyzing the energy conveyed by the biopotential signal. For example, the EMG signal is often rectified and then low-pass-filtered to yield a signal proportional to the force generated by a muscle. The full-wave rectifier can be bypassed through SW3.

The signal is then buffered by IC10 in Figure 2.11 before it is optically isolated from recording instruments connected at output connector J2. Galvanic isolation ensures that the source of biopotential signals (e.g., a patient) is not exposed to dangerous currents leaked from power lines through the subject's heart. This function is implemented through a Hewlett-Packard HCNR201 high-linearity analog optocoupler. This optocoupler includes one LED and two photodiodes, the output photodiode and an input photodiode designed to receive the same light intensity from the LED as the output photodiode. The LED current is controlled through a feedback loop so that the current at the input photodiode is proportional to the voltage of the analog signal at the input of IC10. Under the hypothesis that both photodiodes receive the same light intensity, the current at the output photodiode also follows the input analog signal.

Op-amp IC8 and capacitor C43 integrate the difference between the current through R18 and the current through the input photodiode. The output of this integrator drives the LED so that these currents are equalized. Therefore, the current through the input photodiode (which will be equal to the current of the output photodiode) is equal to the input analog signal divided by R18. The analog signal is recovered at the output through the bandwidth-limited current-to-voltage converter implemented by IC9, R15, and C44. The output of the integrator of the input stage drives the LED through a constant-transconductance stage implemented by Q1, R16, R14, R13, and R17.

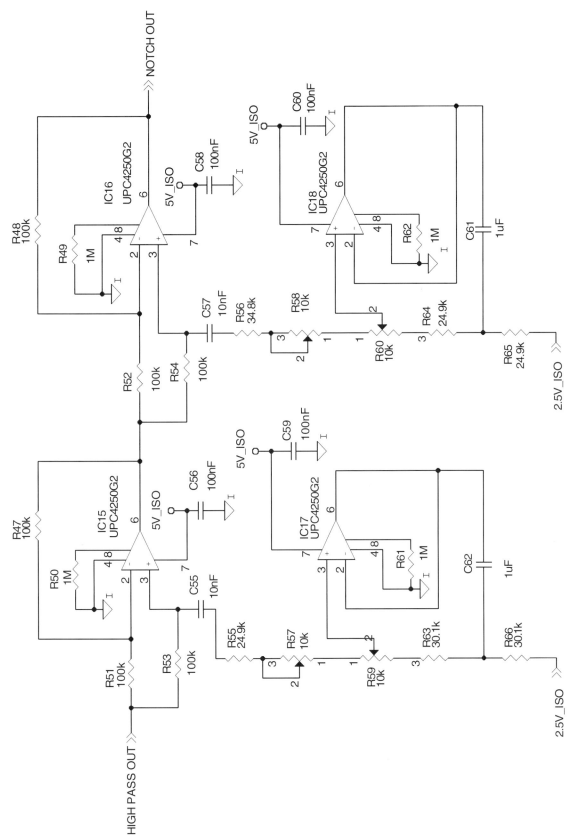

Figure 2.9 Two notch filters are used to reduce power line interference that may be picked up by the amplifier of Figure 2.7. The filter built around IC15 and IC17 has a notch at around 50 Hz, while the other (built around IC16 and IC18) has a notch at 60 Hz.

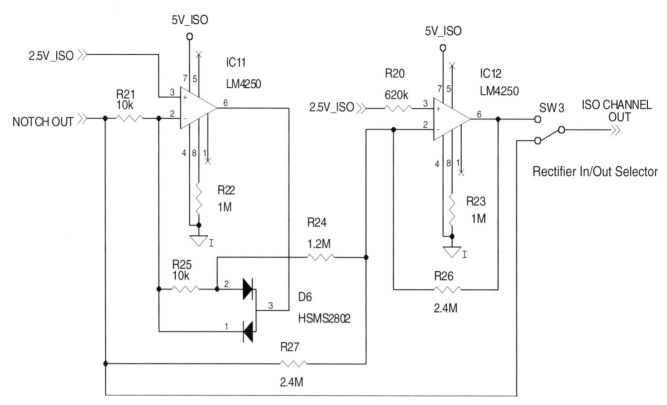

Figure 2.10   This circuit can be used to full-wave-rectify the signal at the output of the notch filters of Figure 2.9. This signal-processing operation is often used in electromyography (EMG) to yield a signal proportional to the force generated by a muscle. The full-wave rectifier can be bypassed through SW3.

Power for the circuit is supplied by the two isolated dc/dc converters (IC2 and IC3) shown in Figure 2.12. 5V_ISO feeds the parts of the circuit that are in galvanic connection with the biopotential source, while +5V_DATA_ACQ feeds the analog signal isolator's output circuit. IC1 buffers the output of the resistor divider formed by R2 and R3 to generate a synthetic analog ground halfway between the isolated ground and the isolated supply (5V_ISO). The dc/dc converters operate from a single +15-V power supply. An appropriate choice is a switched ac/dc medical-grade power supply, manufactured by Condor, model GLM75-15.

ACTIVE FILTERS

Passive filters have their advantages: They consist of simple components, use no gain elements, and require no power supply. Their noise contribution to biopotential signals is limited to the thermal noise from their resistive components, and they can be used in applications which require them to handle large currents and voltages. Despite this, there are many applications which require filter functions that a passive filter could not achieve without the use of inductors or which would become cumbersome and impractical because of the interactions between successive stages. This is where active filters come to the rescue. Active filters use op-amps, along with resistors and capacitors, to implement the desired filter function. Op-amps can be used to simulate the characteristics of an inductor but without the bulk and expense. In addition, higher-order filters can be implemented with ease because cascaded stages have little interaction with each other.

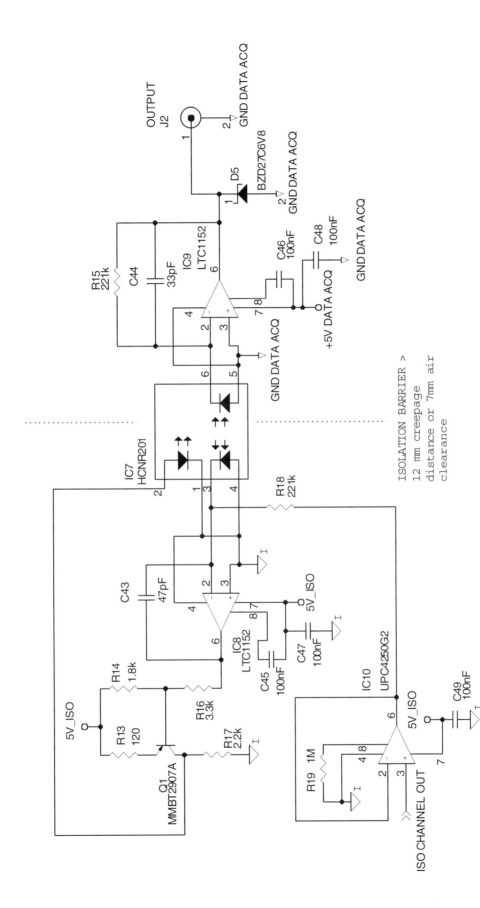

Figure 2.11 The signal at the output of Figure 2.10 is buffered by IC10 and then optically isolated from recording instruments connected at output connector J2. Galvanic isolation to protect the subject from dangerous leakage currents is done through a Hewlett-Packard HCNR201 high-linearity analog optocoupler.

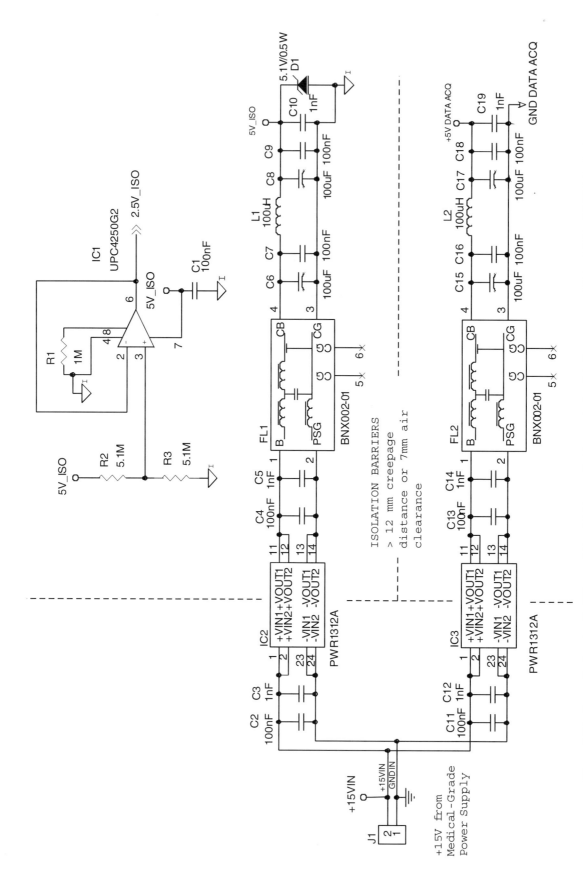

Figure 2.12 Power for the circuit of Figures 2.7 to 2.11 is supplied by two isolated dc/dc converters (IC2 and IC3). 5V_ISO feeds the parts of the circuit that are in galvanic connection with the biopotential source, and +5V_DATA_ACQ feeds the analog signal isolator's output circuit. IC1 generates a synthetic analog ground. The dc/dc converters should be operated from a medical-grade 15-V power supply.

Many good books and articles have been written on the design of active filters, and we will not try to duplicate their efforts. In our view, the books with the most practical approach for the experimentalist are:

- D. Lancaster, *Active Filter Cookbook*, Synergistics Press, 1995.
- P. Horowitz and W. Hill, *The Art of Electronics*, 2nd ed., Cambridge University Press, New York, 1989.
- H. M. Berlin, *The Design of Active Filters, with Experiments*, Howard W. Sams, Indianapolis, IN, 1974.

Designing active filters is not difficult. There are a number of free software packages that will take your input parameters and provide you automatically with a schematic diagram and calculate capacitor and resistor values for specific filter implementations. This doesn't mean that the programs will do everything for you. You still have to decide what type of *filter response* and *implementation* suit your application.

Filter response refers to the shape of a filter's transfer function. Everyone's first approximation to filtering physiological signals is to assume a frequency-domain rectangular passband containing the spectral components of interest while excluding potential interference sources. However, real-world filters do not yield a perfect step in the frequency domain. In fact, to produce such a response would require an infinite number of poles (implemented through an infinite number of amplifiers, resistors, and capacitors) and would result in a filter that is inherently unstable in the time domain. Because of these reasons, real-world filters make use of stable approximations to a perfect step in the frequency domain. Some of the most common filter responses are the Butterworth, Chebyshev, and Bessel. Each of these filter responses has advantages and disadvantages, and it is the designers task to find a suitable compromise that best fits the task at hand. Table 2.5 summarizes the frequency- and time-domain characteristics of these filters, and Figure 2.13 shows the magnitude and phase responses for fourth-order Chebyshev, Butterworth, and Bessel transfer functions with a -3-dB cutoff frequency of 30 Hz.

The Butterworth response (also known as *maximally flat*) is nearly flat in the passband and rolls off smoothly and monotonically. In addition, it has virtually no ripple in either the passband or the stopband. For these reasons, many designers regard the Butterworth filter transfer function as the best compromise between attenuation and phase response for general-purpose applications. This transfer function is certainly the most commonly used in the design of analog biopotential signal filters. Despite this, applications that require a precise estimation of phase shift are better served by Bessel filters, since its phase shift is linear, a property that is not shared by Butterworth or Chebyshev filters.

The next step to designing a filter is to select a suitable implementation. Here again, a compromise has to be made to achieve the desired filter transfer function with real-world analog components. The most common active filter topologies are described below.

TABLE 2.5   Characteristics of Some Common Filter Transfer Functions

Transfer Function	Frequency-Domain Characteristics		Time-Domain Characteristics	
	Ripple	Stopband	Phase	Group Delay
Chebyshev	Equal ripple flat	Steep	Poor	Poor
Butterworth	Smooth	Moderate	Moderate	Moderate
Bessel	Maximum smoothness	Weak	Very flat	Very flat

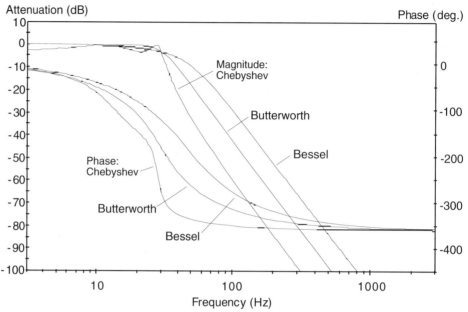

Figure 2.13   Real-world filters do not yield a perfect step in the frequency domain. Some of the most common filter responses are the Butterworth, Chebyshev, and Bessel. Each of these filter responses has advantages and disadvantages, and it is the designer's task to find a suitable compromise that best fits the task at hand from phase- and amplitude-response graphs such as this one for fourth-order filters with a -3-dB cutoff frequency of 30 Hz.

1. Sallen–Key topology (also known as the *voltage-controlled voltage-source topology*) uses an op-amp as a gain block. Because of this, the Sallen–Key configuration is relatively independent of op-amp specifications and requires the op-amp's bandwidth only to extend slightly beyond the filter stopband frequency. The Sallen–Key topology features good phase response, but its frequency response and Q are sensitive to the gain setting.

2. Multiple-feedback topology uses op-amps as integrators that need a minimum loop gain of 20 dB (open-loop gain 10 times the closed-loop gain) to avoid Q *enhancement*, making it difficult to get high-Q performance. However, this filter configuration is relatively insensitive to passive-component values.

3. State-variable topology uses op-amps as amplifiers and integrators, which again need a minimum loop gain of 20 dB. In addition, the op-amps need a frequency response that is flat to beyond the stopband frequency. Despite this, state-variable filters provide independent control over gain, cutoff frequency, Q, and other parameters but require more passive components. A very nice feature of this topology is that the same circuit yields low-pass, high-pass, and bandpass response.

4. Impedance-converter topology (also known as *frequency-dependent negative-resistance* topology) requires op-amps with a minimum loop gain of 20 dB at the resonant negative resistance frequency. Multiple op-amps are needed, and use of dual-packaged devices is recommended for matched performance in each leg. FET-input op-amps are used because of their low bias currents. Although the impedance-converter approach requires more components, it is relatively insensitive to variations in their values.

Since biopotential signal filtering applications commonly have bandwidths limited to the audio range, the biggest trade-off is often the number of op-amps versus the level of control that a designer has over the filter. For a person inexperienced with the design of active filters, we recommend that you try the two and three op-amp topologies that will allow you more ability to "tweak" the end result. In addition, a good way of designing well-behaved filters is to base them on one of the various active filter building blocks offered by analog IC vendors. For example, Burr-Brown (now part of Texas Instruments) offers the UAF42, a universal active filter that can be configured for a wide range of low-pass, high-pass, and bandpass filters. It implements filter functions through a state-variable topology with an inverting amplifier and two integrators. The integrators include on-chip 1000-pF capacitors trimmed to 0.5%. This solves the difficult problems of obtaining tight-tolerance low-loss capacitors. The UAF42 is available in 14-pin DIP and SOL-16 surface-mounted packages.

Burr-Brown's free DOS-compatible FilterPro program lets you design Butterworth, Chebyshev, and Bessel filters, enter the desired performance, and then obtain the passive values required. You can force the program to use the nearest 1% resistors, set some resistor values, enter realistic or measured capacitor values, and then plot the actual gain/phase versus frequency performance. Similarly, Microchip's Windows-based FilterLab lets you design Sallen–Key or multiple-feedback low-pass filters with either Butterworth, Chebyshev, or Bessel responses using their MCP60x family of single-supply op-amps.

Maxim also offers a line of *state-variable filter ICs*, the MAX274 and MAX275. These ICs have independent cascadable second-order sections that can each implement all-pole bandpass or low-pass filter responses, such as Butterworth, Bessel, and Chebyshev, and is programmed by four external resistors. The MAX274 has four second-order sections, permitting eighth-order filters to be realized with center frequencies up to 150 kHz. The MAX275 has two second-order sections, permitting fourth-order filters to be realized with center frequencies up to 300 kHz. Both filters operate from a single +5-V supply or from dual ±5-V supplies. A free DOS-based filter design program is available from Maxim to support the development of applications based on the MAX274 state-variable filter IC.

State-variable filter realizations have the distinct advantage that they provide simultaneous low-pass, bandpass, and high-pass outputs from the same filter circuit. In addition, the filter parameters are independent of each other. For example, the cutoff frequency of the active-feedback state-variable filter circuit of Figure 2.14 is given by

$$f_C = \frac{1}{2\pi(R3)(C1)}$$

where R3 = R4 and C1 = C2. As shown in the ac-sweep PSpice simulation analysis of Figure 2.15, this filter yields simultaneous low-pass and high-pass responses with a -3-dB cutoff frequency f_C and a bandpass response centered at the same frequency. In this example, the resistor values selected for R4 and R6 give the filter a cutoff frequency of approximately 50 Hz. The Butterworth response on a state-variable filter gives it a value $Q = -3$ dB and an in-band gain of the bandpass filter equal to Q ($= 0.707$), making all curves cross at the same point.

Since the cutoff frequency of a state-variable filter depends on the value of two resistors (R3 and R4 in the prior example), it is relatively easy to design a tunable filter by substituting these resistors by two tracking variable resistors. The filter can also be made to have a cutoff frequency that is proportional to a control voltage by using circuits that present a variable resistance as a function of an input voltage.

Although FETs and variable transconductance amplifiers can be used as voltage-dependent resistors, better results are easier to achieve using analog multipliers in series with a resistor as the control elements. The circuit of Figure 2.16 shows how R3 and R4 of the circuit of Figure 2.14 have been replaced by two Analog Devices AD633 precision analog multipliers. The transfer function of the AD633 is given by

$$V_{out} = \frac{(x_1 - x_2)(y_1 - y_2)}{10\,\text{V}} + z$$

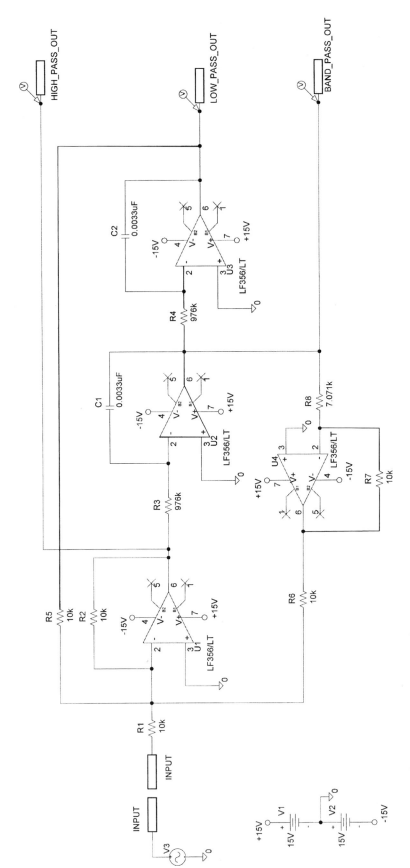

Figure 2.14 State-variable filter realizations have the distinct advantage that they provide simultaneous low-pass, bandpass, and high-pass outputs from the same filter circuit at a −3-dB cutoff frequency $f_c = 1/2\pi(R3)(C1)$.

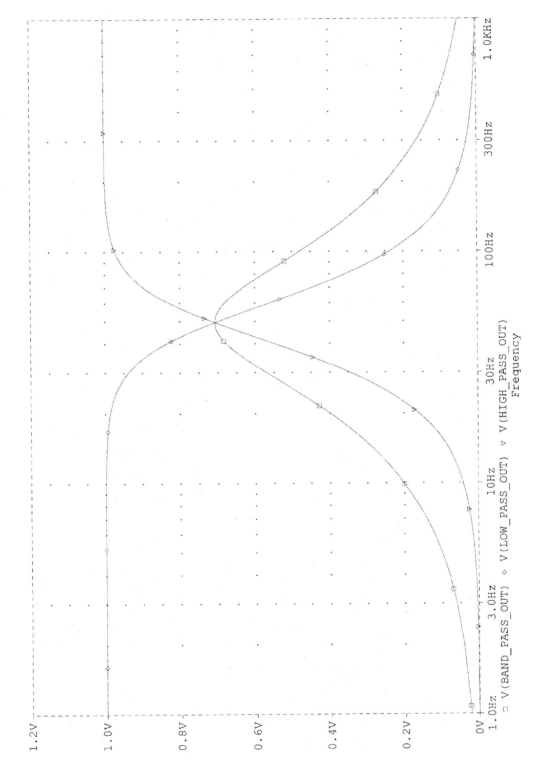

Temperature: 27.0

□ V(BAND_PASS_OUT) ◇ V(LOW_PASS_OUT) ▽ V(HIGH_PASS_OUT)
Frequency

Figure 2.15 A PSpice simulation ac sweep of the state-variable filter of Figure 2.14 yields simultaneous low- and high-pass responses with a −3-dB cutoff frequency f_c and a bandpass response centered at the same frequency. In this example, the resistor values selected for R4 and R6 give the filter a cutoff frequency of approximately 50 Hz.

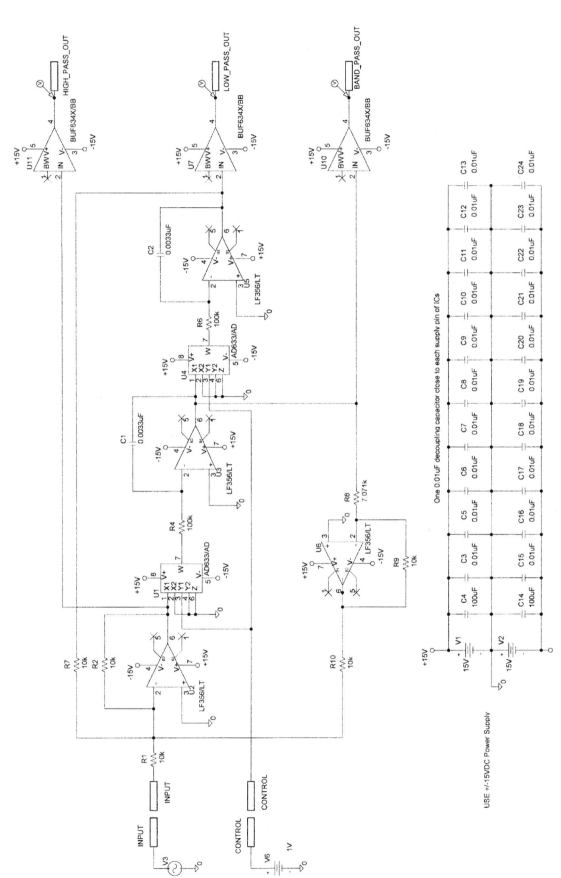

Figure 2.16 Since the cutoff frequency of a state-variable filter depends on the value of two resistors, it is possible to design a voltage-controlled tunable filter by substituting these resistors by two circuits that present a variable resistance as a function of an input voltage. For example, Analog Devices' AD633 precision analog multipliers can be configured such that the apparent impedance between x_1 and their output is proportional to the control voltage at y_1. This makes the cutoff frequency of the high- and low-pass filter outputs as well as the center frequency of the bandpass section be $f_c = $ (control voltage)$/20\pi(R3)(C1)$.

where z is an offset input. In the tunable filter of Figure 2.16, the AD633 is configured such that the apparent impedance between x_1 and the output is proportional to the control voltage at y_1. This makes the cutoff frequency of the high- and low-pass filter outputs, as well as the center frequency of the bandpass section,

$$f_c = \frac{\text{control voltage}}{20\pi(R3)(C1)}$$

PSpice simulation results shown in Figure 2.17 demonstrate the effect of varying the control voltage presented to the y_1 inputs of the AD633s. Although the circuit is shown set up for PSpice simulation, it can be built using real components. Output buffering using Burr-Brown BUF634 buffers make this a very useful stand-alone lab instrument that can be used to filter amplified biopotential signals selectively prior to recording. A digitally programmable version of the tunable filter can be made by substituting two multiplying D/A converters for the AD633s. In this case, the control voltage is replaced by a digital control word supplied to the input of the D/A converters.

50/60-Hz NOTCH FILTERS

Probably the most common problem in the detection and processing of biopotential signals is power line interference. Sixty hertz (50 Hz in Europe) and its harmonics manages to creep into low-level signals despite the use of differential amplification methods and active body potential driving which attempt to eliminate common-mode signals. Unfortunately, 50/60 Hz falls right within the band where biopotentials and other physiological signals have most of their energy. The usual solution to reject unwanted in-band frequencies is the notch filter.

As shown in Figure 2.18, simple implementation of a notch filter known as a *twin-T filter* requires only three resistors and three capacitors. If C1 = C3, C2 = 2C1, R1 = R3, and R2 = R1/2, the notch frequency occurs where the capacitive reactance equals the resistance ($X_C = R$) and is given by

$$f_{\text{notch}} = \frac{1}{2\pi(R1)(C1)}$$

As such, the twin-T notch filter works by phase cancellation of the input signal. When the phase shift in the two sections is exactly $+90$ and -90, the tuned frequency is canceled completely. Signals passed by the filter will experience some distortion since the twin-T notch shifts the phase of low-frequency components ($<f_{\text{notch}}$) by -90 and high-frequency components ($>f_{\text{notch}}$) by $+90$. The insertion loss of the filter will depend on the load that is connected to the output, so the resistors should be of much lower value than the load for minimal loss. The depth and width of the response can be adjusted somewhat with the value of R2 and by adding some resistance across the capacitors.

Twin-T notch filters can achieve very good suppression at their center frequency. However, the use of precise and tightly matched components is extremely important to yield a deep notch at the required frequency. The depth of the notch is defined as the output signal ratio between an out-of-notch component and a component at the notch frequency. In practice, a twin-T notch built with tightly matched components can yield pretty good notch-frequency attenuations. Passive notch filters can be built into small enclosures and placed between equipment stages. For example, the notch filter of Figure 2.19 was built inside a Pomona Electronics model 2391 box, which comes with BNC connectors on each end, making it easy to place it at the input of oscilloscopes and signal recorders.

For most practical applications, however, an op-amp needs to be added to the twin-T network to increase its notch depth as well as to make it insensitive to the impedance of

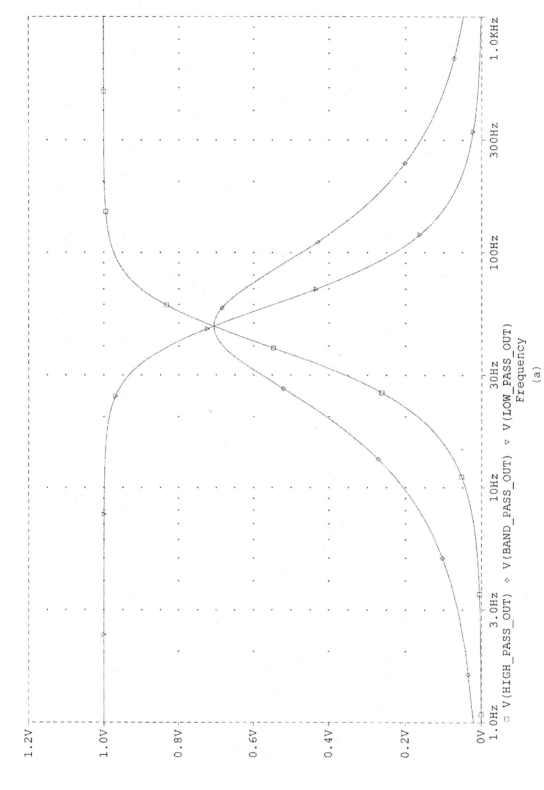

Figure 2.17 PSpice simulation results of the voltage-tunable state-variable filter of Figure 2.16 demonstrate the effect of varying the control voltage presented to the y_1 inputs of the AD633s: (*a*) a cutoff frequency of 50 Hz at 1-V shifts to (*b*) 300 Hz at 6 V.

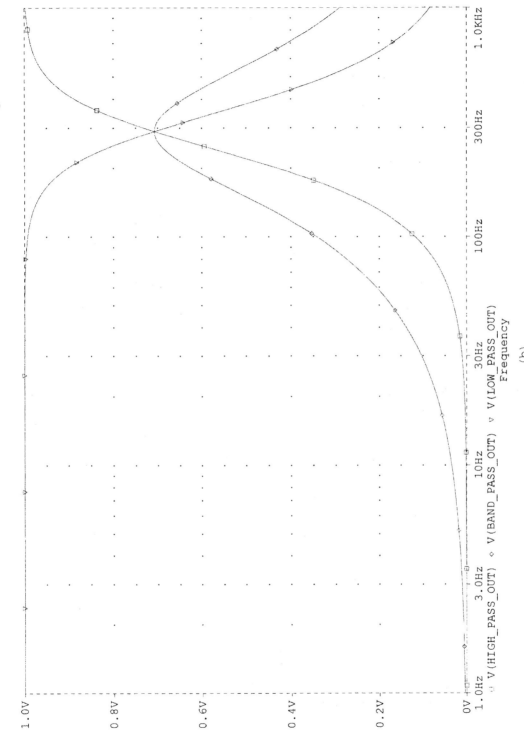

Figure 2.17 (*Continued*)

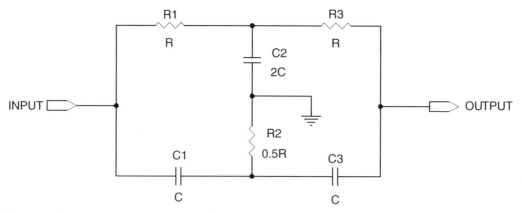

Figure 2.18 The notch frequency of the simple twin-T filter is $f_{notch} = 1/2\pi(R1)(C1)$ if C1 = C3, C2 = 2C1, R1 = R3, and R2 = R1/2. However, the use of precise and tightly matched components is extremely important to yield a deep notch at the required frequency.

Figure 2.19 Passive notch filters can be built into small enclosures with BNC connectors on each end and placed between equipment stages.

the output load. The circuit of Figure 2.20 shows how a unity-gain follower bootstraps the network in an active twin-T notch filter. Since output of the op-amp presents a very low impedance, the notch frequency and depth are not changed. However, the Q value of the filter increases proportionately to the level of signal that is fed back to the junction of R2 and C2 (the point that is grounded in a passive twin-T notch filter).

A very high Q value is not always desirable to filter power line interference. The power line frequency in many countries deviates quite a bit from the nominal 50 or 60 Hz. A second op-amp can be added as shown in Figure 2.21 to control the Q value of the filter. Here, the amount of feedback that is provided to the R2/C2 junction is set by potentiometer R4. An op-amp is needed to buffer the feedback signal to ensure a constant low impedance at the R2/C2 junction so that the notch frequency and depth do not change as a function of the potentiometer setting.

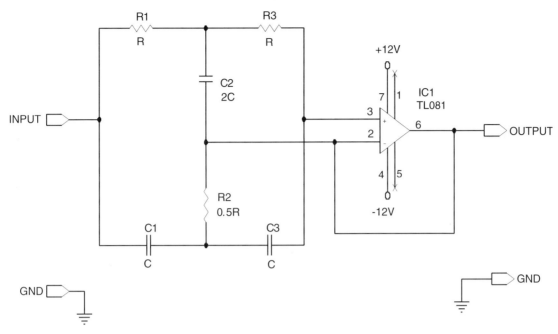

Figure 2.20 An op-amp in unity-gain configuration can be added to the twin-T network to increase its notch depth as well as to make it insensitive to the impedance of the output load. Since the output of the op-amp presents a very low impedance, the notch frequency and depth are not changed. However, the Q value of the filter increases proportionately to the level of signal that is fed back to the twin-T.

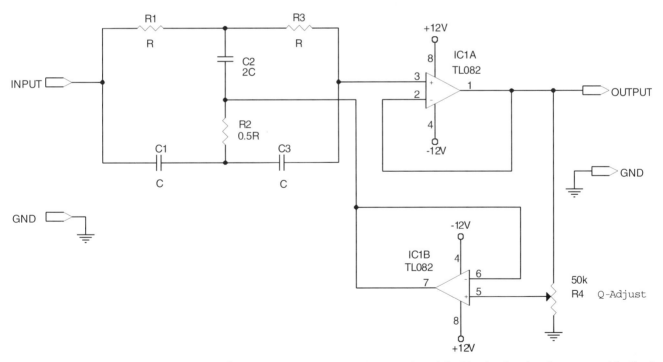

Figure 2.21 An op-amp can be added to the filter of Figure 2.20 to control the Q value of the filter by changing the amount of feedback that is provided to the twin-T. The op-amp is needed to buffer the feedback signal to ensure a constant low impedance at the twin-T so that the notch frequency and depth do not change as a function of the potentiometer setting.

Another popular notch filter is the *gyrator filter*. This is the type of notch filter that was implemented for the biopotential amplifier of Figure 2.9. For the sake of clarity, one notch filter section has been redrawn, simplified, and relabeled in Figure 2.22. In essence, the properties of an inductor are simulated by a simple op-amp circuit called a *synthetic inductor or gyrator*. The impedance at the noninverting input of the synthetic inductor op-amp

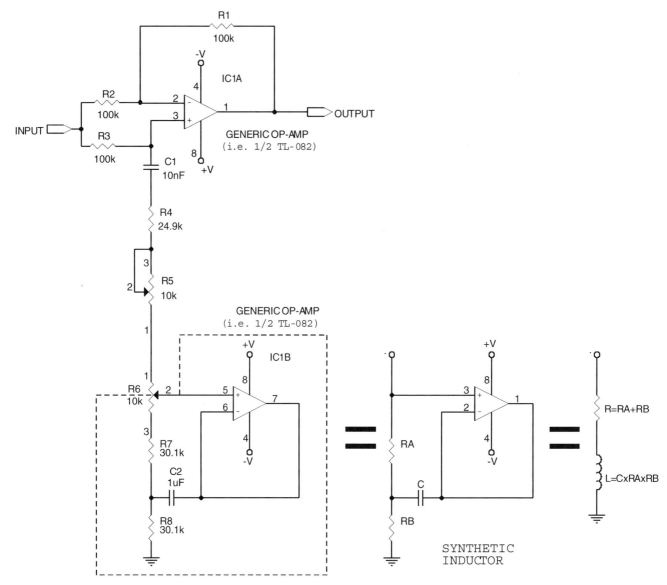

Figure 2.22 The properties of an inductor are simulated in this notch filter by a simple op-amp circuit called a synthetic inductor or gyrator. The impedance at the noninverting input of the synthetic inductor op-amp is equivalent to that of an inductor $L = CR_A R_B$ in series with a loss resistance $R = R_A + R_B$. For the values shown in the circuit, the equivalent inductance has an approximate value of 900 H, yielding a notch frequency of

$$f_{notch} = \frac{1}{2\pi\sqrt{LC}} = \frac{1}{2\pi\sqrt{(900\,\text{H})(10\,\text{nF})}} = 53\,\text{Hz}$$

The output of IC1A tracks the filter's input except for signals close to the notch. The depth of the notch is controlled via R5, while the exact notch center frequency can be trimmed via R6.

is equivalent to that of an inductor $L = CR_A R_B$ in series with a loss resistance $R = R_A + R_B$. For the values shown in the circuit, the equivalent inductance has an approximate value of 900 H. When this inductor is placed in series with a 10-nF capacitor (C1), ac signals reaching the noninverting input of op-amp IC1A are shunted to ground via the series resistances (R4 + R5 + R6 + R7 + R8) only for frequencies close to

$$\frac{1}{2\pi\sqrt{LC}} = \frac{1}{2\pi\sqrt{(900\,\text{H})(10\,\text{nF})}} = 53\,\text{Hz}$$

The output of IC1A tracks the filter's input except for signals close to the notch. The depth of the notch is controlled via R5 (lower value = deeper notch), while the exact notch center frequency can be trimmed via R6.

Yet another approach to the design of a notch filter is to combine a low- and a high-pass filter to yield a filter that excludes only notch frequencies from its bandpass. You will recall from our earlier discussion that a state-variable filter produces simultaneous high-pass, low-pass, and bandpass outputs. Looking at the intersection of the low- and high-pass outputs of the state-variable filter shown in Figure 2.15, it is easy to see how an additional op-amp configured to sum the high-pass output with the low-pass output would yield a signal notched at the common cutoff frequency. The circuit of Figure 2.23 shows a notch filter implemented using a Burr-Brown UAF42 state-variable filter IC. This IC incorporates precision 1000-pF capacitors for the op-amp integrators and an auxiliary op-amp that is used to sum the low- and high-pass outputs. As such, all that is needed to implement a notch filter with this IC are five external resistors. The notch frequency is set via R1 and R2 (where R1 = R2) and is given by

$$f_{notch} = \frac{1}{2\pi(\text{R}1)(1000\,\text{pF})}$$

Whatever notch filter you chose to use, you must remember that the notch filter will not only remove the power line interference but will also take away parts of the signal of interest. In addition, the notch filter may introduce nonlinear phase shifts in frequency components within the filter's passband.

Take, for example, applications that require very subtle analysis of the ECG signal. Arbitrary removal of power line frequency signals may not pose a problem for standard ECG signals since the main frequency components of P-, R-, and T-waves are far below 60 Hz. However, when ECGs are examined for small variations that are indicative of scar tissue due to previous myocardial infarction, removal of power line interference has to be done with utmost care not to eliminate or distort the *ventricular late potentials*, microvolt-level (1 to 20 μV) waveforms that are continuous with the QRS complex, last into the ST segment, and occupy a relatively wide frequency band (40 to 200 Hz) that peaks exactly within the range 50 to 60 Hz.

HARMONIC ELIMINATOR

Unfortunately, power line interference is not limited to 50 or 60 Hz. Fluorescent lights, dimmers, and other nonlinearities introduce powerful components at the harmonics of the power line frequency. A number of independent notch filters at 60, 120, 180 Hz, and so on (or their 50-Hz counterparts), could be cascaded to yield a comb filter to eliminate power line interference at the main frequency and its harmonics. However, an *n-path filter* is a better way than this of implementing a comb filter. This filter implementation generates the necessary poles by switching a sequence of capacitors in synchronism to the power line fundamental frequency.

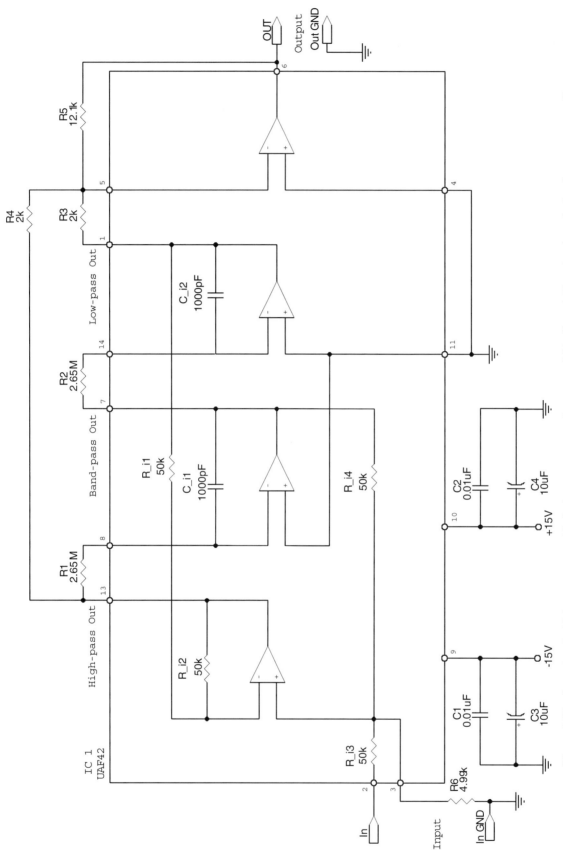

Figure 2.23 Summing the high-pass output with the low-pass output of a state-variable filter yields a signal notched at the common cutoff frequency. The Burr-Brown UAF42 state-variable filter IC incorporates precision 1000-pF capacitors for the op-amp integrators and an auxiliary op-amp that can be used to sum the low- and high-pass outputs. All that is needed to implement a notch filter with this IC are five external resistors. The notch frequency is

$$f_{\text{notch}} = \frac{1}{2\pi(R1)(1000\ \text{pF})} \qquad \text{if } R1 = R2$$

The simplified circuit of Figure 2.24 can be used to explain the process by which such a filter operates. Here, eight capacitors are switched to ground in synchronism with the power line, making this circuit an eight-path filter. A power line sample obtained from transformer T1 and coupled through R8 and C14 is fed to a 74HC4046 phase-locked loop (PLL). A PLL consists of a phase comparator, a loop filter, and a voltage-controlled oscillator. In this application, the phase comparator compares the phase and frequency of the power line sample against those of the internal oscillator and adjusts the oscillator frequency so that it equals some exact multiple of the incoming reference. A 74HC161 is clocked at 16 times the power line frequency by the output of the PLL's oscillator, generating eight digital addresses addresses (on outputs QB–QD) that divide the power line cycle evenly into eight segments. A 74HC4051 is driven by the digital sequence, selecting which one of eight capacitors is connected to ground throughout the power line cycle. During the time that one such capacitor is connected to ground, it samples the filter's input signal (the biopotential signal to be comb-filtered, not the power line reference) with a time constant given by R2. The output of the eight-path filter is inverted by IC1C and added to the original input signal to yield the comb response. Please note that IC1D is used to generate a virtual ground at 2.5 V. However, a split power supply would work equally well by substituting ground by −2.5 V, +2.5 V by ground, and +5 V by +2.5 V.

An intuitive explanation of the comb filter's transfer function is that the capacitors of the n-stage filter charge to a portion of the difference between the current signal voltage and the voltage integrated over the same time segment on previous power line cycles. Each capacitor can be assumed to store an average of the signal at the specific time segment of the power line cycle. Since the components of interest in the signal to be filtered are in most cases uncorrelated to the power line, the capacitors store only a sample of signal components locked to the power line frequency (i.e., the power line fundamental and its harmonics along with any other correlated noise). The output signal is then cleaned from repetitive power line noise when the power line–locked average is subtracted from the input signal.

This type of filter has unique advantages over fixed-frequency notch filters. First, the filter automatically adapts the frequency of its notches to whatever power line frequency is present at the reference port. Second, the filter does not affect the signal when no interference is present. Last, power line–frequency biopotential signal components not locked to the power line are not affected since this filter excludes only signals that maintain a phase lock to the power line for a number of power line cycles. In addition, unlike the notch filters described earlier, changes in component values in this circuit have minimal effect on the filter's response, making it maintenance free.

This filter is not a continous-time system. The use of switched capacitors makes this a sampled system that is bound by Nyquist's sampling theorem. This limits the theoretical bandwidth of the filter to one-half the sampling frequency. Since the signal is sampled eight times during the power line cycle, the theoretical bandwidth of the filter is four times the reference frequency, making it possible to reject only the fundamental, first, and second harmonics (60, 120, and 180 Hz for a 60-Hz reference). Higher bandwidth with rejection of higher harmonics requires increasing the number of sampling capacitors. For example, a 16-path filter at 60 Hz would have a theoretical bandwidth of 480 Hz.

The comb filter of Figure 2.24 needs additional components to reject more noise than that which it introduces by its switching action. This is usually accomplished through various bandpass filters placed at the output of the n-stage filter as well as at the output of the summing amplifier. A family of ready-made universal eliminator modules based on this principle is available from Electronic Design & Research Inc. Models EDR-82534 and EDR-82534A are adaptive comb filters designed specifically for integration within medical instrumentation. These modules comprise a 64-path filter and support a signal bandwidth of dc to 500 Hz (EDR-82534) or dc to 1200 Hz (EDR-82534A). Figure 2.25 shows the pinout and connection to the module. The module's internal block diagram is

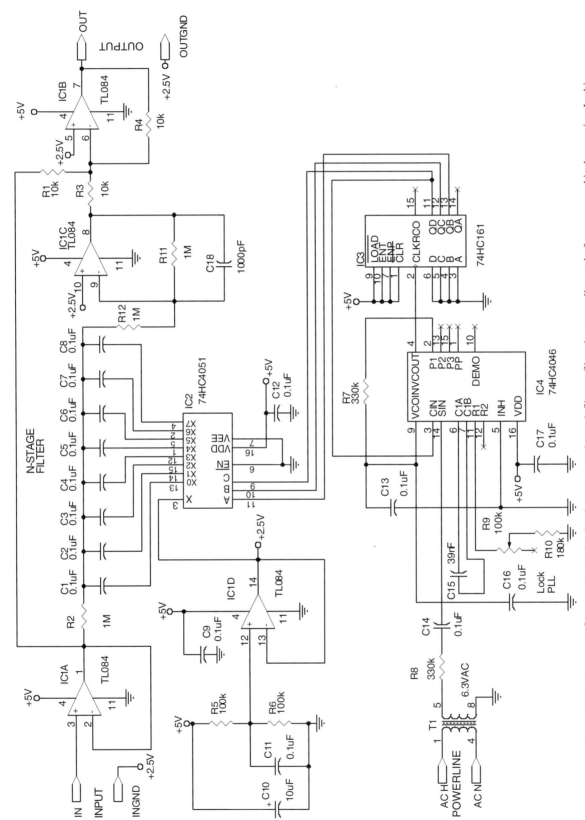

Figure 2.24 An *n*-path filter can be used to implement a comb notch filter to filter the power line main frequency and its harmonics. In this eight-path filter, eight capacitors are switched to ground in synchronism with the power line. The capacitors charge to a portion of the difference between the current signal voltage and the voltage integrated over the same time segment on previous power line cycles. Since the signal of interest is uncorrelated to the power line, the capacitors only store a sample of signal components locked to the power line frequency. The output signal is then cleaned from power line noise when the power line–locked average is subtracted from the input signal.

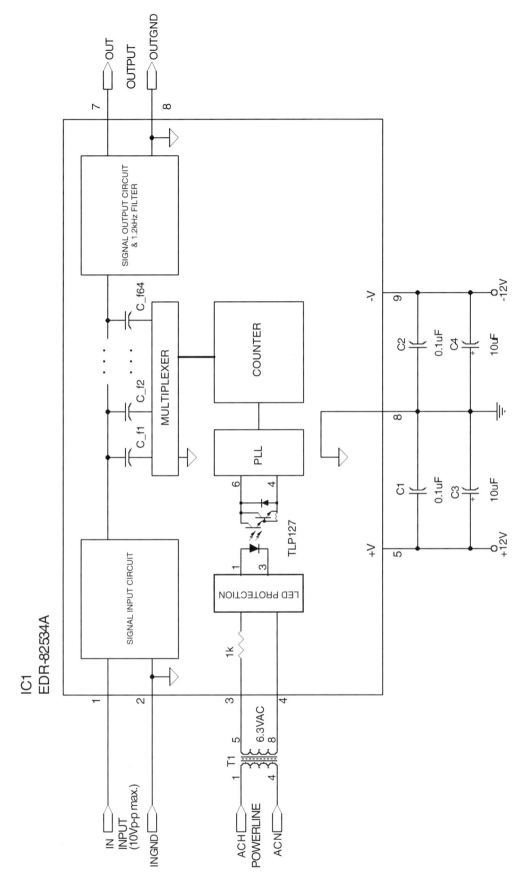

Figure 2.25 Electronic Design & Research Inc.'s EDR-82534 and EDR-82534A are adaptive comb filters designed specifically for integration within medical instrumentation. These modules comprise a 64-path filter and support a signal bandwidth of dc to 500 Hz (EDR-82534) or dc to 1200 Hz (EDR-82534A).

75

presented by the authors for reference purposes only and is not necessarily what Electronic Design & Research places within their "black boxes." No external components (besides power supply and reference signal source) are needed to use these modules. These modules are definitely recommended for applications that can tolerate their size (3.9 in. $\times$ 2 in. footprint, 1 in. height) and cost ($221 for the EDR-82534 and $324 for the EDR-82534A in single units).

SWITCHED-CAPACITOR FILTERS

Lately, many designers have opted for switched-capacitor filters as substitutes for continuous-time active filters. Switched-capacitor filters allow sophisticated, accurate, and tunable analog filter circuits to be manufactured without resistors. The advantage of this is that resistors take up considerable room on integrated circuits and that it is next to impossible to maintain a consistent absolute resistance value from manufacturing run to manufacturing run.

In Chapter 1 we showed how a sampling capacitor C_S, which switches continuously between an input voltage and a hold capacitor C_H with a clock period T, acts as a low-pass filter with a -3-dB cutoff frequency defined by the capacitor ratio

$$\frac{C_H}{C_S} = \frac{1}{2\pi f_{-3dB}T}$$

Capacitor ratios are much easier to maintain from batch to batch of an IC than are precise resistor values. In addition, resistor values that would be prohibitive for integration can easily be synthesized through the resistor equivalence of the switched capacitor. Finally, equivalent resistor values can be tuned simply by changing the switching frequency.

Commercial switched-capacitor ICs based on the same principle offer complete or nearly complete high-order filters in small, inexpensive packages. By switching the capacitor at around 100 times the corner frequency, these filters can attain a good approximation of theoretical performance. Switched-capacitor ICs are available as complete filters or as universal building blocks that require few external capacitors or resistors. Driving clocks may be internal or external to the filter itself. Varying clock frequency permits programming filters "on the fly."

If you are not sure which filter transfer function will work best in your application, switched-capacitor filters can help you try out various possibilities without rewiring your circuit. This is because switched-capacitor manufacturers offer filters with the various transfer functions in pin-compatible packages. For example, the Maxim MAX290 family of low-pass filters offers interchangeable chips that implement Bessel, Butterworth, and elliptic-response transfer functions.

Switched-capacitor filters do have disadvantages. For one, since a switched-capacitor filter is a sampling device, it can result in aliasing errors. Frequency components near and above half the sampling frequency must be eliminated to ensure accuracy. In addition, the output of a switched-capacitor filter usually needs to be low-pass filtered with a continuous-time filter to eliminate clocking signals that always manage to feed through.

The use of switched-capacitor filters can present other traps to the designer of biopotential amplifiers. This is because high-speed clock signals can easily couple to the high-impedance inputs and ground lines. Furthermore, the internal amplifiers within switched-capacitor filter ICs can generate noise and harmonic distortion on processed biopotential signals. Regardless of the precautions that one may take in the design, continuous-time active filters end up being at least 20 to 40 dB quieter than their switched-capacitor counterparts.

With these warnings in mind, let us look at some of the most popular switched-capacitor filter choices for processing biopotential signals.

1. Maxim MAX280. This IC is a fifth-order all-pole low-pass filter with no dc error, making it an excellent choice for processing low-frequency biopotential signals. As shown in Figure 2.26, the filter IC uses an external resistor and capacitor to isolate the fourth-order filter implemented within the IC from the dc signal path. The external resistor and capacitor are used as part of the filter's feedback loop and also form one pole for the overall filter circuit. The values of these components should be chosen such that

$$f_{\text{cutoff}} = \frac{1.62}{2\pi(R1)(C1)}$$

where R1 is usually chosen to be around $20\,\text{k}\Omega$.

The chip's internal four-pole switched-capacitor filter is driven by an internal clock that determines the filter's cutoff frequency. For a maximally flat amplitude response, the clock should be 100 times the desired cutoff frequency. The internal oscillator runs at a nominal frequency of $140\,\text{kHz}$ that can be modified by connecting an external capacitor (C2) between pin 5 and ground. The clock frequency is given by

$$f_{\text{clock}} = 140\,\text{kHz}\left(\frac{33\,\text{pF}}{33\,\text{pF} + C2}\right)$$

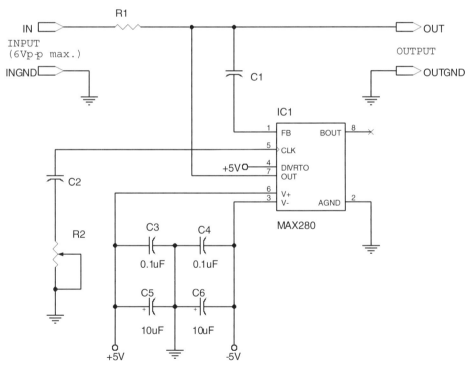

Figure 2.26 The Maxim MAX280 IC is a fifth-order all-pole low-pass filter with no dc error, making it an excellent choice for processing low-frequency biopotential signals. It uses an external resistor and capacitor to isolate the fourth-order filter implemented within the IC from the dc signal path. The internal switched capacitor filter is driven by an internal clock that determines the filter's cutoff frequency. If pin 4 is tied to V+, the filter has a cutoff frequency ratio of 100:1; when tied to ground, 200:1; and when tied to V−, of 400:1.

A series resistor (R2) can be added to trim the oscillation frequency. In this case, the new clock frequency is given by

$$f_{\text{clock}} = \frac{f_{\text{clock}}^{\text{R2}=0}}{1 - 4(\text{R2})(\text{C2})f_{\text{clock}}^{\text{R2}=0}}$$

where $f_{\text{clock}}^{\text{R2}=0}$ is the oscillator frequency when R2 is not present (obtained through the prior equation). Pin 5 of the MAX280 can also be driven from an external clock. In addition, the state of pin 4 must be taken into account to determine the effective clock frequency. If pin 4 is tied to V+, the filter has a cutoff frequency ratio of 100:1; when tied to ground, 200:1; and when tied to V-, 400:1.

2. Maxim MAX29x. This is a family of easy-to-use eighth-order low-pass filters that can be set up with corner frequencies from 0.1 to 25 kHz (MAX291/MAX292) or 0.1 Hz to 50 kHz (MAX295/MAX296). The MAX291 and MAX295 filters provide a Butterworth response, while the MAX292 and MAX296 yield a Bessel response. The clock frequency-to-cutoff frequency ratio for the MAX291 and MAX292 is 100:1; a ratio of 50:1 is used in the MAX295 and MAX296. As shown in Figure 2.27, an external capacitor is used to set the clock frequency of an internal oscillator according to

$$f_{\text{clock}}(\text{kHz}) = \frac{10^5}{3\text{C1}(\text{pF})}$$

Of course, the internal clock can also be driven externally. The MAX29x family has an internal uncommitted op-amp that has its noninverting input tied to ground. This op-amp can be used to build a continuous-time low-pass filter for prefilter antialiasing or clock attenuation at the switched-capacitor's filter output.

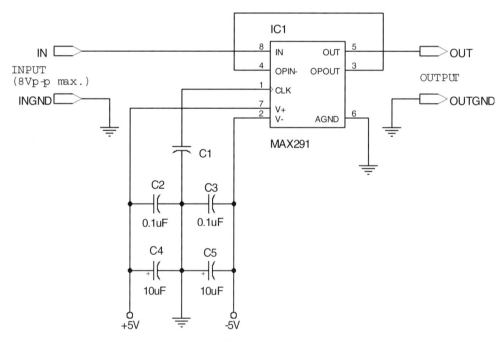

Figure 2.27 Maxim MAX29x ICs are eighth-order low-pass filters that can be set up with corner frequencies from 0.1 to either 25 or 50 kHz with either a Butterworth or a Bessel response. The clock frequency-to-cutoff frequency ratio for the MAX291 and MAX292 is 100:1; a ratio of 50:1 is used in the MAX295 and MAX296. An external capacitor is used to set the clock frequency of an internal oscillator according to $f_{\text{clock}}(\text{kHz}) = 10^5/3\text{C1}(\text{pF})$.

3. National Semiconductor MF4 (or Texas Instruments TLC04). This IC is an easy-to-use, fourth-order Butterworth low-pass filter. The ratio of the clock frequency to the low-pass cutoff frequency is set internally to 50 : 1. A Schmitt trigger clock input stage allows two clocking options, either self-clocking (via an external resistor and capacitor) for stand-alone applications, or for tighter cutoff frequency control an external TTL or CMOS logic compatible clock can be applied. The MF4-50 applies a 50 : 1 clock to cutoff frequency ratio, while the MF4-100 applies a 100 : 1 ratio. The clock frequency of the circuit shown in Figure 2.28 is given by

$$f_{\text{clock}} \approx \frac{1}{1.69(\text{R}1)(\text{C}1)}$$

4. National Semiconductor and Maxim MF10. This IC consists of two independent and general-purpose switched-capacitor active filter building blocks. Each block, together with an external clock and three to four resistors, can produce various second-order functions. Each building block has three output pins. One of the outputs can be configured to perform either an all-pass, high-pass, or notch function; the remaining two output pins perform low-pass and bandpass functions. The center frequency of the low-pass and bandpass second-order functions can be either directly dependent on the clock frequency, or they can depend on both clock frequency and external resistor ratios. The center frequency of the notch and all-pass functions depends directly on the clock frequency, while the high-pass center frequency depends on both the resistor ratio and the clock. Any of the classical filter transfer functions can be implemented by selecting the right component values. The specific design methods and equations for the MF10 are available in the datasheet for these products as well as in National Semiconductor Application Note 307.

Other useful universal switched-capacitor filters are Linear Technologies' LTC1068 and the LTC1562 as well as Maxim's MAX 265 and MAX 266. Maxim also offers two microprocessor-programmable universal switched-capacitor filters: the MAX261 and

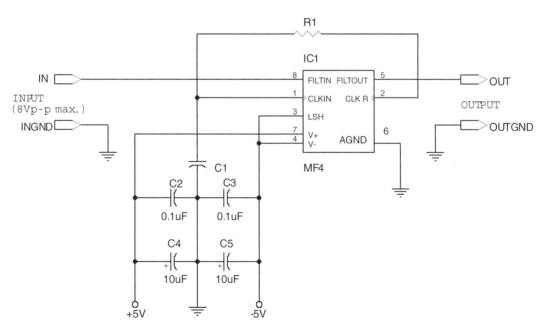

Figure 2.28 National Semiconductor's MF4 IC is a fourth-order Butterworth low-pass filter with a clock-to-cutoff frequency ratio of 50 : 1. The clock frequency is set via a capacitor and a resistor according to $f_{\text{clock}} \approx 1/1.69(\text{R}1)(\text{C}1)$.

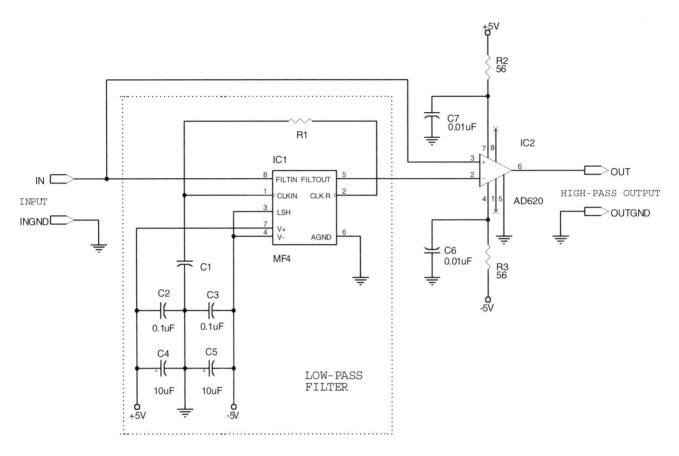

Figure 2.29 An instrumentation amplifier can be used to convert a low-pass filter into its high-pass counterpart. In this example, the input signal is fed to the AD620 instrumentation amplifier both directly and through the low-pass filter. This results in a high-pass function, since the output of the amplifier is the original signal minus the attenuated low-frequency components.

MAX262. One thing that you won't find with ease is a high-pass version of the low-pass filter chips (such as the MAX280 or the MF4). However, there is an alternative to using a universal filter IC. Figure 2.29 shows how an instrumentation amplifier can be used to convert a low-pass filter into its high-pass counterpart. The input signal is fed to the AD620 instrumentation amplifier both directly (to the noninverting input) and through the low-pass filter (into the inverting input). At low frequencies, both inputs of the instrumentation amplifier see the same signal since the low-pass filter passes the signal unaffected. The output of the amplifier should thus be zero. At high frequencies, however, signals are attenuated by the filter and the amplifier outputs the difference. This results in a high-pass function, since the output of the amplifier is the original signal minus the attenuated low-frequency components. If you decide to use this technique with a switched-capacitor filter, take precautions to avoid clock noise from affecting the performance of the filter. If possible, filter the amplifier's power input lines (the figure shows these lines filtered through R2/C7 and R3/C6), and use a continuous-time low-pass filter at the output of the switched-capacitor low-pass filter.

SLEW-RATE LIMITING

Sometimes, artifacts that obscure or distort a biopotential signal have significant spectral components within the bandpass of interest. However, there are times when the morphology of the artifact is sufficiently different from the signal of interest as to allow for its automatic identification and removal. A good example of this are the artifacts produced by cardiac pacemakers on the ECG signal. The Association for the Advancement of Medical Instrumentation (AAMI) recommends designing ECG equipment assuming the ECG waveform of Figure 2.30. The highest slew rate for an ECG can be estimated by dividing the maximum peak amplitude within the AAMI range of 0.5 to 5 mV and dividing it by the minimum rise time of the QR interval within the AAMI range of 17.5 to 52.5 ms. This gives a maximum slew rate of 5 mV/17.5 ms = 0.28 V/s for a worst-case ECG pulse. The slew rate of a pacing pulse is much higher than this, making it possible to design a slew-rate filter that limits the rate of change in the signal rather than a specific frequency band.

This slew-rate filter technique is also useful in other applications that require limiting artifacts from fast transients. For example, it can be used to filter large artifacts produced by magnetic-resonance imaging (MRI) equipment, allowing a patient's vital signs to be monitored during procedures involving this imaging technique. Removing "pop" artifacts (fast transients caused by movement) from EEG recordings and limiting stimulus artifacts in devices designed to measure nerve conduction are also possible through the use of slew-rate limiters.

A simple slew-rate limiting filter is shown in Figure 2.31. This filter, designed by Williams [1998], is a simple op-amp buffered RC low-pass filter modified by the addition of a bidirectional diode clipping network. Whenever the input voltage to op-amp IC1 differs from its output voltage by at least one forward diode drop, the forward-biased diodes will conduct. Under these conditions, the voltage at the output and noninverting input of the op-amp are equal, causing the voltage across R2 to remain approximately constant at one forward diode drop, $V_{\text{F DIODE}}$.

A constant voltage across R2 forces a constant current that charges C1 linearly rather than exponentially. If R2 >> R1, the maximum slope ($\Delta V_{\text{out}}/\Delta$time) of the signal at the output is given by

$$\text{slope}_{\max} = \frac{\Delta V_{\text{out}}}{\Delta \text{time}} = \frac{V_{\text{F DIODE}}}{(R2)(C1)}$$

Input signals with a slope higher than this cause a constant rate of change at the output equal to this limit. A signal segment with a slope lower than this limit passes through the

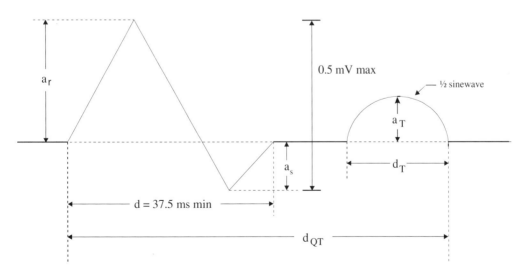

Figure 2.30 The AAMI stereotype ECG waveform has a maximum slew rate of 0.28 V/s, which allows separating the ECG's signal components from artifacts with higher slew rates, such as those from pacing and MRI.

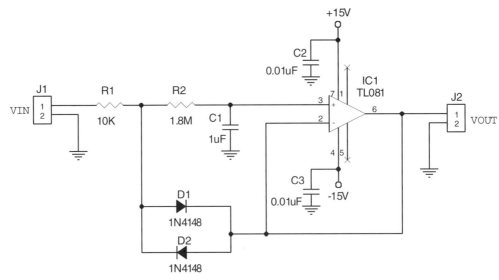

Figure 2.31   A simple slew-rate-limiting filter is based on a simple op-amp buffered *RC* low-pass filter modified by the addition of a bidirectional diode clipping network. Whenever the input voltage to op-amp IC1 differs from its output voltage by at least one forward diode drop, the forward-biased diodes will conduct. In this region, R2 forces a constant current that charges C1 linearly rather than exponentially, thus limiting the slope of signals at the output of the filter.

circuit without attenuation. With the values shown in the schematic diagram, signals with a slope of less than 0.33 V/s pass unaffected, allowing ECGs to be filtered from fast artifacts without distorting their worst-case 0.28-V/s slopes.

J. Moore has proposed a different slew-rate limiter that can be used to filter ECG signals from fast transients. The following circuit is described by Moore [1991] in U.S. Patent 4,991,580 as part of an ECG recorder that is immune to artifacts induced by MRI equipment. When the patient is placed inside the bore of a magnetic-resonance imager, the strong time-varying magnetic fields produced by the MRI system can induce voltage spikes on the ECG leads with an amplitude of 65 mV and a duration of 0.5 ms.

In the circuit of Figure 2.32, when the output of op-amp IC1A is positive, diodes D2 and D3 are forward biased and diodes D1 and D4 are reverse biased. Under these conditions, zener diode D5 is in series with diodes D1 and D4, cathode positive and anode negative. If the op-amp's output voltage exceeds the D5's breakdown voltage plus two diode forward voltage drops, the voltage at the junction between R1 and R2 will be limited to the zener voltage plus the two diode drops. With a 6.2-V zener, the limiting voltage is approximately 7.4 V. Negative voltage swings will have a similar effect, placing the zener diode in series with D1 and D4, limiting the negative swing at the junction between R1 and R2 to −7.4 V.

In operation, the voltage past the rectified-zener limiting bridge is converted by R2 to a current. Since the voltage at the input of the resistor is limited to ±7.4 V, the current flowing through the resistor will be within the range ±0.74 μA. This current charges capacitor C3. The change in voltage across this capacitor is then limited by its capacitance since $dV_C/dt = I_C/C$. For a 1-μF capacitor, and with a current of no more than ±0.74 μA, the slew rate of the signal buffered by IC1B is limited to ±7 V/s. The slew-rate limiting applied to the biopotential signal is calculated by dividing 7 V/s by the gain of the stages preceding the slew-rate limiter. For example, if the front-end biopotential amplifier has a gain of 21, the slew-rate limiting referred back to the biopotential amplifier's input signal is 0.33 V/s.

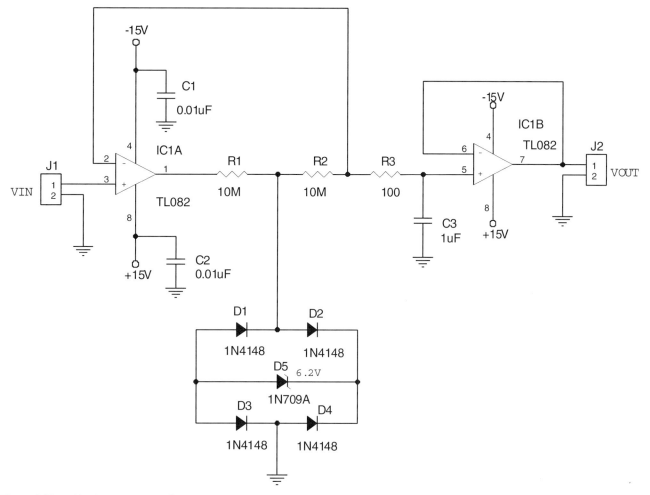

Figure 2.32   This slew-rate limiter filters fast transients such as MRI artifacts from ECG signals that can be used to filter ECG signals from fast transients. The circuit around IC1A acts as a precision-rectified zener diode. R2 converts the voltage of the bridge into a current that is limited by the zener voltage and which charges C3. The change in voltage across this capacitor is limited by its capacitance, thus limiting the slew rate of signals at the input of buffer IC1B.

ECG AMPLIFIER WITH PACEMAKER PULSE DETECTION AND ARTIFACT REJECTION

Patients with cardiac pacemakers may have difficulties, especially just after implantation of the pacemaker and lead system, from failure of the pacing system to properly sense the heart's intrinsic signals or to evoke heartbeats (capture the myocardium). This may reflect problems in the electronics, the leads, the placement of the leads, or the myocardium itself. The most catastrophic event is failure of the pacemaker to capture the heart when the patient's intrinsic rate is slow or nonexistent. To ensure that pacing therapy is being delivered in an appropriate manner, patients must be monitored to determine if the pacemaker is functioning properly and if appropriate benefit is being derived from the therapy. Specifically, it is important to know how often and why a pacemaker is activated, whether it is firing at appropriate times, and whether it is, in fact, capturing the heart to produce a heartbeat in response to the electrical stimulation.

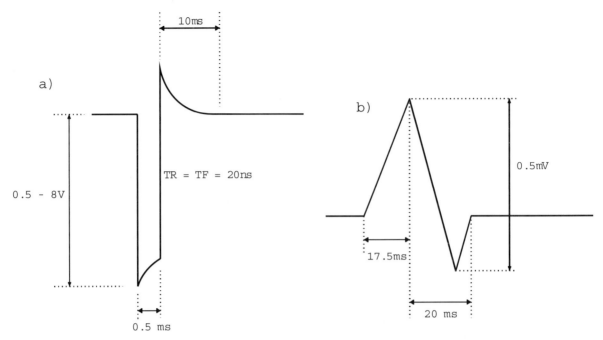

Figure 2.33   A typical pacing pulse consists of a main pulse that stimulates the heart and a discharge phase used to deplete the charge on the capacitive coupling generated by the delivery of the pacing pulse. Simple biopotential amplifiers unduly distort ECG signals from pacemaker patients because of the large amplitude difference between (a) pacing pulses and (b) the heart's intrinsic signals.

As shown in Figure 2.33a, a typical pacing pulse consists of two components, a main pulse and a discharge phase. The main pulse, which is used to stimulate the heart, is characterized by its narrow width, sharp rise and fall, and large variation in amplitude. The actual shape of the pacing pulse depends on the output coupling design of the pacemaker. The discharge phase is used to deplete the capacitive coupling generated by the delivery of the pacing pulse charge built up between the heart's tissue and the pacemaker's electrodes. The shape and size of the discharge phase is a function of the energy content of the pacing pulse and the amount of capacitive coupling.

The problem with using a simple biopotential amplifier to diagnose patients with pacemakers is the large voltage ratio between the artifact caused by the pacing pulse and the true ECG signal. When this ratio is large, the ECG signal baseline will be shifted to the discharge baseline of the pacer artifact signal, thus distorting the ECG. A large artifact signal can also produce amplifier overloads, preventing observation of the heart's electrical activity following pacing. To overcome this problem, ECG amplifiers designed to follow up on patients receiving pacing therapy have the means to separate and separately process the ECG and pacing pulse artifact signals, and augment the pacer artifact by providing a uniform pacing pulse artifact.

Figures 2.34 and 2.35 illustrate a four-lead ECG amplifier useful for evaluating patients implanted with a cardiac pacemaker. In this circuit, op-amps IC1B, IC1C, and IC1D buffer the biopotential signals detected by electrodes placed in the right arm, left arm, left leg, and chest of the patient. IC1A buffers the ECG signal detected from an electrode placed on one of the precordial electrodes.

The buffered left-arm, right-arm, and left-leg signals are summed by resistors R8, R9, and R10 prior to being buffered by IC2B to derive the *Wilson central terminal* (WCT) *potential*, which is considered to be the reference potential for ECG recording from the

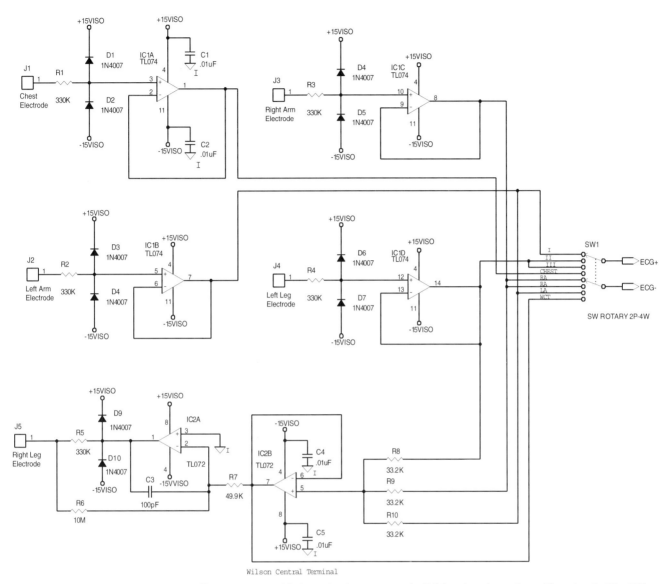

Figure 2.34 This four-lead ECG amplifier separates and independently processes the ECG and pacing pulse artifact signals. The Wilson central terminal is synthesized by R8–R10 together with IC2B and is used for right-leg driving after being inverted IC2A. SW1 selects the lead to be amplified by instrumentation amplifier IC3 of Figure 2.35.

chest electrode. The level of this signal is closely related to the common-mode potential seen by the limb electrodes. As such, it is used to reduce common-mode interference by driving the right-leg electrode through inverting amplifier IC2A.

SW1, a two-pole four-position rotary switch with insulated shaft and 5-kV contact-to-case insulation rating selects which buffered signals are presented to the inverting and non-inverting inputs of instrumentation amplifier IC3 of Figure 2.35. The *bipolar limb leads* are obtained as follows:

- *Lead I:* tracing of the potential difference generated by the heart between the left and right arms, where the left arm (L) is the noninverting input and the right arm (R) the inverting input

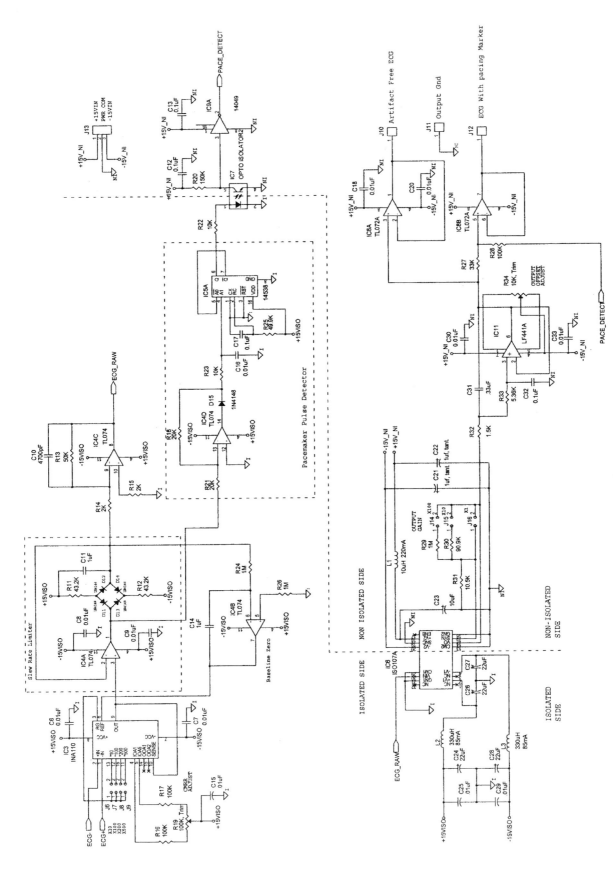

Figure 2.35 The slew-rate limiter and pacing pulse detector implemented around IC4A and IC4D split the amplified signal into the ECG and pacing pulse components. It then restores the baseline and isolates the artifact-free ECG signal and augments the pacing artifact by providing a uniform pacing pulse indicator.

- *Lead II:* tracing of the potential difference between the left leg (F) and the right arm, where the left leg is the noninverting input
- *Lead III:* tracing of the potential difference between the left leg and the left arm, where the left leg is the noninverting input

A *unipolar chest lead* is obtained by connecting the chest electrode to the noninverting input, while the WCT is connected to the inverting input. The specific chest lead (V1–V6) is obtained by placing the chest electrode on specific anatomical landmarks according to international standards:

- V1   in the fourth intercostal, at the right side of the sternum
- V2   in the fourth intercostal, at the left side of the sternum
- V3   halfway between V2 and V4
- V4   in the fifth intercostal, left mid clavicular
- V5   in the same horizontal plane as V4, halfway between V4 and V6 (anterior axially)
- V6   in the same horizontal plane as V4, left mid axially

V1 and V2 are thus placed above the right ventricle, V3–V6 above the left ventricle.

The output of IC3 contains an amplified version of the true ECG signal and the pacing pulse artifact. This signal is separated into the two components via the slew-rate limiter and pacing pulse detector circuits implemented around IC4A and IC4D. Once a certain slew rate has been exceeded by the output of IC3, the slew-rate limiter limits the excursions of the output signal to a specified rate of change. When the input voltage to IC4A is zero, the current flowing through R11 and R12 is the same, and thus the output of the slew-rate limiter is zero (neglecting possible op-amp offset). When a positive-going signal appears at the noninverting input terminal of IC4A, the output of IC4A will go positive, and current through R12 will be sourced by the op-amp. At the same time, current will be sourced by capacitor 46. As long as slewing-rate conditions given by the values of R11 = R12 and C10 are not exceeded, the slew-rate limiter's output voltage tracks IC4A's input voltage. However, if a pacing pulse artifact is presented to the noninverting input of IC4A, the slew-rate limit imposed by the time constant of R11 = R12 and C10 is exceeded and capacitor C10 will charge (or discharge) at a rate limited by the value of R11 = R12. At the time of the slew rate limiting, the output signal's slew rate will be limited to the charge or discharge rate of capacitor C10. Soon after the slew rate of the input signal falls under the slew-rate limit, op-amp IC4A returns from saturation, and the output of the slew-rate limiter tracks IC4A's noninverting input. Once the ECG signal has been cleaned by the slew-rate limiter from the pacing pulse artifact, baseline zeroing is accomplished by feeding the inverted baseline level (derived by inverting and heavily low-pass filtering the slew-rate limiter's output) to the reference pin of IC3.

Pacing pulse detection exploits the fact that the output of IC4A rails when the slew rate of its input exceeds the set limit. IC4D rectifies IC4A's output. Whenever the slew-rate limit is violated (which is assumed to happen only whenever a pacing pulse is present), a positive pulse appears at the trigger input of nonretriggerable one-shot IC5. The artifact-free ECG signal is amplified further via IC4C before it is presented to a Burr-Brown ISO107 isolation amplifier. Every time a pacing pulse is detected, an optically isolated 5-ms pulse is added to the isolated and filtered ECG signal via R27 and R28 before being buffered by IC8B.

The circuits in galvanic contact with the patient are powered via ISO107, which generates isolated ±12 V when powered from ±12 V. L2, C24, C25, and C26 form a pi filter to clean the isolated +12-V power line generated by ISO107 from switching noise. An identical network is used to filter the negative isolated supply rail. A pi filter formed by L1,

C21, and C23 is used to decouple the positive input power rail of IC6 so that switching noise within IC6 does not find its way into the postisolation circuits.

SCRATCH, RUMBLE, CLICK, AND POP

Our good British friend and colleague André Routh has held many positions as a biomedical engineer and scientist at various cardiac pacemaker and diagnostic device companies for over 20 years. In the tradition of British electronics aficionados, his hobby is hi-fi, especially when it's done using valves (vacuum tubes). André's expertise is in biomedical signal processing, and some time ago he implemented a specialized filter for cardiac signals based on the way in which audio professionals filter surface noises from old records (remember vinyl?). In the hi-fi field, these circuits are known as scratch or "click and pop" filters. Another device, the rumble filter, cuts down low-pitched noises, such as vibration from the phonograph motor.

The circuit of Figures 2.36 to 2.39 is a high-pass filter with a cutoff frequency of 0.5 Hz for baseline offset rejection and a low-pass filter with a cutoff frequency of 500 Hz for noise rejection and antialias filtering. The gain of this filter unit can be varied continuously in three ranges, from 0 to a maximum of 100. What makes this filter so unique is its rapid recovery from transient overloads. This is achieved through an input "blanking" scheme that can reject impulses such as pacing spikes. The high-pass filter with its long time constant can cause output saturation when faced with a step-type input or slow decay process such as a defibrillation shock. An output overload detection circuit changes the time constant of the high-pass filter. The cutoff frequency is increased to approximately 50 Hz until the output is within the allowable range for the signal acquisition equipment that follows the filter.

Typically, this filter circuit would be used between an isolated biopotential amplifier and an instrumentation tape recorder or data acquisition unit. The output of the filter is monitored on a LED bar graph display which has a VU meter response with fast attack and slow decay times. Additionally, there are two LEDs to indicate either input or output overloads. In a typical application, an input signal arrives via J2 from the output of an isolated biopotential amplifier. R1, D1, and D2 form an input signal clamp that limits at approximately 8.2 V. R1 and C1 form a low-pass filter with a cutoff frequency of 5.3 kHz. R2 is a passive attenuator whose wiper is connected to the noninverting input of IC1A. The gain of this op-amp is defined by R3 and R4. R4 should be selected to have a resistance equal to R2, so that when R2 is set to maximum the voltage at IC1A's output is equal to the input voltage at J2.

IC1B and associated components form a second-order Butterworth (maximally flat amplitude) low-pass filter. With the components selected, the cutoff frequency calculated is 509 Hz. Front-panel SW1 can be used to bypass the low-pass filter. IC2 is a quad CMOS transmission gate (digitally controlled analog switch). Under normal conditions (i.e., with no input overload), IC2A is closed and IC2B is open, allowing the signal at the wiper of switch SW1 to feed the second-order Butterworth high-pass filter built around IC1C (Figure 2.37). With the components selected, the cutoff frequency is 0.49 Hz. As long as there is no output overload, analog switches IC2C and IC2D are open, and as such, under normal conditions they do not interfere with the nominal operation of the high-pass filter. Front-panel switch SW2 can be used to bypass the high-pass filter.

IC1D is configured as a switchable gain amplifier. The gains available are ×1, ×10, and ×100, depending on the setting of front-panel switch SW3. The low-pass action of C6 depends on the gain of the amplifier stage. At ×100 the cutoff frequency is approximately 4.8 kHz; at ×10 it is about 53 kHz. Potentiometer R13 in conjunction with resistor R14 allows the output offset to be nulled. R18 and C7 form a low-pass filter with a cutoff

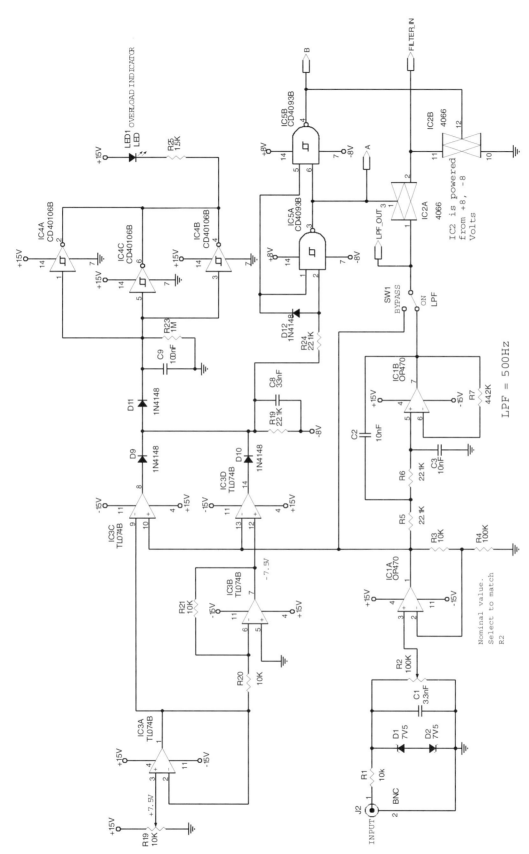

Figure 2.36 This ECG signal filter, designed by André Routh, uses the principles of hi-fi "pop and click" filters to ensure fast recovery from transients without distorting the signal of interest. This is the low-pass filter portion of the instrument. In addition, this figure includes part of the control circuit that enables the unique functionality of this filter is built around IC3, which is configured as a window comparator. With no input overload, IC2A presents the low-pass output to the high-pass filter of Figure 2.37. When the input saturates, the input of the high-pass filter is grounded.

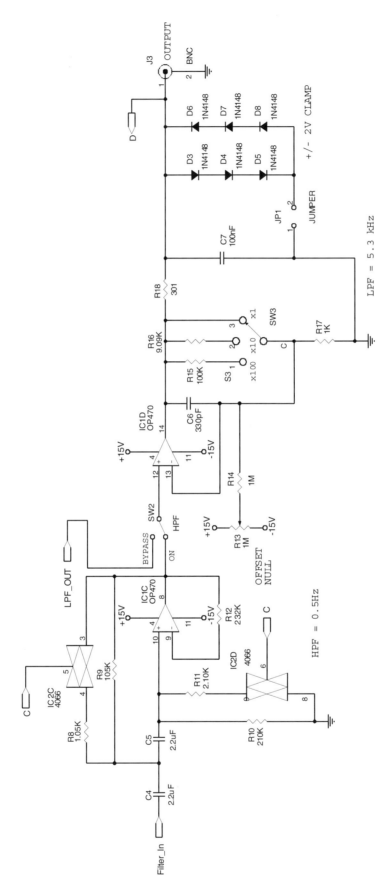

Figure 2.37 Under normal conditions, this filter has a high-pass cutoff of 0.5 Hz. When a transient overload is detected, the high-pass cutoff frequency is increased to approximately 50 Hz and the gain falls to zero until the output is within the allowable range for the signal acquisition equipment that follows the filter.

frequency of 5.3 kHz. When jumper JP1 is closed, the output swing is clamped at approximately ± 2 V by diodes D3–D8.

The unique functionality of this filter comes into action when an overload occurs. IC3 in Figure 2.36 is configured as a window comparator. Potentiometer R19 is adjusted to $+7.5$ V. IC3B is a unity-gain inverter whose output will be at -7.5 V. IC3C and IC3D function as comparators. With no input overload, both IC3C's and IC3D's outputs are low (i.e., close to -15 V). If the input voltage rises above $+7.5$ V, IC3C will switch and D9 will be forward biased. If the input falls below -7.5 V, IC3D will switch and D10 will be forward biased. D9, D10, and R19 form a wired OR gate. D11 prevents the inputs of IC4A, IC4B, and IC4C from receiving voltages below 0 V. C9, R23, and the hysteresis of the Schmitt action of IC4 form a one-shot that acts as a pulse stretcher, causing LED1 to illuminate. R24 and D12 clamp the positive voltage reaching IC5A to $+8$ V. IC5A and IC5B are connected as series inverters whose outputs, labeled A and B, drive transmission gates IC2A and IC2B. Note that A is also ANDed via D13 into IC5C (Figure 2.38).

On receipt of an input overload, IC2A will open and IC2B will close, thereby disconnecting the high-pass filter from the overload. This action is enhanced when the low-pass filter is in the circuit (i.e., SW1 is not set to bypass) because the low-pass filter introduces a delay to the signal. The net effect is that IC2A opens and IC2B closes *before* the overload signal has emerged at the output of the low-pass filter.

The circuit's output (marked D in Figure 2.37) is fed to the precision full-wave rectifier of Figure 2.38 configured around IC6A and IC6B. For positive half-cycles, D14 is forward biased and the circuit behaves like a cascaded unity noninverting amplifier. C10 ensures stability. For negative half-cycles, IC6A and D15 form a precision half-wave rectifier whose output (at D15 anode) is fed to the unity-gain inverter formed by IC6B/R26/R28. The net result is a fast full-wave rectifier that requires only two matched resistors (R26 and R28).

The output of the full-wave rectifier is fed to a comparator formed by IC6D. R30, D16, and D17 limit the comparator output to the ± 8 V range for application to NAND gate IC5C. The output from IC5C is inverted by IC5D, whose output is labeled C. Under normal conditions (i.e., not output overload or input overload), C is low. If an overload occurs, the magic of the circuit kicks in, C goes high, thereby enabling transmission gates IC2C and IC2D. R8 and R11 are switched in parallel with R9 and R10, respectively, when IC2C and IC2D close. The cutoff frequency of the high-pass filter of Figure 2.37 formed around IC1C is increased to approximately 50 Hz and the output of the filter rapidly settles toward 0 V as C4 and C5 discharge.

The rest of the circuit is used to drive a TSM39168 VU LED bar graph display. The output from the full-wave rectifier is fed to the precision half-wave rectifier formed by IC6 and D18. R32, R33, and C11 form an attack/decay circuit where the attack time (i.e., charge time of C11) is determined predominantly by R32/C11 and the decay time by R33/C11. The bar graph display of Figure 2.39 has an integrated driver circuit with VU meter calibration. The scale factor is set by R33 and R34. The transition from the green LEDs to the orange LEDs occurs at 1 V. The tenth LED (red) is activated when the voltage reaches 1.5 V. The output from this last LED is taken to a comparator formed by IC7A and resistor divider R34 and R35. The comparator output feeds the one-shot formed by D19, R36, and C13 and the Schmitt inverters IC4D, IC4E, and IC4F. The stretched pulse illuminates LED2 to indicate that the output level is too high.

A voltage of ± 15 V for the filter unit is supplied via connector J1. C15, C16, C19, and C20 from input filters for the ± 15-V supply to reduce crosstalk between channels. IC8 and IC9 are low-power ± 8-V regulators to supply the analog transmission gates (IC2) and the quad NAND gate (IC5). The $+5$-V power for the bar graph display has a separate ground circuit and is filtered by C23, R38, and C12 to prevent noise from being fed into the amplifier stages.

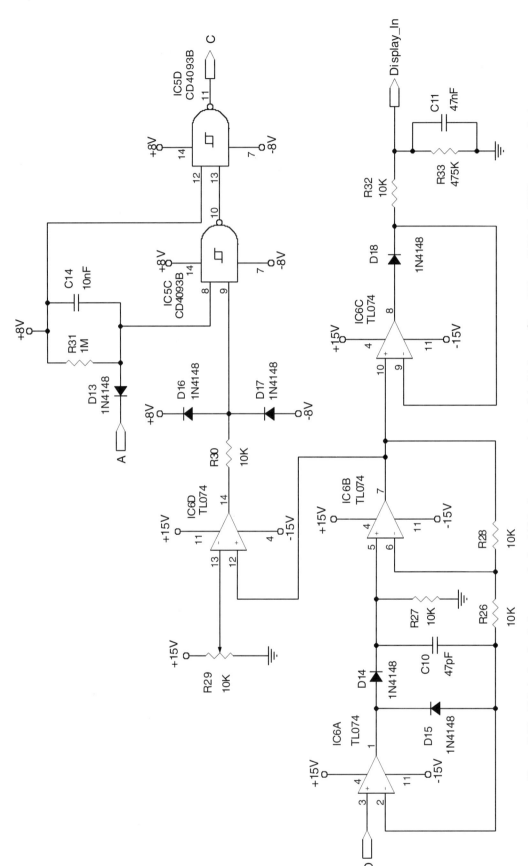

Figure 2.38 This is the circuit that controls the shift in frequency and gain of the high-pass filter of Figure 2.37. The HPF output (marked D) is fed to a precision full-wave rectifier configured around IC6A and IC6B, the output of which is fed to a comparator formed by IC6D. Under normal conditions (i.e., not output overload or input overload), C is low. If an overload occurs, C goes high, thereby increasing the HPF cutoff to 50 Hz and changing its gain to zero.

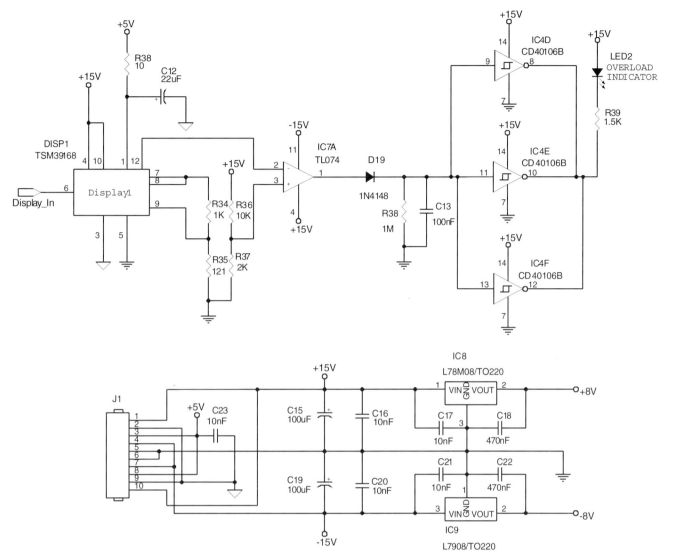

Figure 2.39 Signal levels are indicated by a TSM39168 VU LED bar graph display. The transition from the green LEDs to the orange LEDs occurs at 1 V. The tenth LED (red) is activated when the voltage reaches 1.5 V. LED2 indicates that the output level is too high. A voltage of ± 15 V power for the filter unit is supplied via connector J1. IC8 and IC9 are low-power ± 8-V regulators to supply the analog transmission gate (IC2) and the quad NAND gate (IC5). The +5-V power for the bar graph display has a separate input.

More recently, André designed an intracardiac electrogram signal amplifier that implements a similar transient-elimination technique to enable detection of evoked-response events from the heart soon after the delivery of pacing or defibrillation pulses. The amplifier of Figure 2.40 receives a digital BLANK command just before (e.g., 1 ms) the transient event (e.g., a pacing or defibrillation pulse) is delivered to the heart. Under normal conditions, the electrogram signal detected via intracardiac electrodes (from pacing leads or a mapping catheter) is presented to an INA128 instrumentation amplifier (IC3) via accoupling capacitors C3 and C6. Resistors R4 and R8, together with C5 and C7, low-pass the input signals. The output of the instrumentation amplifier is filtered by the second-order Butterworth high-pass filter built around IC1B.

When a BLANK signal is delivered to the biopotential amplifier circuit, the inputs of the instrumentation amplifier are shorted to the analog signal ground via analog switches

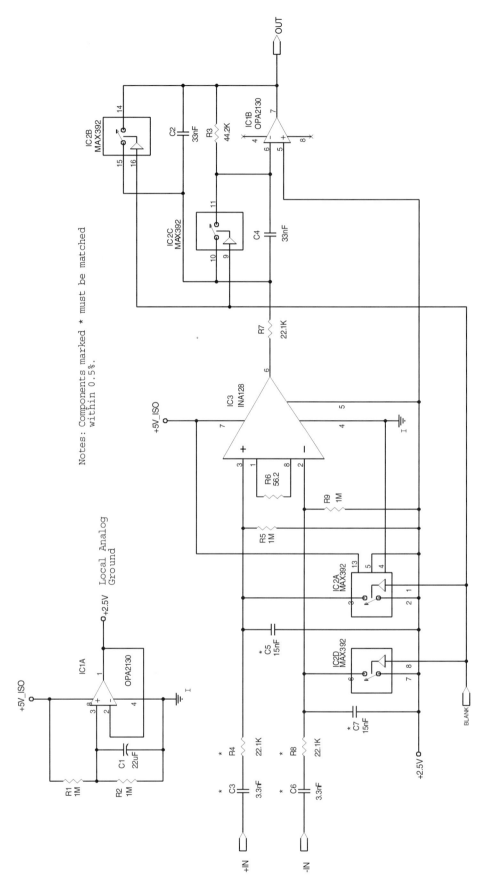

Figure 2.40 This intracardiac electrogram signal amplifier enables detection of evoked-response events from the heart soon after the delivery of pacing or defibrillation pulses. A digital BLANK command is supplied just before the stimulus pulse is delivered to the heart. This command shorts the inputs of the instrumentation amplifier and changes the configuration of IC1B to an amplifier with a gain of zero. Recovery occurs within a few milliseconds after the blanking signal is released.

94

IC2A and IC2D. At the same time, the capacitors of the high-pass filter are shorted, which changes the configuration of IC1B to an inverting amplifier with a gain of zero. After the stimulus subsides, the blanking signal is released, and the amplifier recovers within milliseconds to allow detection of evoked or intrinsic potentials. Note that the circuit is powered from a single +5-V supply. IC1A is used to buffer the 2.5-V voltage produced by resistor divider R1 and R2 and filtered by C1. The 2.5 V at the output of IC1A is used as an analog reference potential for the circuit.

REFERENCES

Koide, F. T., Quick Estimate of Signal Bandwidth, *Electronic Design*, 115–116, December 16, 1996.

Moore, J. C., Method of Improving the Quality of an Electrocardiogram Obtained from a Patient Undergoing Magnetic Resonance Imaging, U.S. patent 4,991,580, 1991.

Suesserman, M., Bootstrapped, AC-Coupled Differential Amplifier, U.S. patent 5,300,896, 1994.

Williams, A., Special Low-Pass Filter Limits Slope, *Electronic Design*, May 25, 1998.

3

DESIGN OF SAFE MEDICAL DEVICE PROTOTYPES

Military downsizing, government cutbacks, and corporate reengineering had the opposite effect on the medical industry as they did on all other areas of technology. As R&D budgets shrank, early-generation technologies that had long been considered obsolete for space, security, and military applications suddenly found a thriving environment in the development of new medical devices. For example, cruise-missile tracking technology has been adapted to steer powerful x-ray beams to destroy brain and spinal-cord tumors precisely without the need of surgery; software used for interpreting spy-satellite images has been used as the basis for detecting subtle but highly virulent breast cancers that were often overseen in visual interpretations of mamographic images; and miniaturized high-energy capacitors developed for portable laser weapons may make it possible to build smaller and lighter automatic implantable cardiac defibrillators.

Many other examples of these dual-use technologies continue to appear with no end in sight, making medical electronics one of today's fastest-growing and most promising technology-based industries. Fortunately for the entrepreneurs among us, prototypes of many new medical instruments can still be developed in a garage-turned-laboratory without the need for esoteric technologies recycled from multibillion-dollar satellite and weapons programs. Rather, a fresh idea, a personal computer, and some simple interface circuitry is all it may take to start the next revolution in medical care.

Despite how simple or complex a medical electronic instrument prototype may be, however, safety must be the primary objective throughout the development effort. Becoming intimately familiar with electrical safety standards is probably the most important thing that a newcomer to the field can do, because the dangers involved in interfacing with the human body are often counterintuitive to an otherwise knowledgeable engineer. For example, did you know that a 60-Hz current of barely 10 μA flowing through the heart has the potential of causing permanent damage and even death?

The objective of this chapter is to introduce the basics of designing and constructing electrically safe medical instrument prototypes. We first present an overview of electrical safety compliance requirements, proceed to look at a number of circuits that enable safe interfacing with medical electronics, then review safety testing methods, and finally, show

Design and Development of Medical Electronic Instrumentation   By David Prutchi and Michael Norris
ISBN 0-471-67623-3   Copyright © 2005 John Wiley & Sons, Inc.

the construction of a number of useful test instruments suitable for assessing the electrical safety of medical electronic instruments.

STANDARDS FOR PROTECTION AGAINST ELECTRICAL SHOCK

It has been a long time now that medical electronic devices left the realm of experimentation and were transformed into irreplaceable tools of modern medicine. This widespread use of a very diverse variety of electronic devices compelled countries to impose regulations that ensure their efficacy and safety. In the United States, the Food and Drug Administration (FDA) is responsible for the regulation of medical devices. In the European Union (EU), a series of directives establishes the requirements that manufacturers of medical devices must meet before they can obtain *CE marking* for their products, to authorize their sale and use. In addition, however, individual nations of the EU may impose local regulations through internal regulatory bodies. Other countries, including Canada, Japan, Australia, and New Zealand, have their own regulations, which although similar to the harmonized European and U.S. standards, have certain particulars of their own.

Safety standards are sponsored by organizations such as the American National Standards Institute (ANSI), the Association for Advancement of Medical Instrumentation (AAMI), the International Electrotechnical Commission (IEC), and Underwriters' Laboratories, Inc. (UL), among many others. These standards are written by committees comprised of representatives of the medical devices industry, insurance industry, academia, physicians, and other users in the medical community, test laboratories, and the public. The purpose of creating these broad-spectrum committees is to ensure that standards address the needs of all parties involved in the development, manufacture, and use of medical devices. Thus, through a consensus process, emerging standards are deemed to capture the state of the art and are recognized at national and international levels.

In general, safety regulations for medical equipment address the risks of electric shock, fire, burns, or tissue damage due to contact with high-energy sources, exposure to ionizing radiation, physical injury due to mechanical hazards, and malfunction due to electromagnetic interference or electrostatic discharge. The most significant technical standard is IEC-601, *Medical Electrical Equipment*, adopted by Europe as EN-60601, which has been harmonized with UL Standard 2601-1 for the United States, CAN/CSA-C22.2 601.1 for Canada, and AS3200.1 and NZS6150 for Australia and New Zealand, respectively.

According to IEC-601, a possible risk for electrical shock is present whenever an operator can be exposed to a part at a voltage exceeding $25\,V_{RMS}$ or $60\,V$ dc, while an energy risk is present for circuits with residual voltages above $60\,V$ or residual energy in excess of $2\,mJ$. Obviously, the enclosure of the device is the first barrier of protection that can protect the operator or patient from intentional or unintentional contact with these hazards. As such, the enclosure must be selected to be strong enough mechanically to withstand anticipated use and misuse of the instrument and must serve as a protection against fires that may start within the instrument due to failures in the circuitry.

Beyond the electrical protection supplied by the enclosure, however, the circuitry of the medical instrument must be designed with other safety barriers to maintain leakage currents within the limits allowed by the safety standards. Since patient and operator safety must be ensured under both normal and single-fault conditions, regulatory agencies have classified the risks posed by various parts of a medical instrument and have imposed specifications on the isolation barriers to be used between different parts. The first type of part is the *accessible part*, a part that can be touched without the use of a tool. Touching in this context not only assumes that contact is made with the exterior of the enclosure or any exposed control knob, connector, or display, but that it could be made accidentally: for example, by poking a finger or pencil through an opening in the enclosure. In fact, most standards define rigid and articulated probes that must be used to verify the acceptability of enclosure openings.

The second type of part is the *live part*, a part that when contacted can cause the leakage current to ground or to an accessible part of the equipment to exceed the limits established by the standard. One form of live part is the mains part, defined as a circuit connected directly to the power line.

The third type of part comprises *signal-input and signal-output parts*, referring to circuits used to interface a medical instrument to other instruments: for example, for the purposes of displaying, recording, or processing data. The fourth and most critical part of a medical instrument is that which deliberately comes into physical contact with the patient. Such a part, called an *applied part*, may include a number of patient connections which provide an electrical pathway between it and the patient. The patient circuit comprises all patient connections as well as all other parts and circuits of the medical instrument that are not electrically isolated from these connections.

The level of electrical shock protection provided to patients by the isolation of applied parts classifies them as follows:

- *Type B*: applied parts that provide a direct ground connection to a patient
- *Type BF* (the F stands for "floating"): indicates that the applied part is isolated from all other parts of the equipment to such a degree that the leakage current flowing through a patient to ground does not exceed the allowable level even when a voltage equal to 110% of the rated power line voltage is applied directly between the applied part and ground
- *Type CF*: similar to type BF, but refers to applied parts providing a higher degree of protection, to allow direct connection to the heart

The use of F-type applied parts is preferable in all cases to type B applied parts. This is because patient environments often involve simultaneous use of multiple electronic instruments connected to the patient. In any case, type B applied parts are prohibited whenever patient connections provide either low-impedance or semipermanent connections to the patient (e.g., through recording bioelectrodes as in ECG or EEG, or for stimulation of tissues, such as TENS). Furthermore, all medical electrical equipment intended for direct cardiac application (e.g., intracardiac electrophysiology catheters, invasive cardiac pacing) must contain only CF-type applied parts. Additionally, the applied parts of instruments for cardiac diagnosis and therapy are often designed to withstand the application of high-voltage high-energy shocks, such as those used for cardiac cardioversion and defibrillation.

These classifications have more than academic purpose. The standards provide the designer with clear indications regarding the minimal level of circuit separation and the application of insulation between these parts to accomplish acceptable levels of isolation. As such, *insulation* is not only defined as a solid insulating material applied to a circuit, but also to spacings that establish creepage distances and air clearance between parts. The separation of two conductive parts by air alone constitutes a *clearing distance*, while the separation of conductive parts on a nonconductive plane (e.g., tracks on a printed circuit board) is a *creepage distance*. The minimum separation distance between elements of two parts is determined by the working voltage between parts as well as by the insulation rating required to afford protection against electrical shock.

A basic insulation barrier is applied to live parts to provide basic protection against electrical shock. For example, its use applies to the separation between a live part and an accessible conductive part that is protected by connection to ground. Supplementary insulation is an independent insulation barrier applied in addition to basic insulation in order to provide protection against electrical shock in the event of failure of the basic insulation. Double insulation and reinforced insulation provide protection equivalent to the use of both basic and supplementary insulation.

TABLE 3.1   Spacings (Millimeters) Required to Provide Various Levels of Insulation between Parts of a Medical Device

Ac Voltage:		125 V	250 V	380 V
Dc Voltage:		150 V	300 V	450 V
Basic insulation (between	Air clearance	1	1.6	2.4
parts of opposite polarity)	Creepage distance	2	3	4
Double or supplementary	Air clearance	1.6	2.5	3.5
insulation	Creepage distance	3	4	6
Double or reinforced	Air clearance	3.2	5	7
Reinforced insulation	Creepage distance	6	8	12

Figure 3.1 and Table 3.1 present a partial view of how to achieve the minimal required insulation ratings between parts. Although these are only a subset of all possibilities contemplated by the standards, they certainly provide a very practical reference for the designer.

LEAKAGE CURRENTS

Evidently, the purpose of the various isolation barriers is to ensure that leakage currents are maintained within safe values even when a single-fault condition occurs. Three types of leakage currents are defined within the standards:

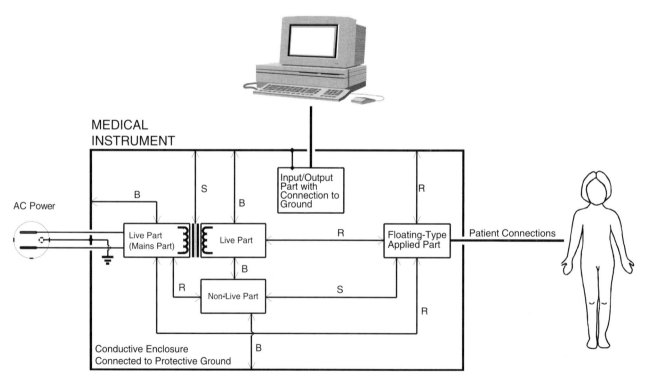

Figure 3.1   Some of the requirements for insulation between the parts of a medical instrument. Insulation types: B, basic; S, supplementary; R, reinforced.

1. *Ground leakage current:* current flowing from all mains parts through or across the insulation into the protective ground conductor of the grounded power cord
2. *Enclosure leakage current:* total current flowing from the enclosure and all accessible parts (but excluding applied parts) through an external conductive connection other than the protective ground conductor to ground or another part of the enclosure
3. *Patient leakage current:* current flowing from the applied part by way of the patient to ground, or flowing from the patient via an F-type applied part to ground; originates from the unintended appearance of voltage from an external source on the patient

It must be noted, however, that these leakage currents must not be confused with currents generated intentionally by the medical device to produce a physiological effect on the patient, or used by the applied part to facilitate measurement without producing a physiological effect. Examples of patient intentional currents are those used for the stimulation of nerves and muscle, cardiac pacing and defibrillation, and cutting and cauterization with radio frequency. Patient auxiliary currents are used to bias the front-end amplifiers designed to detect biopotentials, to enable the measurement of impedance of living tissues, and so on.

In the standards, the terms *voltage* and *current* refer to the root-mean-square (RMS) values of an alternating, direct, or composite voltage or current. Remember that by definition, the RMS value of an alternating voltage V across a resistor R equals the direct voltage, causing the same dissipation level in R. For a sinusoidal waveform, the RMS voltage V_{RMS} is related to the peak-to-peak voltage $V_{p\text{-}p}$ by

$$V_{p\text{-}p} = V_{RMS}\sqrt{2} \approx 1.414 V_{RMS}$$

A similarly corresponding definition applies to the value of an RMS current. In the case of composite (ac + dc) signals, the RMS value is calculated from

$$V_{RMS} = \sqrt{V_{dc}^2 + V_{ac_{RMS}}^2}$$

As shown in Table 3.2, allowable patient leakage and auxiliary currents are defined for both normal and single-fault conditions, assuming that the equipment is operating at maximum load and that the supply is set at 110% of the maximum rated supply voltage. *Single-fault conditions* are defined as conditions in which a single means of protection against a

TABLE 3.2   Some Allowable Values of Continuous Leakage and Patient Auxiliary Currents under Normal and Single-Fault Conditions (Milliamperes)

| | Equipment Type | | | | | |
| | B | | BF | | CF | |
Condition	Normal	Single fault	Normal	Single fault	Normal	Single fault
Ground leakage current	0.5	1	0.5	1	0.5	1
Enclosure leakage current	0.1	0.5	0.1	0.5	0.1	0.5
Patient leakage current	0.1	0.5	0.1	0.5	0.01	0.05
Patient leakage current (with power line voltage on the applied part)	—	—	—	5	—	0.05
Dc patient auxiliary current	0.01	0.05	0.01	0.05	0.01	0.05
Ac patient auxiliary current	0.1	0.5	0.1	0.5	0.01	0.05

safety hazard in the equipment is defective or a single external abnormal condition is present. These include interruption of the supply by opening the neutral conductor as well as interruption of the protective ground conductor. Patient leakage current between an F-type applied part and ground assumes that an external voltage equal to 110% of the maximum rated supply voltage is connected directly to the applied part. For battery-powered equipment, the external voltage that is assumed to be connected to the F-type applied part is 250 V. In addition, it must be noted that grounding of the patient is considered to be a normal condition.

The allowable leakage current levels have been set as a compromise between achievable performance and overall risk. Although a 60-Hz current as low as $10\,\mu A$ flowing through the heart may cause ventricular fibrillation (a disorganized quivering of the lower chambers of the heart muscle that quickly leads to death) under highly specific conditions, the probability of such an event is only 0.2%. Under more realistic clinical conditions, however, a $50\text{-}\mu A$ current flowing from a CF-type applied part through an intracardiac catheter has an overall probability of just 0.1% of causing ventricular fibrillation. This probability is very similar to that of causing fibrillation due to the irritation caused by mere mechanical contact of the catheter with the heart wall. Obviously, for equipment that does not come in direct contact with the heart, allowable leakage currents have been increased up to the point where even under single-fault condition, the probability of causing ventricular fibrillation is no higher than 0.1%, even though the actual current may be perceptible to the patient.

DESIGN EXAMPLE: ISOLATED DIFFERENTIAL ECG AMPLIFIER

Let's use a simple circuit as an example to illustrate the various considerations regarding the safe design of a medical instrument. Figure 3.2 presents the schematic diagram of a simple biopotential amplifier intended to detect a differential ECG signal through surface ECG electrodes. In the circuit, signals picked up by electrodes attached to the patient's skin are amplified by IC1, a Burr-Brown INA110 instrumentation amplifier IC. The gain of the front-end stage is programmable between unity and 500 by jumpers JP2–JP5. Since IC1 is dc-coupled, care must be exercised in the selection of gain so that the amplifier is not saturated by dc offset voltages accompanying the biopotential signal. For example, to use this circuit as part of a surface ECG amplifier, the gain must be calculated to cope with offset potentials of up to $\pm300\,mV$. In general, IC1's gain should be kept low so that dc coupling does not result in its saturation. Potentiometer R1 is used to trim the input offset to IC1. R1–R3 can be omitted from the circuit for most applications that do not require extreme dc precision.

Direct connection of IC1's inputs to patient electrodes is possible since the amplifier uses a maximum bias current of 50 pA, and the FDH300 low-leakage diodes used to protect the inputs of IC1 contribute no more than an additional 1 nA each to the patient auxiliary current. The total 54 nA maximum is well under the allowed 0.01 mA auxiliary current for CF-type applied parts. If the application permits it, however (e.g., if the skin–electrode interface has a sufficiently low impedance), it is a good idea to add resistors larger than 300 kΩ in series with the patient connections. These resistors would effectively limit the auxiliary current flowing through the patient to less than 0.05 mA in case a fault in IC1 or in D1–D4 short-circuit the patient's connection with one of the isolated power rails.

Depending on the biopotential signal being amplified, either dc or ac coupling are required. For dc coupling, IC1 is referenced to the isolated ground plane I_{G1}, which also serves as the patient common input. Since the INA110 has FET inputs, bias currents drawn through input source resistances have a negligible effect on dc accuracy. However, a return path must always be provided to prevent charging of stray capacitances which may saturate the INA110. If this amplifier would be needed to amplify completely floating sources

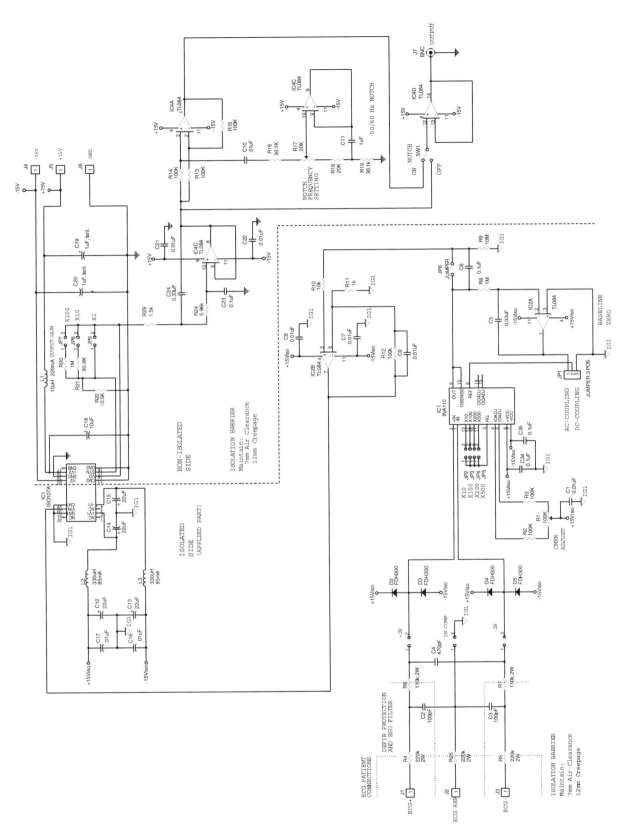

Figure 3.2 A universal differential biopotential amplifier serves as an example of designing medical electronic instruments for safety. An INA110 instrumentation amplifier acts as an impedance buffer and provides gain to weak electrical activity detected by electrodes on the patient's body. For use as a single-lead ECG amplifier, a protection network is used to limit currents during defibrillation or electrosurgery procedures. Galvanic isolation for the amplified signal is provided by an ISO107 isolation amplifier IC. This IC also provides isolated power to operate the circuitry of the applied part.

or capacitively coupled sensors, a 10-MΩ resistor to the isolated ground plane from each input should be used. When ac coupling is desired, IC4A, R8, and C5 are used to offset IC1's reference to suppress a baseline composed of components in the range dc to 0.48 Hz. Also for ac coupling, any remaining baseline at IC1's output may be eliminated by a high-pass filter (1.59 Hz at -3 dB) formed by C6 and R9.

IC1's output signal is amplified by IC4b. Notice that the gain of this stage is fixed at 101. Galvanic isolation is provided by IC3, a Burr-Brown ISO107 isolation amplifier IC. This type of IC resembles an operational amplifier but is designed with an internal isolation barrier between its input and output pins. The ISO107's signal channel has a small-signal bandwidth of 20 kHz and provides an isolation barrier rated at a continuous 2500 V. In addition to providing a signal channel across the isolation barrier, the ISO107 has an internal dc/dc converter which powers the isolated side of the ISO107 circuitry and provides isolated power (± 15 V at ± 15 mA typical) for the rest of the circuitry of the applied part (i.e., IC1 and IC2). The isolation rating of the barrier for the dc/dc converter is the same as that for the signal channel. In total, the 60-Hz leakage current through IC3 does not exceed 2 μA with 240 V ac applied across its isolation barrier. The output gain of IC3 is selected through jumpers JP7–JP9 to provide gains of 1, 10, or 100. IC3's output is then low-pass filtered by IC4C. With the component values shown, the filter has a cutoff frequency of 300 Hz. You may recalculate these values to match the bandwidth required by your application.

In one position of SW1, the filter's output is buffered directly by IC4D and presented to the output of the biopotential amplifier. In the other position, SW1 redirects the output of IC4C to a tunable-frequency notch filter before being buffered by IC4D. This makes it easy to eliminate 50/60-Hz power line hum that may have been picked up through common-mode imbalances between the differential patient connections.

As shown up to this point, patient leakage and auxiliary currents have been kept within allowed limits by virtue of appropriate selection of the components for the circuit. However, appropriate layout and interconnection are as important in ensuring a safe design. To do so, every conductive point belonging to the isolated portion of the circuit must be separated from every conductive point in the nonisolated side of the circuit by the required air clearance and creepage distances corresponding to reinforced insulation at the rated working voltage. The layout of a prototype instrument that incorporates this ECG amplifier is shown in Figure 3.3.

Since there is a 30-mm separation between the closest pins across the ISO107 isolation barrier, and considering that the internal isolation barrier is rated at a continuous 2500 V at 60 Hz, the standards would consider this barrier to be equivalent to 1000 V ac–rated reinforced insulation. This separation would also be needed between all other isolated and nonisolated points of the circuit. Most commonly, a biopotential amplifier is operated in environments where the power line voltage is the highest potential of concern and has a maximum rated value of 240 V$_{RMS}$. According to Table 3.1, this would require an air clearance of 5 mm and an 8-mm creepage distance. Remember that these distances also apply to the separation of any point on the isolated side and any conductive fastening means in connection with any nonisolated part of a medical instrument.

Amplifying the electrical activity produced by the heart introduces a number of additional requirements addressed by the front-end protection circuit shown in Figure 3.2. Physicians conducting electrophysiological diagnosis and therapy of conditions involving the heart assume the possibility of ventricular fibrillation during a procedure. Reverting fibrillation back into a normal rhythm driven by the sinus node of the heart involves briefly forcing high current through the heart. To overcome tissue resistivity, this implies the delivery of a high-energy, high-voltage pulse.

Typical external defibrillators deliver this pulse by discharging a 32-μF capacitor charged up to 5000 V dc through a 500-μH inductor directly into paddle electrodes placed on the chest of the patient, who may be assumed to act as a 100-Ω resistor. A sizable fraction of

Figure 3.3 Layout of an instrument that incorporates the ECG amplifier of Figure 3.2. Note that the gap in components and conductors which forms the insulation barrier is traversed only by the ISO107 isolation amplifier. An external ELPAC model MED113TT medical-grade power supply is used to power the circuit from the 120 V ac power line.

the defibrillation pulse may appear at the ECG recording electrodes as well as between the isolated patient ground and the power line ground. The front-end protection circuit places 330-kΩ resistors (R4 in series with R6, R25, and R5 in series with R7) in series with the patient leads to limit the peak defibrillator input current to under 10 mA. For this application, 2-W carbon-composition high-voltage-rating resistors are chosen, to withstand the several dozen watts of instantaneous power that may be dissipated during each defibrillation pulse.

Since voltages close to the full 5000-V defibrillator capacitor initial voltage could appear across these resistors, care must be taken to ensure that current does not find an alternative path by producing a spark or by creeping across the printed circuit. The insulation required to withstand the peak voltage of the defibrillator pulse should be chosen to be a minimum air clearance of 7 mm and a minimum creepage distance of 12 mm. This separation would also apply to the isolation barrier between the applied part and all other parts of the medical instrument.

A second consideration must be made for equipment that may be used in the operating room. Here the applied part of the instrument may be exposed to very strong RF currents coming from an electrosurgery (ESU) unit used for either cauterizing wounds or cutting tissue. Usually, continuous-wave or gated damped sinusoids are applied between a large-area electrode on the patient's back and the scalpel electrode. Through RF heating, tissues are cut and blood is coagulated, causing small ruptured vessels to close. The RF component of the ESU waveform typically is within the range 200 kHz to 3 MHz, and power levels into a 500-Ω load range from 80 to 750 W. Open-circuit voltages range from approximately 300 V and can be as high as 9 kV.

If the circuit of Figure 3.2 were used in the presence of ESU, the path of RF leakage current would probably be from the ESU electrodes into one or more of the device's patient electrodes, through the coupling capacitance of the ISO107, through the stray capacitance of the power supply transformer, into the power line, and back through the stray capacitance of the ESU generator's power transformer. To deal with these RF currents, medical electronic equipment often includes filters that attenuate RF signals before they can be detected by the circuit's nonlinearities. In our front-end protection circuit, RF appearing at the ECG+ and ECG− electrodes is sinked to the isolated ground by C2 and C3. C4 is used to eliminate any remaining RF that can be demodulated differentially by D1–D4 or IC1's circuitry. Here again, currents driven by very high RF voltages must not find alternative paths such as corona discharge or creepage, and for these reasons, appropriate spacings must be observed.

STAND-ALONE ANALOG ISOLATORS

The previous design example demonstrated the use of the ISO107 isolation amplifier embedded within a circuit to provide a signal path across the isolation barrier. In many cases, having instead a self-contained general-purpose isolation module like that of Figure 3.4 can simplify the design of prototype and experimental equipment. The circuit diagram for such a module is shown in Figure 3.5. This module was designed as a stand-alone isolation board to protect subjects connected to isolated biopotential amplifiers from lethal ground fault currents as well as those originating from defibrillator pulses. The heart of the module is IC1, an Analog Devices' 284J isolation amplifier. This device meets leakage standards of $2\,\mu A$ maximum at 115 V ac, 60 Hz. This performance results from the carrier isolation technique, which is used to transfer signals and power across the isolation barrier, providing a maximum isolation of $2500\,V_{RMS}$ at 60 Hz for 1 minute, and $\pm 2500\,V_{p\text{-}p}$ maximum continuous ac, dc, or 10-ms pulses at 0.1 Hz.

Figure 3.4 A stand-alone signal isolator can be built using an Analog Devices' 284J isolation amplifier. The input voltage range for this module is ± 5 V differential at unity gain. However, this module can also be used for the direct low-level amplification of biopotential signals with a low input noise $10\,\mu V_{p\text{-}p}$, medium input impedance $10^8\,\Omega$, and high CMR (110-dB inputs to output, 78-dB inputs to guard). The module can generate isolated power for input circuitry, such as biopotential signal buffer preamplifiers or instrumentation amplifiers.

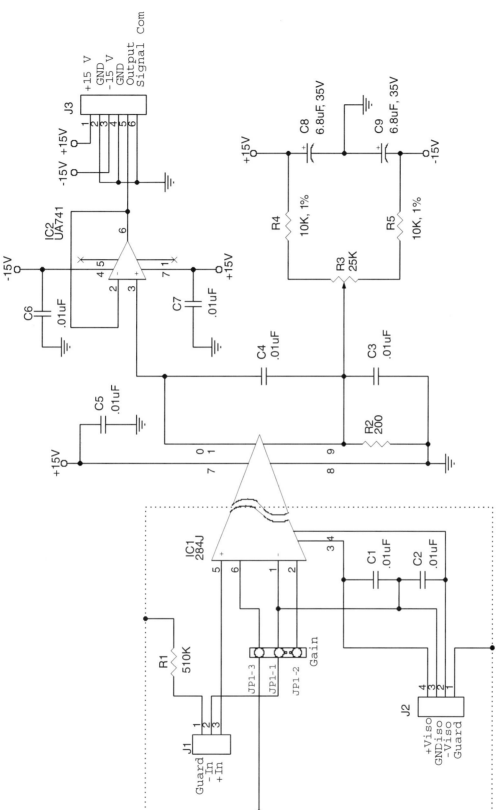

Figure 3.5 This isolation amplifier meets leakage standards of 2.0 μA at 115 V ac, 60 Hz, providing a maximum isolation of 2500 V_{RMS} at 60 Hz for 1 minute, and ±2500 V_{P-P} maximum continuous ac, dc, or 10-μs pulses at 0.1 Hz.

Bipolar input signals, present at J1-3, and referenced to isolated ground at J1-2, are introduced into the isolated signal input of IC1. IC1's gain can be set in the range 1 to 10 by changing the value of a resistor connected between JP1-1 and JP1-2 according to

$$\text{gain} = 1 + \frac{100\text{k}\Omega}{10.7\text{k}\Omega + R_i(\text{k}\Omega)}$$

To preserve high CMR, this resistor and all connections to it must be guarded with a shield connected to JP1-3. Best performance is achieved by placing a shorting jumper between JP1-1 and JP1-2 and operating the circuit at a gain of 10. Leaving JP1 open results in unity gain.

The bandwidth supported by the 284J is dc to 1 kHz (small signal), dc to 700 Hz (full power $G = 1$), and dc to 200 Hz (full power $G = 10$). IC1's output is buffered by IC2, a unity-gain buffer, in order to drive low-impedance loads connected between the module's output at J3-5 and nonisolated ground at J3-6. Trimmer R3 is used to zero the output offset voltage over the gain range. IC1's output is low-pass filtered to roll off noise and output ripple. Cutoff (-3 dB) of the low-pass filter is given by

$$f_{-3\text{dB}}(\text{Hz}) = \frac{1}{2\pi \times C4(\text{F}) \times 1000\,\Omega}$$

Use of a 1-μF capacitor results in a cutoff frequency of approximately 160 Hz.

The input voltage range for this module is ± 5 V differential at unity gain. However, this module can also be used for the direct low-level amplification of biopotential signals with a low input noise 10 $\mu V_{\text{p-p}}$, medium input impedance of $10^8\,\Omega$, and high CMR (110-dB inputs to output, 78-dB inputs to guard). Differential measurement of biopotential signals is achieved between J1-3 (noninverting input) and J1-2 (inverting input), while CMR is optimized by connecting J1-1 to a distant reference electrode. If the module is used as a biopotential amplifier, the leads to the electrodes should be low-loss low-capacitance coaxial cables, whose shields are connected to J1-1. This module should be operated using a symmetric ± 15 V regulated power supply (J3-1 $= +15$ V, J3-2 $=$ nonisolated ground, J3-3 $= -15$ V). Dual ± 8.5 V dc at 5 mA of isolated power are provided at J2. These lines may be used to power floating input circuitry such as biopotential signal buffer preamplifiers or instrumentation amplifiers.

THREE-PORT ISOLATION

Most isolation amplifier ICs on the market that contain an internal dc/dc converter to power the isolated side of the amplifier (as well as support circuitry) are labeled as input or output isolation amplifiers. This refers to the direction in which power is sent across the isolation barrier. An *input isolation amplifier* thus powers the isolated input side of the amplifier through an internal isolating dc/dc converter while operating its output side from the same source that powers the dc/dc. Conversely, an *output isolation amplifier* uses the power directly to supply the input side of the amplifier and the dc/dc converter, while the output of the dc/dc is used to power the isolated output stage of the amplifier. Since most isolation amplifiers with internal dc/dc converters have additional power capacity, input isolation amplifiers typically make extra power available on the isolated input side for driving external signal conditioners or preamplifiers, while an output isolation amplifier makes extra power available on the isolated output side for driving external loads.

Although most applications in medical devices require input isolation amplifiers, there are cases in which a signal generated by the device is required to be sent to a floating applied part. For example, a stimulus waveform may be generated using a D/A on a nonisolated side of the device and delivered to a stimulation circuit using an output isolation amplifier. An interesting isolation amplifier that can fulfill both roles is Analog Devices' AD210. This component has a true three-port design structure, which permits it to be applied as an input or output isolator, in single- or multichannel applications. In the AD210, each port (input, output, and power) remains independent, with $2500\,V_{RMS}$ (continuous) and $\pm 3500\,V_{p-p}$ isolation between any two ports.

In the circuit of Figure 3.6, an AD210BN is used to power and isolate a general-purpose instrumentation biopotential amplifier front end. Biopotential signals detected by electrodes connected to input connector J1 are dc-coupled to the inputs of an AD620 instrumentation amplifier via input resistors R5 and R8. Low-leakage diodes D1–D4 are used to protect the AD620 from high-voltage transients. Resistors R3 and R9 provide a dc path for bias currents whenever the amplifier is used to detect signals from capacitive sensors.

IC2 is used to buffer the offset voltage set by trimmer potentiometer R10. The offset level is fed directly to the reference pin of the instrumentation amplifier. Gains of $\times 1$, $\times 2$, $\times 10$, and $\times 100$ are selected through jumpers JP1–JP3. The output of the instrumentation amplifier is fed to the input of the AD210 isolation amplifier. The output of the isolation amplifier is filtered by the active low-pass filter formed around IC4. Isolated $\pm 15\,V$ to power IC1 and IC2 is generated by the AD210. The AD210 also produces a separate, isolated $\pm 15\,V$ which is used to power IC4. A single 15 V at 80 mA supply is all that is needed to power the complete circuit, thanks to the AD210's three-port feature. Figure 3.7 shows an implementation of this circuit where the various insulation barriers can be seen clearly.

ANALOG SIGNAL ISOLATION USING OPTICAL ISOLATION BARRIERS

High performance usually comes at a high price, and the ISO107, 284J, and AD210BN are no exceptions. The unit price for each of these is over $100, making their use prohibitive in many low-cost designs as well as in instruments that involve a large number of analog signals crossing the isolation barrier. In these cases, analog isolators can be built using low-cost optoisolators as isolation channels. Optoisolators or optocouplers operate by emitting and detecting modulated light. An input current drives a light-emitting device internal to the optocoupler, and an internal photodetector drives the output circuitry. Optoisolators usually consist of an LED and a phototransistor which are galvanically isolated from each other and are located opposite each other in a lighttight package.

The simplest form of optical isolation for an analog signal is implemented by the circuit of Figure 3.8. In this circuit, an input voltage is converted by IC1 and transistor Q1 to a proportional current to drive a LED. The light output of the LED is proportional to the drive current between $500\,\mu A$ and $40\,\mu A$. However, since the best linear behavior for optical output flux versus input current occurs in the range 5 to $20\,\mu A$, offset is introduced in the voltage-to-current conversion. A 1.2-V reference voltage is generated across reference diode D1. Resistor R5 and potentiometer R6 select the fraction of the reference voltage that should be used as an offset. The voltage divider formed by resistors R1 and R2 and potentiometer R3 are used to scale the input signal.

The LED D2 and a photodiode D3 are mounted across from each other inside a piece of dark PVC pipe. The silicon photodiode, operating as a light-controlled current source, generates an output current that is proportional to the incident optical flux supplied by the LED emitter. Translating the photodiode current back to a voltage is done with series resistor R10. IC2 buffers the voltage across the current-sensing resistor and cancels the offset caused by keeping the LED always illuminating the photodiode. Although operation of this circuit seems straightforward, its performance leaves a lot to be desired, especially as far as stability is concerned. The problem is that the optical output of a LED as a function of drive current is very unstable, changing widely with the LED's age, temperature, and the dynamics of the drive current. For this reason, this circuit is not recommended except for applications where good linearity and precision are not required.

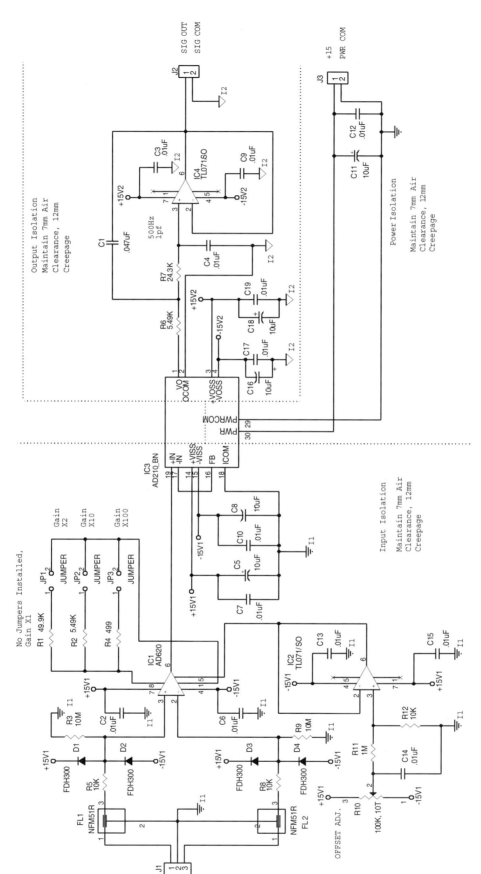

Figure 3.6 In this biopotential amplifier, each port (input, output, and power) of the AD210 isolation amplifier remains independent with 2500 V$_{RMS}$ (continuous) and ±3500 V$_{P-P}$ isolation from any of the other ports.

110

Figure 3.7   This implementation of the general-purpose biopotential amplifier of Figure 3.6 requires a single supply of 15 V at 80 mA, thanks to the AD210's three-port feature.

Figure 3.9 shows a circuit that implements an interesting way of somewhat linearizing the response of an optoisolator while simplifying the circuitry needed to introduce offset to pass bipolar signals. This isolated EEG amplifier is an adaptation of a circuit by Porr [2000]. Here, an Analog Devices AMP01 instrumentation amplifier is used as the high-input-impedance front-end amplifier for the biopotentials collected from EEG scalp-surface electrodes. The gain of this stage is 20. IC1's output is high-pass filtered by C5, C6, and R3 to introduce a −3-dB cutoff frequency of 0.32 Hz. A selectable-gain stage is implemented around op-amp IC2 to boost IC1's output signal approximately 100, 200, 500, or 1000 times. A Sallen–Key second-order low-pass filter built around IC3A is then used to limit the bandpass of the EEG amplifier to approximately 34 Hz.

IC3B drives the LEDs of optoisolators IC4 and IC5. The phototransistor in IC4 is used to set the inverting input of IC3B such that the LED is driven to a point that balances the signal at IC3's noninverting input. When the phototransistor in IC5 is not illuminated, its collector is pulled up to the nonisolated positive supply rail by R21. However, as signals cause IC3B to drive the LED, the phototransistor pulls the collector toward the nonisolated negative supply rail. The isolated signal is high-pass filtered by C18 and R22 and buffered via IC8 before being presented to the output.

LINEAR ANALOG ISOLATION USING OPTOISOLATORS

A way of using an optoisolator for analog signals while maintaining good linearity is first to convert the analog signal into a pulse train of variable frequency (or pulse width), which is then used to drive the optoisolator. At the other side of the optoisolator, the pulse train is demodulated to render the original signal. Another possibility is to place the optoisolator within a servo loop that makes use of the loop's error to convey a high-linearity analog signal. Yet another solution is to convey true-digital data through the optoisolator.

The analog isolator circuit of Figure 3.10 works by pulse-width modulating a pulse train in proportion to the input voltage. As shown in Figure 3.11, an input signal is presented to

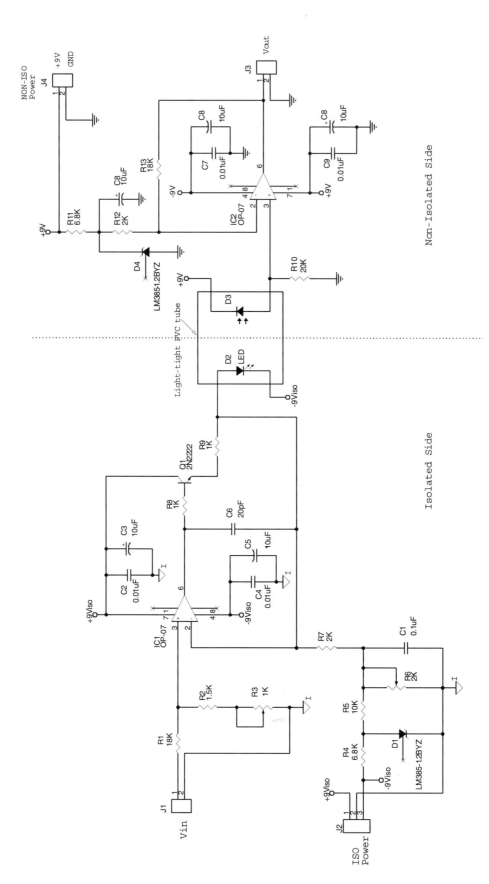

Figure 3.8 The simplest form of optical isolation for an analog signal is implemented by driving an LED with a current proportional to the input signal. Offset is introduced to operate the LED in the range where optical output flux versus input current is almost linear. A photodiode mounted across the LED inside a piece of dark PVC pipe generates an output current that is proportional to the incident optical flux supplied by the LED.

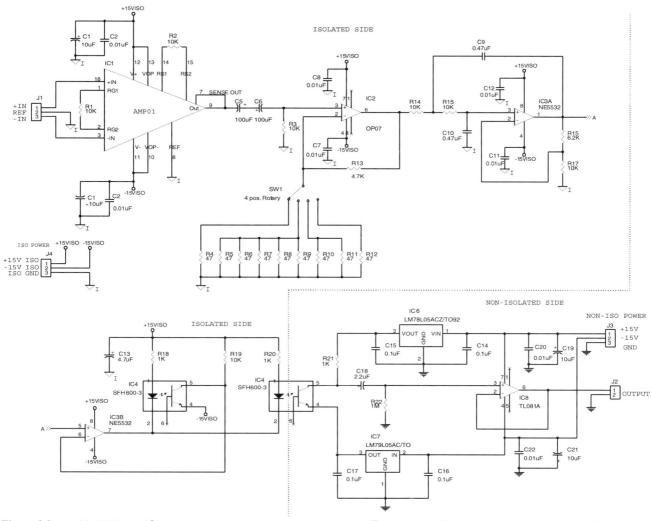

Figure 3.9 In this EEG amplifier, an extra optoisolator is used to introduce offset automatically to pass bipolar signals and to linearize the response of the isolation amplifier. IC3B drives the LEDs of optoisolators IC4 and IC5. The phototransistor in IC4 is used to set the inverting input of IC3B such that the LED is driven to a point that balances the signal at IC3's noninverting input. When the phototransistor in IC5 is not illuminated, its collector is pulled up to the nonisolated positive supply rail by R21. However, as signals cause IC3B to drive the LED, the phototransistor pulls the collector towards the nonisolated negative supply rail.

the differential amplifier implemented by IC1A. This differential amplifier has unity gain and is used to add an offset voltage that is set via potentiometer R3. The input signal is then low-pass filtered through R6 and C6 to the dc–65 Hz modulation bandwidth, which can be supported by the pulse-width modulator implemented around IC1B. The pulse-train output from the pulse-width modulator drives an LED inside optocoupler IC2 via transistor Q1. The Optek OPI1264 optoisolator used in this circuit was selected because it has a very respectable 10-kV isolation rating (although its UL file rates it at 3500 V ac for 1 minute).

At the phototransistor end of IC2, the light pulses generated by the LED are received and converted into a replica of the original pulse train through the *RC* high-pass filter formed by R13 and C8. The circuit centered on IC3A acts as a pulse-train demodulator to reconstruct the original modulating signal, which, depending on the characteristics of the input signal, may have an offset. Finally, the low-pass filter implemented around IC3B removes any remnants of the pulse train's carrier.

Figure 3.10   In this analog signal isolator, the analog signal is converted into a pulse train of variable pulse width, which is then used to drive the optoisolator. At the other side of the optoisolator, the pulse train is demodulated to render the original signal. The large cylinder at the center of the board is an Optek OPI1264 optoisolator through which isolation levels of 10 kV can be obtained.

A similar circuit can be built using an integrated voltage-to-frequency converter. The circuit of Figure 3.12 is an application suggested by Analog Devices for its ADVFC32 integrated circuit. This chip is an industry-standard, low-cost monolithic voltage-to-frequency (V/F) converter or frequency-to-voltage (F/V) converter with good linearity and operating frequency up to 500 kHz. In the V/F configuration, positive or negative input voltages or currents can be converted to a proportional frequency using only a few external components. For F/V conversion, the same components are used with a simple biasing network to accommodate a wide range of input logic levels.

In an analog isolator circuit, an input signal in the range 0 to 10 V drives IC1, an ADVFC32 configured as a V/F converter. Input resistor R1 and offset resistor R2 have been selected such that a 0-V input causes IC1 to oscillate at 50 kHz, while a 10-V input yields an output of 500 kHz. IC2 is a bandgap voltage regulator used to generate the offset voltage reference. IC3, a high-frequency optoisolator with a high isolation voltage rating, is used to transmit the frequency-modulated signal generated by IC1 across the isolation barrier to IC4, an ADVFC32 configured as a F/V converter. Integration for the V/F function is provided by an internal op-amp that has C12 within its feedback loop. This capacitor defines the frequency response of the isolation amplifier circuit. With 1000 pF, the bandwidth of the amplifier is dc to 3 kHz. The output of the V/F is offset to reproduce the input range 0 to 10 V by summing a current produced through R9 and R10 by a second bandgap reference IC5. Prior to exiting the circuit, any remaining carrier is filtered via R6 and C13, which form a 3-kHz low-pass filter.

In the new generation of *analog optocouplers* that have appeared on the market, the LED they use has widely variable electrical-to-optical transfer characteristics, just as in any other optocoupler. However, these optocouplers have a second photodetector, which is used as part of a feedback loop to stabilize the LED's optical output. The circuit of Figure 3.13 is very similar to the simple optical analog isolator of Figure 3.8. The main difference is that IC1's output is not simply a current proportional to its input voltage. Rather, part of the LED's optical flux is detected by the second photodiode and used to provide feedback to the op-amp current source. Since the stability of a photodiode is not usually a concern, and since the characteristics of both photodiodes are closely matched during manufacture, the

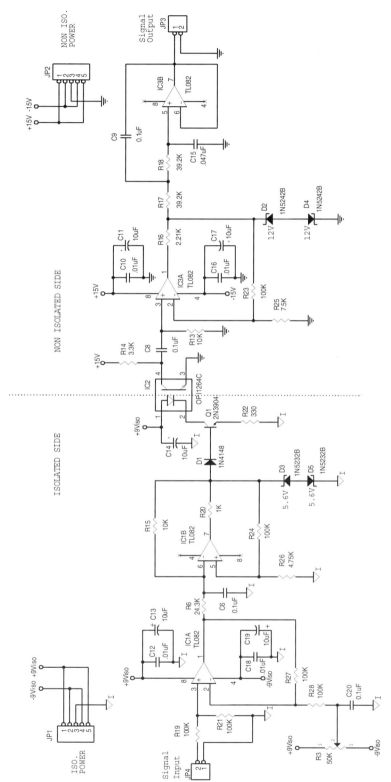

Figure 3.11 In this pulse-width modulation optical isolation amplifier, the input signal is first offset by IC1A. IC1B implements the pulse-width modulator, which drives the LED of an Optek OPI1264 optoisolator through transistor Q1. The pulse train at the optoisolator's is demodulated via IC3A and low-pass filtered by IC3B.

115

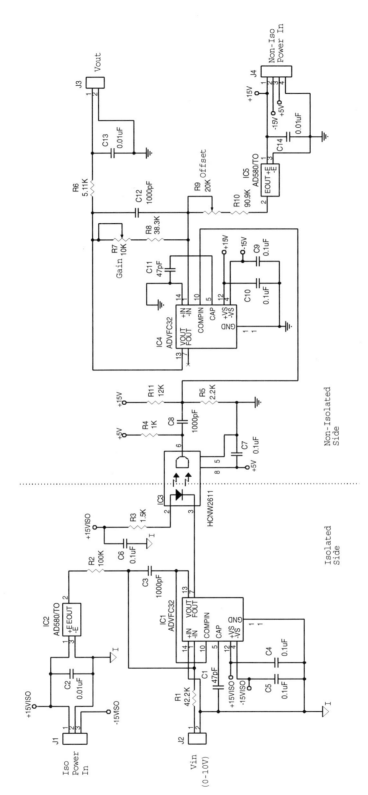

Figure 3.12 An optical isolation amplifier can be built using a voltage-to-frequency (V/F) converter that drives the optoisolator and a frequency-to-voltage (F/V) converter to recover the original signal. The ADVFC32 can be used to implement both functions. An input signal in the range 0 to 10 V causes the LED inside the optoisolator to be flashed at a frequency between 50 and 500 kHz.

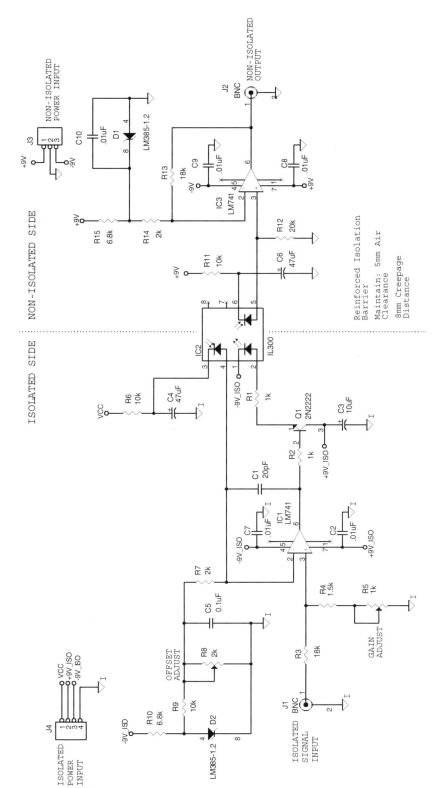

Figure 3.13 Unlike the open-loop isolation amplifier of Figure 3.8, the output of IC1 in this closed-loop circuit is not simply a current proportional to its input voltage. Rather, part of the LED's optical flux is detected by the second photodiode and used to provide feedback to the op-amp current source. Since the stability of a photodiode is not usually a concern, and since the characteristics of both photodiodes are closely matched during manufacture, the transfer characteristic of the complete circuit is highly linear regardless of LED nonlinearity, aging, temperature, or LED drive current dynamics.

117

transfer characteristic of the complete circuit is highly linear regardless of LED nonlinearity, aging, temperature, or LED drive current dynamics.

In this circuit, the LED driver is prebiased by introducing a negative offset voltage into the inverting input of op-amp IC1. The feedback photodiode sources current to R7, which is translated into the control voltage for the servo loop. The op-amp will drive transistor Q1 to supply LED current to force sufficient photocurrent to keep the voltage at the noninverting input of IC1 equal to that at its inverting input. The output photodiode is connected to a noninverting voltage follower amplifier. The photodiode load resistor R11 performs the current-to-voltage conversion. IC3 is used to buffer R11 as well as to zero-offset the output signal's baseline.

Note that none of the optocoupler-based isolators described above generate their own isolated power. Once an isolation dc/dc converter has to be added, and considering the limited performance of these circuits, applications that require only one or two isolated signal channels may be better off using commercial, self-contained analog isolation ICs. Isolators like the optocoupler circuits shown are most often found in multichannel applications, where the cost of a large number of self-contained isolation ICs would be prohibitive.

DIGITAL ALTERNATIVE TO SIGNAL ISOLATION

The large majority of modern medical electronic instruments make use of either an embedded microcomputer or an external PC for control, data processing, and display. This implies that in most cases, an analog-to-digital converter is used at some point within the instrument to support data acquisition functions. The circuit of Figure 3.14 places the A/D converter in direct connection with the applied part of the medical instrument, and relays digital rather than analog signals across the isolation barrier. This alternative over analog signal isolation has the advantage that the additional noise, nonlinearity, and complexity of the latter can be avoided by translating signals to digital format early in the process. Furthermore, optoisolators for high-speed digital signals are inexpensive and widely available. In addition, serial data formats can be used to minimize the number of digital signals that must be communicated concurrently through the isolation barrier.

Many modern high-end medical instruments make extensive use of this philosophy. If you have the opportunity, examine the circuit schematics of one of today's electrocardio-

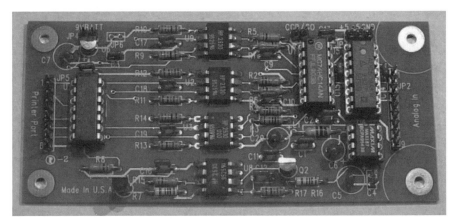

Figure 3.14   The majority of modern medical electronic instruments make use of either an embedded microcomputer or an external PC for control, data processing, and display, meaning that an A/D converter is used at some point within the instrument to support data acquisition functions. This circuit places the A/D converter in direct connection with the applied part of the medical instrument, and relays digital rather than analog signals across an optical isolation barrier.

graphy instruments. You will probably find an elegant design comprising instrumentation amplifiers for each lead followed directly by an A/D converter and optical isolation leading to a DSP microprocessor. Often, the complete applied part is contained within a "medallion" to which the patient leads are directly connected, and digital signals to and from the embedded microcomputer are relayed through optical fiber.

The sample circuit of Figure 3.15 is not as complex as those of high-end commercial instruments but provides a very simple and convenient interface between analog-output applied parts and most PCs on the market. Instead of connecting to the computer's

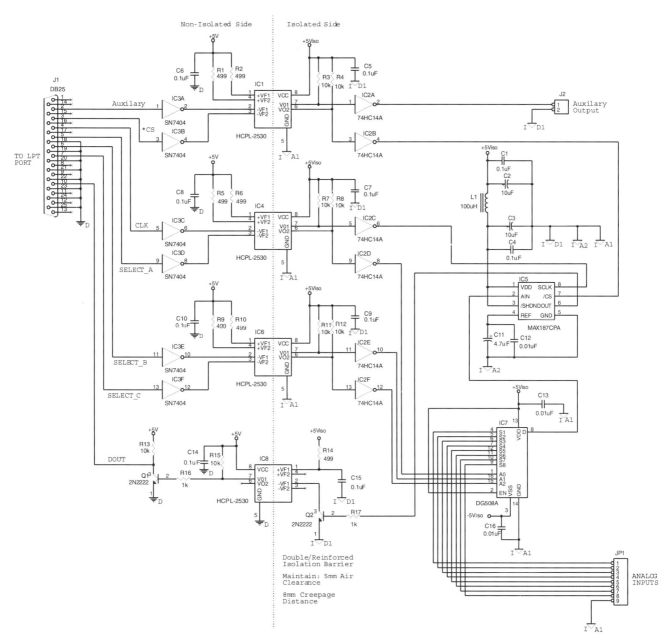

Figure 3.15 A MAX187 low-power 12-bit A/D converter IC forms the core of the isolated A/D converter module. A two-wire serial interface conveys data from the MAX187 through optoisolators back to the PC through the printer port. The PC also controls an isolated signal multiplexer that allows one of eight analog signals to be presented to the input of the A/D.

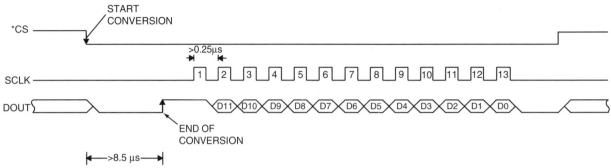

Figure 3.16 Data acquisition and serial protocol timing for the MAX187. A/D conversion is initiated by a falling edge on the *CS line. After conversion, data are read out in serial format, shifted from the sequential-approximation register on each falling-edge transition of SCLK.

expansion bus, it plugs into one of the parallel printer ports, which is used as a serial I/O for an eight-channel A/D converter. As shown in the diagram, at the heart of the circuit is a MAX187. This IC is one of Maxim's single-chip A/Ds featuring a 12-bit 8.5-μs successive-approximation converter, 1.5-μs track-and-hold, on-chip clock, precision 4.096-V reference, $\frac{1}{2}$LSB nonlinearity, and high-speed three-wire serial interface.

One of eight isolated analog signals to be measured is presented to the analog input line AIN of the MAX187 by way of a DG508A analog multiplexer IC. Voltages within the range 0 to 4.096 V can be converted by the A/D into distinct digital codes for every 1 mV of change. MAX187's A/D conversion initiation and data-read operations are controlled by the *CS (chip select) and SCLK (serial clock) lines. As shown in Figure 3.16, an A/D conversion is initiated by a falling edge on the *CS line. At this point, the track-and-hold holds the input voltage and the successive-approximation process begins. The start of conversion is acknowledged by the MAX187 changing the state of the DOUT line from high impedance to the low state. After an internally timed 8.5-μs conversion period, the end of conversion (EOC) is signaled by the DOUT line going high.

Once a conversion is completed, data can be obtained in serial format, shifted from the sequential-approximation register on each falling-edge transition of SCLK. Since there are 12 bits, a minimum of 13 falling-edge pulses are required to shift out the A/D result. Isolation between the PC's printer port and the MAX187 is provided by IC4–IC7. Bits 1 and 2 of the PC's LPT 8-bit output port (hex address 378 for LPT1:) are toggled by software to implement the control portion of the MAX187 serial protocol. Bit 6 of the printer status port register (hex address 379 for LPT1:) is used to receive the serial data from the MAX187. Bits 3–5 of the output port are used to control the analog signal multiplexer. Bit 0 of the output port is an auxiliary line that may be used in the control of the applied part's circuitry.

Power for the MAX187 must be supplied from a patient-contact-rated isolated power supply capable of delivering ±5 V. A pi filter formed by C1–C4 and L1 ensures a clean supply to the A/D. In addition, you may notice that two separate isolated ground planes, one analog and one digital, are shown in Figure 3.15. Ideally, the signal ground plane, used as the reference for the analog input signal, should be constructed to shield the analog portions of the A/D and signal multiplexer: namely, the input network and voltage reference filtering and decoupling capacitors. The analog and digital ground planes should be connected at a single point, preferably directly to the isolated ground line supplying the circuit.

SOFTWARE FOR THE ISOLATED A/D

The sample program that follows is for driving the isolated A/D converter from the printer port of a PC. The program flow starts by initializing the ports. Notice that use of the standard

LPT1: is assumed, and you may need to change the output port and status port locations to suit your specific installation.

```
' ATODSAFE.BAS is a QuickBasic sample program to acquire data using
' the 8-channel, isolated 12-bit A/D converter. The use of LPT1:
' is assumed.
'
' Printer port locations
' ---------------------------
CONST prinop = &H378 ' Printer Output Port (could be &H278 or &H3BC)
CONST prinstat = &H379 ' Printer Status Port (could be &H279 or &H3BD)
'
' Define control pin locations
' -------------------------------------------
CONST aux = 1, notcs = 2, sclk = 4, sela = 16, selb = 32, selc = 64
'
' Initialize
' ----------
OUT prinop, 0    ' clear printer port
CLS              ' clear screen
INPUT "Please input channel to acquire "; chan
' determine control bits for desired channel
selas = (chan AND 1) * sela
selbs = (chan AND 2) / 2 * selb
selcs = (chan AND 4) / 4 * selc
'
' Acquisition and display control
' -------------------------------
start:
SCREEN 2                          ' CGA graphics mode 640×200
GOSUB acquire                     ' determine first display point
y = INT((4.096 − vout) * 45) + 10 ' compute position of starting
point
start1:
CLS                               ' refresh screen
LOCATE 2, 2: PRINT "4.096V";      ' place y-axis labels
LOCATE 7, 2: PRINT "3.000V";
LOCATE 13, 2: PRINT "2.000V";
LOCATE 19, 2: PRINT "1.000V";
LOCATE 25, 2: PRINT "0.000V";
PSET (60, y)                      ' place first sample
FOR x = 60 TO 640                 ' horizontal sweep
   GOSUB acquire                  ' acquire a sample
   y = INT((4.096 − vout) * 45) + 10 ' compute position on screen
   LINE −(x, y)                   ' display data
   IF INKEY$ <> " " THEN GOTO progend ' press any key to escape
NEXT x                            ' next sample
GOTO start1                       ' start a new screen

progend:
'
```

```
' Leave program
' -------------
OUT prinop, 0                              ' clear printer port
SCREEN 0                                   ' return to text mode screen
END


acquire:
'
' Acquisition loop
' ----------------
OUT prinop, selas + selbs +                ' keep CS' deasserted
selcs + notcs
convert:
OUT prinop, selas + selbs + selcs          ' convert by asserting CS'
loop1:
bit = (INP(prinstat) AND 64) / 64          ' read status port and filter DOUT
IF bit = 0 THEN GOTO loop1                 ' wait for EOC signal
dat = 0                                    ' clear A/D accumulator
FOR clocknum = 11 TO 0 STEP -1             ' clock 12 bits serially
    OUT prinop, selas + selbs +            ' clock pulse rising edge
    selcs + sclk
    OUT prinop, selas + selbs + selcs      ' clock pulse falling edge
    bit = (INP(prinstat) AND 64) / 64      ' read status port and filter DOUT
    dat = dat + (2 ^ clocknum) * bit       ' accumulate from bit 11 to bit 0
NEXT clocknum                              ' next bit
OUT prinop, selas + selbs +                ' one more clock to reset A/D
selcs + sclk
OUT prinop, selas + selbs + selcs
OUT prinop, selas + selbs +                ' deassert CS'
selcs + notcs
vout = dat * .001                          ' translate A/D data to Volts
RETURN
```

The program is presented only as an example and for the sake of simplicity will run the A/D converter as fast as the PC is able to drive its lines. Sampling rate control could be implemented either by inserting for-to loops adaptively to introduce controlled delay between samples [Prutchi, 1996] or by controlling the acquisition process from interrupts generated by high-resolution hardware timing [Ackerman, 1991–1992; Schulze, 1991].

The actual acquisition subroutine starts by setting up the multiplexer while keeping *CS deasserted. Conversion for the channel selected is then initiated by asserting *CS and polling for the end-of-conversion signal before attempting to read the conversion data. At this point, the A/D accumulator variable is cleared, and each of the 12 bits is clocked-in serially. The value of each bit is read from the status port and is multiplied by the decimal value of its binary position before being accumulated. Finally, one more clock pulse is inserted to reset the A/D, the *CS line is deasserted, and A/D data are translated to volts.

ISOLATED ANALOG MULTIPLEXER

The number of channels that can be acquired through the isolated A/D can be expanded by using additional multiplexers. Figure 3.17 shows a PCB that implements a 64-channel

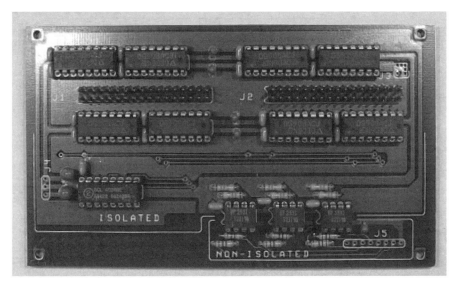

Figure 3.17 The number of channels that can be acquired through the isolated A/D can be expanded by using additional multiplexers. This circuit implements a 64-channel analog signal multiplexer with isolated control.

analog signal multiplexer with isolated control. Through its use, up to 64 high-level analog signals may be connected selectively to an A/D or other acquisition and recording instrumentation. Signals may originate from biopotential amplifiers, physiological sensors, or a combination of both.

The channel addressing is controlled digitally through a parallel asynchronous input. Optical isolation is implemented onboard to maintain patient-contact-rated isolation between the analog electronics and the digital control. Additional multiplexer circuits may be stacked to increase the channel capability of a system. In PC-based data acquisition systems, this board may be used to collect analog signals from up to 64 channels through a single analog input of the A/D. The digital control is easily obtainable from a parallel output port available in most every PC data acquisition board.

Typical multichannel biopotential signal acquisition applications include topographic brain mappers, body potential mapping (BPM) ECG, the recording of surface array EMG signals, and so on. The signal range for the DG508 multiplexers in the circuit of Figure 3.18 is ± 12 V with a bandwidth of at least 50 kHz. The in–out resistance of a selected channel is less than $600\,\Omega$. Scanning of an array can be done at a maximum frequency of 1000 channels/s. Channel selection is accomplished via optoisolators, which can be driven directly by TTL logic. The circuit requires ± 12 V isolated power to operate.

POWER SUPPLIES

As Figure 3.1 showed, having reinforced insulation between the applied part and every other part of the medical instrument does not mean that similarly strong insulation is not needed between a mains part and other live or nonlive parts besides the applied part. This implies that although you may be using a component which itself powers the applied part across an appropriate isolation barrier (e.g., ISO107, 284J), the instrument's power supply must still meet the same requirements as a safety isolation transformer.

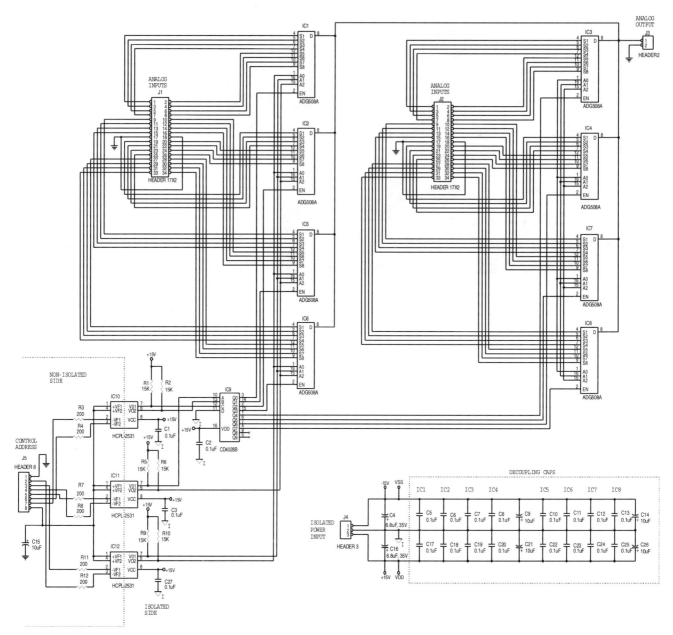

Figure 3.18   Using this circuit, up to 64 high-level analog signals may be connected selectively to an A/D or other acquisition and recording instrumentation. Signals may originate from biopotential amplifiers, physiological sensors, or a combination. Channel addressing is digitally controlled through a parallel, optically isolated asynchronous input. Additional multiplexer circuits may be stacked to increase the channel capability of a system.

To accomplish the required levels of isolation, medical instruments are often designed to incorporate a safety extra-low-voltage (SELV) transformer to derive their operating power from the power line. This type of transformer supplies a voltage under 25 V ac through an output winding that is electrically separated from ground and the body of the transformer by at least basic insulation, and which is separated from the input winding by at least double insulation or reinforced insulation.

In addition, SELV transformers for medical equipment usually have an electrostatic shield that is tightly wound over the insulation of the primary windings. This shield reduces capacitive coupling between primary and secondary windings, thus reducing leakage currents at the power line frequency. The shield is coated with reinforced insulation to create a reinforced insulation barrier between the primary and secondary windings. The core itself is isolated from the windings by supplementary or reinforced insulation.

Another convenient alternative for powering medical instruments is the use of batteries. This substitution not only ensures inherently low leakage currents but can make the equipment highly portable. Considering that you may need to travel all the way to South America, Eastern Europe, or Asia to run your first tests, the independence provided by a battery-operated power supply is certainly a welcome blessing for an evaluation prototype. Whatever the choice in power supply, it is generally a good idea to purchase it as an approved OEM (original equipment manufacturer) assembly. This helps you concentrate your efforts on the core of your instrument rather than having to deal with the headaches of designing and constructing supplies that perform as required by the safety standards.

Along the same philosophical lines, designing an instrument to make use of preapproved components can help considerably to receive and maintain safety approval once you embark on the production and sale of a medical product. You can still use components that have not been certified by a U.S. Nationally Recognized Testing Laboratory (NRTL, or its equivalent in other countries); however, the assured continuity of safety performance will have to be investigated for each device to be used. This is complicated further by the fact that once you receive approval for your product, any change in any component will require requalification of the complete assembly. Finally, keep in mind that safety standards usually impose special performance characteristics for certain components, such as power cords, switches, line filters, fuse holders, optoisolators, CRTs and displays, and printed circuit boards.

ADDITIONAL PROTECTION

Regardless of how carefully you designed your instrument, absolute safety cannot be guaranteed in the real world. Despite all the safety testing and evaluation required by the FDA, medical device manufacturers still pay a premium for insurance to protect themselves from exposure against liability. For this reason, it often happens that additional or redundant hardware to ensure safety beyond the minimum requirements is cost-effective, since it will bring concomitant savings in insurance costs due to reduced risk.

Being extra conservative is especially important at the prototyping stage, since as an entrepreneur you probably do not have the legal and financial umbrella of a large corporation to protect you against an unintentional mishap. Our personal preference is to introduce, at the very least, an additional but independent layer of protection against electrical shock at the patient interface. A very practical method to accomplish this is to use Ohmic Instruments' Iso-Switch patient-lead fault interruptors. These devices, shown in Figure 3.19, are two-lead semiconductor devices that can be placed, almost transparently, in series with every patient connection to break the patient circuit in case an overcurrent fault develops.

As shown in the V–I plot of Figure 3.20, an Iso-Switch patient-lead fault interruptor rated at $\pm 10 \, \mu A$ acts as a 40-kΩ resistor. Once the trip point of the Iso-Switch is exceeded, the device presents a negative-slope resistance of magnitude equal to that of the positive slope within the trip boundaries. The trip time under an overcurrent condition is very fast, typically $10 \, \mu s$. Once the device trips, the resistance of the Iso-Switch increases to

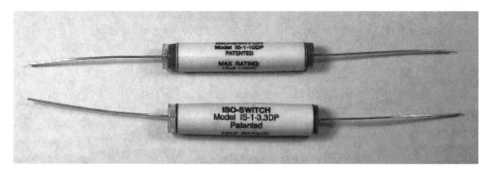

Figure 3.19 Ohmic Instruments' Iso-Switch patient-lead fault interruptors are two-lead semiconductor devices that can be placed, almost transparently, in series with every patient connection to break the patient circuit in case an overcurrent fault develops. Various models are available for different applications, with trip currents ranging from ±1 μA to ±100 μA. Various operating voltage ranges are also offered: ±325 V for 115-V ac instruments, ±720 V for 220-V ac equipment, and the DP (defibrillation-proof) series that withstands pulses of ±5 kV for 10 ms.

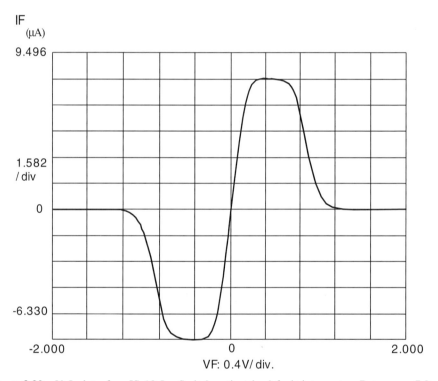

Figure 3.20 V–I plot of an IS-10 Iso-Switch patient-lead fault interruptor. Between −7.9 and +7.9 μA (rated ±10 μA), the device acts as a 40-kΩ resistor. Beyond the trip point, the resistance of the Iso-Switch increases to approximately 1000 MΩ at the maximum absolute operating voltage of 325 V.

approximately 1000 MΩ at the maximum absolute operating voltage for the device. Once the overload is removed, the device resets itself automatically.

Various Iso-Switch models are available for different applications, with trip currents ranging from ±1 μA to ±100 μA. Various operating voltage ranges are also offered: ±325 V for 115-V ac instruments and ±720 V for 220-V ac equipment; and the DP series designed for applications where defibrillation protection is desired, withstands pulses of up to ±5 kV for 10 μs. At prices in the range $29 to $34 each (one patient connection), the added protection provided by these fault interruptors is certainly affordable for evaluation prototypes, and sometimes even for the final design.

TESTING FOR COMPLIANCE

Although the presentation of medical device electrical safety standards above is by no means intended to replace the actual standards in scope or in content, it hopefully introduced many of the most important design requirements covered by the major standards. With medical equipment, however, designing solely for compliance is not sufficient. The consequences of a malfunctioning device can be so serious that testing to ensure proper performance is of utmost importance. As such, construction standards are only one aspect covered by applicable safety standards. The other major aspect defined by the standards comprises performance requirements.

Within this second category, standards specify the multiple tests which are applicable to diverse types of equipment and to identify in detail the criteria for compliance. In fact, almost every construction requirement is linked to a certain performance requirement which defines the testing that is performed to verify the acceptability of the construction. In essence, the electrical tests described by the standards are designed to probe insulation, components, and constructional features which could lead to a safety hazard under either normal or single-fault conditions.

Ground Integrity

Since the enclosure of the medical device is the first barrier of protection against the risks of electrical shock, the first test to be conducted is one to assess the integrity of the protective ground which guards a metallic enclosure and any other grounded exposed part. UL standard 2601-1 establishes that the impedance between the protective ground pin in the power plug and each accessible part which could become live in case the basic insulation failed should be less than 0.1 Ω. The standard also requires that the test be conducted by applying a 50- or 60-Hz ac current with an RMS value of 10 to 25 A for 5 seconds. Despite this, however, a reasonable approximation of this measurement can be obtained by using the 1-A dc current supplied by the circuit of Figure 3.21. Resistance is then assessed by measuring the voltage across the grounding path.

Here, op-amp IC2 and power FET Q1 form a voltage-to-current converter that is driven by a reference voltage set by R6 to maintain a 1-A constant current on a conductor connected between the ground terminal of J5 and connector J4. Power for the circuit is derived from three alkaline D cells, providing a maximum voltage compliance of approximately 4.5 V. Because full-range operation of the circuit is accomplished by driving the gate of Q1 well above its source-to-drain voltage, IC2 is operated from 12 V generated by charge pump IC1.

You may notice that J2 and J3 are labeled to be connected to a *Kelvin probe*. This type of test probe separates the point through which current is introduced from that through which the voltage across the unknown resistance is measured. The use of such a technique is required for low-resistance measurements because it effectively excludes the resistance of the test leads and avoids the voltage measurement errors that are often introduced by high-current-density concentrations on the current-injection terminals. As shown in Figure 3.22, a large alligator clip (e.g., Radio Shack 270-344) can easily be converted into a Kelvin probe by replacing the standard metallic axis by a nylon bolt with nylon spacers to isolate the jaws from each other and by covering the ends of the inner spring with a

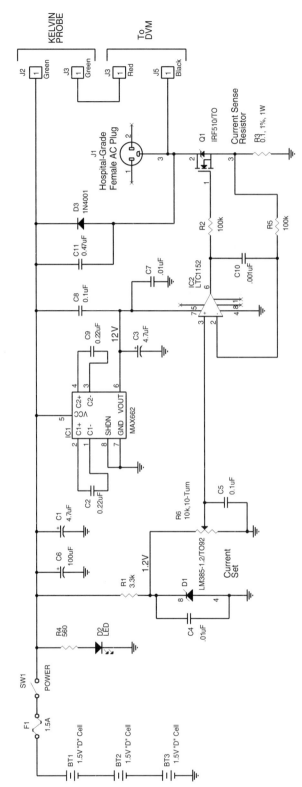

Figure 3.21 This simple adapter permits the measurement of milliohm resistances with any digital voltmeter. The circuit operates by generating a 1-A constant current on the unknown resistance of the ground path between the ground terminal of J1 and connector J2. A voltmeter connected across the power cord ground conductor measures resistance on a scale of 1 V/Ω.

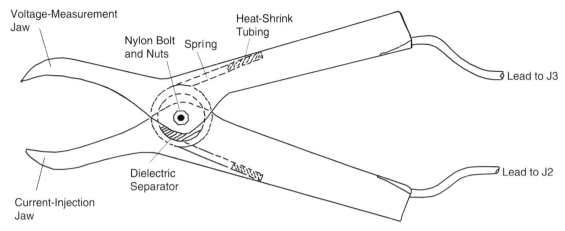

Figure 3.22 An alligator clip can be converted into a Kelvin probe by replacing the standard metallic axis by a nylon bolt with nylon spacers to isolate the jaws from each other. For the same reason, the ends of the inner spring must also be insulated.

suitable insulator. Two separate leads are then used to connect to the jaws of the probe, one to inject current and the other to sense voltage.

Once the circuit and the probe are assembled, calibrate the adapter to produce exactly 1 A. Plug the power cord of the instrument under test to the hospital-grade ac plug J1, and clip the Kelvin probe to an exposed conductive point of the case that is supposed to be protectively grounded. A digital voltmeter connected between J4 and J5 will directly read the protective ground resistance on a scale of 1 V/Ω. It must be remembered, however, that the measurement of resistance provided by this instrument only approximates the impedance test intended by the standards. The discrepancy between the methods is especially evident for high-power circuits, since a dc measurement of resistance does not convey any information regarding the inductive component of impedance. Moreover, dc ohmmeters are usually fooled by the polarized interface that results when an oxidation layer forms between connections in a defective ground system. This last concern may be alleviated by running the test once again but with the current injection polarity reversed. Nonlinear polarization indicating oxidation must be suspected if resistance measurements taken with opposite current injection polarities do not agree to a high degree. Failing this test is an immediate show-stopper. Before proceeding with any further testing, you must locate the faulty connection responsible for compromising the integrity of the protective ground.

Measuring Leakage and Patient Auxiliary Currents

Leakage and auxiliary current tests are the most important tests to establish the electrical safety of a medical electronic instrument. These are also the tests that are most commonly failed during safety approval submissions as well as during the periodic tests that hospitals conduct to ensure the safety of medical electronic devices throughout their service life. In the case of medical electronic instruments, measurements of leakage and auxiliary currents are taken using a load that simulates the impedance of a human patient. The *AAMI load* is a simple *RC* network that presents an almost purely resistive impedance of 1 kΩ for frequencies up to 1 kHz. As shown in Figure 3.23*a*, this load constitutes the core of the current measuring device. If a 1-μA current is forced through the AAMI load at different frequencies, the high-impedance RMS voltmeter within the measuring device would read the values presented in the graph of Figure 3.23*b*.

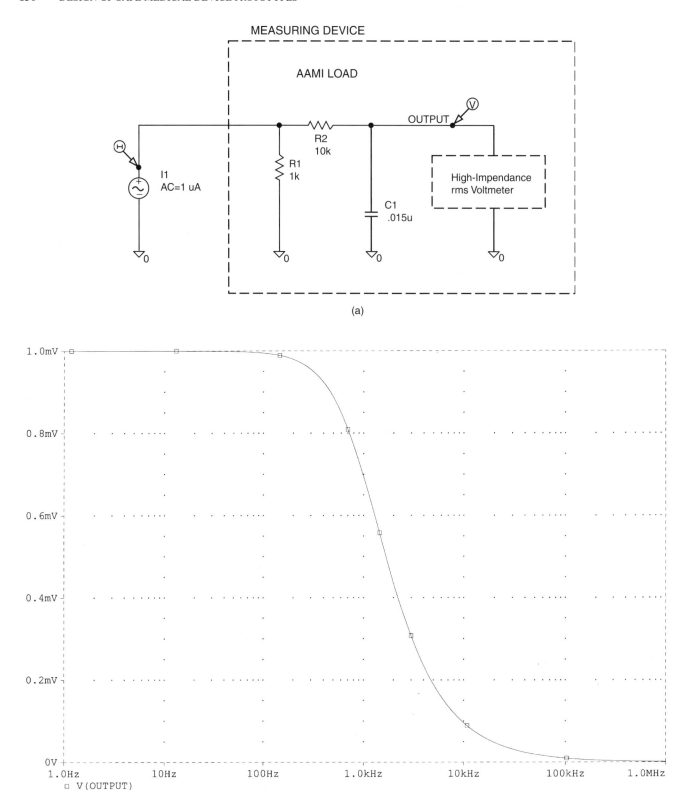

(a)

(b)

Figure 3.23 Leakage and auxiliary currents are measured using a load that simulates the impedance of a human patient. (*a*) The AAMI load is the core of the current-measuring device defined by safety standards harmonized with EN-601-1. (*b*) This load presents an impedance of approximately 1 kΩ for frequencies up to 1 kHz.

The frequency-response characteristics of the AAMI load have been selected to approximate the inverse of the risk current curve as a function of frequency. In turn, this risk current curve was derived from strength/frequency data for perceptible and lethal currents. These data showed that between 1 and 100 kHz, the current necessary to pose the same risk increases proportionately to 100 times the risk current between dc and 1 kHz. Since insufficient data exist above 100 kHz, AAMI decided not to extrapolate beyond 100 kHz, but rather, to maintain the same risk current level corresponding to 100 kHz.

Actual current measurements should be conducted after preconditioning the device under test in a humidity cabinet. For this treatment, all access covers that can be removed without the use of a tool must be opened and detached. Humidity-sensitive components which in themselves do not contribute significantly to the risk of electrocution may also be removed. Next, the equipment is placed in the humidity cabinet containing air with a relative humidity of 91 to 95% and a temperature t within the range $+20$ to $+32°C$ for 48 hours (or 7 days if the instrument is supposed to be drip-proof or splash-proof). Prior to placing it in the humidity cabinet, however, the equipment must be warmed to a temperature between t and $t + 4°C$.

The measurement should then be carried out 1 hour after the end of the humidity preconditioning treatment. Throughout this waiting period and during testing, the same temperature t must be maintained, but the relative humidity of the environment must only be 45 to 65%. Testing should be performed with the equipment's on–off switch in both conditions while connected to a power supply set at 110% of the maximum rated supply voltage. When operational, the maximum rated load must be used. As mentioned in the first part of this chapter, allowable patient leakage and auxiliary currents are defined for both normal and single-fault conditions. *Single-fault conditions* are defined as conditions in which a single means of protection against a safety hazard in the equipment is defective or a single external abnormal condition is present. Specific single-fault conditions that must be simulated during testing include interruption of the supply by opening the neutral conductor as well as the interruption of the protective ground conductor. Patient leakage current between an F-type applied part and ground assumes that an external voltage equal to 110% of the maximum rated supply voltage is in direct connection with the applied part. For battery-powered equipment, the external voltage that is assumed to be connected to the F-type applied part is 250 V.

Leakage current tests are conducted as shown in Figure 3.24a–d, with the device's power switch in the on and off conditions and creating the single-fault conditions specified in the figure. If the enclosure or a part thereof is made of insulating material, a piece of metal foil 20 cm $\times$ 10 cm applied to the nonconductive part of the enclosure must be used as the protectively grounded enclosure connection. The metal foil is wrapped on the surface of the insulating enclosure, simulating the way in which a human hand could act as a capacitively coupled electrode.

The connections for measuring patient auxiliary currents are shown in Figure 3.24e. Here, the current flowing between each patient connection and every other patient connection is measured under normal and single-fault conditions. For this test, the measuring instrument should be capable of differentiating the ac components from the dc components of the RMS current reading. As you can see from Table 3.2, different ac and dc auxiliary current levels are permitted to flow through the patient, depending on the use intended for the equipment.

Versatile Microammeter

Figures 3.25 to 3.28 present the schematic diagrams of a versatile instrument for the measurement of leakage and auxiliary currents. The core of the circuit is an AAMI load that converts a leakage or auxiliary current into a voltage waveform with a factor of 1 V/mA

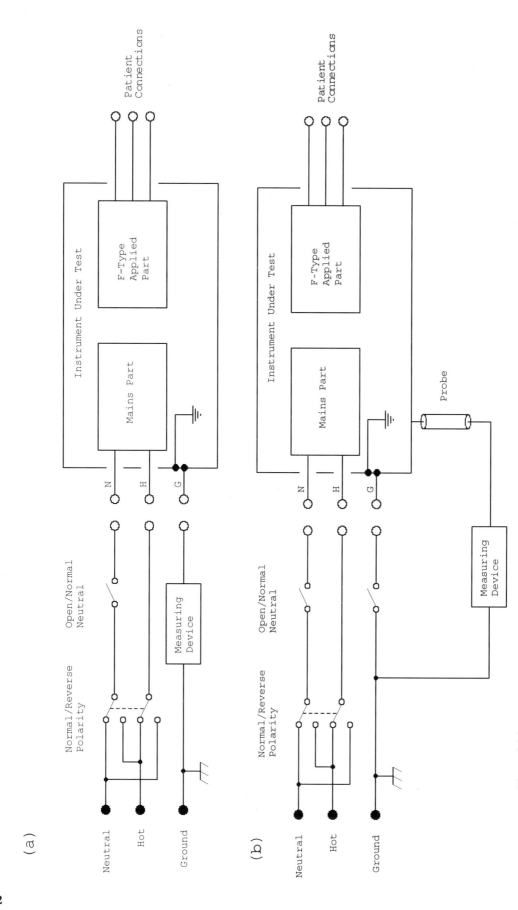

Figure 3.24 Leakage current measurements are obtained with the device's power switch in the on and off conditions as well as while inducing single-fault conditions in the circuit as shown. (*a*) Ground leakage current is measured between the protective ground conductor of the grounded power cord and ground. (*b*) The enclosure leakage current measured is the total current flowing from the enclosure and all accessible parts through an external conductive connection other than the protective ground conductor to ground. (*c*) Patient leakage current is measured between each and all patient connections and protective ground. (*d*) A second test for leakage current involves applying power line voltage to the applied parts. (*e*) The patient auxiliary current is measured between every patient connection and all other patient connections.

132

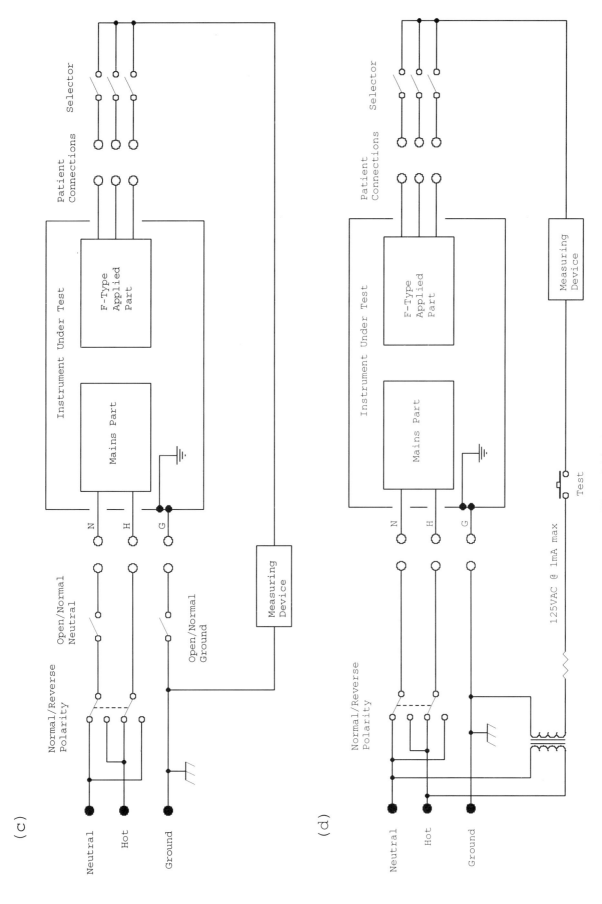

Figure 3.24 (*Continued*)

133

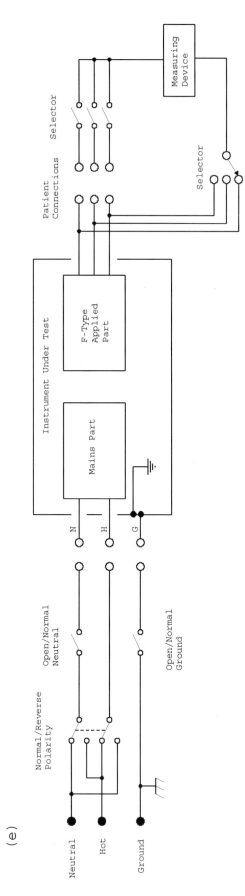

Figure 3.24 (*Continued*)

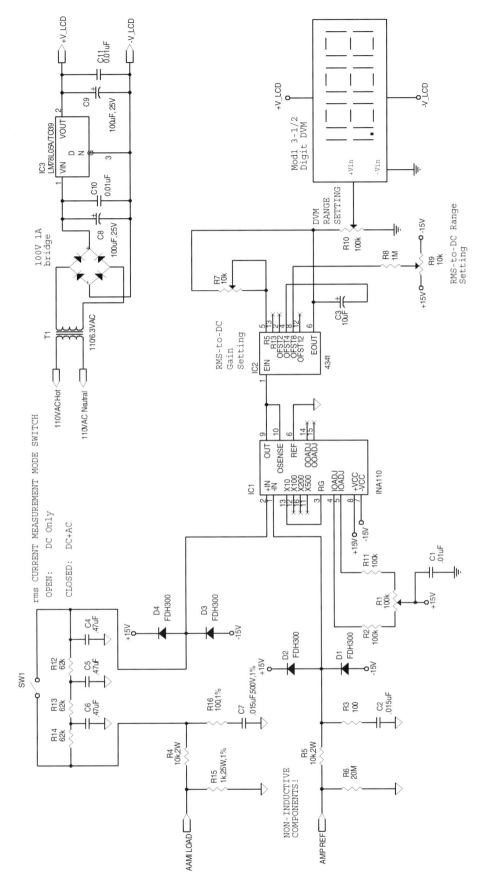

Figure 3.25 An AAMI load converts currents into voltage signals that are amplified by an instrumentation amplifier. The RMS value of the amplified signal is converted by IC2 into a dc voltage that is measured by a $3\frac{1}{2}$-digit DVM module.

135

for frequencies under 1 kHz. The output of the AAMI load is amplified with a gain of 10 by instrumentation amplifier IC1. The RMS value of the amplified signal is then computed by IC2, a true-RMS-to-dc converter IC, and displayed by a $3\frac{1}{2}$-digit DVM module. Note that some modules need to be supplied from a power supply that floats in relation to the reference potential of the voltage to be measured. If necessary, use the circuit shown in the insert of Figure 3.25 to supply floating power to the DVM module.

R1 must be set to produce a reading of zero with no current flowing through the AAMI load. Potentiometers R7 and R9 should be set to produce a reading of 1999 counts on the DVM for a 1.999-mA dc current through the AAMI load. In this same circuit, notice that an *RC* low-pass filter network can be interposed between the AAMI load and the voltage amplifying circuit. This serves to allow the measurement of the dc component of a patient auxiliary current.

The circuit presented in Figure 3.26 makes it possible easily to configure a measurement setup to conduct the various leakage and auxiliary current tests required by the safety standards. In this circuit, switch SW3 selects the connection of the AAMI load to measure either a patient current, the current on the protective ground pin, or the enclosure leakage current. Switch SW4 is used to select the source of a patient current among the different possible combinations of patient connections. Also through this circuit, power line–level voltages can be applied to the patient connections by way of relay K1. The figure shows the connection distribution suitable for testing a 12-lead ECG; however, the connections of any other applied part can be substituted. The connectors used for establishing connection to the leads of an applied part must be selected carefully so as not to contribute themselves to the measured leakage or auxiliary currents. A good choice are Ohmic Instruments' 301PB ECG binding posts, which can accommodate either the standard snap-ons or the pin tips used for establishing connection to ECG patient electrodes.

The construction of the AAMI load, its power supply, and its switching network deserves special attention, since stray coupling within the circuit can render measurements useless. This is because wiring and/or PCB traces that are part of the AAMI load and the voltmeter may be coupled capacitively to ground or the power line. Some forethought in the layout will avoid a lot of aggravation later when you try to calibrate the instrument. The circuit of Figure 3.27 controls the power supplied through J13 to the device under test. SPDT relays with contacts rated for 125 V ac at 20 A are used to reverse the power line polarity at J13 as well as to cause open-ground and open-neutral single-fault conditions.

Three neon lamps are used to indicate that the measurement instrument is powered as well as to verify that the ac plug from which power is obtained is wired correctly. Normal and fault conditions are shown in Table 3.3. In addition, the figure shows the 115-V ac isolation transformers and the voltage-divider network formed by R17–R19, which are used to generate 125 V ac for measuring the patient leakage current with applied power line voltage. Notice that R20 limits the current that can be delivered by this circuit to approximately 1 mA.

Finally, Figure 3.28 presents the dc power supply section of the measurement instrument. Linear regulators are used to generate the various power levels required by the microammeter. In addition, R24–R26 are used to derive an ac signal to balance the measurement circuit during application of 125 V ac to the patient connections. When measuring currents, place the equipment under test on a nonconductive bench, away from grounded metal surfaces. In addition, make sure that all external parts of an applied part, including patient cords, are placed on a dielectric insulating stand (e.g., a polystyrene box) approximately 1 m above a grounded metal surface.

Finally, a word of caution: Be very careful when using this instrument! Remember that unrestricted power line voltages are used to power the device under test, making the risk of electrocution or fire very real. Moreover, single-fault conditions forced during testing

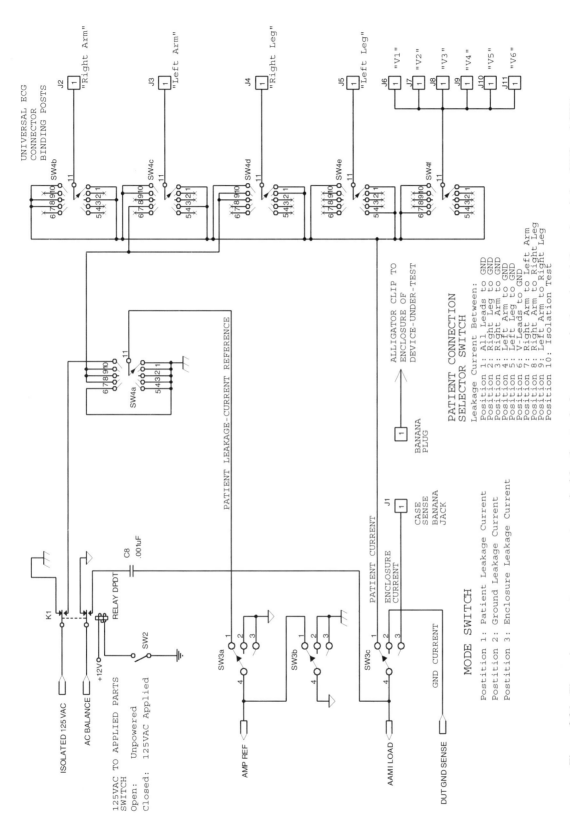

Figure 3.26 The various current measurements required by the standards are conveniently selected through switches SW3 and SW4. Patient connections are shown for a 12-lead ECG; however, the connections of any other applied part can be substituted. In addition, power line voltage can be applied to the patient connections to conduct the live patient leakage current test.

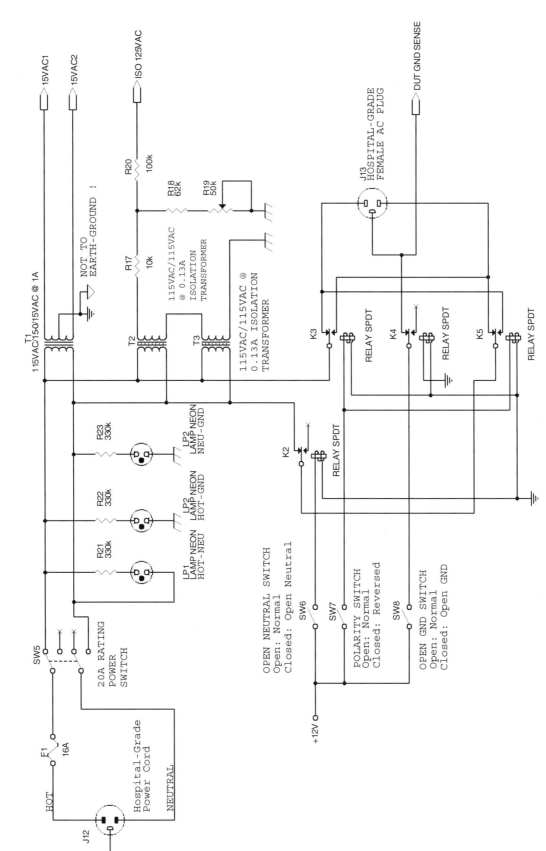

Figure 3.27 Power-control relays are used to reverse power line polarity and cause single-fault conditions for analyzing the instrument under test. Two 115-V ac isolation transformers and the voltage-divider network formed by R17–R19 are used to generate the 125 V ac for measuring the patient leakage current with applied power line voltage.

TABLE 3.3   Normal and Fault Conditions Detected by Three Neon Lamps Used in the Circuit of Figure 3.27 to Verify That the AC Plug from Which Power Is Obtained Is Wired Correctly

Indicator Light			Condition
LP1 (Hot–Neutral)	LP2 (Hot–Ground)	LP3 (Neutral–Ground)	
Off	Off	Off	Instrument off or open hot
Off	On	Off	Open neutral
Off	On	On	Hot–ground reversed
On	Off	Off	Open ground
On	Off	On	Hot–neutral reversed
On	On	Off	Correct, or ground–neutral reversed

may result in enclosure of the device under test becoming live, threatening anyone who would come in touch with it accidentally. In addition, since power line–level voltages can be injected into patient connections and the associated power system, never conduct these tests in the vicinity of a patient or on a power system branch that is used to power medical electronic instruments connected to patients.

HiPot Testing

Once compliance with current leakage limits is established, high-potential application testing, commonly known as *hiPot testing*, is used to assess the suitability of the insulation barriers between isolated parts of a medical instrument. In essence, very high voltage is applied differentially between the parts separated by the isolation barrier under test. As shown in Table 3.4, the test voltage is dependent on the voltage U to which the barrier is subjected under normal operating conditions at the rated supply voltage. While high voltage is applied, current is monitored to ensure that no arc breakdown occurs. HiPot testers have internal circuitry that automatically disconnects the high-voltage supply across the insulation under test whenever current exceeds a preset threshold value.

Although slight corona discharges are allowed by the standards, excessive RMS leakage current measurement is not sufficient for reliable detection of dielectric breakdown. Rather, milliampere-level current spikes or pulses should be monitored, since these are indications of the type of arc breakdown that occurs on insulation prior to catastrophic and destructive breakdown. Here again, the test voltage is supposed to be within the rated operating frequency for the instrument under test, and measured magnitudes refer to their ac RMS values. Despite this, the technique is sometimes modified by applying the dc equivalent to the peak-to-peak amplitude of the ac RMS voltage required. This obviously reduces current leakage between parts, since capacitive and inductive coupling disappear, leaving a current signal that directly conveys information about insulation breakdown processes at the peak of the dielectric stress.

As an example of applying this technique, if the highest-rated supply voltage for an instrument is 125 V ac, the standard requires that testing of basic insulation be conducted at 1000 V_{RMS}. The peak-to-peak voltage of the ac test signal required would thus be 1000 $V_{RMS} \times 1.41 = 1410$ V_{P-P}. As such, 1410 V dc would be applied, for example, between a wire connecting the hot and neutral of the power cord and another wire attached to the protective ground connection of the instrument. Similarly, the insulation barrier between an F-type applied part and any other point of the instrument of the example is required to be tested at 3000 V_{RMS}, which corresponds to a 4230 V dc voltage for the modified test. Breakdown of the insulation would then be indicated by current spikes that would appear while the high voltage is applied between a point that ties together all patient

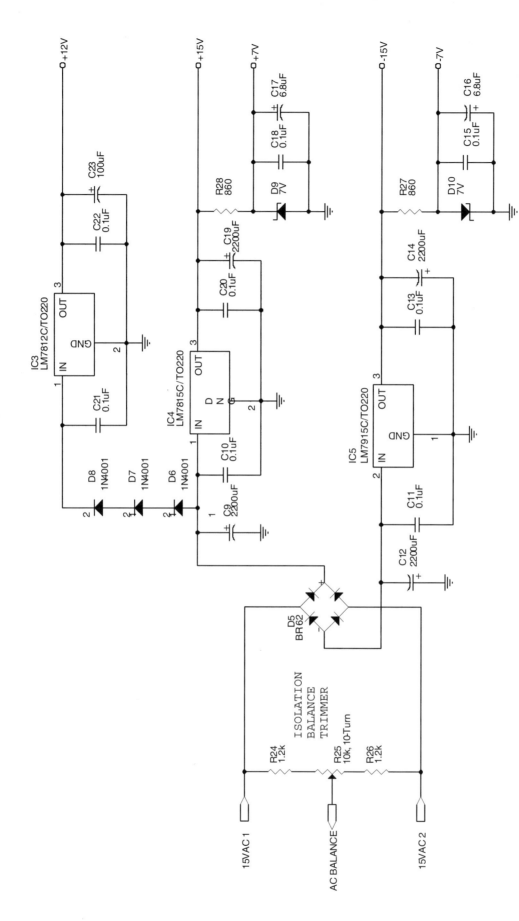

Figure 3.28 Linear regulators generate the various voltages required by the ammeter's circuitry. In addition, R24–R26 are used to derive an ac signal to balance the measurement circuit during application of 125 V ac to the patient connections.

TABLE 3.4   Some Voltages Used for HiPot Testing of Insulation Barriers

Insulation Type	HiPot Test Voltages for Reference Voltage U (V)				
	$U\,50$	$50 < U\,150$	$150 < U\,250$	$250 < U\,1000$	$1000 < U\,10,000$
Basic	500	1000	1500	$2U + 1000$	$U + 2000$
Supplementary	500	2000	2500	$2U + 2000$	$U + 3000$
Reinforced	500	3000	4000	$2(2U + 1500)$	$2(U + 2500)$

connections and a point that ties all nonisolated I/O lines and the hot, neutral, and ground connections of the power cord.

It must be noted, however, that despite its convenience, this method is not always accepted by regulatory bodies as a reasonable substitute for the tests specified by the standards. In any case, make sure that the hiPot tester that you use can detect the precatastrophic breakdowns, since otherwise, the insulation in your instrument may break down, delivering an instantaneous lethal level of current.

HiPot testing should be conducted after the equipment is preconditioned in a humidity cabinet. As before, all access covers that can be removed without the use of a tool must be detached. Humidity-sensitive components which in themselves do not contribute significantly to the risk of electrocution may also be removed. In addition, however, voltage-limiting devices (e.g., spark-gap transient-voltage suppressors, Iso-Switches) in parallel with insulation to be tested can be removed if the test voltage would make them become operative during the hiPot.

HiPot tests of the various insulation methods must be conducted with the instrument in a humidity cabinet. For each test the voltage should be increased slowly from zero to the target potential over a 10-second period and then kept at the required test level for 1 minute. If breakdown does not occur, tripping the hiPot tester's automatic shutoff, the test is completed by lowering the voltage back to zero over a 10-second period.

Finally, it must be noted that the standards do not except battery-powered equipment from hiPot testing. Instead, the reference voltage U is set to be 250 V. Fully or partially nonconductive enclosures are not excluded either. In these cases, the same metal-foil method used for current leakage testing must be used, being careful that flashover does not occur at the edges of the foil at very high hiPot test levels.

Most of the circuitry inside commercial hipot testers is really used to control high-voltage source and detecting currents that exceed a set threshold. If the $1000 or so needed to buy a low-end hipot tester are outside your budget, you may conduct design-time tests using the circuit shown in Figure 3.29. Here, variac T1 is used to change the supply voltage to the primary of a high-voltage transformer T2. We used a surplus transformer with a ground-referenced secondary rated for a maximum output of 5 kV at 5 mA. Not any high-voltage transformer should be used for this application. A unit with good voltage regulation is needed. Avoid using neon-light transformers because these are built to provide a constant current to the load. Under unloaded conditions, they will present a voltage of 9 to 15 kV at the load.

The high voltage applied to the device under test starts as 117 V ac. This voltage is controlled using a Crydom solid-state relay. If the AC_ENABLE signal is a logic high, +12 V dc is applied to the control inputs of the Crydom solid-state relay, and if SW2, the ARM key switch, is in the on position, 117 V ac will be applied to the inputs of variac T2. T1, a Magnatek-Triad step-up transformer, is fed directly by the variac's output. The high voltage produced by T2 is then applied to the device under test through a touch-proof banana connector as shown in Figure 3.30.

The current-trip circuit of Figure 3.31 acts as a milliammeter-controlled relay that disconnects the supply to the variac if the isolation barrier under test should fail. Ac current

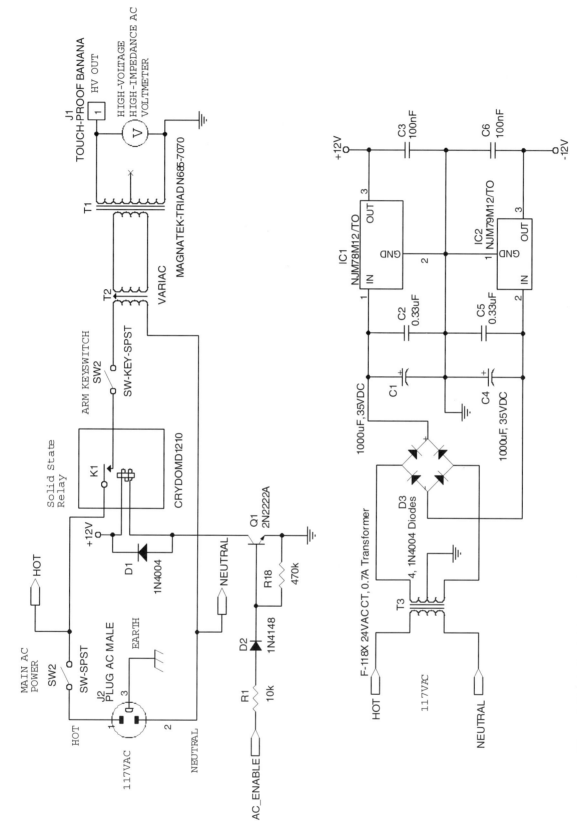

Figure 3.29 A simple hiPot tester for design time can be constructed using a variac to change the supply voltage to the primary of a high-voltage transformer. The variac's supply is controlled through a Crydom solid-state relay.

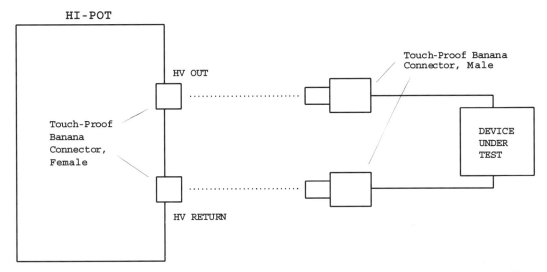

HI-POT

HV OUT

HV RETURN

Touch-Proof
Banana
Connector,
Female

Touch-Proof Banana
Connector, Male

DEVICE
UNDER
TEST

Figure 3.30   The high voltage produced by T2 of Figure 3.29 is applied to the device under test through a touch-proof banana connector. The current that leaks through the DUT's barriers is returned to the hiPot's sense resistor via a second touch-proof connector.

flowing through the device under test flows from the HV OUT terminal, J1, and through J3, the HV RETURN, and then through R9, the current-sensing resistor to ground. The direction of the current flow is reversed on the opposite phase of the ac waveform applied to the device under test. The voltage that appears across R9 is proportional to the current flow through the device under test. A current of 1 mA flowing through the device under test appears as 7 V RMS across R9. This voltage is buffered by IC1A and full-wave rectified by IC1B and IC1C. The peak of the current-sense voltage across R9 is compared to the dc voltage level from R13 by IC1D. If the peak current-sense voltage is greater than the voltage from R13, the output from IC1D will be +12 V dc, a logic level high. This logic high causes IC2B to be reset, setting the AC_ENABLE output low, switching off K1 and the high voltage to the device under test. IC3A provides a power-on reset circuit, so the control flip-flop, IC2B, always powers on in a reset or off condition. Switch SW3 provides a way to stop the test by resetting the control flip-flop manually to the off state.

To start a test, first the reset input to the control flip-flop is a logic low, meaning that the comparator output is a logic low, and SW3 is open. SW4 is pressed and a logic level high is input to the clock input of the control flip-flop. The Q output will switch to a logic high state. This high output on AC_ENABLE will switch K1 on, applying high voltage to the device under test. As the control knob on the variac is increased, the high voltage to the device under test is increased. If the current flowing through R9 (i.e., the current leaking through the device under test) increases above the level set by R13, the output of IC1D switches to a high state, resetting the control flip-flop and switching K1 off, disconnecting the high voltage from the device under test. Because the reset input to the flip-flip has priority over the clock input to the flip-flop, the Q output cannot be turned on anytime the current-sense voltage is greater than the trip level set by R13. This is done for safety purposes, so that if there is an overcurrent condition, the high voltage must first be reduced to the point where the overcurrent condition is removed before the test can be restarted by pressing SW4. IC3B and IC3D are connected in parallel to provide enough current to drive the high-voltage-indicator LED, D9.

Utmost care should be exercised when using a hiPot tester. High voltage below the current-trip threshold can give a nasty or lethal shock. In addition, dielectric breakdown carries the associated risk of ignition. Be prepared to deal with emergencies. In addition, consider

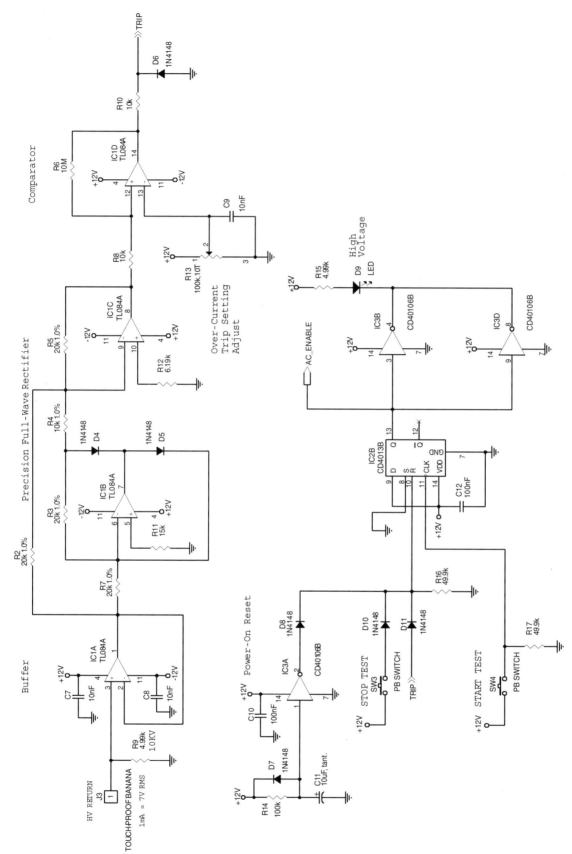

Figure 3.31 The current-trip circuit acts as a milliammeter-controlled switch that disconnects the supply to the variac if the isolation barrier under test should fail.

that dielectric breakdown or single-fault conditions may result in any part of the device under test becoming electrified at very high voltage, threatening anyone who came close to it.

Testing for Other Risks

You will probably be able to get by just by being able to pass the tests outlined above as long as evaluation of the prototype is conducted on a very limited number of patients while under the supervision of a physician. As long as the instrument is built solidly enough to inspire confidence, we have seldom encountered situations where an engineering evaluation prototype would be required to pass the battery of tests specified by the standards to ensure compliance with the mechanical and labeling requirements demanded for prerelease or commercial products. However, there are other very realistic risks in a clinical environment, and you should make sure that you will not cause undue interference or harm through mechanisms other than leakage currents.

For an evaluation prototype, you should at least test for and verify that the following conditions are met:

- The equipment has been designed to minimize the risk of fire and explosion. Safety standards typically limit temperature rises allowed for components as well as defining enclosure requirements for containing fires within the instrument. Whenever possible, select materials to be compliant with UL-94V (*Flammability of Plastic Materials for Parts in Devices and Appliances*). In addition, if the device is supposed to operate in areas where flammable anesthetics or oxygen-enriched atmospheres are used (e.g., operating rooms, hospital rooms), special requirements must be met to ensure that these explosive atmospheres are not ignited.

- If intentional sources of ionizing radiation are present, the equipment must be evaluated by the Center for Devices and Radiological Health (CDRH). If components are used that may generate ionizing radiation which is not used for a diagnostic or therapeutic purpose (e.g., from CRTs), you must ensure that the exposure at a distance of 5 cm from any accessible part of the equipment, averaged over a 10-cm^2 area, is less than 0.5 milliroentgen per hour. Devices that make use of ultraviolet radiation and lasers should also be investigated to ensure safety. In addition, devices that make use of ultrasound and RF emissions are also regulated in specific substandards of the IEC-601-2 series.

- The equipment must not emit electromagnetic interference (EMI), which may cause other equipment to malfunction. The device should also be designed to be immune to electromagnetic interference, power line "glitches," and electrostatic discharge. We deal with these issues in Chapter 4.

- Applied parts designed to come in contact with the biological tissues, cells, or body fluids of a patient must be assessed as to their biocompatibility. Diligent prudence should be applied in the selection of materials, making sure to test for biological effects, including cytotoxicity, sensitization, irritation, intracutaneous reactivity, and so on.

- The medical device and all peripheral equipment should be fitted with the means necessary to disconnect them from a patient immediately and completely should an emergency arise. In addition, connectors used on patient connections are identified clearly and uniquely and must be of a type that cannot be plugged into the power line accidentally or form an electrical path to any point when they are not connected to the equipment.

In addition to establishing reasonable safety, take the time to educate clinical staff members who support your evaluations as to the potential dangers posed by an instrument, regardless of how remote and unlikely they may be. Remember that if something goes wrong, it is the clinical staff who will have to save the patient's life!

CONCLUDING REMARKS

Applying the principles and requirements described by the safety standards is important for even the first engineering evaluation prototype of a medical electronic device. This is because the biomedical equipment department of any hospital hosting the preclinical trials will demand that the device passes, at a bare minimum, all electrical safety tests. Furthermore, since we live in a litigious society, it is a good idea to maintain good records showing that careful consideration was given to the standards and regulations all the way from the beginning of the design.

As you can appreciate from even the brief overview of limited scope presented in this chapter, the safety and performance requirements for medical devices are many and very stringent. However, we believe that these requirements are not enforced by regulatory agencies with the intent of discouraging the advancement of the medical sciences. Rather, it is our perception that applicable standards and regulations are there to help the designer develop a product that provides true benefit to the patient at the same time that it reduces foreseeable risks. We encourage you to pursue data that clearly demonstrate clinical efficacy for an idea that you may have for a medical product. In addition, we urge you to realize that to be approved, new medical technology must absolutely be based on solid physiological and technical grounds. Armed with this information, and if you can adapt to a changing regulatory environment, we are convinced that a very receptive audience of investors eagerly awaits to back your idea.

REFERENCES

Ackerman, B., High-Resolution Timing on a PC, *Circuit Cellar INK*, 24, 46–49, December 1991–January 1992.

Association for the Advancement of Medical Instrumentation, *Safe Current Limits for Electromedical Apparatus*, ANSI/AAMI Standard ES1, 1993.

IEC-601-1, *Medical Electrical Equipment—Part 1: General Requirements for Safety*, 1988; Amendments 1, 1991, and 2, 1995.

IEC-601-1-1, *Medical Electrical Equipment—Part 1: General Requirements for Safety*; Section 1: Collateral Standard: Safety Requirements for Medical Electrical Systems, 1992.

IEC-601-2-27, *Medical Electrical Equipment—Part 2: Particular Requirements for the Safety of Electrocardiographic Monitoring Equipment*, 1994.

Porr, B., EEG Preamp: Anti Alias Filter and Isolation Circuit, `www.cn.stir.ac.uk/~bp1/eegviewer/preamp/`, 2000.

Prutchi, D., LPT:Analog!—a 12-Bit A/D Converter Printer Port Adapter, *Circuit Cellar INK*, 67, 26–33, February 1996.

Schulze, D. P., A PC Stopwatch, *Circuit Cellar INK*, 19, 22–23, February–March 1991.

UL-94V, *Flammability of Plastic Materials for Parts in Devices and Appliances*, UL-2601-1, *Standard for Safety: Medical Electrical Equipment—Part 1: General Requirements for Safety*, 2nd ed., 1997.

4

ELECTROMAGNETIC COMPATIBILITY AND MEDICAL DEVICES

Have you heard about the wheelchair that moved on its own every time a police car passed by? No, it's not part of a joke. This actually happened, and several people were seriously injured when radio signals from the two-way communications equipment on emergency vehicles and boats, CB, and amateur radios interfered with proper operation of the control circuitry of powered wheelchairs, sending some off curbs and piers. Similar reports of improper operation of apnea monitors, anesthetic gas monitors, and ECG and EEG monitors due to electromagnetic interference prompted government agencies to look carefully at these occurrences and establish regulations by which equipment must possess sufficient immunity to operate as intended in the presence of interference.

Complying with these regulations is not easy. The technologies involved in modern circuit design have considerably blurred the boundaries between the digital and analog worlds. Suddenly, multihundred megahertz and even gigahertz clocks became commonplace in high-performance digital circuits, making it necessary to consider every connection between components as an RF transmission line. At the same time that the need for higher performance pushes designers toward high-speed technology, the marketplace is demanding more compact, lighter, and less power-hungry devices. With smaller size, analog effects again enter into consideration, because as components and conductors come into close proximity, coupling between circuit sections becomes a real problem.

Obviously, self-interference within a circuit must be eliminated to make the product workable, but this still does not make the product marketworthy. This is because strict regulations concerning electromagnetic compatibility are now being enforced around the world in an effort to ensure that devices do not interfere with each other. In the United States the FCC regulates the testing and certification of all electronic devices that generate or use clock rates above 9 kHz [Dash and Strauss, 1995]. In principle, the FCC's charter is to protect communications from unwanted electromagnetic interference (EMI). In the European Common Market, on the other hand, an electromagnetic compatibility (EMC) directive is now in effect, which not only establishes requirements against causing undue interference to radio and telecommunications equipment, but also institutes requirements

Design and Development of Medical Electronic Instrumentation   By David Prutchi and Michael Norris
ISBN 0-471-67623-3   Copyright © 2005 John Wiley & Sons, Inc.

by which equipment must possess sufficient immunity to operate as intended in the presence of interference [Gubisch, 1995].

Regulatory bodies around the world have developed standards and regulations covering both emissions and immunity that designers must take very seriously. Failure to comply with EMI and EMC regulations can have a serious impact on everyone associated with a product, starting with the designer, through the manufacturer, the marketing and distribution network, and extending even to customers. The consequences of noncompliance include halting manufacturing and distribution, levying fines, and the publication of public notices of noncompliance to warn potential customers and other agencies. These considerations become especially important in the case of medical equipment, since it often involves sensitive electronics that can be affected adversely by electromagnetic interference, leading to potentially serious hazards to patients and health-care providers.

The European Community regulates emissions and immunity of medical devices through the EN-60601-1-2 standard (*Medical Electrical Equipment—Part 1: General Requirements for Safety*; Section 2: Collateral Standard: Electromagnetic Compatibility— Requirements and Tests) as well as the EN-55011 standard (*Limits and Methods of Measurement of Radio Disturbance Characteristics of Industrial, Scientific and Medical Radio Frequency Equipment*). In EN-60601-1-2, pass/fail criteria are defined by the manufacturer. As a result, the manufacturer may chose to classify a failure mode that does not pose a hazard to the patient as a "pass." In the United States, the FDA is adopting many of the IEC-60601-1-2 requirements but is imposing restrictions on a manufacturer's ability to adopt pass/fail criteria. The FDA prescribes that a passing result corresponds to maintaining clinical utility. In addition, there are discrepancies between the immunity levels recommended by European authorities and the FDA. Because of these differences in opinion, as well as because the standards are relatively new, changes occur often, and we advise engineers to keep updated on the latest versions.

Assuring compliance with the rules involves an extensive series of tests. The EMI and EMC standards enforced by the various regulatory agencies clearly define the construction of test sites as well as the test procedures to be followed. Even a fairly spartan facility capable of conducting these tests ends up costing over $100,000 just to set up, and for this reason, most companies hire an outside test lab at the rate of $1500 to $3000 per day to conduct testing. Considering how fast charges can accumulate during testing, it is obviously not a smart move simply to hire a test lab and wait for the results. Rather, designers should familiarize themselves with the relevant EMI and EMC standards and make sure that compliance requirements are considered at every stage in the design process.

In this chapter we present the major EMI/EMC requirements for medical devices, look at the theory of how circuits produce EMI, and describe some low-cost tools and methods that will allow you to identify and isolate the sources of EMI that inevitably make it into a circuit.

EMISSIONS FROM MEDICAL DEVICES

The FCC's main concern with RF emissions from electronic devices is possible interference with communications devices such as commercial radio and TV receivers. From the point of view of agencies regulating medical devices (in the United States the FDA), the concern about unintentional electromagnetic emissions extends to the way in which they could interfere with diagnostic or therapeutic medical devices. Note the word *unintentional*, since these standards do not apply directly to medical devices that intentionally generate electromagnetic signals (e.g., telemetry ECG transmitters, electrosurgery equipment, magnetic resonance imagers) which require special emissions that limit exemptions at specifically allocated frequency bands.

EN-60601-1-2 sets forth requirements for emissions based on the CISPR-11 standard developed by the International Electrotechnical Commission's Special Committee on Radio Interference (CISPR). These requirements address both radiated emissions (i.e., electromagnetic interference coupled to victim receivers over wireless paths), as well as conducted emissions (i.e., electromagnetic interference coupled to power lines and other conductors) from medical equipment.

Emission limits are set based on the type of setting the device will be used. Class A requirements are the least stringent and apply to medical devices intended to operate in areas where receivers are not usually present. Class B requirements apply to equipment that may operate in close proximity to radio and TV receivers, such as in a patient's home or hospital room.

Conducted emissions are tested below 30 MHz, while radiated emissions are tested above 30 MHz. Although both emission mechanisms overlap, regulatory agencies set this boundary because low-frequency interference is primarily conducted (since low frequencies do not radiate very efficiently without intentionally designed antenna elements) and high-frequency interference is primarily radiated (since high frequencies are conducted poorly through inductive wires).

RADIATED EMISSIONS FROM DIGITAL CIRCUITS

As they operate, digital circuits constantly switch the state of lines between a high-voltage level and a low-voltage level to represent binary states. As shown in Figure 4.1a, the resulting time-domain waveform on any single line of a digital circuit can thus be idealized as a train of trapezoidal pulses of amplitude (either current I or voltage V) A, rise time t_r, fall time t_f (between 10 and 90% of the amplitude), pulse width τ (at 50% of the amplitude), and period T.

The Fourier envelope of all frequency-domain components generated by such a periodic pulse train can be approximated by the nomogram of Figure 4.1b. The frequency spectrum is composed mainly of a series of discrete sine-wave harmonics starting at the fundamental frequency $f_0 = 1/T$ and continuing for all integer multiples of f_0. The nomogram identifies two frequencies of interest. The first is f_1, above which the locus of the maximum amplitudes rolls off with a $1/f$ slope. The second, f_2, is the limit above which the locus rolls off at a more abrupt rate of $1/f^2$. These frequencies are located at

$$f_1 = \frac{1}{\pi t} = \frac{0.32}{\tau}$$

and

$$f_2 = \frac{1}{\pi t} = \frac{0.32}{t}$$

where t is the faster of (t_r, t_f).

The envelope of harmonic amplitude (in either amperes or volts) is then simplified to

$$[V \text{ or } I] = \begin{cases} 2A(\tau + t) & f < f_1 \\[2mm] \dfrac{0.64A}{Tf} = -20 \, \text{dB/decade roll-off} & f_1 \leq f < f_2 \\[2mm] \dfrac{0.2A}{Ttf^2} = -40 \, \text{dB/decade roll-off} & f_2 \leq f \end{cases}$$

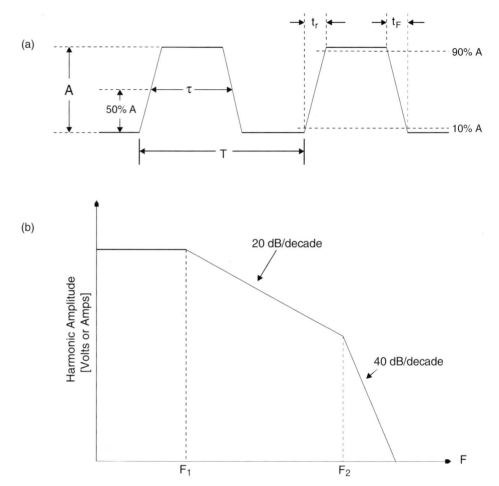

Figure 4.1   A pulse train with the characteristics shown in (*a*) produces a spectrum with an envelope that can be approximated by the nomogram of (*b*).

For nonperiodic trains, the nomogram must be modified to account for the broadband nature of the source. To do so, a nomogram of the spectral density envelope of the signal can be defined for a unity bandwidth of 1 MHz by

$$(V \text{ or } I)\left[\frac{\text{dBV}}{\text{MHz}} \text{ or } \frac{\text{dBA}}{\text{MHz}}\right] = \begin{cases} 6 + 20\log(A\tau) & f < f_1 \\ 20\log(A) - 4 - 20\log[f(\text{MHz})] & f_1 \le f < f_2 \\ 20\log\dfrac{A}{\tau(\mu s)} - 14 - 40\log[f(\text{MHz})] & f_2 \le f \end{cases}$$

Depending on its internal impedance, a circuit carrying such a pulse train will create in its vicinity a field that is principally electric or magnetic. At a greater distance from the source, the field becomes electromagnetic, regardless of the source impedance. If there is a coupling mechanism, which can be either conduction or radiation, some or all of the frequency components in the digital pulse train's spectrum will be absorbed by some "victim" receiver circuit.

To illustrate the magnitude of the problem, imagine a medical instrument's main circuit board, consisting of a CPU, some glue logic, and memory ICs, that has been housed in an unshielded plastic case. Let's assume that at any given time, a number of these ICs are toggling states synchronously, at a frequency of 100 MHz, for instance. Furthermore, assume that the total power switched at any given instant during a synchronous transition is approximately 10 W. Now, in a real circuit, efficiency is not 100%, and a small fraction of these 10 W will not do either useful work or be dissipated as heat by the ICs and wiring, but rather, will be radiated into space. Assuming a reasonable fraction value of 10^{-6} of the total switched power at the fundamental frequency, the power radiated is 10 μW.

Now, let's assume that an FM radio is placed at a distance of 5 m from the device. The field strength E produced by the 10 μW at this distance may be approximated by the formula

$$E = \sqrt{\frac{30 \text{ radiated power (W)}}{\text{distance (m)}}} = \sqrt{\frac{30 \times 10 \times 10^{-6}}{5}} = 3.46 \frac{\text{mV}}{\text{m}} = 70.79 \frac{\text{dB}\mu\text{V}}{\text{m}}$$

Considering that the minimum field strength required for good reception quality by a typical FM receiver is approximately 50 dBμV/m, the radiated computer clock would cause considerable interference to the reception of a radio station in the same frequency. In fact, interference caused by the computer of this example may extend up to 50 m or more away!

From the past discussion, it is easy to conclude that a first method for reducing radiated emissions is to maintain clock speeds low as well as to make rise and fall times as slow as possible for the specific application. At the same time, it is desirable to maintain the total power per transition to the bare minimum. Transition times and powers depend primarily on the technology used. As shown in Table 4.1, the ac parameters of each technology strongly influence the equivalent radiation bandwidth. In addition, the voltage swing, in combination with the source impedance and load characteristics of each technology, determines the amount of power used and thus the power of radiated emissions on each transition. Figure 4.2 shows how the selection of technology plays a crucial role in establishing the bandwidth and power levels of radiated emissions that will require control throughout the design effort.

Another problematic circuit often found in medical devices is the switching power supply. Here, high-power switching at frequencies of 100 kHz and above produce significant harmonics up to and above 30 MHz, requiring careful circuit layout and filtering. Fully

TABLE 4.1 The Most Popular Logic Families Have Very Different Timing and Driving Parameters, Resulting in Radiated Emissions Spectra with Different Characteristics

Technology	Minimum Voltage Swing (V)	Minimum Transition Time t (ns)	Typical Bit Pulse Width τ (ns)	Equivalent Bandwidth (MHz)	Single-Load Input Capacitance (pF)	Output Source Impedance (Low/High) (Ω)
5-V CMOS	5	70	500	4.5	5	300/300
12-V CMOS	12	25	250	12	5	300/300
HCMOS	5	3.5	50	92	4	160/160
TTL	3	8	50	4	5	30/150
TTL-S	3	2.5	30	125	4	15/50
TTL-LS	3	5	50	65	5.5	30/160
TTL-FAST	3	2.5	25	125	4.5	15/40
ECL	0.8	2	20	160	3	7/7
GaAs	1	0.1	2	3200	1	N/A

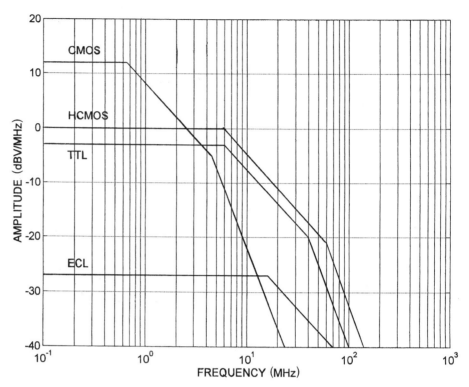

Figure 4.2   The characteristic voltage spectrum envelope of emissions by every logic gate is highly dependent on the technology being used.

integrated filters are available for dc power lines. For example, Figure 4.3 shows the way in which muRata BNX002 block filters are used to filter the raw dc power outputs produced by two C&D Technologies' HB04U15D12 isolating dc/dc converters. In the circuit, each dc/dc (IC1 and IC2) produces unregulated 24 V ($\pm$12 V if the center-tap common is used), which is isolated from the +15 V dc power input by an isolation barrier rated at 3000 V dc (continuous, tested at 8 kV, 60 Hz for 10 s). The outputs of the dc/dc converters are filtered via filters FILT1 and FILT2, which internally incorporate multiple EMI filters implemented with feed-through capacitors, monolithic chip capacitors, and ferrite-bead inductors. Each of these filters attenuates RF by at least 40 dB in the range 1 MHz to 1 GHz. C1/C4 and C7/C10 are used to reduce ripple, and the circuits following these capacitors are linear regulators that yield regulated $\pm$24 V at 50 mA to the applied part for which this isolation power supply was designed.

Another filter worth mentioning is muRata's PLTxR53C common-mode choke coil. This family of modules is ideal for suppressing noise from a few megahertz (1 to 5 MHz, depending on the model) to several hundred megahertz (10 MHz to 1 GHz, depending on the model) from dc power supplies. This module is useful in suppressing noise radiated from the cable connecting a device to an external wall-mounted or "brick" ac adapter.

ELECTROMAGNETIC FIELDS

EMI standards establish that radiated-emissions test measurements should be performed at a distance of 10 to 30 m, depending on the device's classification. For compliance testing, the device under test should be placed on a nonconductive table 0.8 m above a ground

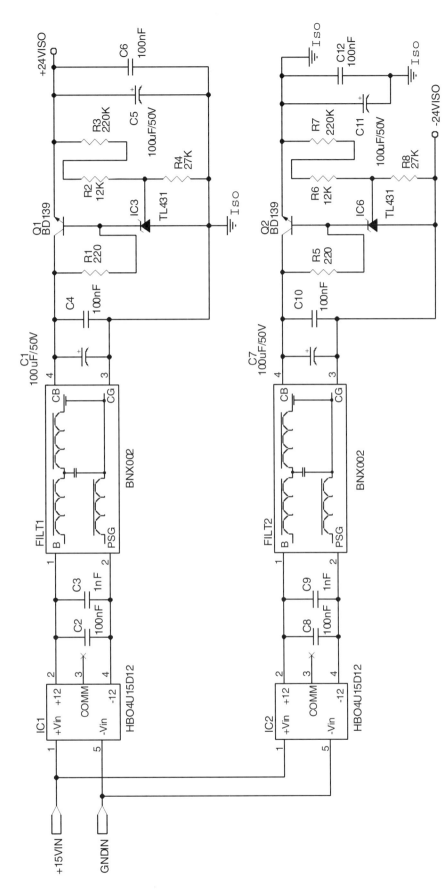

Figure 4.3 BNX002 block filters by muRata are used to filter the raw dc power outputs produced by two isolating dc/dc converters. These filters internally incorporate multiple EMI filters implemented with feed-through capacitors, monolithic chip capacitors, and ferrite-bead inductors.

TABLE 4.2   EN-55011 Radiated Emissions Limits for Group 1 Devices[a]

| Frequency (MHz) | Test Distance (m) | Field Strength (dBµV/m) | |
		Class A	Class B
30–230	10	40	30
230–1000	10	47	37

[a]The lower limits apply at the transition frequency.

plane. The table is typically centered on a motorized turntable that allows 360° rotation. A measurement antenna is positioned at a distance of 10 to 30 m as measured from the closest point of the device under test. The radiated emissions are maximized by configuring and rotating the device under test as well as by raising and lowering the antenna from 1 to 4 m. A spectrum analyzer with peak detection capabilities is used to find the maxima of the radiated emissions during the testing. Then, final measurements are taken using a spectrum analyzer with quasi-peak function with a measurement bandwidth of 120 kHz. The test setup is shown in Figure 4.4. The limits for radiated emissions per EN-55011 for group 1 devices are presented in Table 4.2.

In reality, devices to be tested are not usually taken directly to the open-field test site. Rather, they are first scanned for potentially offensive radiated emissions in a small shielded room. The compliance testing is then conducted in the 10-m open-field test site, paying special attention to peak emissions detected in the shielded room. This is almost a practical necessity, because open-field test sites, even when located far from large metropolitan areas, are still inundated by human-made RF signals. As an example, Table 4.3 shows results obtained recently when testing an implantable-device programmer for radiated emissions. The specific frequencies selected for testing were identified the night before taking the device to an open-field test site in the middle of Texas's hill country. Figure 4.5 shows the device being tested at the open-field site. The device sits atop a motorized turntable. A biconical antenna can be seen placed 10 m away from the device under test. At the 10-m distance specified for the tests, radiated emissions have their electric-field

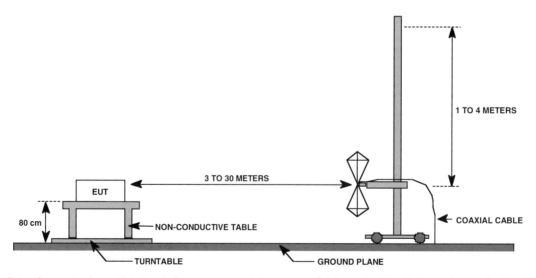

Figure 4.4   Setup for conducting radiated-emission measurements in an open-field test site. The device under test is placed on a nonconductive table 0.8 m above a ground plane, and the distance between the device and the antenna is 10 m.

TABLE 4.3   EN-55011 Sample Worksheet for Testing Radiated Emissions[a]

Measurement distance: 10 m
Antenna polarization: vertical
Detector function: quasi-peak

Frequency (MHz)	EUT Direction (deg)	Antenna Elevation (m)	Recorded Level (dBμV)	Amplifier Gain (dB)	Antenna Factor (dB/m)	Cable Loss (dB)	Corrected Level (dBμV/m)	Limit (dBμV/m)	Margin (dB)
112.7	0	1.0	30.9	27.1	12.2	1.6	17.6	40	−22.4
130.0	270	1.0	42.8	27.0	11.8	1.7	29.3	40	−10.7
60.0	270	1.0	38.5	27.3	8.9	1.0	21.1	40	−18.9
200.0	270	1.0	28.5	26.7	10.8	2.1	14.7	40	−25.3
298.9	150	1.0	30.2	26.5	13.9	2.6	20.2	47	−26.8
400.0	90	3.5	30.3	27.2	15.3	3.0	21.4	47	−25.6
77.0	210	1.0	33.0	27.3	6.3	1.2	13.3	40	−26.7
110.0	300	1.0	39.0	27.1	12.1	1.6	25.6	40	−14.4

[a]Corrected level = recorded level + antenna factor + cable loss. The frequencies of interest were selected during a precompliance scan of the device in a shielded room.

Figure 4.5   A prototype implantable-device programmer is being tested at an open-field test site. The device sits atop a motorized turntable. A biconical antenna is placed 10 m away from the device under test.

E and magnetic-field **H** vectors orthogonal to each other but in the same plane. Under these conditions, electromagnetic propagation occurs as a plane wave.

If the test probe is brought closer and closer to the device under test, however, the nature of the electromagnetic field changes. Near the source of the radiation, the field produced is mostly a function of the impedance of the source. If the field is generated by a circuit carrying high current and low voltage, the field will be mostly magnetic in nature. If on the other hand, the field is produced by an element placed at high voltage but carrying little or no current, the field will be mostly electric in nature. This is the domain of the near field, while the plane wave is in the domain of the far field.

The ideal generator for a magnetic field, or H-field as it is known, is thus a circular loop of area $S(\text{m}^2)$ carrying an ac current I of wavelength λ. It should be noted that although a static field is generated by a dc current and can be calculated with the method to follow, static H-fields do not cause radiated emissions and are thus disregarded for EMI purposes. If the loop size is smaller than the observation distance D, the magnitudes of the **E** and **H** vectors can be found using the solutions derived from Maxwell's equations. In the near field, the simplified values for these magnitudes are

$$E(\text{V/m}) = \frac{Z_0 IS}{2\lambda D^2}$$

and

$$H(\text{A/m}) = \frac{IS}{4\pi D^3}$$

where Z_0 equals the impedance of free space, 120π or $377\,\Omega$. Inspecting these equations, we find that in the near field, H is independent of λ and decreases drastically with the inverse of the cube of the distance. At the same time, the electric field increases as frequency increases and falls off with the inverse of the square of distance.

The wave impedance may be defined as the division of E by H in an electromagnetic version of Ohm's law:

$$Z_{\text{wave}}(\Omega) = \frac{E(\text{V/m})}{H(\text{A/m})}$$

Thus,

$$Z_{\text{wave}} = \frac{Z_0 2\pi D}{\lambda}$$

where $D < 48/f(\text{MHz})$. In the far field [i.e., $D > 48/f(\text{MHz})$], on the other hand, both E- and H-fields decrease as the inverse of the observation distance as described by

$$E(\text{V/m}) = \frac{Z_0 \pi IS}{\lambda^2 D} \qquad H(\text{A/m}) = \frac{\pi IS}{\lambda^2 D}$$

which maintains a constant impedance equal to Z_0, allowing direct calculation of radiated power density in W/m² simply by multiplying E and H. Notice that E and H, and thus power, increase with the square of frequency. This shows, once again, that limiting the bandwidth of radiated signals by a pulse train is of utmost importance in controlling EMI.

The region dividing the near field from the far field is called the *transition region* [i.e., at $D \approx 48/f(\text{MHz})$]. In it, abrupt transitions occur on the near-field characteristics until a smooth blending leads to the far-field characteristics. Electromagnetic fields can also be created by passing an alternating current through a straight wire dipole, just as in a radio antenna. In this case, the near-field electric and magnetic vector amplitudes are

$$E(\text{V/m}) = \frac{Z_0 Il\lambda}{8\pi^2 D^3}$$

and

$$H(\text{A/m}) = \frac{Il}{4\pi D^2}$$

where l is the dipole length in meters. In contrast with the near-field H of a loop which falls with the inverse of D^3, the near-field H of a dipole falls off as $1/D^2$. Similarly, the near-field E of a dipole falls off as $1/D^3$, in contrast to the near-field E of a loop that falls as $1/D^2$. The wave impedance of emissions radiated by a dipole is also affected differently by frequency:

$$Z_{\text{wave}} = \frac{Z_0 \lambda}{2\pi D}$$

Compare this equation with the equation describing Z_{wave} in the near field. The change in wave impedance as a function of frequency in the case of a dipole is inverse to that of a loop.

In the far field, the behavior of the E- and H-fields is again similar to that of electromagnetic radiation from a loop; that is, they decrease as the observation distance increases as described by

$$E(\text{V/m}) = \frac{Z_0 Il}{2\lambda D} \qquad H(\text{A/m}) = \frac{Il}{2\lambda D}$$

Beyond the transitional point, the wave impedance again remains constant at the value of Z_0. The result of a constant impedance in the far field means that the ratio of E to H components remains constant regardless of how the field was generated.

Of course, real-life circuits are neither ideal open wires nor perfect loops, but rather, hybrids of these two. In a simplified form, as shown in Figure 4.6, a more realistic model of a circuit which radiates electromagnetic emissions can assume that an ac voltage source

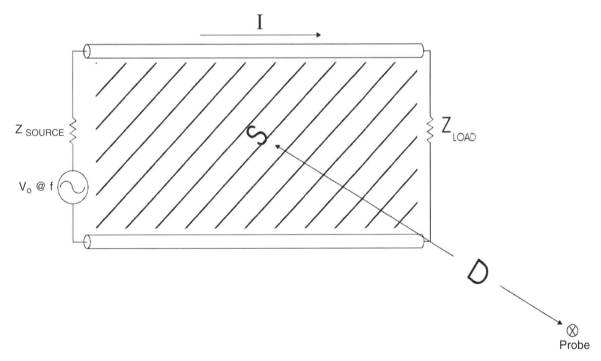

Figure 4.6   A simplified but realistic model of a circuit that radiates electromagnetic emissions. In it, an ac voltage source causes the flow of a current I in a rectangular loop enclosing an area S. The voltage seen by the load depends on the source and load impedances.

causes the flow of a current I in a rectangular loop enclosing an area S. The source impedance is Z_{source}, and the impedance of the load is Z_{load}, resulting in an overall equivalent impedance of $Z_{circuit} = Z_{source} + Z_{load}$.

In the near field, the electric- and magnetic-field vector magnitudes are given by

$$E(\text{V/m}) = \frac{V_0 S}{4\pi D^3}$$

where $Z_{circuit} \geq 7.9D(\text{m}) \cdot f(\text{MHz})$, or

$$E(\text{V/m}) = \frac{0.63 S l f(\text{MHz})}{D^2}$$

where $Z_{circuit} \leq 7.9D(\text{m}) \cdot f(\text{MHz})$, and

$$H(\text{A/m}) = \frac{IS}{4\pi D^3}$$

In the far field, the electric- and magnetic-field vector magnitudes are given by

$$E(\text{V/m}) = Z_0 H = \frac{0.013 V_0 S[f(\text{MHz})]^2}{D Z_{circuit}}$$

and

$$H(\text{A/m}) = \frac{35 \times 10^{-6} IS[f(\text{MHz})]^2}{D}$$

The second lesson of controlling radiated emission leaps out from these equations-keep the area enclosed by loops carrying strong time-varying currents to the minimum possible. Similarly, traces carrying high voltages should be kept as short as possible and be properly terminated.

Besides directing our attention to the parameters affecting radiated emissions, these equations are very useful when designing for compliance with EMI requirements. As exemplified by Figure 4.7, near- and far-field ballpark estimates of EMI can be obtained from known circuit parameters for a large number of common circuit topologies.

PROBING E- AND H-FIELDS IN THE NEAR FIELD

The main reason why EMI standards establish that testing should be performed in the far field is that as demonstrated above, a constant impedance in the far field causes the ratio of E to H components to remain constant regardless of how the field was generated. This means that measurements can be reproduced with reliability and standardized methods of testing can be defined with ease. From the past equations, however, it would seem possible to establish a quantitative correlation that would allow far-field estimates from near-field measurements. Unfortunately, in practice, this is not the case. Near-field measurements are extremely dependent on the exact geometry of the source, the position of the near-field probe, and the interaction between the probe and the source to accomplish the exact measurements necessary for calculating the behavior of the radiation in the far-field region.

Although not applicable to predict the outcome of compliance tests, near-field measurements can nevertheless be very useful to the designer in locating potential sources of radiated emissions. Here, near-field qualitative measurements with simple instruments

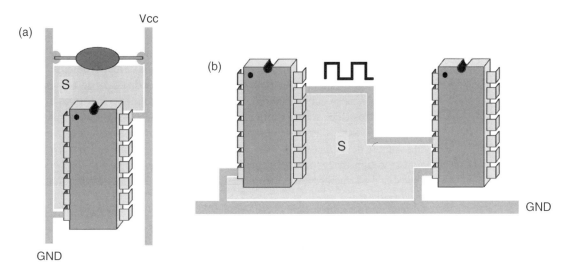

Figure 4.7 Simple differential-mode radiating circuit configurations are created when an ac current flows on a current path that forms a loop enclosing a certain area *S*. (*a*) Transient power demands of an IC are supplied by a decoupling capacitor, causing brief, strong currents that circulate on a loop formed by the supply-bus PCB tracks. (*b*) Fast digital signals driving low-impedance inputs form EMI-radiating loops when current returns through distant ground paths.

can accurately pinpoint sources of EMI and identify their basic characteristics. In essence, if a strong E-field is detected from a certain circuit section but a relatively weak H-field is sensed at the same point, the culprit can usually be traced to a train of high-voltage pulses on a long wire, an unterminated line, or a trace driving a high-impedance load. Conversely, if the H-field is strong but the E-field probe detects little activity, the source of EMI is most probably a looplike circuit through which strong currents circulate. Examples of such situations are PCB tracks carrying strong currents, inductors in switching power supplies, and eddy currents induced in metal enclosures by strong fields inside the case.

Since the same equations used to describe emission of radiation are applicable to the reception of emissions, it is apparent that a small loop of wire can act as a near-field probe which is mostly sensitive to H-fields. E-fields, on the other hand, would then be detected, preferably by a short exposed wire. Measurements can then be taken with either a wideband ac voltmeter or a spectrum analyzer. Even a simple single-turn wire loop at the end of a coax cable can be a very effective H-field probe. With this arrangement, maximum output from the probe is recorded when the loop is in immediate proximity and aligned with a current-carrying wire. This directionality is very useful for pinpointing the exact source of a suspicious signal.

The diameter of the loop makes a large difference on H-field measurements [Kraz, 1995]. The area enclosed by the loop influences the sensitivity of the probe, since it determines the number of magnetic flux lines that are intercepted to produce a detectable signal. A larger loop will obviously develop a larger voltage at the input of the voltmeter or spectrum analyzer. On the other hand, larger loops have inherently larger self-inductance and equivalent capacitance than small loops. As inductance increases, the network formed with the complex impedance of the measurement setup resonates at lower frequencies, beyond which the probe cannot be used. Moreover, larger loops make it much more difficult to identify the exact source of an interfering signal, because their size does not allow them to pick up radiations selectively from single lines when a multitude of the latter are clustered close together. Coils with multiple turns can be used to increase the sensitivity without appreciably increasing the physical size of the coil. However, this solution will again result in reduced spectral response due to increased self-inductance. It is clear, then, that loop geometry must be chosen for every specific instance based on a solution of compromise. In general, it is a good idea to keep a variety of probes handy to tackle different problems.

Another convenient H-field probe can be constructed similar to ac tongs. In this case, a magnetically permeable material concentrates the magnetic flux lines created by the circuit under test. The resulting magnetic flux is then detected by a coil with multiple turns. If the tongs were to enclose completely the conductor through which a current is flowing, the voltage developed across the coil would be proportional to the vector sum of the currents through the conductor. This is, of course, impractical for the needs of sniffing H-fields, and a structure with open-ended tongs is more suitable for probing a circuit without modifying it.

The probe can be built as shown in Figure 4.8, using a small ferrite bead (e.g., 0.1 in. thick, 0.3 in. in outside diameter) that has been sectioned in half. The actual construction depends on the actual ferrite that you select, but in general, 40 to 50 turns of thin enameled copper wire provide suitable sensitivity. The terminals of the coil should be soldered to the center and shield of coaxial cable. After insulating the central conductor connection, a portion of the braid should be used to cover the assembly, thus E-field-shielding the coil. The assembly can then be mounted at the end of a small plastic tube and embedded within a glob of epoxy. For the prototype probe, a virtually flat bandwidth was measured from around 600 kHz to approximately 10 MHz.

Better bandwidth can be achieved by using a VCR magnetic head instead of the ferrite assembly. Video heads are designed for broadband detection of magnetic fluctuations, and for this reason they can be used for sniffing H-fields from 2 MHz up to approximately 120 MHz with relatively flat response. To construct the probe, carefully remove one of the

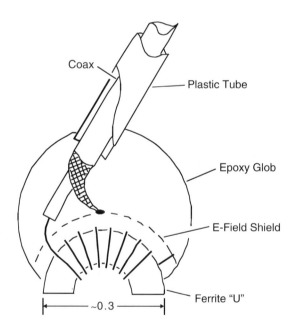

Coax

Plastic Tube

Epoxy Glob

E-Field Shield

Ferrite "U"

~0.3

Figure 4.8 A useful H-field probe can be constructed from a small ferrite bead that has been sectioned in half. Approximately 40 turns of thin enameled wire are used to detect the magnetic flux concentrated by the ferrite. A small portion of the coax cable braid is used as an E-field shield for the coil. The assembly is mounted at the end of a small plastic tube that serves as a handle and embedded in a glob of epoxy.

magnetic heads from a discarded drum. Even a very worn-out head will work well in this application. Very soiled heads should be cleaned with a swab and pure alcohol. Degaussing will also help improve the sensitivity of an old head. All other aspects of constructing and using this probe are the same as for the ferrite-bead probe.

For E-fields, the simplest near-field probe is a coax cable in which a short segment of the center conductor extends beyond the braid at the unterminated end of the coax. Similar to the loop probe, a longer wire will pick up a stronger signal at the expense of specificity and bandwidth. In general, the length of the wire should be selected so that measurements can be performed with a sensitivity of approximately 3 mV/m. At this level, potentially problematic emissions can be identified without causing undue concern about low-level emissions.

Constructing the ideal H- or E-field probe for a specific job may take some trial and error, since the effort of electromagnetic modeling required for proper design is most probably an overkill for most applications. One test that you may nevertheless want to perform on a probe is to determine the existence of resonances within the desired spectral range. To conduct the test, a wideband probe should be connected to an RF generator set up to track the tuning frequency of a wideband spectrum analyzer. The probe under test should be located in close proximity to the emitting probe and connected to the input of the spectrum analyzer. The limit of the useful bandwidth of a probe is the point at which the first abrupt resonance appears.

Before even plugging the spectrum analyzer to the power line, however, the first step in conducting a near-field EMI study should be to draw a component placement diagram of the assembly to be probed. The diagram should indicate circuit points identified in the mathematical circuit harmonic analysis as potential sources for EMI radiation. Only after this preliminary work has been done should bench testing begin. A coarse near-field sweep should be conducted at relatively high gain to identify EMI hot spots in the assembly. A technique that works well is to log the frequencies at which strong components appear when scanning the unit under test. Detailed scanning using a more discriminating probe can then concentrate on the hot spots to identify the culprit circuit generating offending emissions.

A very valuable source of clues for future troubleshooting can be built along the way by printing the spectral estimate at each point in which measurements strongly agree or disagree with the circuit's harmonic analysis. In any case, keep detailed and organized notes of the near-field scans, since these will certainly prove to be invaluable when attempting quick fixes while the clock is running at the far-field compliance-testing facility.

BARE-BONES SPECTRUM ANALYZER

While an ac voltmeter can provide an indication of the field strength to which a probe is exposed, it does not provide any indication of the spectral contents of an emission. A spectrum analyzer is a tool that certainly cannot be beaten in the search for offending signals. Unfortunately, spectrum analyzers are often beyond the reach of a designer on a tight budget. For near-field sniffing, however, even the crudest spectrum analyzer will do a magnificent job.

Figure 4.9 shows a simple home-brewed adapter to convert any triggered oscilloscope into a spectrum analyzer capable of providing qualitative spectral estimates with a bandwidth of 100 kHz to 400 MHz. As shown in Figure 4.10, a voltage-controlled TV tuner IC1 forms the basis of the spectrum analyzer. Most any voltage-controlled tuner will work, and you may be able to get one free from a discarded TV or VCR printed circuit board. The connection points and distribution vary from device to device, but the pinout is usually identified by stampings on the metallic can of the device.

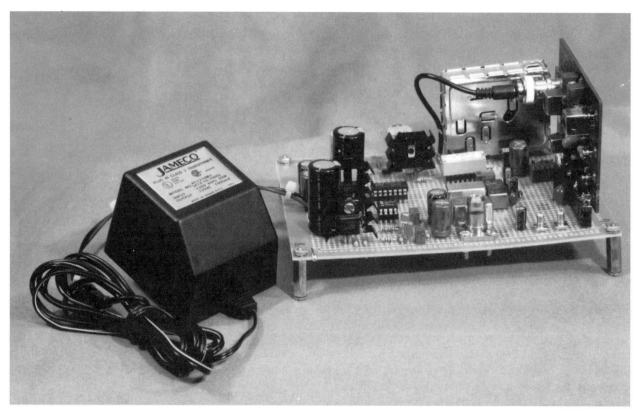

Figure 4.9   A simple circuit can be used to convert any triggered oscilloscope into a 100-kHz to 400-MHz spectrum analyzer suitable for near-field EMI sniffing.

Varactor-controlled TV tuners receive signals on their RF input at a frequency determined by the voltage applied to the VTUNE input. With power applied to the UHF section of a tuner, typical control voltages between 0 and 32 V span a frequency range of approximately 450 to 850 MHz. The sensitivity of the tuner can be adjusted through the AGC input. The output of the tuner is a standard 45-MHz IF. However, 450 to 850 MHz is not a range that is directly applicable to the large bulk of EMI sniffing work. For this reason, the more appropriate range 100 kHz to 400 MHz is up-converted to the tuner's input range through a circuit formed by IC4–IC7 (manufactured by Mini-Circuits). Here, signals from the probe are low-pass filtered by IC6 and injected into the IF port of a TUF-2 mixer. The LO input of the mixer is fed with the output of IC4, a self-contained voltage-controlled oscillator tuned to 450 MHz by potentiometer R20. The RF port of the mixer outputs signals with frequency components at the sum and difference between the IF input and the LO frequency. This output is high-pass filtered by IC5 to ensure that only up-converted components are fed to the tuner input.

Sweeping the tuner across its range is accomplished by a sawtooth waveform that spans approximately 1 to 31 V. The basic sawtooth is generated by IC3 and Q1 and buffered by IC2B. The span of the sawtooth is set by attenuator R10, while the center of the sweep is adjusted by introducing an offset on IC2A by means of R13. The output of IC2A is amplified by transistor Q2. Q2 should be selected for a gain of 50 or less. The final span and linearity of the sweep is adjusted in three ranges by R1, R2, and R3.

The IF output of the tuner is attenuated to a level suitable for processing by R7, R8, and R9. The actual value of the resistors for this attenuator must be selected based on the output level of the specific tuner that you use. Then, in the circuit of Figure 4.11, IC8, a NE/SA605

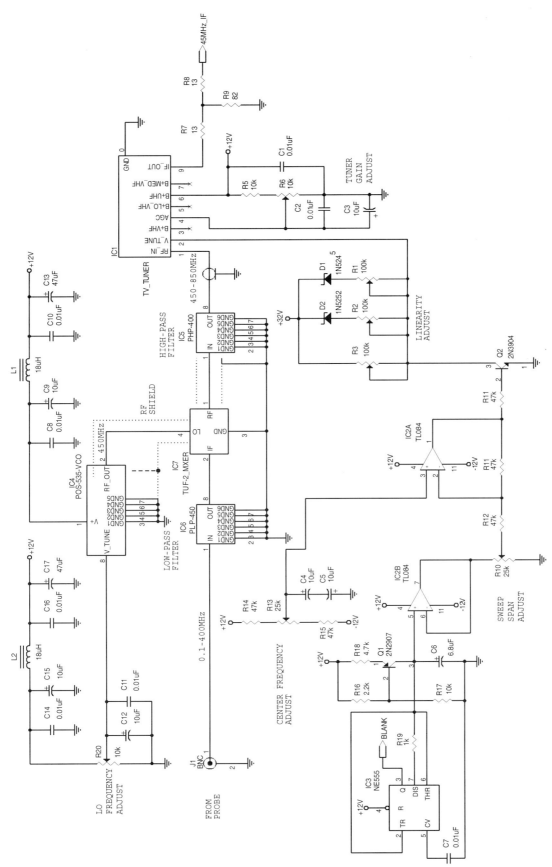

Figure 4.10 A varactor-based TV tuner is the heart of a simple spectrum analyzer. Signals of 100 kHz to 400 MHz from a sniffing probe are up-converted to the 450- to 850-MHz UHF band, where the tuner can be swept by a sawtooth waveform. The tuner produces a 45-MHz intermediate frequency that can be processed to derive the input signal spectrum. Direct connection of the probe to the tuner input extends the range of the spectrum analyzer to the high-VHF/UHF region (450 to 850 MHz).

163

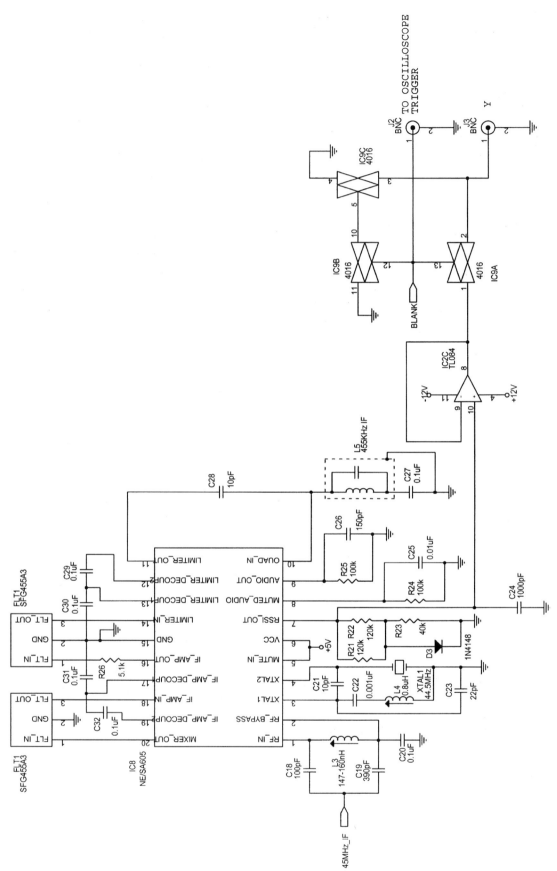

Figure 4.11 The intermediate-frequency output of the tuner is detected by IC8, a single-chip IF processor. The received signal strength indicator (RSSI) output as a function of the sawtooth signal driving the tuner is a logarithmic representation of the spectrum of the signal picked up by the probe.

single-chip IF processor, takes care of detecting the signal and producing a logarithmic output of signal strength.

In this portion of the circuit, the 45-MHz IF signal is coupled to the input of a RF mixer internal to IC8 by way of a tuned circuit formed by C18, C19, and L3. The LO input of this mixer is fed from a 44.5-MHz crystal-controlled oscillator. The resulting 455-kHz IF is filtered by two ceramic filters, FLT1 and FLT2. An internal received signal strength indicator (RSSI) circuit is used as a detector and linear-to-logarithmic converter. The RSSI output, as a function of the sawtooth signal driving the tuner, is thus a logarithmic representation of the spectrum of the signal picked up by the probe. RSSI is a current signal that requires conversion to a voltage by the network formed by R21–R23 and D3. C24 low-pass filters the RSSI output to produce a smooth display, and IC2C acts as a buffer and impedance transformer for the current-to-voltage converter. Finally, IC9A blanks the output during retrace.

Figure 4.12 presents the power supply circuit for the adapter. Most of the circuitry, including the up-converter, tuner, and sawtooth generator, is powered by ± 12 V; $+5$ V powers the IF processor. The $+32$ V to drive the tuner's varactors is obtained by multiplying the 12 V ac input to $+48$ V, reaching the desired voltage through IC10, an LM317 adjustable linear regulator.

To operate the spectrum analyzer, the Y output of the adapter is connected to the vertical input of the oscilloscope, and the TRIGGER output is connected to the trigger synchronization input of the scope. The horizontal frequency of the oscilloscope is set such that one full sweep caused by the sawtooth fits the full graticule on the oscilloscope's screen. Fine-tuning can be accomplished either by trimming the time base of the scope or by adjusting the value of R18 appropriately. Alternatively, a two-channel oscilloscope can be operated in the X–Y mode by injecting the sawtooth available at pin 7 of IC2A to the appropriately scaled X-axis channel.

The comb generator circuit of Figure 4.13 can be used for calibrating the adapter. The circuit is simply a TTL-compatible 40-MHz crystal-controlled oscillator module feeding a synchronous binary counter. It is called a comb generator because the spectral pattern of any of its outputs resembles an ordinary hair comb with its prongs pointing up. Because these spectral components occur at harmonic multiples of the fundamental square-wave frequency selected, it follows that the frequency difference between consecutive "prongs" must be the same as the value of the fundamental frequency of the square wave.

Figure 4.14 presents the pattern obtained when the 20-MHz comb output of the generator is probed by a commercial spectrum analyzer. Ac coupling was accomplished through a series-connection 15-pF capacitor, and termination to ground of the output side of this capacitor was performed through a 50-Ω noninductive resistor. This is the gold standard against which the adapter should be calibrated.

Start testing the adapter by setting the sawtooth generator to vary the voltage at the VTUNE input of the tuner between approximately 1 and 31 V. Initially, set R6 to apply 2.5 V dc to the AGC pin of IC1. The up-converter LO frequency should be adjusted to 450 MHz by trimming R20. 9.6 V dc at the VTUNE input of IC4 will typically result in the desired LO frequency. L4 should be trimmed to achieve stable oscillation of the 44.5-MHz IF LO oscillator. With a 40-MHz comb applied to the input of the adapter through a 15-pF coupling capacitor and with 50-Ω termination, adjust L3 to obtain an approximation of the expected 40-MHz comb pattern on the oscilloscope. After achieving a satisfactory display for the 40-MHz comb, calibrate the linearity of the adapter using a 20-MHz comb by first trimming R3 to produce equal spacing between spectral lines throughout the lower third of the display. Then linearize the midrange by trimming R2, and finally, the high range by trimming R1.

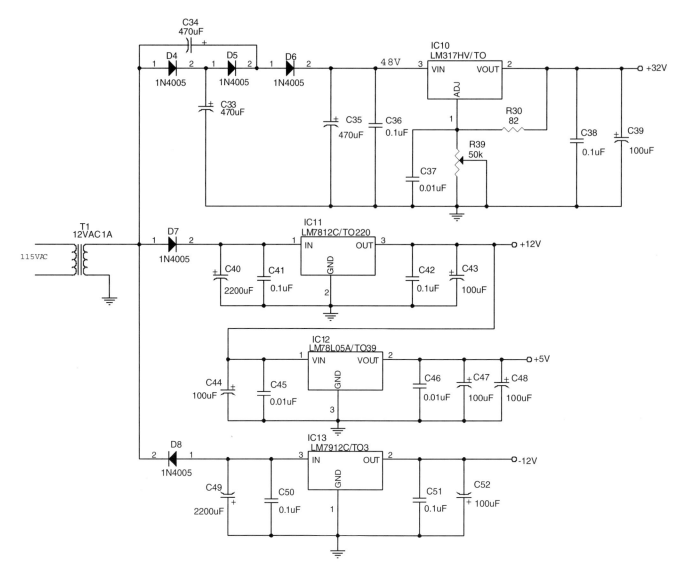

Figure 4.12 Dc power for the various circuits of the spectrum analyzer is derived from a single 12-V ac input. A voltage of +12 V powers most of the circuitry, including the up-converter, tuner, and sawtooth generator; +5 V powers the IF processor. Sweeping the tuner across its 450 to 850 MHz range requires up to +32 V to drive its varactors. A voltage of −12 V is used as bias to ensure that sweeping can be accomplished within any desired portion of the full range.

CONDUCTED EMISSIONS

Conducted emissions measurements are made to determine the line-to-ground radio noise from each power-input terminal of a line-powered medical device. Measurements are taken using a line impedance stabilization network (LISN). A spectrum analyzer and a quasi-peak adapter with a measurement bandwidth of 9 kHz are typically used to record the conducted emissions. As shown in Figure 4.15, tests are performed in a shielded room.

A LISN is a passive RCL network that connects between the ac power line and the device under test. The purpose of the LISN is to present a standard line impedance to the device under test regardless of local power line impedance conditions. The LISN also

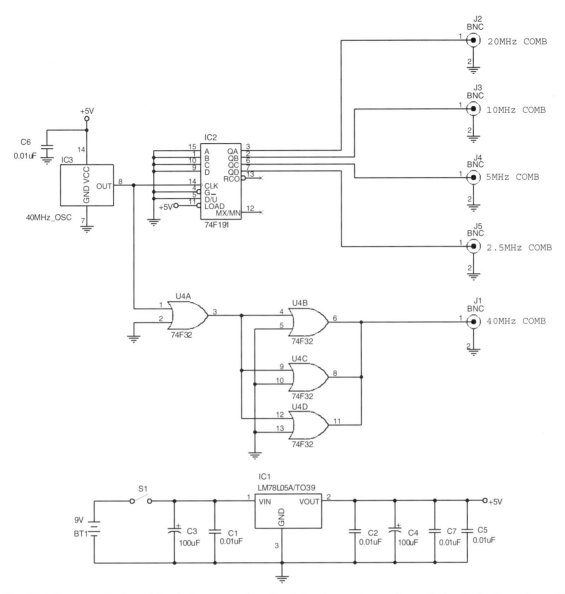

Figure 4.13 High-frequency clocks and fast logic generate broadband signals extending well into the hundreds of megahertz. This generator produces various comb patterns which are useful in the calibration of spectrum analyzers.

isolates the device under test from unwanted interference signals on the power line and provides a test point to probe emissions conducted from the device under test toward the power line. Figure 4.16 presents the circuit for a 50 Ω/50 μH LISN following the definition of standard CISPR-16-1. This circuit provides a 50-Ω output impedance for measurement of RF emissions produced by the device under test. This impedance was selected because theoretical and empirical data have shown that the power circuitry statistically looks like a 50-Ω impedance to standard electronic equipment, and RF test equipment is typically designed for 50-Ω input. The bandwidth is typically determined by the operating frequency of the potential victims of the device under test's conducted emissions. For the majority of medical devices, emission measurements are carried out from 150 kHz to 30 MHz. This ensures that devices do not interfere with VLF or HF radio communication systems and other electronic devices operating at these frequencies.

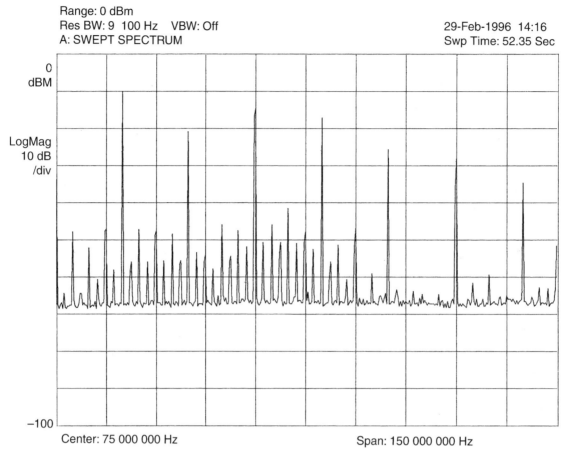

Figure 4.14   The spectral pattern obtained from the output of the comb generator can be used as a frequency ruler because it presents strong spectral lines at every harmonic of the fundamental square wave. Notice the similarity between the envelope formed by the spectral components of this 20-MHz comb and the nomograms of Figure 4.2.

Each of the 50-μH inductors and 1-μF capacitors form an unbalanced filter. The inductors must be of sufficiently large wire gauge to carry the full ac current demanded by the device under test with less than a 2-V drop. Conducted emissions are then measured using a spectrum analyzer with quasi-peak detection. Measurements are taken between hot to ground and then between neutral to ground. A 50-Ω resistor needs to be connected across the 1-kΩ resistor, which is not connected to the spectrum analyzer's 50-Ω input. Switch SW1 accomplishes phase selection and automatic shunting of the LISN leg not being observed.

Note that the LISN established by the standards presents a 1-μF capacitance between the hot line and the LISN and device under the test's safety ground. A ground fault could lead to potentially lethal currents to operators in contact with the LISN or the device under test. For this reason, it is advisable to wire the LISN's ground terminal permanently to ground.

A safer way of running design-time tests is to use a LISN made from a modified power protector designed to filter power line glitches prior to supplying power to computers and other electronic equipment. The circuit for this LISN is shown in Figure 4.17. The input connector and power cord, circuit breaker F1, and neon light are found almost universally in power protector strips. You may also leave any MOVs that you find in the power strip.

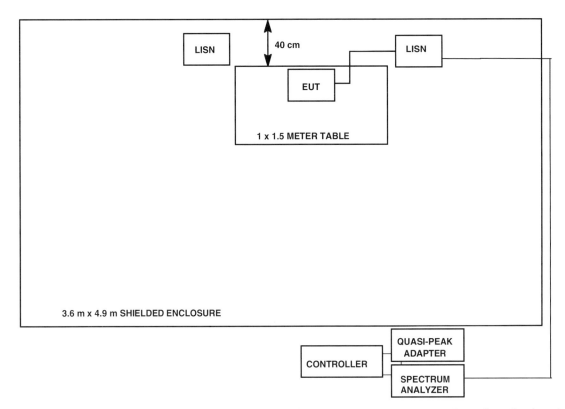

Figure 4.15   Setup for performing conducted emissions testing in a shielded room. Measurements are taken using a line impedance stabilization network (LISN). A spectrum analyzer and a quasi-peak adapter with a measurement bandwidth of 9 kHz are typically used to record the conducted emissions.

Modify the filter circuitry and configure it as shown in the schematic diagram. You may remove all except one of the power outlets to accommodate the components. The power outlet used to connect to the device under test (J2) would be one of the power strip's original power outlets.

Emissions radiated from this LISN are coupled inductively to a spectrum analyzer. L3 is a single loop of No. 14 stranded insulated wire that exits and reenters the power strip's enclosure. The H-field probe made of a VCR head, described above, would then be used to pick up conducted emissions. Although this LISN does not yield results identical to those of the standards, it makes it easy to detect emissions conducted by the device under test into the power line. In addition, although conducted emissions tests should be performed on both phases (hot and neutral) of the power line, most offensive units reveal themselves with just the hot-to-ground measurement provided by this LISN.

When the device is prepared for testing, the power cord in excess of the distance is folded back and forth, forming a bundle 30 to 40 cm long in the approximate center of the cable. Power supply cords for any peripheral equipment should be powered from an auxiliary LISN. Excess interface cable lengths should be bundled separately in a noninductive arrangement at the approximate center of the cable with the bundle 30 to 40 cm in length. The emissions conducted are maximized by varying the operating states and configuration of the device under test. The limits for conducted emissions per EN-55011 for group 1 devices are shown in Table 4.4. As an example, Table 4.5 shows the results we obtained recently when testing an implantable-device programmer for conducted emissions.

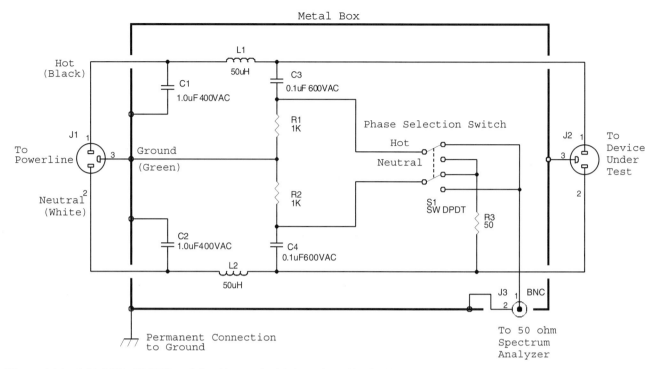

Figure 4.16 A 50 Ω/50 μH LISN as defined by standard CISPR 16-1. This circuit provides a 50-Ω output impedance for measurement of RF emissions produced by the device under test. Conducted emission measurements are carried out from 150 kHz to 30 MHz.

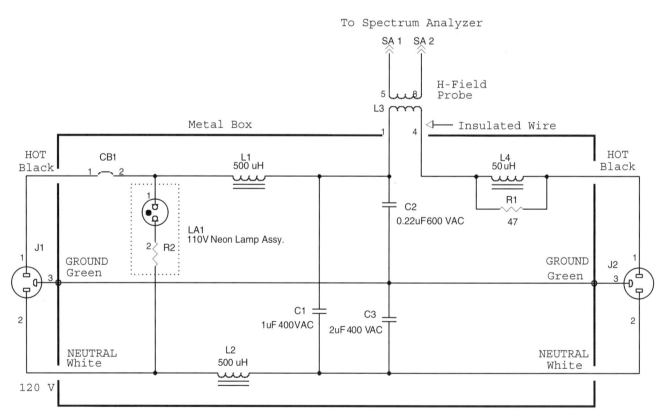

Figure 4.17 A safer way of running lab tests is to use a LISN in which radiated emissions are coupled inductively to a spectrum analyzer. L3 is a single loop of No. 14 stranded insulated wire which is coupled to an H-field probe made of a VCR head.

TABLE 4.4   EN-55011 Conducted Emissions Limits for Group 1 Devices*[a]*

| | Maximum RF Line Voltage (dBµV) | | | |
| | Class A | | Class B | |
Frequency (MHz)	Quasi-peak	Average	Quasi-peak	Average
0.15–0.5	79	66	66, decreasing with log of frequency to 56	56, decreasing with log of frequency to 46
0.5–5.0	73	60	56	46
5.0–30.0	73	60	60	50

*[a]*The limit decreases linearly with the logarithm of the frequency in the range 0.15 to 0.5 MHz. The lower limits apply at the transition frequencies.

TABLE 4.5   EN-55011 Sample Worksheet for Testing Conducted Emissions

Line measured: phase

Frequency Input (MHz)	Reading input (dBµV)	Correction Factor (dB)	Correction Reading (dBµV)	Limit (dBµV)	Margin (dB)	Detector Function
0.257	45.5	0.8	46.3	66	−19.7	Peak
0.323	42.7	0.7	43.4	66	−22.6	Peak
0.195	41.9	1.0	42.9	66	−23.1	Peak
0.385	38.8	0.6	39.4	66	−26.6	Peak
0.451	34.8	0.6	35.4	66	−30.6	Peak
20.100	57.3	1.8	59.1	60	−0.9	Peak
18.100	56.0	1.5	57.5	60	−2.5	Peak
18.500	53.1	1.6	54.7	60	−5.3	Peak
19.000	50.8	1.7	52.5	60	−7.5	Peak

SUSCEPTIBILITY

It is really surprising that regulatory agencies around the world took so long to take steps that would protect patients and health-care providers from EMI-induced medical-device failures. Although the military and aviation industries had been developing hardware immune to EMI for many years, it was only in 1994 that the FDA started taking action by warning the medical device industry about their concerns regarding EMI-induced failures, as well as by making specific recommendations for immunity levels for critical devices. In 1998, European agencies advanced this process by making it mandatory for medical devices to comply with a fairly comprehensive EMC standard to be marketable in Europe. Today, most nations which require medical devices to comply with EMC requirements make use of standards based on the EN-60601-1-2 immunity requirements shown in Table 4.6, which cover electrostatic discharge (ESD), radio-frequency interference (RFI), and a variety of power disturbances. IEC-60601-1-2 itself cites extensively the test methods and immunity levels of the basic IEC-61000-4 series of standards.

At the time of this writing, EN-60601-1-2, pass/fail criteria are ultimately defined by the manufacturer. This is because the current immunity criteria for medical products are defined in this standard as "equipment and/or system continues to perform its intended function as specified by the manufacturer or fails without creating a safety hazard." A safety hazard is then defined as a "potentially detrimental effect on the patient, other persons, animals, or the surroundings arising directly from equipment." As a result, the manufacturer

TABLE 4.6 EN-60601-1-2 Immunity Requirements

Susceptibility	Test Requirements	Relevant Standard
Electrostatic discharge (ESD)	3-kV contact to conductive accessible parts and coupling planes; 8-kV air discharge to nonconductive accessible parts	IEC-61000-4-2
Radiated emissions (EMIs)	3 V/m, 26 MHz to 1 GHz, modulated at passband or 1 kHz	IEC-61000-4-3
Conducted emissions	Test from 150 kHz to 80 MHz into power line to 3 V/m; bulk current injection for patient cables	IEC-61000-4-6
Power line voltage dips, interruptions, and variations	100% dropout for $\frac{1}{2}$ cycle, 60% sag for 5 cycles, 30% sag for 25 cycles; low-powered equipment maintains clinical utility, high-powered equipment remains safe	IEC-61000-4-11
Electrical fast transients (EFTs)	1 kV at power line for plug-connected equipment; 2 kV at power line for permanently installed equipment; 0.5 kV for signal lines longer than 3 m	IEC-61000-4-4
Surge	1 kV differential mode at power line, 2 kV common mode at power line; signal lines not tested	IEC-61000-4-5
Magnetic fields	10 A/m at power line frequency	IEC-61000-4-8

may chose to classify a failure mode that does not pose a risk to the patient (or other surrounding targets) as a *pass*. For example, the manufacturer may pass a device that fails to start operating or stops operating when exposed to EMI threats as long as these modes do not pose a threat to the patient. A common classification for device performance during testing is as follows:

- *Criteria level A:* normal performance within equipment specifications
- *Criteria level B:* degradation or loss of function or performance which is self-recoverable when the interfering signal is removed
- *Criteria level C:* degradation or loss of function or performance that requires system reset or operator intervention when the interfering signal is removed

In the United States, the FDA is adopting many of the IEC-60601-1-2 requirements but imposes restrictions on the manufacturer's ability to adopt pass/fail criteria. The FDA prescribes that a passing result corresponds to maintaining clinical utility. Some of the same concerns are also being adopted for revised versions of IEC-60601-1-2 and drafts show that the failure criteria will change in focus from the safety hazard to specified performance compliance. Because of the fast-changing nature of this field, we present this section as a primer on the issue of immunity to EMI, but strongly advise you to keep updated on the latest versions of applicable standards.

Susceptibility to Electrostatic Discharge

Do you know what is the potential difference between you and the doorknob before you shout "ouch!" on those winter days that electrostatic charges seem to love? 6 kV! But there are also occasional 15-kV discharges—the kind that make you hope that someone else will open the door for you. Regardless of the specific number of kilovolts, such discharges suffice to fry many static-sensitive ICs. However, the primary concern with ESD events is the large amount of RF energy they convey over an extremely wide band of frequencies. IEC-61000-4-2 considers ESD between a human being and a medical device as the primary source of ESD-related failures. Testing is done by delivering 3-kV discharges directly to a device's exposed conductive components and 8-kV air discharges to parts that may be recessed.

The human body model of Figure 4.18 represents the discharge from the fingertip of a standing person delivered to a device, modeled here by a 150-pF capacitor discharged through a switching component and a 330-Ω series resistor into the device under test. This model, which hasn't changed much since it was developed in the nineteenth century, was originally used to investigate explosive gas mixtures in mines. An ESD simulator is not more than an instrument implementing this model. A high-voltage power supply is used to charge a 150-pF capacitor via a charging resistor. The capacitor's charge is then delivered to the device under test by way of a 330-Ω resistor. The switch may be a vacuum relay, a high-voltage semiconductor switch, or a spark gap.

Construction of the ESD simulator should enable it to generate a discharge waveform with the parameters shown in Table 4.7. Obviously, the most critical design consideration is being able to generate the ESD with a rise time of 0.7 to 1 ns. The frequency content of such an ESD waveform is flat to around 300 MHz before it begins to roll off, so it contains significant energy at 1 GHz and above. Short rise time is so important because it is the pulse's dV/dt as well as the dI/dt that it causes which allow ESD to induce currents and voltages in a device's circuits which lead to failures. Suppose, for example, that a PCB track within the medical device's circuit has an inductance of 10 nH/cm. If current from an ESD event is directly or indirectly coupled to that PCB track, the voltage induced along a length l of that track will be given by

$$V = 10(\text{nH/cm}) \cdot l(\text{cm}) \frac{dI}{dt}$$

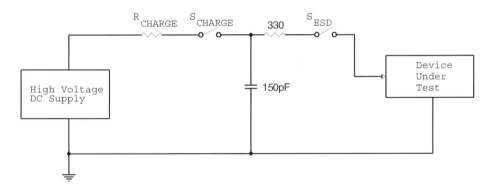

Figure 4.18 The human body model of ESD represents discharge delivered to a device from the fingertip of a standing person. It is modeled by a 150-pF capacitor discharged through a switching component and a 330-Ω series resistor into the device under test.

TABLE 4.7   IEC-801-2 (1991) ESD Waveform Parameters

Severity Level	Voltage (kV) (±5%)	First Peak Current of Discharge (A) (±10%)	Rise Time (ns)	Current at 30 ns (A) (±30%)	Current at 60 ns (A) (±30%)
1	2	7.5	0.7–1	4	2
2	4	15	0.7–1	8	4
3	6	22.5	0.7–1	12	6
4	8	30	0.7–1	16	8

Since even a modest ESD event can develop a current of 10 A in 1 ns, the voltage induced across 1 cm of PCB track will be as high as 100 V! Similarly, high currents can flow through capacitances on and across circuit components. Current flowing through a capacitor is given by

$$I = C\frac{dV}{dt}$$

If an ESD event causes a change in potential of 1 kV within 1 ns, the current flowing through an unprotected input with 10-pF capacitance would be as high as 10 A.

Testing for Immunity against ESD   Two ESD testing techniques are used to check medical devices. The first is air discharge; the second is contact discharge. Testing by air discharge consists of charging the ESD simulator to the required test voltage and slowly moving the simulator's discharge electrode toward the device under test until discharge occurs. This is very similar to what happens when a charged human approaches a device. However, test results obtained through this technique are notoriously unrepeatable, since the tester's rate of approach, exact angle of approach, conditions of the air around the device, and other variables influence the magnitude and path the discharge will take through the device under test.

The contact test technique was developed in an attempt to improve repeatability. In this test, the discharge electrode of the ESD simulator is held in contact with a metallic surface on the device under test when the discharge switch closes. The actual discharge occurs within the ESD simulator in a controlled environment, and the current can be injected at the same contact point each time. The test requires an unpainted conductive contact area on the device under test. As such, this test applies only for devices that have a conductive surface from which paint can be removed and is not applicable when no metallic surfaces are directly accessible.

Testing to EN-61000-4-2 involves delivering air discharges of up to ±8 kV (using an 8-mm round tip to simulate a human finger) to everything nonmetallic that is normally accessible to the operator. Contact discharges of up to ±3 kV (using a sharp tip that is touched against the product before the discharge) are applied to operator-accessible metal parts. Test voltages are increased gradually from low values, often using the settings 25%, 50%, 75%, and then 100% of the test voltage. This is because ESD failures are sometimes seen to occur at lower voltages but not at the maximum test level. The highest test level on an ESD test is not necessarily the one most likely to cause a failure.

It must be noted that the contact test is more severe than the air-discharge test. This is because the former yields faster rise times than the latter. In turn, faster rise times yield higher bandwidth for the EMI generated by the ESD event. An 8-kV air discharge is in the same category as a 6-kV contact discharge, and a 15-kV air discharge is as severe as an 8-kV contact discharge. Note the nonlinear relationship. European regulatory agencies are considering increasing the 3-kV contact test level to 6 kV, so keep yourself up to date with the standards.

Despite the simplicity of the human discharge model, ESD simulators are not all that simple, and commercial units are certainly expensive. However, for development-time testing meant to give you a good "gut feeling," there are some simple alternatives to buying a fully compliant ESD test system. Tiwari [1996] proposed modifying a piezoelectric type of kitchen gas lighter as a fast-static-charge generator which can produce an ESD-like discharge through air.

As shown in Figure 4.19, the modification involves removing the gas reservoir and replacing the gas feed line by a pin which extends beyond the gas lighter's tip. When the handle is squeezed and the tip of this makeshift ESD gun is placed in close proximity (e.g., $\frac{1}{4}$ in.) to a conductive member of the device under test, a spark jumps, conveying

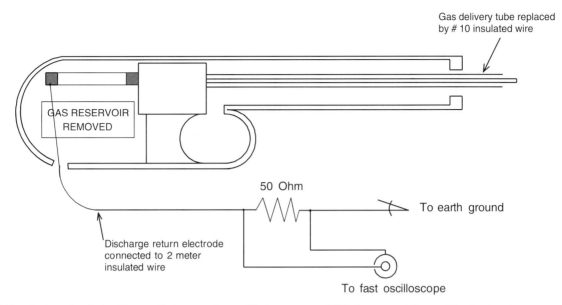

Figure 4.19 A piezoelectric-ignition gas lighter can be modified to generate ESD-like events. The modification involves removing the gas reservoir and replacing the gas feed line by a pin that extends beyond the gas lighter's tip. Typical ignition piezoelectric crystals can generate discharges conveying approximately 0.2 μC within a total pulse of 100 ns to 1 μs. To assess the current and waveform delivered by the discharge, use a 50-Ω resistor in series with the ESD gun's ground terminal as a current shunt.

approximately 0.2 μC within a total ESD pulse of 100 ns to 1 μs. To assess the current and waveform delivered by the discharge, use a 50-Ω resistor in series with the ESD gun's ground terminal as a current shunt.

A simulator which produces waveforms that are closer to a professional unit compliant with IEC-801-2 [1991] can be built for under $100 using surplus high-voltage components. In the circuit of Figure 4.20, a TDK model PCU-554 dc-to-ac inverter is used to drive a Cockroft–Walton quintupler. The dc-to-ac inverter is originally sold as a cold-cathode fluorescent lamp driver for LCD screen backlighting and may be substituted by any similar part capable of delivering at least 1.2 kV$_{RMS}$ at 10 mA. The module produces a high-voltage output that is proportional to its dc input. Dc power for the module is supplied by a variable power supply built around IC1, a LM317 adjustable voltage regulator. The PCU-544 operates well for input voltages in the range 1.5 to 5 V.

The output polarity of the Cockroft–Walton multiplier depends on the way in which its diodes are oriented. Since ESD standards call for testing with discharges of both polarities, the multiplier was designed to yield either positive or negative output. If the high-voltage ac output of the dc-to-ac inverter is connected to point A of the voltage multiplier and point B is connected to ground, the output at point D will be positive. If, however, point C receives the high-voltage ac and point D is connected to ground, point B will be negative. The multiplier can be built on a piece of perfboard, with square-pin connectors at points A, B, C, and D. Ideally, the multiplier assembly should be potted in RTV silicone rubber. This board can then be disconnected from the main circuit and turned around to change polarity.

Switching C5, the ESD model capacitor, between the output of the voltage quintupler and the output is accomplished by K1, a vacuum relay. Vacuum relays are much better at generating fast-rise-time waveforms than most other switches (e.g., firing thyratrons) and yield more reproducible waveforms than those of spark gaps. Vacuum relays can be expensive (a few hundred dollars), and it is better to search the inventory of electronic surplus stores such

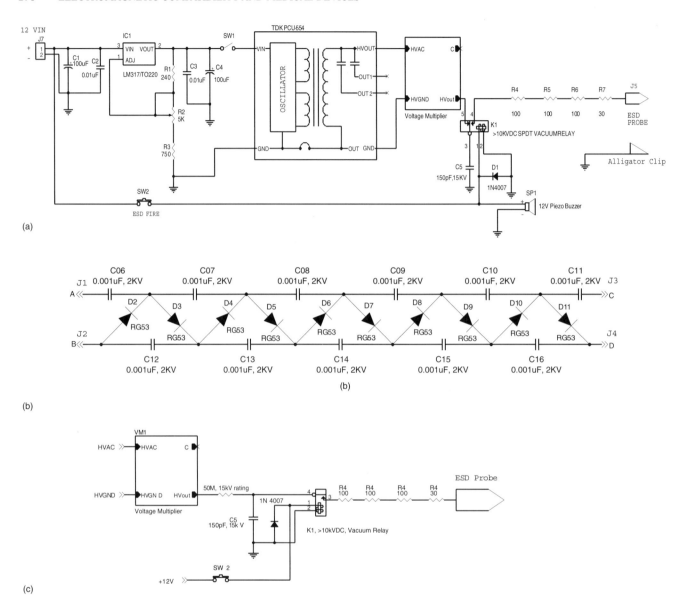

Figure 4.20 A simulator which produces waveforms that are closer to a professional unit compliant with IEC801-2 (1991) can be built using surplus high-voltage components. (*a*) A cold-cathode fluorescent lamp driver and a quintupler produce the high voltage to charge the ESD model capacitor. A vacuum relay transfers the charge to the device under test via the ESD resistor network. (*b*) The quintupler should be built as a stand-alone module, making it possible to reverse the ESD polarity. (*c*) If necessary, a SPST vacuum relay can replace the SPDT unit.

as Surplus Sales of Nebraska and Fair Radio Sales for a suitable SPDT relay with at least a 10-kV contact rating and a 12-V dc coil. If you cannot find a suitable SPDT relay, you can use a 50-MΩ resistor (with at least a 15-kV rating) to charge the 150-pF capacitor constantly, and use a SPST vacuum relay to deliver the ESD to the ESD resistor network.

Use carbon-composition (noninductive), high-voltage resistors for R4–R7 and build the high-voltage discharge path with the shortest possible lead lengths. This will ensure low path inductance and fast ESD pulse rise times. When pushbutton switch SW2 energizes K1, high voltage will be present at the ESD probe. A piezoelectric buzzer is used to warn the user that the probe is potentially charged. To comply with the standard, the probe must have specific dimensions. However, good results are obtained using a $\frac{3}{8}$-in. smooth rounded-head bolt as the probe tip for air discharge. For contact-discharge tests, use the pointed edge of a $\frac{1}{4}$-in. steel nail.

During operation, the discharge return ground cable of the generator must be connected to earth ground. The ground cable should be at least 2 m long and have insulation rated at 12 kV or more. Dc power for the ESD gun can be obtained from a 12-V battery pack or a 12-V dc adapter with a current rating of at least 400 mA. Finally, make a calibration dial to be placed around the shaft of potentiometer R2 by measuring the voltage across C5 using a high-impedance high-voltage probe and a digital voltmeter.

The standards call for a ground reference plane to be placed in the floor of the laboratory. The plane should be an aluminum or copper sheet no thinner than 0.25 mm, covering an area no smaller than 1 m², and projecting at least 0.5 m beyond all sides of the device under test. This is the ground reference to which the ESD simulator should connect. The plane must also be connected to the protective earth ground. The device under test should be placed on a nonconductive test table 0.8 m high. All nonconductive construction (e.g., all wood) is necessary because metal objects in the table construction would distort the RF fields radiated by the ESD event field. The table should be placed no closer than 1 m to the walls of the laboratory or any other metallic object. A 1.6 m × 0.8 m metallic sheet horizontal coupling plane covered with a 0.5-mm insulating support is placed between the tabletop and the device under test. This coupling plane must be connected to the reference ground plane via two 470-kΩ resistors in series.

Indirect Injection of ESD Fields   Since ESD events generate large amounts of RFI, it is not always necessary for the ESD event to happen between a charged body and the medical device itself. A discharge between two bodies in the vicinity of the medical device may suffice to cause a failure. For this reason, IEC-61000-4-2 specifies that testing shall also be done by generating EMI fields through ESD between the ESD simulator and the isolated horizontal coupling plane, as well as between the ESD simulator and an isolated vertical coupling plane. The vertical coupling plane is effectively an antenna of dimensions 0.5 m × 0.5 m that is placed on the horizontal coupling plane but is isolated from it. An ESD generator is then placed in the center of the vertical edge, and at least 10 impulses of either polarity are applied. The vertical coupling plane must also be connected to the reference ground plane via two 470-kΩ resistors in series.

Susceptibility to Radiated Electromagnetic Interference

IEC-61000-4-3 specifies a modulated RFI test of 3 V/m as representative of the radiated electromagnetic interference that may be caused on a medical device by wireless communication equipment. For critical equipment such as life-support devices, 10 V/m is used for testing. Currently, tests should be performed at frequencies of 26 MHz through 1 GHz with 1 kHz at 80% amplitude modulation, but there is serious talk about extending the upper limit to 3 GHz. The frequency band is covered in steps of 1% of the fundamental frequency. For frequencies of 26 to 200 MHz, a biconical transmit antenna is commonly used. For frequencies above 200 MHz, a double-ridged horn transmit antenna is the popular choice.

As shown in Figure 4.21, the testing is usually performed in a shielded enclosure with anechoic material placed throughout the enclosure to minimize reflections. The transmit antenna is typically located 3 m from the device under test. An isotropic field strength meter is placed inside the room at a location physically close to the device under test and used as a secondary indication of the field strength. Testing is performed utilizing linearly polarized antennas, with the device under test exposed to both vertically and horizontally polarized fields on each of four sides. In addition to the frequency sweep, the device under

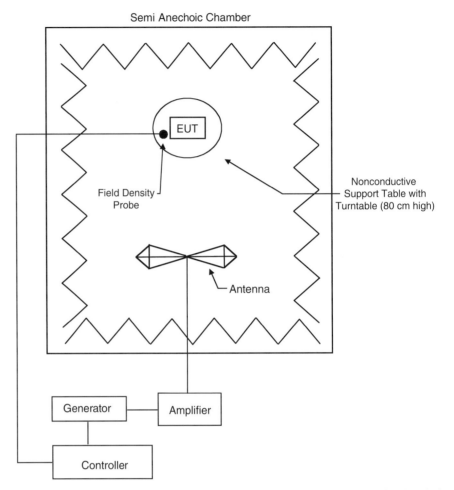

Figure 4.21 Setup for assessing the susceptibility of a medical device to radiated emissions. Testing is performed in a shielded enclosure with anechoic material placed throughout the enclosure to minimize reflections. The transmit antenna is typically located 3 m from the device under test. An isotropic field strength meter is placed inside the room to yield an indication of the field strength.

test is also exposed to a 3-V/m field of 900 MHz modulated with a 200-Hz square wave and modulated with 50% duty cycle in both vertical and horizontal polarizations.

Sometimes, additional test and monitoring equipment is needed to generate test signals and to evaluate the performance of the device under test. Figure 4.22 shows the experimental setup used to test the RFI susceptibility of a prototype implantable-device programmer. The implantable device programmed by this device is meant to interact with the patient's heart. Although the implantable device itself was not the subject of this specific test, it had to be in communication with the programmer so that the performance of the programmer could be evaluated while being exposed to the 3-V/m RFI. In addition, since the programmer has an ECG input, a patient simulator had to be connected to the programmer during the tests. The patient simulator as well as the implantable device were placed under an aluminum foil shield. A shielded closed-circuit TV camera relayed the image from the programmer's computer screen to those who were monitoring the device outside the shielded room.

Because of the amount of EMI generated, there is no easy legal way of conducting this test outside a shielded room. As such, the common engineering practice is to apply good design practice and then cross fingers when running the test at a qualified facility. Design-lab testing

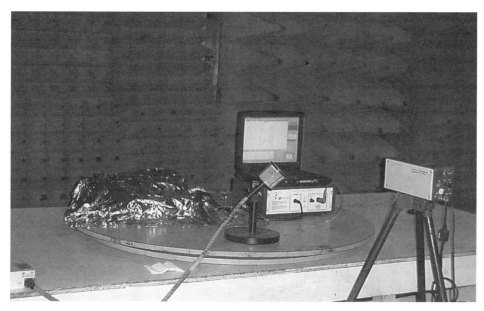

Figure 4.22   A prototype implantable-device programmer is being tested to assess its susceptibility to radiated EMI. An implantable device and a simulator need to interact with the device under test to assess its behavior. These test accessories are placed under the aluminum foil shield. A TV camera relays the image from the programmer's computer screen to the control station outside the shielded room.

will probably become more popular in the future. Since cellular telephones and handheld transceivers can produce field strengths above 3 V/m, regulatory agencies are considering increasing the EMI field level to 10 V/m for all medical electronic equipment. Passing 10 V/m will be a very difficult challenge for the designers of sensitive patient-connected devices!

A beefed-up indirect-injection ESD test can serve as the basis for a test to give a rough indication of a device's susceptibility to radiated EMI. This is the way in which the military test equipment hardened against electromagnetic pulses (EMP) generated either by nuclear explosions. A cheap wideband EMI generator, albeit not nearly as powerful as that used to test for EMP susceptibility, can be built using a high-voltage generator that charges a capacitor and releases its energy into an antenna. The trick is to produce a very fast rise time (less than 1 ns, if possible) and a relatively long total duration (100 ns or more). One way of doing this is shown in Figure 4.23. The core of this wideband EMI generator is Blumlein's pulse generator. The capacitances of two transmission lines are charged by a high-voltage power supply via a series charging resistor R_{charge}. When charging, the transmission lines are effectively in parallel because inductor L_{bypass} does not present any substantial impedance to low-frequency signals. When a certain voltage is developed across the transmission line, the spark gap breaks down, effectively shorting one end of transmission line 1. This causes a very fast pulse to appear across the wideband antenna. Blumlein generators are often used to power nitrogen lasers, ground-penetrating radar, and other instruments that require sharp, high-voltage pulses.

A traveling-wave TEM horn antenna can be used to radiate the pulse generated by the Blumlein generator toward the device under test. A traveling-wave TEM horn consists of a pair of triangular conductors forming a V structure in which a spherical TEM-like mode wave propagates along the axis of the V. The schematic diagram for an experimental wideband generator circuit is shown in Figure 4.24. Here, a push-pull oscillator drives a TV flyback. The original primary of the flyback transformer is not used. Instead, new primaries are made by winding two sets of four turns each of insulated No.18 wire around the exposed core of the flyback transformer. Feedback for the oscillator is obtained through an additional coil of four turns of No.24 wire wound around the core.

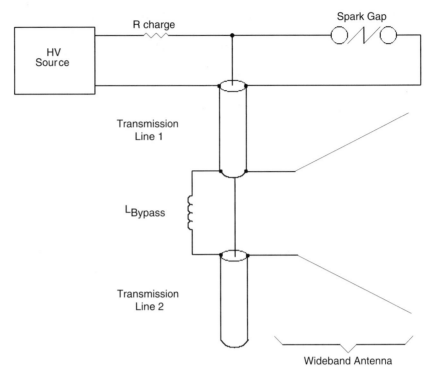

Figure 4.23 A beefed-up indirect-injection ESD test can serve as the basis for wideband assessment of a device's susceptibility to radiated EMI. The transmission lines of a Blumlein pulse generator are charged by a high-voltage power supply via R_{charge}. When a certain voltage is developed across the transmission line, the spark gap triggers, causing a very fast pulse to be delivered to a wideband antenna.

Applied at the input of the flyback driver, 12 V should produce 15 to 20 kV dc at the output of the flyback's tripler. This high voltage is used to charge two transmission line capacitors Z1 and Z2 which are etched on a double-sided 0.4-mm-thick copper-clad PCB as shown in Figure 4.25. The TEM horn antenna is formed from two truncated triangular pieces of single-sided PCB and edge-soldered to the Blumlein generator board. The spark gap is simply a copper or bronze U shape with a bolt and nuts that permit the discharge gap width to be adjusted.

Susceptibility to Conducted Electromagnetic Interference

EMI susceptibility tests for medical devices conducted according to IEC-61000-4-6 involve injecting RF voltages onto the power line and bulk RF currents into other signal cables. A current probe is clamped around the entire cable bundle, and radio-frequency

> **Warning!** This is a dangerous device! It produces high voltages that can cause very painful or lethal electrical shocks. In operation, the spark gap produces significant levels of ultraviolet radiation, which must be shielded to prevent eye damage. In addition, spark discharges can ignite flammable or volatile atmospheres. Finally, the EMI levels generated by this circuit are certainly above what the FCC likes to see dumped into the atmosphere. Thus, this generator should only be operated inside a properly shielded room.

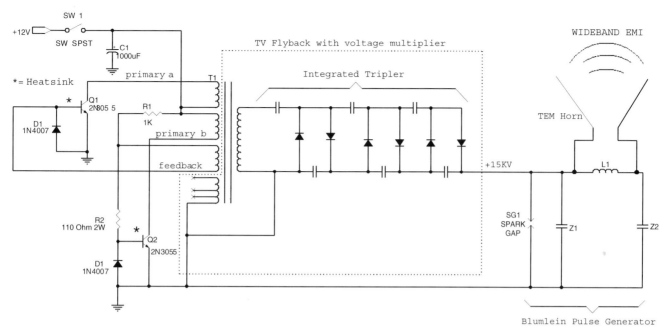

Figure 4.24   In this experimental wideband generator, a push-pull oscillator drives a TV flyback to produce 15 to 20 kV dc. This high voltage is used to charge two transmission line capacitors, Z1 and Z2, which are etched on a double-sided PCB.

energy is injected. The frequency range for immunity tests is 150 kHz to 80 MHz, and the injected RF has an amplitude of 3 V. Figure 4.26 shows a typical test setup. The device under test is placed in the approximate center and 10 cm above a reference ground plane, and is powered and operated in a normal configuration. Injection of RF into the ac power leads is performed with the coupling network shown in Figure 4.27. Testing of signal input leads is performed via a current clamp on the leads.

Susceptibility to Fast Power Line Transients

IEC-61000-4-4 deals with the immunity that devices must present against repetitive fast transients that may be induced, for example, by inductive disconnects on the power line circuit from which the medical device is powered. Electrical fast transients (EFTs) are caused any time that gaseous discharge occurs (a spark in air or other gas), the most common being the opening of a switch through which current is flowing. As the switch is opened, arcing occurs between the contacts: first at low voltage and high frequency while contacts are close together, and later at a higher voltage and lower frequency as the contacts separate.

Figure 4.28 shows the experimental setup to test for susceptibility to EFT. The device under test is placed in the approximate center of a reference ground plane and is powered and operated under worst-case conditions. Throughout the test, the device under test is observed for any indications of erratic operation. Transients are applied to the power leads through the use of a coupling/decoupling network. In this network, 33-nF capacitors couple the high-voltage pulses from the EFT burst generator between ground and the live and neutral lines of the device under test's power input. The network also includes a filter to prevent the high-voltage pulses from coupling into the real power line.

The device under test is subjected to 1-kV discharges to the ac power input leads. Each pulse should reach 900 V by 5 ns ± 30% and should spend no more than 50 ns ± 30% above 500 V. The burst of pulses is delivered with a 5-kHz repetition rate. Both positive and negative polarity discharges are applied. For each discharge sequence the duration is

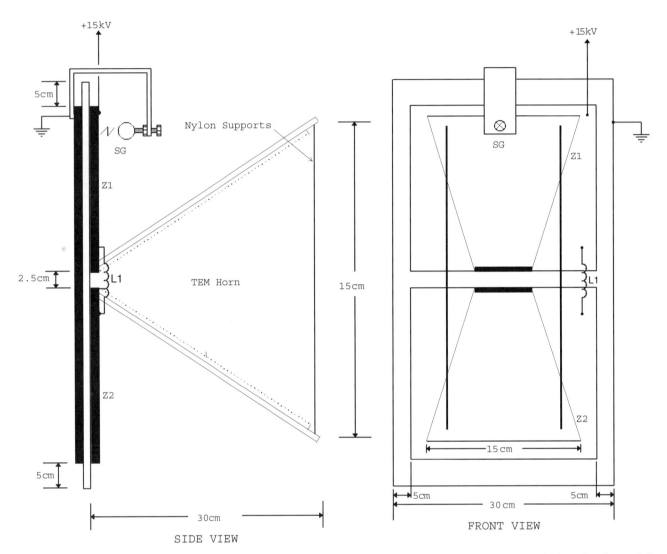

Figure 4.25   Construction details for the Blumlein generator. The TEM horn antenna is formed from two truncated triangular pieces of single-sided PCB and edge-soldered to the Blumlein generator board. The spark gap is in a bronze U shape with a bolt that permits the discharge gap to be adjusted.

1 minute with a 1-minute pause between sequences. EFT is capable of inducing EMI within the device under test over a 60-MHz bandwidth.

A capacitive coupling clamp is used to couple bursts onto signal, data, I/O, and telecommunications lines. The coupling clamp is really a 1-m-long metallic plate that couples the EFT to signal lines without galvanic connection. This plate is suspended by insulator blocks 10 cm over a reference ground plane. An EFT generator is a relatively complex piece of equipment. However, for design-time testing, Guettler [1999] proposed a simple line-disturbance simulator which generates inductive-disconnect transients through the use of a fluorescent-lamp ballast inductor and a modified glow-discharge starter. In the circuit shown in Figure 4.29, a fluorescent-lamp glow-discharge starter is modified by removing its noise-suppression capacitor. In operation, when SW2 is closed, the glow-discharge starter SW3 switches on and off at random. The abrupt current variations through the

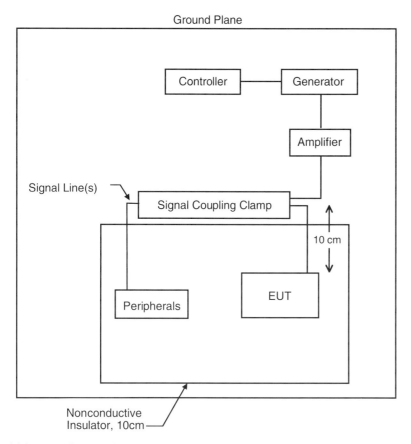

Figure 4.26   Setup for assessing the susceptibility of a medical device to conducted emissions. The test involves injecting RF voltages onto the power line and bulk RF currents into other signal cables. A current probe is clamped around the entire cable bundle, and radio-frequency energy is injected. The frequency range for conducted immunity tests is 150 kHz to 80 MHz.

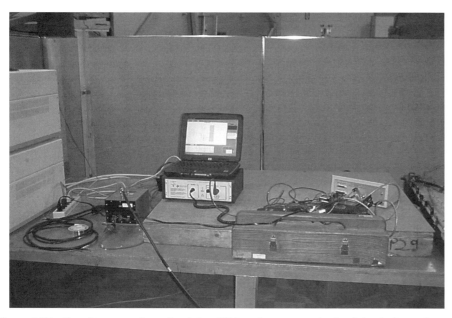

Figure 4.27   Coupling network used to inject RF into the ac power leads of the device under test.

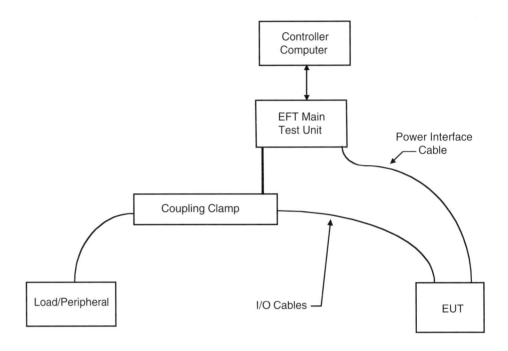

Figure 4.28 Setup for assessing the susceptibility of a medical device to electrical fast transients (EFTs). This test simulates the power line transients that are caused by switching off an inductive load.

fluorescent lamp ballast L5 induce noise at the device under the test's power line input. An *LRC* network filters the transients so that they do not flow back into the real power line.

Susceptibility to High-Energy Power Line Transients

IEC-61000-4-5 deals with large surges that may directly reach the device's power line because of lightning. The surge pulses for this test are much wider than those used in EFT testing. The pulses of IEC-61000-4-5 are tens of microseconds wide, making them able to induce EMI within the device under test over a 300-kHz bandwidth. Figure 4.30 presents a typical setup for testing a medical device's susceptibility to high-energy power line surges. The device under test is placed in the approximate center of a reference ground plane and is powered and operated in a normal configuration. Throughout the test, the device under test is observed for indications that a failure has occurred. Transients are applied to the ac power line leads as well as to I/O lines through coupling/decoupling networks. The surge is applied in the common mode (line to ground) at 2 kV. The surge is then applied in the differential mode (line to line) at 1 kV. A series of six positive and six negative surges are applied with a 1-minute interval between surges.

The impulse generator for these tests uses a 20-μF capacitor charged to the test voltage, which is discharged through a switch into the device under test via a series coupling capacitor and a 40-Ω resistor. In addition, a 50-Ω resistor placed just after the switch is used for shaping the test pulse. For line-to-ground testing, the coupling capacitor has a value of 18 μF. For line-to-line testing the capacitor is 9 μF with a 10-Ω resistor added in series. Just as with the coupling/decoupling networks for EFT testing, a filter is placed in series with the real power line to avoid transients from entering it.

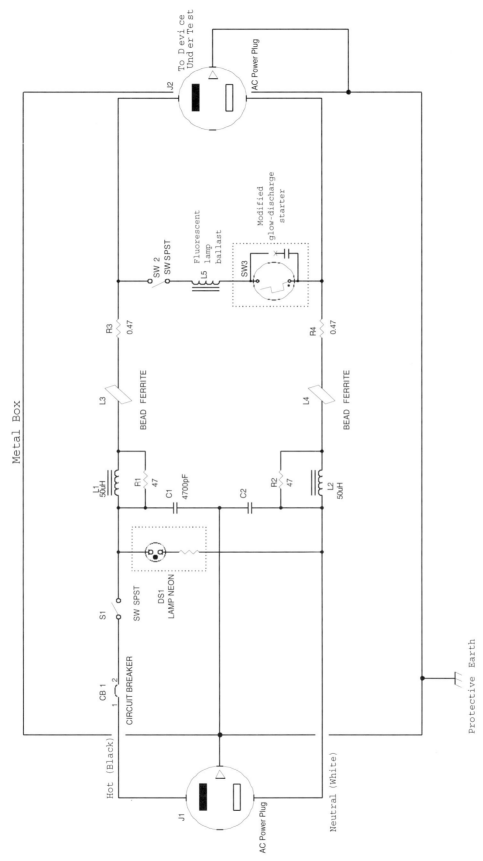

Figure 4.29 This simple line-disturbance simulator generates inductive-disconnect transients like those of the EFT test through the use of a fluorescent-lamp ballast inductor and a modified glow-discharge starter.

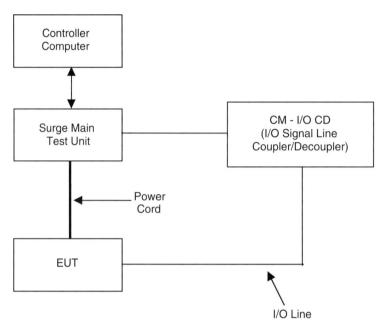

Figure 4.30   Setup for assessing the susceptibility of a medical device to high-energy power line surges. Transients are applied to the ac power line leads as well as to I/O lines through the use of coupling–decoupling networks. This tests simulates the transient variations in power line voltage that may be induced by a lightning strike.

High-energy power line transients are not easy to generate without a pulse generator such as the one described by the standard. Because of the voltages and energies involved, we recommend using a commercial unit. However, if you decide to build your own, follow the component values suggested by IEC-61000-4-5, use conservative ratings, and above all, keep it safe.

Susceptibility to Voltage Dips, Short Interruptions, and Voltage Variations

IEC-61000-4-11 covers power line voltage dips and interruptions. The voltage variations and their duration are as follows:

- 30% Reduction for 10 ms
- 60% Reduction for 100 ms
- >95% Reduction for 5000 ms
- ±10% Voltage variation

For this test, the performance of the device under test is assessed for each combination of test level and duration selected with a sequence of three dips/interruption with intervals of 10 seconds minimum between each test event. Each representative mode of operation of the device under test should be assessed. For each voltage variation, a different pass/fail criterion may be used. However, the equipment must exhibit safe conditions following long outages.

The implication of this test to designers is the need to provide energy reserves in the equipment power supply to maintain operation through brief dips and outages, at least for critical functions. Designers must also ensure that a device cannot power-up in an unsafe mode after a prolonged power outage. A simple way of simulating these conditions during the design phase is to use a variac. Although timing an exact 10 ms for a dip is not an easy manual task, good approximations for these dips and interruptions can be achieved, especially when the variac's output is monitored with an oscilloscope. Remember, however, that power line voltages are present, and an oscilloscope with grounded input channels should not be used without appropriate isolation. As shown in Figure 4.31, a small filament transformer (e.g., 110 V ac/6.3 V ac at 100 mA) provides appropriate isolation for monitoring the variac's output with a grounded oscilloscope.

Susceptibility to Magnetic Fields

IEC-61000-4-8 deals with interference that may be caused on a device by low-frequency magnetic fields, such as those generated by the power lines. These magnetic fields can produce jitter on CRT displays, distortion in amplified signals equipment, or false readings in equipment magnetic or electromagnetic field sensors. As shown in Figure 4.32, the device under test is placed in the approximate center of a referenced ground plane at a height of 10 cm and is powered and operated in a normal configuration. The magnetic field is increased to 10 A/m (approximately 125 mG) and is applied to three axes of the device under test. The field is maintained for a period of approximately 5 minutes for each of three axes, while the device under test is monitored for any indication of erratic operation.

As shown in Figure 4.33, a test system for design-time evaluation is easy to build. Use four pieces of 92-cm-long $\frac{3}{4}$-in. PVC pipe, four $\frac{3}{4}$-in. PVC pipe elbows, and one $\frac{3}{4}$-in. PVC pipe tee to construct the 1 m × 1 m frame for the current loop. Thread No. 18 insulated copper through the loop to form two complete turns. Solder a flexible twisted-pair cable to the loop wires where they exit the pipe. R1 will be used to monitor the RMS current flowing through the coil. The coil is powered by a transformer rated at 24 V at 10 A, which is in turn powered from the power line through a variac.

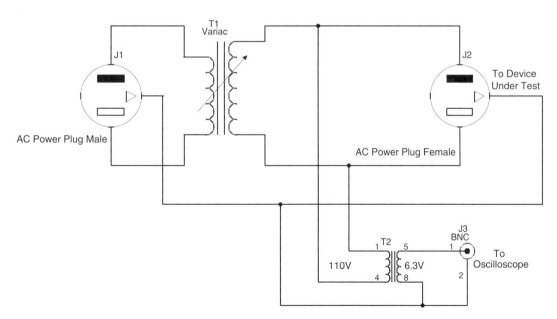

Figure 4.31   A simple way of simulating power line voltage dips and interruptions is to use a variac. Although timing an exact 10 ms for a dip is not an easy manual task, good approximations for these dips and interruptions can be achieved when the variac's output is monitored with an oscilloscope. A small filament transformer provides appropriate isolation for connection with a grounded oscilloscope.

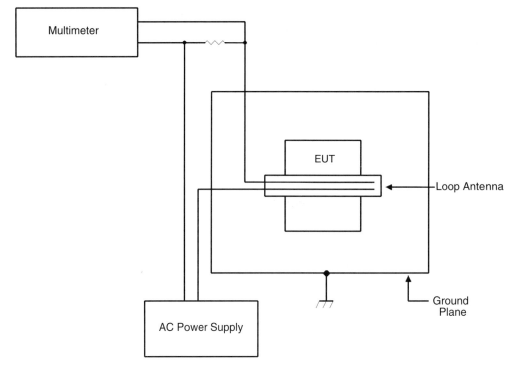

Figure 4.32   Setup for assessing the susceptibility of a medical device to magnetic fields. The magnetic field applied to three axes of the device under test is 10 A/m. This test simulates interference that may be caused on a device by low-frequency magnetic fields, such as those generated by the power lines.

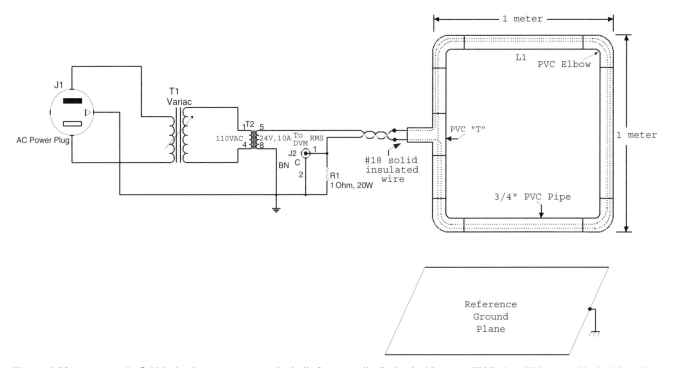

Figure 4.33   A magnetic-field induction generator can be built from a coil of wire inside some PVC pipe. This assembly is driven by a variac–transformer combination to produce the desired magnetic field at the center of the coil.

To use the test system, increase the variac's output until you read 5.55 V$_{RMS}$ across R1. At this point, 4.5 A should be circulating through the two loops of L1. Correcting for the difference in magnetic field produced by a square loop versus that of a circular loop, and applying the Biot–Savart law, the magnetic field at the center of the loop should be $(4.5\,A \times 2)/0.9\,m = 10\,A/m$.

GOOD DESIGN PRACTICES, REMEDIES, AND DUCT TAPE

No cure for EMC problems is better than prevention. Trouble avoidance in EMC is accomplished by considering the emissions and susceptibility aspects of EMC at every stage of the design process. The following important questions must be part of the circuit design, selection of components, and packaging:

- Will this part of the design generate or be susceptible to interference?
- What are the characteristics of the interference?
- At what frequency or frequencies does it occur?
- From where is it most likely to originate?
- Which radiated and/or conducted path(s) can the interference take from source to victim?

Once potential sources of interference are identified, you must decide what to do to reduce their impact. There are four broad solutions to an individual EMC problem:

1. *Prevention*. Eliminate the sources of potential interference.
2. *Reflection*. Keep internally generated signals inside the device and keep external interference outside the device's enclosure.
3. *Absorption*. Use filter networks and filtering materials to absorb interfering signals.
4. *Conduction*. Divert interfering signals to the device's RF ground.

Fortunately, most of the rules and perils are known in the war against EMI. Designing an instrument to pass EMC testing is, in all likelihood, all that will be needed to ensure proper performance under real-world situations. Avoid overdesign. The authors are not aware of a single medical device malfunction attributed to interference by unknown UFO radiation. All you need to do is figure out the potential level of interference that you may encounter, and design within these limits.

Kendall [1998] proposed a simple way of estimating the amount of protection that may be needed in a medical device to counteract an EMI threat. His step-by-step procedure demonstrates how to estimate the protection that needs to be incorporated in the design of an analog comparator with 5-mV sensitivity.

1. Start by identifying the RF threat level. For example, if the applicable standard for your device establishes immunity against radiated interference at 3 V/m, use this level for your calculations.
2. Multiply the threat level by the field uniformity of the test chamber in which the device will be exposed to EMI. A factor of 2 is appropriate for ferrite-lined chambers, while a factor of 4 is typically used for semianechoic chambers. Assuming a ferrite-lined chamber, the uniform field will be $2 \times 3(V/m) = 6(V/m) = 136(dB\mu V/m)$.
3. Account for losses between the source and the victim. A minimum theoretical loss of $-14\,dB$ would happen in the case in which the source and the victim are both

connected to perfectly tuned dipole antennas. However, more severe losses can be assumed for less perfect situations, such as when the length of the signal line under consideration is much shorter than one-fourth wavelength of the offending spectral component for which the threat analysis is conducted. Assume for this example that the line of interest presents an impedance of $100\,\Omega$ at the frequency of interest and that the coupling between the offending source and the signal line under analysis is $10\,dB$ below ideal.

4. Calculate the coefficient to be used for the offending signal at the victim circuit. In the example $136(dB) - 14(dB) - 10(dB)$ yields $112\,dB\mu V$ across the signal line's load impedance of $100\,\Omega$.

5. Convert this level back to linear units: $112\,dB\mu V = 400\,mV$. This will be the voltage induced by the offending source at the frequency of interest on the signal line under analysis. Note that if the load impedance increases, so will the induced voltage. For example, for a 100-kΩ load, the induced voltage will be as high as $2\,V$.

6. Compare the induced voltage levels against typical circuit threshold values at all susceptible frequencies. For example, if $26\,MHz$ is within the bandwidth of the processing circuit connected to the line under analysis, and since the threshold value for this example was chosen to be $5\,mV$, the protection level required for a load impedance of $100\,\Omega$ would be $20\log(400\,mV/5\,mV) = 38\,dB$. For a 100-k$\Omega$ load impedance, the protection need would increase to $52\,dB$.

With this approximation in hand, it is possible to select shielding, grounding, and filtering components that will afford a combined protection that surpasses the estimate by a certain safety margin.

Shields Up!

Shielding and grounding (reflection and conduction) are the primary methods of guarding against EMI entry and exit to and from a circuit. Chances are that you will not build your own enclosure. Rather, you will probably use an off-the-shelf case or hire an enclosure manufacturer to supply you with custom-made enclosures. In either case, look at the enclosure's data sheets for EMC specifications. The authors' preference is to use enclosures which have a conductive cage that is contained completely inside a plastic enclosure without any exposed metallic parts.

If a conductive enclosure is chosen, ensure that the conductive surface is as electrically continuous as possible. For a split enclosure, ensure as good an electrical contact as possible between the parts. Openings in the case that are required for display windows, cooling slots, and so on, must be kept as small as possible. If the size of the opening is larger than $\frac{1}{20}$ of a potential offending EMI component, use transparent grilles to close the RF gap. Finally, ensure that unshielded lines that carry offending signals do not pass directly through a shielded enclosure. Use shielded cables for high-sensitivity inputs.

EMI grounding requires different, sometimes conflicting considerations from those used to protect low-frequency low-level signal lines. The first difference is the issue of single-point versus distributed grounding. Single-point grounding of circuits is a common practice in the design of low-noise electronic circuits because it eliminates ground loops. This assumption is valid only up to a few megahertz. At higher frequencies in the radio spectrum, line inductances and parasitic capacitances become significant elements, voiding the effectiveness of single-point grounding. For example, for the 300-MHz components of an ESD event, a 0.25-cm length of wire or PCB track acts as a one-fourth wavelength antenna, providing maximum voltage at the ungrounded end. As such, any cable that is longer than $\frac{1}{10}$ to $\frac{1}{20}$ of offending spectral components should be grounded at both ends. If this poses a ground-loop problem for low-frequency signals, one end can be coupled to ground through a 0.01-µF capacitor.

Whenever possible, the shield of external cables should be properly terminated to the equipment enclosure. Poor termination, which may be imposed by leakage current and isolation requirements, may result in capacitive coupling of EMI to signal lines. So, by all means, and as long as isolation and leakage requirements permit, bond the cable shield directly to the device's conductive enclosure. Contrary to the suggestions above, when the potential problem is ESD, the effective solution is not to shield with a conductive layer but rather, to insulate. By not allowing an ESD spark to occur at all, there are no bursts of electric and magnetic fields to radiate EMI.

For this purpose, plastic enclosures, plastic knobs and switch caps, membrane keyboards, plastic display windows, and molded lampholders help eliminate ESD discharge points. As a rule of thumb, a 1-mm thickness of PVC, ABS, polyester, or polycarbonate suffices to protect from 8-kV ESD events. The area protected by a nonconductive cover is more difficult to assess because surface contamination by fingerprints and dust attract moisture from the air to form a somewhat conductivity paths through which ESD can creep. During 8-kV testing, an ESD gun can produce sparks that follow random paths over a supposedly nonconductive surface all the way to a metallic part 5 cm away. The same happens on metallic surfaces painted with nonconductive paint, where surface sparks seek pinhole defects on the paint.

A very common design mistake is to assume that 15-kV-rated insulation on LCD displays, membrane keypads, potentiometers, and switches is sufficient to protect circuitry connected to these components. The problem is that although ESD won't go through the insulation, it will creep to the edges of the insulation and hit wiring on the edges of these components. As such, extend the dielectric protection of panel-mounted controls to prevent or at least divert ESD currents from reaching vulnerable internal circuits.

The Real Bandwidth of Signal Lines

Protecting medical devices from EMI is especially difficult because it often involves sensitive electronics that can pick up and demodulate RFI. Interfering signals can be recognized as real features of physiological signals, leading to potentially serious risks to patients and health-care providers. Take, for example, the polling of a cellular phone, which happens at a frequency close to that of the heart's normal rhythm. If detected and interpreted incorrectly, a pacemaker could assume that the EMI bursts are really the heart beating at an appropriate rhythm, causing it to inhibit the delivery of pacing therapy.

A common mistake in the design of medical instrumentation, especially of biopotential amplification and processing stages is to assume that the RF bandwidth of the circuit is limited to the intended operational bandwidth. The limited bandwidth of an op-amp or of a low-pass filter intended to limit the bandwidth of biopotential signals will do little to prohibit pickup and demodulation of RF signals. If not controlled, RFI can easily induce volt-level RF currents in biopotential amplifiers designed to detect micro- or millivolt-level signals. These RF currents will surely find nonlinear paths (e.g., zener protection diodes and parasitic diodes) that demodulate them, yielding high-level in-band signals that obscure, if not completely swamp, real biopotentials.

Besides using proper shielding, one effective solution is to place RF filters on every line connected to the outside world, especially those that convey low-level signals from patient sensors and electrodes to high-input-impedance analog circuits. Figure 4.34 shows the input filters placed in immediate proximity of the signal-input connector of an amplifier used to record intracardiac electrograms. In this circuit, individual NFM51R00P106 single-line chip filters made by muRata are used to shunt RF signals to the isolated ground plane without affecting low-frequency signals. These filters have a nominal −3-dB cutoff frequency of 10 MHz, yielding a minimum attenuation of 5 dB at 20 MHz, 25 dB at

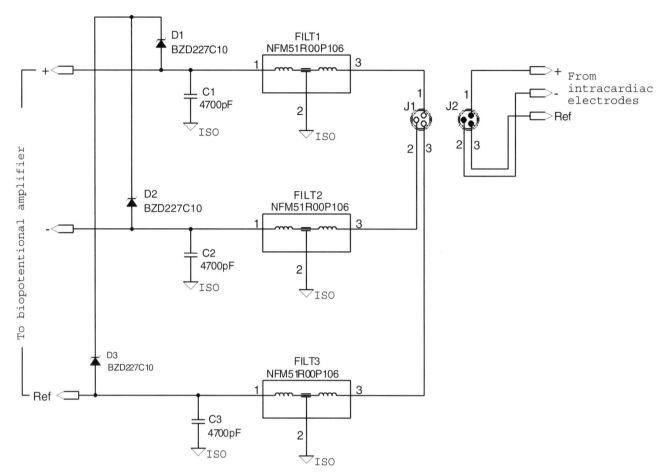

Figure 4.34 RF filters are often needed on every line connected to the outside world. In this intracardiac electrogram signal input circuit, individual single-line chip filters are used to shunt RF signals to the isolated ground plane without affecting low-frequency signals. These filters have a nominal −3-dB cutoff frequency of 10 MHz, yielding a minimum attenuation of 5 dB at 20 MHz, 25 dB at 100 MHz, and 30 dB above 500 MHz. 4700-pF capacitors shunt RF signals that may be induced by electrosurgery. Zener diodes with their cathodes connected to a common point to limit surge voltages that may be caused by defibrillation.

100 MHz, and 30 dB above 500 MHz. The 4700 pF following these filters is used to shunt RF signals that may be induced by electrosurgery tools. The value of 4700 pF was selected for this design to reduce the possibility that electrosurgery currents circulating through the electrodes in contact with the heart would reach a level capable of ablating myocardial tissue. Finally, zener diodes with their cathodes connected to a common point are used to limit the voltage imposed on the inputs of the amplifier by high-voltage transients such as defibrillation.

It is a good idea to filter every line connected to the outside world, even those that carry high-level signals over low-impedance paths (e.g., high-level analog or digital outputs) because if left unfiltered, that line can act as an antenna picking up EMI from outside the device and reradiating it within the device's enclosure. One of the most sensitive and often disregarded lines in a medical device is the reset line. This is because EMI or ESD coupling into a reset line can cause a reset, often leading to temporary interruption of service, which may leave a patient without the support of a life-sustaining therapy until function is restored. For this reason, reset lines and components that can evoke reset events (e.g., reset switches, microprocessor watchdogs, and power supply supervisors) must be thoroughly shielded and decoupled.

Although designers usually protect power supplies from the high-voltage transients that may be encountered under the conditions simulated by EFT and high-energy surges, the EMI component of these events is often forgotten. High-voltage surges may bypass the power supply and associated filters completely and attack the device's circuitry directly. The same filtering recommendations apply then to protecting sensitive analog and digital circuits against the EMI produced by power line surges.

Paying Attention to PCB Layout

Although in earlier sections we advocated maintaining clock speeds low, as well as rise and fall times as slow as possible, some medical device applications really demand lightning-fast processing. Let's digress and imagine what it would be like to own a Ferrari F40, capable of achieving a speed of 200+ mph!—its magnificent turbo-charged V-12 engine purring while cruising down the road at a speed at the limit of human reflexes. Waking up to reality, though, you would seldom (if ever) be able to floor the gas pedal of this marvel. Even if disregarding the legal limit, our roads are just not designed to support much more than half the maximum speed of a loaded sports car. The awesome power of sports engines can be let loose only in special race tracks, constructed with the right materials and slants.

Although you may not consider adding a Ferrari to your estate at this moment, its power does relate to the topic of this chapter, as you are probably using increasingly fast logic and microprocessors in your projects. However, in close resemblance to the sports car analogy, very high bus speeds result in interconnection delays within the same order of magnitude as on-chip gate delays, and for this reason typical PCB design, which considers traces as low-frequency conductors rather than as high-frequency transmission lines, will ensure that such a project turns into a very impressive and expensive paperweight.

Some 20 years ago, while some of us were building microcomputers with 2-MHz Z80s, 8080s, and CDP1802s, engineers designing with ECL technology already faced problems related to the implementation of printed circuit boards, backplanes, and wiring for high-speed logic circuits. Today, however, multihundred megahertz and even gigahertz buses are commonplace, and we face strict regulations on the RF emissions escaping from such wideband sources. For this reason, we should all acknowledge that the utopian idea that digital signals behave as ones and zeros must be replaced by a more realistic approach that involves RF transmission line theory. Through this new approach, printed circuits are designed to convey pulse transmissions with minimal distortion through channels of appropriate bandwidth—no quasi-dc signals anymore! Interestingly, the same PCB layout practices that are useful in the design of high-speed circuitry apply to the design of circuitry with increased immunity against EMI.

Transmission Line Model of PCB Track

PCB design for high-speed logic and RFI immunity demands the use of power and ground planes, and plain double-sided PCBs are not recommended. In the former, a surface stripline track such as that depicted in Figure 4.35 will have an impedance Z_t given by

$$Z_t = \frac{87}{\sqrt{\varepsilon + 1.41}} \ln \frac{5.98h}{0.8w_t + t_t}$$

where ε is the dielectric constant of the PCB dielectric, h the height of the track above the ground or power plane, and w_t and t_t the width and thickness of the track, respectively. A PCB track buried within the fiberglass–epoxy laminate will have its impedance reduced by about 20% compared with that of a surface track.

This PCB track can be modeled as a transmission line [Magid, 1972], and a short pulse applied to one end of this transmission line will appear on the other side, supplying the

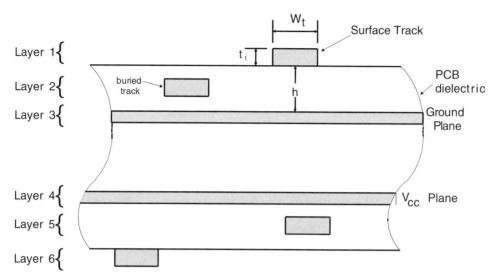

Figure 4.35   The transmission line impedance of a stripline PCB track is affected by its position relative to the ground or power plane as well as by its geometry and by the dielectric constants of the board and the surrounding medium.

load impedance Z_l with a distorted version of the pulse, and presenting an effective delay τ. On a typical surface PCB track, the pulse conduction velocity is approximately 0.15 ns/in. = 0.06 ns/cm, so $\tau = 0.15 l_t$(ns) represents the total delay caused by a track of length l_t (measured in inches).

If the load impedance does not match the track's impedance perfectly, a part of the arriving signal will be reflected back into the transmission line. In general, pulse reflection occurs whenever transmission lines with different impedances are interconnected or when a discontinuity occurs in a single transmission line. For a connection between two transmission lines of impedances Z_0 and Z_1, the reflected voltage V_r is related to the incident voltage V_i through

$$V_r = \frac{Z_1 - Z_0}{Z_1 + Z_0} V_i$$

The ratio $\Gamma = V_r/V_i$, called the *reflection coefficient*, describes what portion of the pulse incident from Z_0 on Z_1 will be reflected back into Z_0. Γ, V_i, and V_r are usually complex quantities because they deal with both the magnitude and phase of the signals that travel along transmission lines.

Using tracks buried within the PCB between the power and ground planes is an effective way of controlling RFI and ESD threats. In contrast with one- or two-sided PCBs, multilayered PCBs with ground and power planes keep ground impedances and loop lengths sufficiently low to avoid circuit tracks from picking up significant levels of RFI. Typically, PCBs with signal tracks sandwiched between ground and power planes are a full order of magnitude less sensitive to RFI than are well-designed two-sided boards.

For nonmedical equipment, the power and ground planes are usually grounded to the chassis. In medical equipment, however, it is usually the most sensitive circuits that need to float, making it impossible to apply classical EMC techniques for proper signal-path decoupling and shielding. For floating circuits that end up being sensitive to EMI, try to form a capacitive grounding path between the power and ground planes of the PCB and the chassis. Of course, the level of coupling must fall well within the amount of enclosure leakage permitted for the floating part.

Pulse Reflection and Termination Techniques

In a typical circuit, a driving logic element and a receiving logic element are connected by a PCB track. In the equivalent circuit shown in Figure 4.36, a pulse with amplitude V_d is injected by the driver logic element, which presents an output impedance Z_d into a PCB track of length l_t and impedance Z_t. The pulse carried by the PCB track is then presented to the input impedance Z_r of the receiving element. If we suppose that a $Z_t = 100\,\Omega$ track carries the pulse, and after looking at the data sheets for a selected 5-V family of high-speed logic, we find that the output and input impedances at our frequency of interest are $Z_d = 50\,\Omega$ and $Z_r = 10\,k\Omega$, respectively, then upon reaching the receiver, a reflected pulse starts traveling back toward the driver with amplitude that can be approximated by

$$V_r = \frac{Z_r - Z_t}{Z_r + Z_t}\,V_d = \frac{10{,}000 - 100}{10{,}000 + 100}(5\,V) = 4.9\,V$$

which assumes a negligible attenuation of the pulse throughout its conduction, and which takes into consideration only the real parts of the variables. This reflected pulse will be rereflected back toward the receiver upon hitting the driver with an amplitude approximated by

$$V_r = \frac{Z_d - Z_t}{Z_d + Z_t}\,V_r = \frac{50 - 100}{50 + 100}(4.9\,V) = -1.63\,V$$

This negative signal will interact with the original incident pulse with a delay equivalent to the time it takes for the pulse to travel back and forth along the track $\tau(ns) = 0.31 t_r$.

Depending on the length of the PCB track, the -1.63-V reflection could distort the leading edge of the pulse so much that it will cause the false detection of a logic-low (Figure 4.37). A different combination of impedances could have caused the reflected pulse

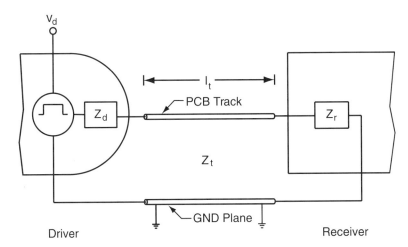

Figure 4.36 The output of a logic element connected through a PCB track to the input of another logic element can be modeled as an ideal voltage step generator that drives a transmission line of impedance Z_t through an output impedance Z_d. The transmission line is then terminated by the receiver's input impedance Z_r.

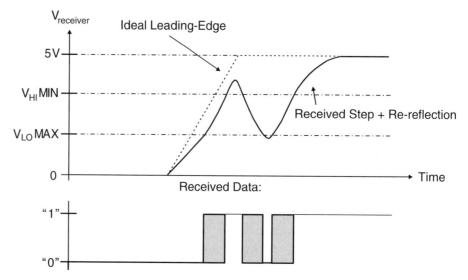

Figure 4.37   A critical length of PCB track could cause the rereflected pulse to distort the leading edge of the pulse so much that it will cause, in this case, the false detection of a logic-low.

to be positive, possibly causing the false detection of a logic-high, or the false activation of an edge-sensitive device. Moreover, a reflected pulse presented to the receiver will cause yet another reflected pulse, which although with far less amplitude, may still be able to cause erroneous operation of a circuit.

Obviously, the solution to the reflected-pulse problem is to match the impedances in the best possible way. This design procedure, called *transmission line termination*, can be accomplished in four different ways: series, parallel, Thévenin, and ac, as shown in Figure 4.38. Series termination is recommended whenever $Z_d < Z_t$ and the line is driving a reduced number of receivers. This technique, which gives good results in most high-speed TTL circuits, consumes negligible power and requires the addition of only one resistor, the value of which is given by

$$R_{term} = Z_t - Z_d$$

The major drawback of the series termination technique as far as logic signal integrity goes is that it increases signal rise and fall times. However, the same is a blessing as far as reducing electromagnetic emissions.

In contrast to series termination, which eliminates pulse reflection at the driver end, all other techniques eliminate reflection at the receiver end of the PCB track. Parallel termination $R_{term} = Z_t$ as well as Thévenin termination $R_{term} = 2Z_t$ techniques consume large amounts of power; however, they provide very clean signals. Ac termination $R_{term} = Z_t$, which uses a small capacitor to couple only ac components to ground, is not as power hungry as the preceding methods but adds capacitive load to the driver and increases the time delay due to its inherent RC constant.

Parallel Path Skew and Track Length Equalization

Parallel transmission over data and address buses requires that all signals arrive at their destination concurrently. Often, however, pulses sent down parallel paths do not arrive at the same time because of differences in the length of these paths. As shown in Figure 4.39,

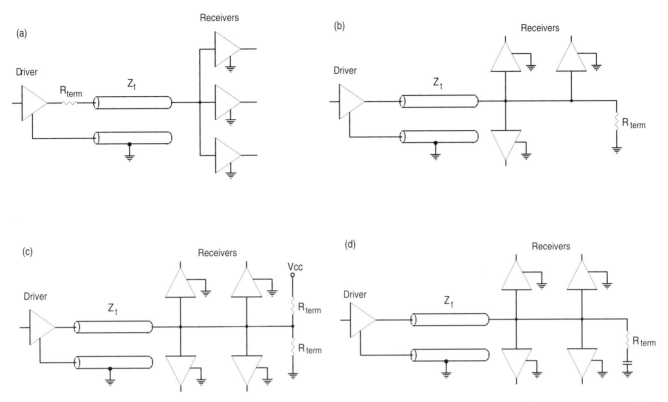

Figure 4.38 Transmission line termination techniques: (*a*) series termination; (*b*) parallel termination; (*c*) Thévenin termination; (*d*) ac termination.

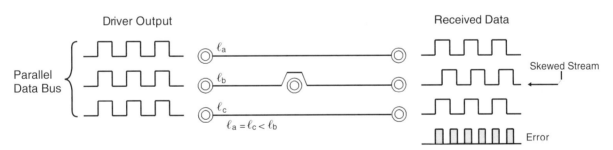

Figure 4.39 Unequal lengths of PCB track on a parallel bus will cause skew between the pulse streams, leading to reception errors.

the skew induced in bit sequences sent along parallel paths of different lengths can cause errors in the communication between circuits, especially when transmitter and receiver circuits are placed in different boards interconnected through backplanes or ribbon cables. The obvious solution is to keep parallel paths as short as possible and ensure equal PCB track lengths for all parallel paths.

Skew also deserves very serious consideration in the design of high-speed microprocessor clock distribution networks. In general, all logic computation during a single clock cycle has to be performed within the very short time left over by the delays suffered by logic signals during that clock period: path delays, setup and latchup times, logic gate propagation delays, and skew. As clock frequencies increase, the time left over for logical computation decreases, up to the point that skew often causes system failures due to incomplete processing during a clock cycle.

For this reason, PCB tracks that distribute the clock must be tuned so that the delay from the clock driver to each load is the same. Whenever possible, the loading on each track carrying the clock should be the same, and in this case skew is minimized by making all tracks the same length. For unbalanced loads, delay times can be tuned through *RC* terminations or by careful adjustment of the track lengths.

Crosstalk and Vulnerable Paths

Crosstalk is the noise induced into a track by the presence of a pulse stream in an adjacent track. In essence, crosstalk is EMI caused by the product on itself. The amount of crosstalk is affected by track spacing, routing, signal direction, and grounding. The major problem with crosstalk arises when the voltages induced on a quiet line are sufficient to be detected as a change in logic state by the receivers of that line. In high-speed systems, the capacitive and inductive coupling between lines is considerable, and crosstalk must be reduced through appropriate design.

First, proper transmission line termination reduces the amount of radiated energy from a driven track, and spurious emissions that nevertheless escape can be shielded through the use of grounded guards. This design consideration is particularly important for lines driven with high-voltage, high-current, and high-frequency signals. Floating lines connected to high-impedance receivers are notably sensitive to crosstalk, and proper shielding, as well as maintaining them at a distance from possible radiating tracks, must be ensured. In addition, it is possible to see from transmission line theory that crosstalk between two adjacent tracks is minimized if the two signals flow in the same direction.

The analysis should be extended to identifying potential coupling paths between signal lines and RFI sources (including ESD) and then taking steps to minimize them through proper placement of PCB tracks and components. For example, shields can be reinforced where transformers and heat sinks are placed, the areas of loops formed by PCB tracks should be minimized, and magnetic coupling paths should be oriented orthogonally.

Finally, remember that components, connectors, and mounting parts that can be accessed from the outer world are very often the paths of entry into a device's circuit for EMI and especially for ESD. Even an exposed metallic screw on an otherwise insulating panel can make it possible for unwanted signals to get into the circuit and cause interference. Common panel-mounted vulnerable parts include membrane keyboards, LEDs, potentiometers, connectors, and switches, together with their mounting hardware.

Analysis of Circuit Board Performance

Although you may consider such tools as a time-domain reflectometer or an RF network analyzer as belonging strictly to a communications lab, these can aid considerably in the design of circuit boards for high-speed and high-immunity applications. These tools are capable of measuring the actual impedances, time delays, and complex reflection coefficients of a circuit. These measurements often show that calculations of these parameters result in very crude estimates that have to be improved on for good circuit performance. In most cases, the iterative process of design will require building and evaluating a test board to determine if the original design considerations were effective. This test board is usually not populated with the actual active components, but the PCB tracks, passive components, sockets, and connectors, as well as the terminated dummy IC packages, form a network of transmission lines that can be analyzed with confidence.

The time-domain reflectometer (TDR) [Strassberg, 1993] of Figure 4.40, which is often used in the troubleshooting of LANs, injects a very sharp pulse V_i into the transmission line under analysis. Then an oscilloscope or a computer fitted with a high-speed A/D receives the reflected pulses. The time delay and shape of the reflected pulses contain the information required to estimate the impedance of the line and its termination (Figure 4.37).

In contrast to TDR, network analyzers such as that in Figure 4.41 operate in the frequency domain and enable exact measurement of the complex reflection coefficient as a function of frequency, measurement of crosstalk between lines, and measurement of phase skew between signals [Montgomery, 1982]. A network analyzer can also be used to identify tracks on which resonance problems could arise and to perform objective crosstalk measurements.

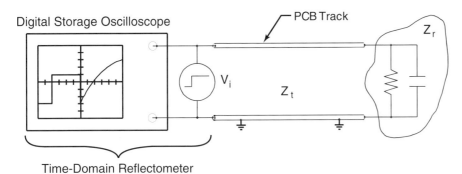

Figure 4.40   In a time-domain reflectometer, a voltage step with very short rise time is injected into a transmission line. After a certain delay, a reflected pulse adds up to the step. The timing and wave-shape of the reflected pulse contain information regarding the characteristics of the transmission line and of the termination.

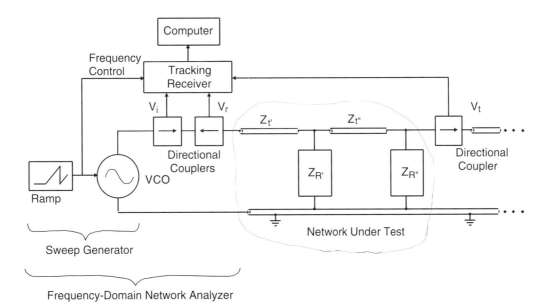

Figure 4.41   In a frequency-domain RF network analyzer, a sweeping sinusoidal signal is injected into the input of the transmission line leading to the subnetwork under test. Directional couplers feed a synchronized RF receiver with samples of the incident, reflected, and transmitted portions of the signal. A computer is used to calculate and display the complex reflection and transmission functions, as well as other relevant parameters.

In fact, ignoring capacitor and inductor resonance is the most common mistake in designing circuits with good EMI/EMC characteristics. Parasitic inductance and capacitance cause components and PCB tracks to resonate, often at frequencies as low as 1 to 30 MHz. For example, many 10-nF decoupling capacitors with $\frac{1}{2}$-in. leads resonate at around 20 MHz. Above the resonant point, capacitors act as inductors and inductors act as capacitors, which typically render filters useless against high frequencies.

Regardless of the usefulness of these RF test tools, their price puts them out of reach for most electronics hobbyists and small engineering firms. But don't be discouraged: Building successful high-immunity and high-speed circuits on a budget is possible by adopting conservative design policies. As you may realize by now, an analog circuit simulator could be as helpful as a digital circuit simulator in the design of your next circuit. For critical circuits, do not assume ideal capacitors and inductors, but rather, consider the RF characteristics of these components in your analysis.

If possible, use ceramic surface-mounted capacitors and short, fat PCB tracks to keep inductance low. As far as inductors are concerned, ferrites perform better than wire-wound types. This is because ferrites have high resonant frequencies and absorb large amounts of energy at resonance. If wire-wound inductors are nevertheless required, place small ferrites in series with the inductor (e.g., mount small ferrite beads on the inductor's leads) to protect the circuit over a wider bandwidth. Of course, make sure that the magnetic flux created by currents flowing through wire-wound inductors does not couple to adjacent components and PCB traces.

Duct Tape and Test-Time Bandages

Despite thorough engineering, surprises do happen during compliance testing. Inevitably, while you are still giving the last touches to your design, the marketing department has already sold a few dozen units to their most prestigious customers and eagerly awaits getting their hands on the very prototype you are testing for the next trade show. "This is not good," you think. Next comes the calculation of how long you can live on your credit cards before finding another job in a distant corner of the world where no one would have heard about your mistake.

In reality, vulnerabilities in well-designed devices can often be patched up at the test site. Understand, however, that any modifications you make to pass the tests will have to be implemented in the actual product. Don't go overboard implementing every possible solution at once. Although adding a capacitor here, a ferrite there, and some shielding to a cable may sound trivial, each of these modifications can turn into a logistic nightmare during production. Moreover, unbudgeted additions multiplied by the number of units to be produced over time may end up reducing the profit margin to intolerable levels.

In addition, and before applying any corrective measures to a medical device prototype, however, make sure that by filtering or shielding a specific line, component, or assembly, you do not violate insulation or leakage requirements. With that said, let's look at the essential elements of a first-aid kit.

Ferrites   Ferrites are the first thing most people use to try to clean the signals from a suspect line. A huge variety of ferrites are available; they come in all sizes and provide attenuation at diverse frequency ranges. Ferrites also come with various installation options: Some require you to pass the cable or lead component through them, while some have clamps that make it easy to retrofit equipment without disconnecting wires. Ferrites are also available to specifically treat differential or common-mode problems. Ferrite diagnostic kits are available from various vendors (e.g., Fair-Rite), and test houses usually have plenty of ferrite beads and chokes available for their customers.

Ferrite is even available in tape form (actually, it is a ferrite-coated tape) and can be applied to PCBs and as a shield on the chassis, enclosure, or other surfaces. Before buying tape specifically for this application, try the magnetic backing sold in arts and craft stores to make refrigerator magnets out of photographs. It works wonders!

Connector Filters   If a signal line on a connector is a suspect for noncompliance, the easiest and cleanest way to apply in-line filtering is either to exchange the original connector for a filtered one or to insert a filtered adapter between the original plug and its target socket. D-type filtered connectors and filtering adaptors are widely available and come with a feedthrough capacitor in the range 50 to 2000 pF with or without ferrite inductors. You should also consider using ferrite plates with holes to match your connectors' pinout configuration.

Connector Shields   Very often, shielded wires are used to protect sensitive or noisy lines, only to terminate at an unshielded connector. Connector backshields can easily be retrofitted onto many cables. Make sure, however, that by connecting the cable's shield to an exposed metallic connector you do not violate insulation requirements.

Capacitors and In-Line Filters   Small capacitors can be tacked on to suspect lines. These are especially useful when dealing with problems that may be occurring on a PCB. It is always a good idea to take an assortment of pF- and nF-range chip capacitors to the test house. If additional filtering is needed, some surgery can be done to a PCB to retrofit in-line filter modules. Beware, however, that the addition of capacitors to the circuit of a medical device may change its leakage characteristics, causing the device to fail leakage and/or hipot safety tests.

Cable Shielding   Cable shielding is the next step in troubleshooting and fixing EMC problems. Mesh with and without zip-on sheaths, as well as conductive foils with or without adhesive backing, are available and can be applied to cables with ease. You will need to decide how thick a shield you use and how you connect the shield to ground.

Enclosure Gaskets   Metallic enclosures are often assumed to be bulletproof barriers against EMI, both incoming as well as outgoing. However, no instrument is perfectly sealed, since cables, displays, and controls couple the inside of the instrument to the outside environment. In fact, a metallic case can sometimes act as a resonator, guiding EMI to (or from) the vulnerable (or offensive) circuit.

Assuming that filtering and shielding of cables and controls has not sufficed to control an EMI problem, the next step is looking for gaps in the enclosure which may require shielding. Here, you can use metallic foils and tapes to improve contact along enclosure seams and determine if a permanent solution could be achieved through the use of conductive EMC gaskets. In addition, temporary application of conductive foils to display windows and ventilation apertures is a very useful diagnostic to determine if conductive transparent screens (e.g., very fine wire mesh) need to be applied to these openings. Many of these shielding materials are available from Chomerics.

If the problem is ESD, consider even tiny gaps or joints in enclosure shields which are weak spots, because they divert very large fast currents as they flow around the enclosure, causing current density hot spots that emit strong EMI through the shield and into the enclosure. In fact, for the very high frequency components of ESD pulses, gaps and joints may act as slot antennas that help get EMI into the enclosure.

On a different front, consider that metallic shields can sometimes be completely transparent to offensive fields that have a dominant magnetic components. In these cases, vulnerable parts of the metallic enclosure may require further shielding with a material that provides a low-reluctance magnetic path for the interference field. The idea is to make the shield attract flux lines to itself and divert the magnetic field away from the sensitive component. The magnetic sheets mentioned above are a good start for solving these problems. If nonmagnetic shields are needed (e.g., to form a shield close to a CRT or a magnetic sensor), you may try one of the shielding alloys produced by Magnetic Shield Corporation. Magnetic Shield sells a $150 engineering kit that includes various 10 in. × 15 in. sheets as well as some braided sleeving made of their CO-NETIC and NETIC alloys. The kit even includes an ac magnetic field probe that can be used with a DVM or oscilloscope to measure magnetic fields from 10 Hz to 3 kHz.

Conductive Spray Paint   Many medical devices are not built with metallic enclosures. If extensive shielding of the case becomes necessary, an alternative to changing the design to use a conductive enclosure is to spray-paint the enclosure using conductive paint. EMI spray paints are available to provide varying degrees of EMI shielding, all the way from light, graphite-based paints to provide mild shielding against ESD through nickel/chrome-loaded sprays that can divert strong magnetic fields away from sensitive components.

Shielding Components   If you got this far down the list of quick patches, chances are that you may need to go back to the drawing board (or more likely, the PCB layout station). Before going back home, though, you may try to shield individual components. You could apply one of various available conductive foils and tapes directly to PCBs (to shield tracks) or to components. There are even precut conductive cardboard boxes that can be used to shield entire sections of a circuit. However, if you need to build a complete village of protective housings on your PCB, it may be worthwhile biting the bullet and going back to the lab to reengineer the product.

CONCLUDING REMARKS

Designing medical equipment that can pass EMI/EMC compliance testing without fixes or delays never happens by mistake. Rather, it involves considering compliance with EMI and EMC regulations from the very beginning of product formulation and design. A head start in the battle against EMI can be obtained by developing a first prototype free of foreseeable trouble. This is possible by carefully selecting the technologies that fulfill the product requirements while minimizing EMI, observing good design and construction practices, and making extensive use of circuit simulation tools. Near-field probing of the first prototype should reveal real-world EMI effects that escaped from the limited view of initial modeling. In addition, RF techniques prove to be essential in the design of circuits that can exploit the power of modern high-speed processors. Correcting any problems through filtering, shielding, or redesign is still inexpensive at this early stage in the design, and the second prototype will already have a good chance of passing compliance testing with minimal rework.

In this chapter we have presented only a few of the ways in which technology selection, circuit design, and layout techniques influence the generation of and immunity against EMI. A discussion of detailed techniques to control EMI by virtue of good design is beyond the scope of this book. However, many books and articles have been published that disclose the secrets of the EMI/EMC world, all the way from Maxwell's equations, through the legalities of regulation, into the tricks of the trade for taming EMI [Mardiguian, 1992; Marshman, 1992; Williams, 1999]. Considering the stiff economical, technical, and legal penalties brought by manufacturing a medical product that does not comply with EMI and EMC regulations, you should be motivated to keep EMC in sight at every turn of the design process.

REFERENCES

Dash, G., and I. Strauss, Inside Part 15—Digital Device Approval, *Compliance Engineering, 1995 Annual Reference Guide*, A11–A18.

EN-55011, *Limits and Methods of Measurement of Radio Disturbance Characteristics of Industrial, Scientific and Medical Radio Frequency Equipment*, 1998.

EN-60601-1-2, *Medical Electrical Equipment—Part 1: General Requirements for Safety*; Section 2: Collateral Standard: Electromagnetic Compatibility—Requirements and Tests, 1998.

EN-61000-4-3, *Electromagnetic Compatibility—Part 4: Testing and Measurement Techniques*; Section 3: Radiated, Radio-Frequency, Electromagnetic Field Immunity Test, 1995.

EN-61000-4-4, *Electromagnetic Compatibility—Part 4: Testing and Measurement Techniques*; Section 4: Electrical Fast Transient/Burst Immunity Test Basic Publication, 1995.

EN-61000-4-5, *Electromagnetic Compatibility—Part 4: Testing and Measurement Techniques*; Section 5: Surge Immunity Test, 1995.

EN-61000-4-6, *Electromagnetic Compatibility—Part 4: Testing and Measurement Techniques*; Section 6: Immunity to Conducted Disturbances, Induced by Radio-Frequency Fields, 1996.

EN-61000-4-8, *Electromagnetic Compatibility—Part 4: Testing and Measurement Techniques*; Section 8: Power Frequency Magnetic Field Immunity Test Basic, EMC Publication, 1993.

EN-61000-4-11, *Electromagnetic Compatibility—Part 4: Testing and Measuring Techniques*; Section 11: Voltage Dips, Short Interruptions and Voltage Variation Immunity Tests, 1993.

Gubisch, R. W., The European Union's EMC Directive, *Compliance Engineering, 1995 Annual Reference Guide*, A55–A64.

Guettler, P., Disturbance Simulator Checks Lines, *EDN*, 152, October 14, 1999.

IEC-60601-1-2, *Medical Electrical Equipment—Part 1: General Requirements for Safety*; Section 2: Collateral Standard: Electromagnetic Compatibility Requirements and Tests, 1993.

Kendall, C. M., Protecting Circuits against Electromagnetic Threats, *Medical Device and Diagnostic Industry*, 78–82, March 1998.

Kraz, V., Near-Field Methods of Locating EMI Sources, *Compliance Engineering*, 43–51, May–June 1995.

Magid, L. M., *Electromagnetic Fields, Energy, and Waves*, Wiley, New York, 1972.

Mardiguian, M., *Controlling Radiated Emissions by Design*, Van Nostrand Reinhold, New York, 1992.

Marshman, C., *The Guide to the EMC Directive*, IEEE Press, Piscataway, NJ, 1992.

Montgomery, D., Borrowing RF Techniques for Digital Design, *Computer Design*, 207–217, May 1982.

Strassberg, D., Time-Domain Reflectometry: In >50-MHz Digital Design, Measurements Are a Must, *EDN*, 65–72, August 19, 1993.

Tiwari, S. S., Simple ESD Gun Tests IC, *EDN*, September 12, 1996.

Williams, T., *EMC for Product Designers*, 2nd ed., Newnes, Oxford, 1999.

5

SIGNAL CONDITIONING, DATA ACQUISITION, AND SPECTRAL ANALYSIS

So far we have seen how to design electrodes, amplifiers, and filters to detect electrophysiological signals that result from electrochemical events occurring in the living body. However, biopotentials are only a small fraction of the signals generated by physiological processes. Think about it for a second—physicians most often assess your health condition by looking at changes in weight, heart and lung sounds, arterial blood pressure, and temperature—none of these involve recording biopotential signals.

Mimicking what your doctor does today, the next mirror in your bathroom, your next bathroom scale, and your future shower and toilet will probably be loaded with sensors that can automatically and unobtrusively measure your temperature, look at skin marks, analyze your body composition, and examine your secretions for the telltale signs of infections or chemical imbalance. A behind-the-scenes computer will scan the sensor data set and suggest that you visit your doctor if it finds suspicious changes.

It will take a while before we develop instruments sophisticated enough to make "Bones" McCoy feel cozy, but the trend of embedding sensors and intelligence into even the simplest medical instruments and home health appliances will continue to grow in the years to come. In fact, a recent study by a major technology-development company demonstrated that the largest emerging markets for the decade of the 2000s will be based on affordable, highly specific, very reliable sensors. This trend has already resulted in the development of many self-contained sensors that incorporate all the necessary front-end electronics. Typically, these only need a power input to produce a high-level output signal proportional to the measured variable. As shown in Table 5.1, many sensors with onboard electronics are already available to measure temperature, pressure, acceleration, gas concentrations, flow, magnetic fields, and so on. Although the robo-doc may not be here yet, these sensors are being designed for incorporation into consumer appliances, and as such they are becoming more available and more affordable by the day.

Design and Development of Medical Electronic Instrumentation   By David Prutchi and Michael Norris
ISBN 0-471-67623-3   Copyright © 2005 John Wiley & Sons, Inc.

TABLE 5.1   Some of the Sensors with Onboard Electronics Widely Available to Measure a Diversity of Physical Variables

Sensor	Manufacturer	Representative Part	Excitation	Range	Output
Temperature	Analog Devices	AD590	4 to 30 V dc	−55 to +150°C	1 mA/°K
Peripheral pressure pulse	UFI	1010	None (piezoelectric)	Typical human pulse pressure from finger	20–40 mV into 1 MΩ
Respiration	UFI	Pneumotrace II	None (piezoelectric)	Typical human chest expansion	20–100 mV into 1 MΩ
Acceleration	Analog Devices	ADXL05	5 V dc	Selectable ±1 g or ±5 g	Selectable 200 mV/g or 1 V/g
Fluid pressure	MSI	MPS3102	10 to 32 V dc	0 to 1000 psi	4 mV/psi
Barometric pressure	Sensortechnics	144SC0811-BARO	7 to 24 V dc	800 to 1100 mbar	16 mV/mbar
Magnetic field	Honeywell	HMC2003	6 to 15 V dc	±2 G	1 V/G
Gas concentration (CO, combustible gases, etc.)	Capteur	Various with driver circuit	5 V dc	Sensor dependent	0 to 5 V range, sensor dependent
Ac or dc current (no contact)	Lem Heme	LTA50P	±15 V dc	0 to 50 A	100 mV/A
Air/gas flow	Honeywell	AWM3300V	10 V dc	0 to 1 L/min	4 V/L/min
Humidity	Thermometrics	RHU-217	5 V dc	30 to 90% RH	33 mV/%RH

UNIVERSAL SENSOR INTERFACE[1]

The device of Figure 5.1 is a simple sensor interface that plugs into the printer port of the PC. With it, it is possible to excite, control, and read a wide variety of sensors directly from a personal computer. This *universal sensor interface* features an eight-channel 12-bit A/D converter with user-selectable input ranges, two 12-bit D/A converters, two 100-μA current sources for direct sensor excitation, and an uncommitted current mirror and multiple digital I/O lines.

A block diagram of the universal sensor interface is shown in Figure 5.2. Note that the circuit can accept and read analog signals from up to eight sensors. The sensor-derived signals are fed through individual prescalers and applied to an eight-channel multiplexer, permitting the user to select a particular sensor-generated signal. At any given time, one of the eight scaled input signals is presented at the output of the multiplexer to a 12-bit A/D converter. The voltage reference for the A/D is provided by a precision 2.5-V reference IC.

Analog outputs are generated by the universal sensor interface through dual voltage-output D/A converters. The resolution of the D/A converters is also 12 bits. To simplify the reading of resistive sensors, the universal sensor interface also includes a precision current-reference IC which incorporates two 100-μA current sources and an uncommitted current mirror. These can be configured to provide constant-current excitation in the range 100 to 400 μA.

Power for the A/D section is obtained directly from the printer port. The D/A converters and current sources are powered from an internal +5-V linear regulator. Whenever the D/A converters or the current sources are needed, an external +9-V to +12-V supply must be provided. This power supply is not required if only the A/D converter section is used. Finally, the digital inputs and outputs of the printer port which are not used by the circuitry within the universal sensor interface are made available to the user as general-purpose I/O lines. These can be used to read switches, provide power to sensors with integrated electronics, or effect real-time control of devices based on information gathered from sensors.

[1]The material in this section appeared originally in *Popular Electronics* [Prutchi, 1999].

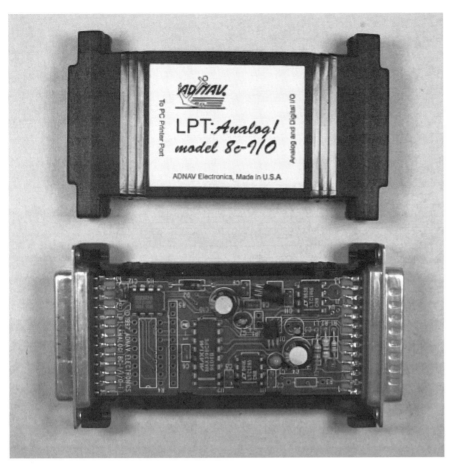

Figure 5.1   The universal sensor interface is a simple device that plugs into the printer port of the PC. It can excite, control, and read a wide variety of sensors directly from a PC.

Unlike ordinary data acquisition setups, the universal sensor interface offers a great degree of flexibility and portability. Virtually all PCs come equipped with a parallel printer port, and reconnecting this data acquisition system to the computer typically does not require delving into the enclosure. Moreover, since the use of the printer port is standardized, the same software will run on any PC without the need for reconfiguration. As long as the signals to be converted are correctly scaled and sampled, the universal sensor interface should have you experimenting with the new generation of sensors in no time at all.

A/D Converter

As shown in Figure 5.3, a Linear Technology LTC1285 integrated circuit (IC1) is at the heart of the A/D circuitry within the universal sensor interface. This IC implements a 12-bit successive-approximation A/D converter, complete with a sample-and-hold and a high-speed three-wire serial interface. A 2.5-V reference voltage is regulated by D1 and supplied to the VREF input of the A/D converter.

The serial interface to the LTC1285 requires only three digital lines. Data-conversion initiation and data-read operations are controlled by the *CS (chip select) and SCLK (serial clock) lines. Data are obtained serially through the DOUT line. As shown in Figure 5.4*a*, an A/D conversion is initiated by a falling edge on the *CS line. At this point, the sample-and-hold stores the input voltage and the successive-approximation process commences.

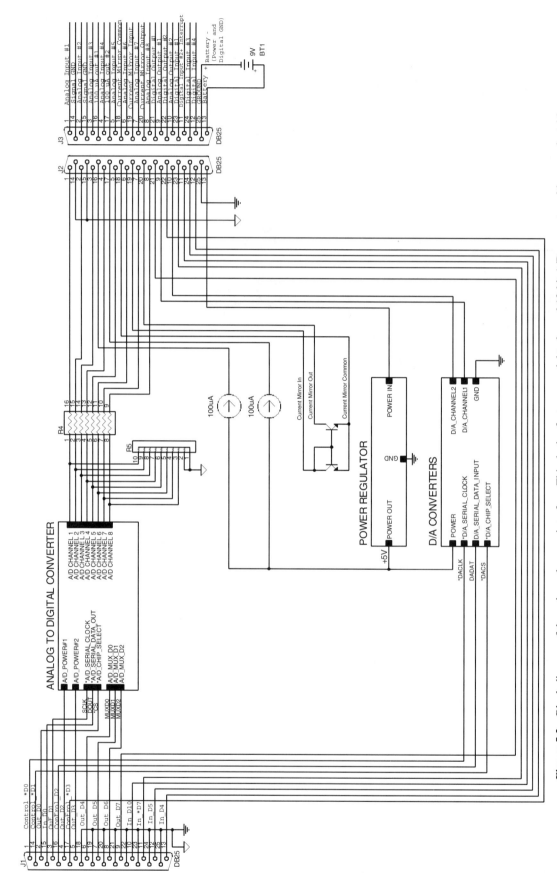

Figure 5.2 Block diagram of the universal sensor interface. This device features an eight-channel 12-bit A/D converter with user-selectable input ranges, two 12-bit D/A converters, two 100-μA current sources for direct sensor excitation, as well as an uncommitted current mirror and multiple digital I/O lines. (From Prutchi [1999]. Reprinted with permission from *Popular Electronics*, June 1999 © Gernsback Publications Inc.)

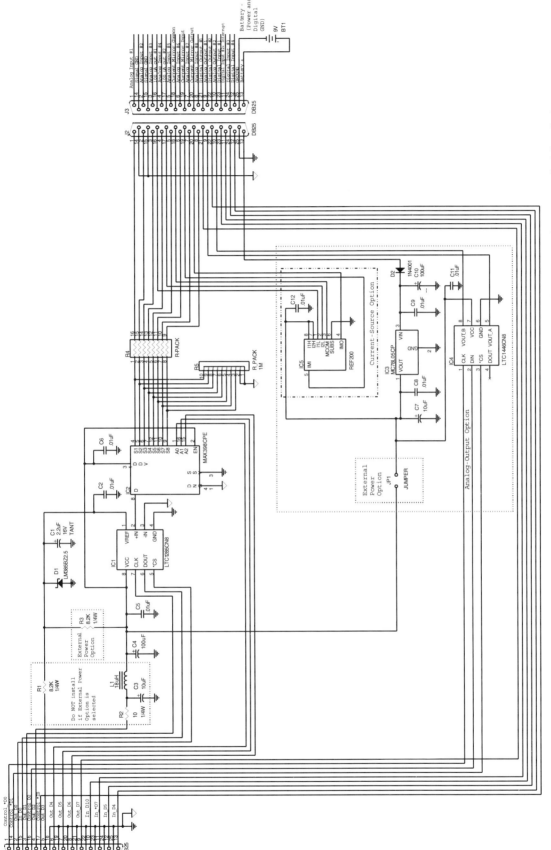

Figure 5.3 Schematic diagram of the universal sensor interface. The micropower components of the A/D section are supplied directly from the printer port. The D/A and current sources require an external power supply. (From Prutchi [1999]. Reprinted with permission from *Popular Electronics*, June 1999 © Gernsback Publications Inc.)

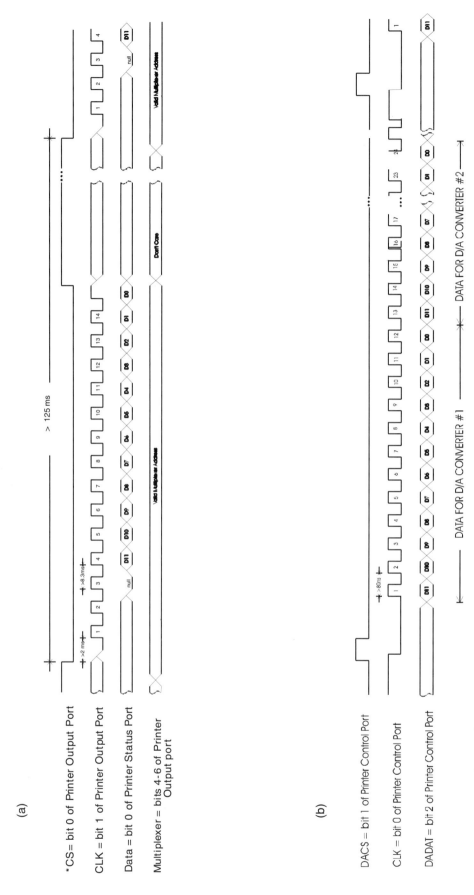

(a)

*CS= bit 0 of Printer Output Port

CLK = bit 1 of Printer Output Port

Data = bit 0 of Printer Status Port

Multiplexer = bits 4-6 of Printer
Output port

(b)

DACS = bit 1 of Printer Control Port

CLK = bit 0 of Printer Control Port

DADAT = bit 2 of Printer Control Port

Figure 5.4 Timing diagram for the serial protocols of the A/D and D/A converters of the universal sensor interface: (*a*) timing diagram for the A/D section; (*b*) protocol for the LTC1446 D/A IC. (From Prutchi [1999]. Reprinted with permission from *Popular Electronics*, June 1999 © Gernsback Publications Inc.)

As conversion takes place, data are obtained in serial format on each falling-edge transition of SCLK after the third clock pulse. Since there are 12 bits, a minimum of 14 falling-edge pulses are required to shift out the A/D result. Bits 0 and 1 of the LPT 8-bit output port (hex address 378 for LPT1:) are toggled by software to implement the control portion of the A/D's serial protocol. Bit 0 of the printer status port register (hex address 379 for LPT1:) is used to receive the serial data from the A/D.

A 3-bit parallel interface using bits 4, 5, and 6 of the LPT 8-bit output port (hex address 378 for LPT1:) controls the analog input multiplexer (IC2) that selects which one of the eight analog inputs is presented to the input of the A/D converter. Power for the A/D converter, the 2.5-V reference, and the multiplexer is supplied directly from the lines controlled by bits 2 and 3 of the parallel printer port.

The following VisualBasic code module (DECLARATIONS.BAS) shows the exact location of the control bits for the universal sensor interface.

```
' This module contains the general declarations required
' to operate the Universal Sensor Interface
'
'
'
' Declare DLL to enable I/O through the printer port
' -----------------------------------------------
'
Public Declare Function Inp Lib "inpout32.dll" _
Alias "Inp32" (ByVal PortAddress As Integer) As Byte
Public Declare Sub Out Lib "inpout32.dll" _
Alias "Out32" (ByVal PortAddress As Integer, ByVal Value As Byte)
'
' Printer port locations
' ----------------------
Global Const prinop1 = &H378      ' Printer Output Port for LPT1
Global Const prinstat1 = &H379    ' Printer Status Port for LPT1
Global Const princont1 = &H37A    ' Printer Control Port for LPT1
Global Const prinop2 = &H278      ' Printer Output Port for LPT2
Global Const prinstat2 = &H279    ' Printer Status Port for LPT2
Global Const princont2 = &H27A    ' Printer Control Port for LPT2
Global Const prinop3 = &H3BC      ' Printer Output Port for LPT3
Global Const prinstat3 = &H3BD    ' Printer Status Port for LPT3
Global Const princont3 = &H3BE    ' Printer Control Port for LPT3
'
' Define control pin locations for LPT:Analog! model 8c-I/O
' --------------------------------------------------------
Global Const notdaclk = 1, notdacs = 2, dadat = 4
' notdaclk is the D/A serial clock line (active low)
' notdacs  is the D/A chip select line (active low)
' dadat    is the D/A serial data input line
Global Const power = 12, notcs = 1, sclk = 2, muxd0 = 16, muxd1 = 32, muxd2 = 64
' power    are the two D/A power supply lines
' notcs    is the A/D chip select line (active low)
' sclk     is the A/D serial clock line
' muxd0    is the A/D input multiplexer data line 0
' muxd1    is the A/D input multiplexer data line 1
' muxd2    is the A/D input multiplexer data line 2
```

```
'
' Define variables for printer port location
' -----------------------------------------
Global prinop    ' Printer Port Output Port location variable
Global prinstat  ' Printer Port Status Port location variable
Global princont  ' Printer Port Control Port location variable
```

The following VisualBasic code module (ACQUIRE.BAS) shows how the serial protocol for the A/D is implemented.

```
Function Acquire (mux) As Single
'
' Function Acquire (mux) executes the A/D protocol on the
' channel specified by mux. The A/D's result in A/D counts is
' returned via the Acquire variable.
'
'
'
' Define variables
' ---------------
Dim clocknum As Integer    ' clock counter variable
Dim dat As Integer         ' accumulated data variable
Dim bit As Integer         ' filtered data bit variable
Dim bit7 As Integer        ' current value of Out D7
' Acquisition loop
' ----------------
bit7 = Inp (prinop) And 2 ^ 7              ' evaluate current value of Out
D7
Out prinop, power + notcs + sclk + mux + bit7  ' power up but keep CS'
deasserted and
                                          ' do not upset bit 7 of Output
port
Out prinop, power + mux + bit7            ' convert by asserting CS', pull
SCLK low
                                          ' without upsetting bit 7 of
                                          ' Output port
For clocknum = 1 To 2                     ' two clock pulses are required
to start
  Out prinop, power + sclk + mux + bit7   ' conversion process
  Out prinop, power + mux + bit7
Next clocknum
dat = 0                                   ' clear A/D accumulator variable
For clocknum = 11 To 0 Step -1            ' clock 12 bits serially
  Out prinop, power + sclk + mux + bit7     ' clock pulse rising edge without
                                            ' upsetting
                                            ' bit 7 of Output port
  Out prinop, power + mux + bit7            ' clock pulse falling edge without
                                            ' upsetting
                                            ' bit 7 of Output port
  bit = (Inp (prinstat) And 8) / 8        ' read Status port and filter bit
                                          ' corresponding
                                          ' to A/D's serial data output
```

```
    dat = dat + (2 ^ clocknum) * bit      ' accumulate from bit 11 to bit 0
Next clocknum                             ' next bit
Out prinop, power + notcs + mux + bit7   ' deassert *CS without upsetting bit 7
Acquire = dat                             ' return 12-bit A/D count
End Function
```

Signal Conditioning

Input signals rarely fit exactly within the universal sensor interface's input range of 0 to 2.5 V. Signals of smaller amplitude than the full range waste resolution, while signals outside the full range will end up being clipped to the limits of the range. Please note that input signals presented to the A/D multiplexer exceeding the range 0 to 2.5 V may cause permanent damage to the circuitry of the interface. Measuring a signal that spans 0 V and a value larger than 2.5 V is easily accomplished. A resistive voltage divider such as that of Figure 5.5a can scale a large unipolar signal to the desired range. For ease of use, the universal sensor interface has onboard locations reserved for resistive voltage dividers. The voltage-divider resistor packs are marked R4 and R5. To use them, first select the appropriate 10-pin single-in-line bussed resistor array for R5. Then select either a DIP resistor pack or individual $\frac{1}{4}$-W precision resistors to be placed on R4.

Logging data from the sensors on a hyperbaric chamber (a pressurized vessel used to study the therapeutic use of high atmospheric pressures) provides a good example of how to select components for R4 and R5. Table 5.2 shows typical ranges for sensors that monitor a small hyperbaric chamber. For this application, the range for analog channels 1 to 4 of the universal sensor interface should be 0 to 5 V, the range for channel 5 should be 0 to 10 V, and the range of channels 6 to 8 should remain at 2.5 V. Assume that an impedance of 10 kΩ is appropriate for all channels.

For this example, R5 can be selected to be a 10-pin bused 10-kΩ resistor pack. A suitable device is a CTS 770-series 10-kΩ, single-in-line conformal 10-pin bused resistor network (Digi-Key part 770-101-10K-ND) or similar (e.g., Jameco 24643). 1% $\frac{1}{4}$-W resistors for R4 are selected using the following formula:

$$2.5\,\text{V} = V_{max}\frac{10\,\text{k}\Omega}{\text{R4} + 10\,\text{k}\Omega}$$

As such, 10-kΩ resistors should be placed between pads pairs [1,16], [2,15], [3,14], and [4,13] of R4 to yield a range of 0 to 5 V for channels 1 to 4. For channel 5, the formula requires a 30-kΩ resistor for a range of 0 to 10 V. Since 30 kΩ is not a standard value, a 30.1-kΩ 1% resistor should be selected and soldered between pads 5 and 12 of R4. Channels 6 to 8 do not require scaling, and thus pad pairs [6,11], [7,10], and [8,9] should be jumpered.

As shown in Figure 5.5b, a small unipolar signal requires just an op-amp-based amplifier to take advantage of the full resolution of the A/D. Signals riding on a median different than the A/D's midpoint (1.25 V) can be offset appropriately by using the circuit of Figure 5.5c. Alternatively, a "quick and dirty" way of introducing offset that can sometimes be used to enable measurement of bipolar signals is to place a 1.5-V battery in series with the signal source.

Current measurements can be obtained by using a suitable shunt. For example, as shown in Figure 5.5d, the popular 4- to 20-mA current loop used to convey information from many industrial instruments and sensors can be converted to a voltage by using a metal-film 124 Ω ± 1% resistor shunt across the input terminals of the universal sensor interface. Since the 4- to 20-mA current will be translated into the range 0.496 to 2.48 V, some measurement resolution will end up being wasted. If the full 12-bit resolution is desired, you may use a 154 Ω ± 1% resistor instead and use an op-amp to introduce a 0.616-V offset to the measurement.

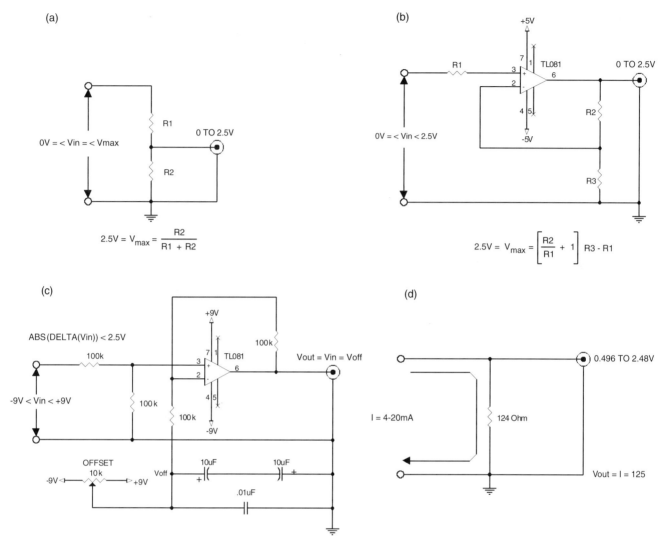

Figure 5.5 Simple circuits can be used to scale different signals to fit within the range 0 to 2.5 V of the universal sensor interface. (*a*) Large unipolar signals can be attenuated by a voltage divider. (*b*) Small unipolar signals can be amplified to make use of the A/D's full resolution. (*c*) Bipolar signals can be converted into unipolar signals by introducing offset. (*d*) 4- to 20-mA current loop signals can be read through a resistive shunt. (From Prutchi [1999]. Reprinted with permission from *Popular Electronics*, June 1999 © Gernsback Publications Inc.)

D/A Converter

Analog outputs are generated by the universal sensor interface through a LTC1446 dual voltage-output D/A converter IC (IC4). The resolution of the D/A converters is 12 bits, providing 1 mV per bit over the range 0 to 4.096 V. There may be an offset error of up to 18 mV (less than 3 mV typical), and the nonlinearity is ±0.5 LSB maximum. The D/A outputs have a maximum current-handling capability of 100 mA, and their maximum source impedance to ground is 120 Ω. In practice, the rail-to-rail buffered outputs can source or sink 5 mA when operating with a 5-V supply while pulling to within 300 mV of the positive supply voltage or ground. The outputs swing to within a few millivolts of either supply rail when unloaded and have an equivalent output resistance of 40 Ω when driving a load to the rails. The D/A buffer amplifiers can drive 1000 pF without going into oscillation.

TABLE 5.2   Some Output Ranges of Sensors Used for Environmental Conditions in a Small Hyperbaric Research Chamber

ADC Channel	Sensor	Output Range (V)
1	Inside oxygen partial pressure	0–5
2	Inside CO_2 partial pressure	0–5
3	Atmospheric pressure	0–5
4	Hyperbaric chamber pressure	0–5
5	Inside relative humidity	0–10
6	Inside ambient temperature	0–1.2
7	Outside ambient temperature	0–1.2
8	Specimen temperature	0–1.2

The D/A converters are powered from a +5-V linear regulator (IC3). Whenever either the D/A converters or the current sources are needed, an external +9- to +12-V supply must be provided. In some cases, powering the A/D section through an external power supply may also be desirable. To do so, do not install R1, R2, C3, and L1. Instead, install R3 and place a jumper on JP1. These changes supply the A/D section of the Universal Sensor Interface from the regulated +5 V supplied by IC3.

The D/A converters are controlled through a serial protocol that requires only three digital lines. Bits 0 to 2 of the LPT control port (hex address 37A for LPT1:) are used for this purpose. As shown in Figure 5.4*b*, data-conversion initiation and data-write operations are controlled by the *DACS (D/A chip select) and DACLK (D/A serial clock) lines. Data are supplied to the D/A converters serially through the DADAT line. The data on the DADAT input are loaded into a shift register (internal to the D/A converter control circuitry) on the rising edge of the DACLK clock. Data are loaded as one 24-bit word where the first 12 bits are for DAC 1 and the second 12 are for DAC 2. For each 12-bit segment the MSB is loaded first. Data from the shift register are loaded into the D/A register when *DACS is pulled high. The clock is disabled internally when *DACS is high. DACLK must be low before *DACS is pulled low to avoid an extra internal clock pulse.

The following VisualBasic code module (DTOA.BAS) shows how the D/A serial control protocol is implemented.

```
Function dtoa(dtoa1 As Integer, dtoa2 As Integer) As Integer
'
' Function DtoA (dtoa1, dtoa2) executes the serial protocol to update D/A
' converters with the values defined by dtoa1 and dtoa2, where
' dtoa1 contains the code for the desired D/A channel 1 output
' dtoa2 contains the code for the desired D/A channel 2 output
'
'
'
' Define variables
' ---------------
Dim bit3 As Integer            ' value of Control *D3
Dim clocknum As Integer        ' clock pulse counter
Dim delay As Integer           ' delay loop dummy variabe
' D/A loop
' --------
bit3 = Inp(princont) And 8     ' evaluate current status of Control *D3
Out princont, notdaclk + bit3       ' set D/A clock line low without
                                    ' upsetting bit 3 of the Control port
```

```
          Out princont, notdaclk+notdacs+bit3  ' assert D/A CS' line without
                                                ' upsetting bit 3 of the Control
                                                ' port
      For clocknum=11 To 0 Step -1              ' clock 12 bits of D/A channel 1
                                                ' from MSB to LSB
      If (dtoa1 And (2 ^ clocknum))=0 Then      ' if current bit converts to 0
          Out princont, notdacs+bit3            ' then do not set data bit high
                                                ' while causing a clock rising edge
                                                ' all without upsetting bit 3 of the
                                                ' Control port
          For delay=1 To 5: Next delay
          Out princont, notdacs+notdaclk+bit3  ' return clock line to low
                                                ' without upsetting bit 3 of the
Control port
          Else                                  ' if, on the other hand, current
                                                ' bit converts to one
          Out princont, notdacs+notdaclk+dadat+bit3   ' then set data bit high
                                                         ' while
                                                ' maintaining the 'CS and clock
                                                ' and do not
                                                ' upset bit 3 of the Control
                                                ' port
                                                ' lines low
          For delay=1 To 5: Next delay    ' introduce delay to comply with IC's
                                                ' timing requirements
          Out princont, notdacs+dadat+bit3  ' cause clock rising edge to
                                                ' clock data bit in without
                                                ' upsetting
                                                ' bit 3 of the Control port
          For delay=1 To 5: Next delay    ' introduce  delay  to  comply  with
                                                IC's
                                                ' timing requirements
          Out princont, notdacs+notdaclk+dadat+bit3 ' return clock line to
                                                      ' low without
                                                      ' upsetting bit 3 of the
                                                      ' Control port
          End If
      Next clocknum                             ' next bit
      For delay=1 To 5: Next delay
      For clocknum=11 To 0 Step -1              ' clock 12 bits of D/A channel 2
                                                ' in the same way
          If (dtoa2 And (2 ^ clocknum))=0 Then
             Out princont, notdacs+bit3
             For delay=1 To 5: Next delay
             Out princont, notdacs+notdaclk+bit3
          Else
             Out princont, notdacs+notdaclk+dadat+bit3
             For delay=1 To 5: Next delay
             Out princont, notdacs+dadat+bit3
             For delay=1 To 5: Next delay
             Out princont, notdacs+notdaclk+dadat+bit3
          End If
```

```
For delay = 1 To 5: Next delay
Next clocknum
Out princont, bit3          ' update D/A data by deasserting 'CS
                            ' without upsetting bit 3 of the Control port
                            ' under this condition, clock and
                            ' data states are of no relevance
dtoa = 0                    ' dummy return variable = 0
End Function
```

Current Sources

A Burr-Brown REF200 IC is used to make two 100-μA current sources and an uncommitted current mirror available on the universal sensor interface. Constant-current sources are highly useful for the excitation of resistive sensors. Figure 5.6 shows various ways in which the current source outputs and the current mirror can be connected to obtain the following:

- Two 100-μA sources
- One 200-μA source
- One 300-μA source
- One 400-μA source

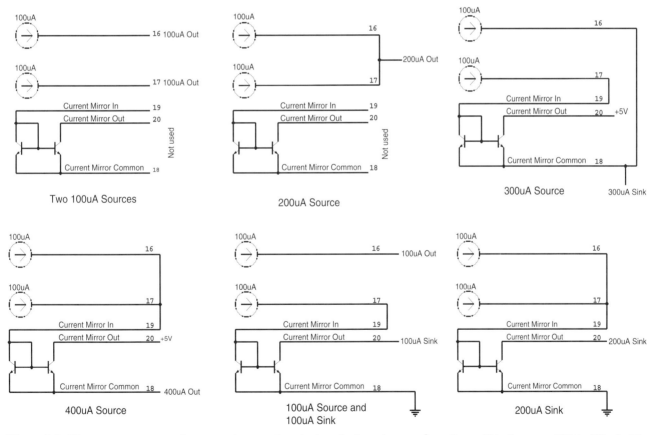

Figure 5.6   The current sources and current mirror can be wired to obtain various configurations within the range 100 to 400 μA. (From Prutchi [1999]. Reprinted with permission from *Popular Electronics*, June 1999 © Gernsback Publications Inc.)

- One 100-μA source and one 100-μA sink
- One 200-μA sink

The accuracy of these sources and sinks is typically better than ±1%, and the voltage compliance is 3.6 V.

Digital I/O

The digital inputs and outputs available on the analog and digital I/O connector of the Universal Sensor Interface are direct pass-throughs of the unused I/O lines from the printer port. The four digital input lines are acquired through bits 4, 5, 6, and 7 of the LPT status port (hex address 379 for LPT1:). Please note that in the standard parallel port, bits 1 to 3 of the status byte do not convey line state information. In addition, digital input 2 (bit 6 of the status port) can be used to drive interrupt-driven acquisition programs. The two digital output lines are controlled through bit 3 of the LPT control port (hex address 37A for LPT1:) and bit 7 of the LPT output port (hex address 378 for LPT1:).

The following VisualBasic code module shows how to read and filter the four digital inputs available on the universal sensor interface.

```
status = Inp(prinstat)                    ' Acquire data from digital inputs
din1 = (status And 128) / 128             ' evaluate digital input #1
If din1 = 1 Then din1 = 0 Else din1 = 1   ' invert din1
din3 = (status And 32) / 32               ' evaluate digital input #2
din2 = (status And 64) / 64               ' evaluate digital input #3
din4 = (status And 16) / 16               ' evaluate digital input #4
```

Construction of the Universal Sensor Interface

The universal sensor interface can be constructed on a small prototyping board using point-to-point wiring techniques. To achieve good A/D performance, build separate analog and digital ground planes and join these at a single point close to the ground connection on J1. If possible, build the universal sensor interface on a four-layer PCB. The entire assembly can be made to fit neatly in a 3-in. hood enclosure with DB25 connectors at the ends.

Software for the Universal Sensor Interface

Software for the universal sensor interface is available in the universal sensor interface directory at the Wiley ftp site associated with this book. This directory contains a number of data acquisition utilities as well as drivers and program examples. The program examples are thoroughly commented and should be self-explanatory.

Figure 5.7 presents the control panel for the VisualBasic (v5.0) application project LPT8_DVM.VBP. This is an example of how to develop a virtual instrument to acquire analog and digital data as well as to control the D/A and digital outputs of the Universal Sensor Interface. A 32-bit I/O DLL (Dynamic Link Library) is used to allow input and output operations to be performed on the printer port under the command of VisualBasic. If 16-bit operation is required, modify the programs to make use of the 16-bit CUSER2.DLL file. LPT8_LOGGER.VBP is basically the same as LPT8_DVM.VBP, but a file dialog has been added to make it possible to log acquired data directly to disk.

In addition, the archive also contains the following examples in QuickBasic:

1. LPTAN8.BAS is a simple program for driving the universal sensor interface A/D. Notice that use of the standard LPT1: is assumed, and you may need to change the output

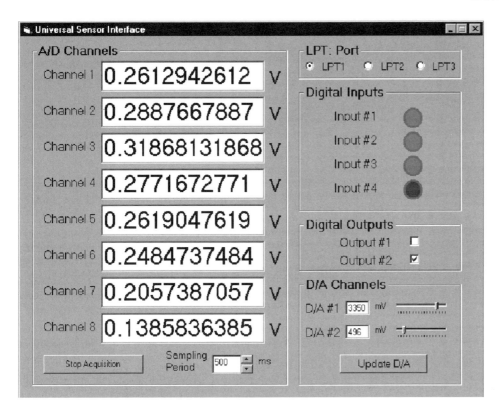

Figure 5.7 Virtual instruments such as this eight-channel DVM based on the universal sensor interface can easily be built using VisualBasic. (From Prutchi [1999]. Reprinted with permission from *Popular Electronics*, June 1999 © Gernsback Publications Inc.)

port and status port locations to suit your specific installation. This program will acquire and display data on the screen at the maximum rate supported by your computer when running QuickBasic.

2. ACQUIRE8.BAS implements an eight-channel oscilloscope/four-channel logic analyzer. In this program, the sampling rate is regulated by inserting for–to loops to introduce delay between samples. The number of loops required to reach the correct delay is based on a calculation of the time that it takes the computer to complete a single data acquisition and display operation, as well as of the delay introduced by the addition of a for-to loop. The performance of this program depends on the processor speed of the computer, the resources available to the processor, the access speed to the video card, and so on. Some typical acquisition rates achieved by this program are shown in Table 5.3.

3. LPT8FAST.BAS is the same as ACQUIRE8.BAS, but provisions have been made to allow acquired data to be recorded on disk. The compiled version (LPT8FAST.EXE) of this program is intended to be run from a bootable diskette in which CONFIG.SYS first initializes a RAM drive. In contrast to writing to a hard drive, writing to a RAM drive is virtually instantaneous, so LPT8FAST stores data in real time on the RAM drive, and only then copies the generated file onto the desired file on the hard drive. Disk writing is done in ASCII format, allowing you to import acquired data directly into virtually any of the most popular application programs (spreadsheets, data analysis software, etc.).

4. The LPT8FAST.EXE utility is meant to be executed from a bootable diskette drive. To create the bootable diskette, first format a new diskette and transfer into it the system files, your DOS COMMAND.COM, HIMEM.SYS, and RAMDRIVE.SYS files. Then

TABLE 5.3 Representative Maximum Sampling Rates Achieved by the Compiled Version of ACQUIRE8 to Run the Universal Sensor Interface as an Oscilloscope/Logic Analyzer

Acquisition/Computer	1 Analog Channel + 4 Digital Inputs (samples/s)	8 Analog Channels + 4 Digital Inputs (samples/s)
486, 66 MHz	834	205
Pentium, 133 MHz	2260	523
Pentium, 166 MHz	2314	561
Pentium, 200 MHz	2766	743

copy the LPT8FAST.EXE file under the root directory of the diskette. Boot your computer from the diskette you created. This should take you to the utility's setup dialog. When asked for it, enter the path and file name under which the program will store acquired data (e.g., C:\MYDATA.TXT). Within this utility, the computer will calculate the maximum sampling rate that can be achieved by this program using the Universal Sensor Interface in conjunction with your computer. You may then select any sampling rate up to the maximum rate calculated for your computer. In addition, you may need to select a different drive letter than the default D: for the RAM drive created by the bootup process. Pay attention to the letter assignment made for the RAM drive when RAMDRIVE.SYS runs (just after booting from the diskette that you created). After entering the desired acquisition setup, data will be acquired into the RAM drive and displayed simultaneously in the computer screen. Stop the data acquisition process by pressing any key. Upon doing so, the acq8.txt file on the RAM drive will be saved to the file you specified.

This program is intended only as an example of implementing the serial protocol required to collect data from the universal sensor interface through the PC printer port, and major enhancements could be made to it. First, QuickBasic imposes a major limit on data acquisition speed. Even running on a 200-MHz Pentium PC, the single-channel sampling rate is limited to about 2.7 kHz. True compilers with in-line assembly (e.g., C++) can increase the effective sampling rate to the vicinity of the 7.5-kHz maximum sampling rate supported by the A/D in the universal sensor interface. Another improvement that can be made is to eliminate the insertion of delay loops to control timing, and instead, to control the acquisition process from interrupts generated by high-resolution hardware timing [Schulze, 1991; Ackerman, 1992].

5. ATOD_SL8.BAS is similar to LPTAN8.BAS, but acquisition is regulated through the TIMER command. Through it, data frames are acquired at desired intervals in the range 1 to 86,400 seconds (1 second to 24 hours between samples). Data frames are acquired and stored on an array. After acquisition is complete, the array is stored to a file on disk.

6. DTOA.BAS is a simple program that implements the serial protocol to write values to the D/A converters.

Signal Conditioning

Sensors that do not produce an output voltage directly can also be measured by the universal sensor interface. For example, Figure 5.8 shows the typical setup for reading a resistive sensor. Here a thermistor (Radio Shack Catalog 271-110) is excited by one of the 100-μA sources. The voltage developed across the thermistor by the constant-current excitation is related to temperature. THERMOM.BAS, a QuickBasic program also in the accompanying disk, is used to read temperature in °C and °F using this circuit. Other resistive sensors, such as piezoresistors, resistive position indicators, resistive humidity sensors, photoresistors, and so on, can be measured in the same way.

The circuit of Figure 5.9 shows how the universal sensor interface can be used for reading sensors that require automatic baseline cancellation. For example, a strain-gauge load cell

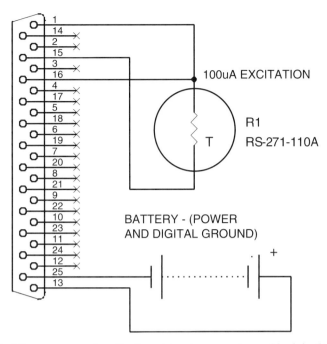

Figure 5.8 Resistive sensors such as the thermistor shown can be read by injecting a known current into the sensor and measuring the voltage developed across the sensor. (From Prutchi [1999]. Reprinted with permission from *Popular Electronics*, June 1999 © Gernsback Publications Inc.)

Warning!   Be extremely careful when using the universal sensor interface to control line-powered devices. Life-threatening voltages and currents are present in these circuits. Isolate and fuse all circuitry properly on the power line path.

---

from an inexpensive digital kitchen scale can be excited from an op-amp (IC1A) driven by D/A converter 2. When the scale is not loaded, the output from the cell is a voltage that floats somewhere in the range of 0 to the drive voltage generated by A/D converter 2. This output voltage is slightly amplified by IC1B (gain = 1.78). IC1C is used to cancel the load cell's base-line level by offsetting the output of IC1B by the amount established by D/A converter 1. The signal developed by a real load is further amplified by IC1D before being fed to analog input 1. Resistor R4 and zener diode D1 protect the A/D input from output voltages produced by IC1D, which may be beyond the acceptable input range. SCALE.BAS is the QuickBasic program that implements an auto-zeroing digital scale by controlling the sensor's output offset.

The Universal Sensor Interface is also ideal for implementing low-cost process or environmental control systems. For this type of application, the device's digital outputs are often required to switch ac loads such as heaters, lamps, or motors. As shown in Figure 5.10, the digital outputs of the universal sensor interface can be used to control an ac load connected to the power line through an opto-isolated triac. The digital output signal is used to drive a 2N2222 transistor, which in turn controls the LED inside the optocoupler (IC1). The opto device selected for this application is a zero-voltage switching triac which is used to drive a power triac (Q1) capable of handling the necessary switching power. A device that can be used to switch light loads is the Q4004.

Digital inputs can also be isolated. The circuit of Figure 5.11 shows how a PS2506 can be used to sense an ac or dc voltage safely to signal a digital input line. R1 should be selected to produce no more than ±20 mA of current through the LED of the PS2506.

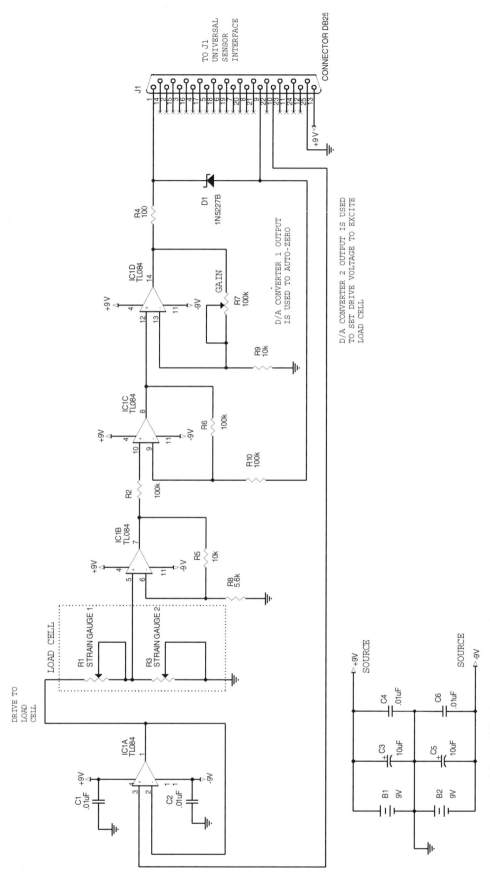

Figure 5.9 A very precise scale can be built for the PC from the load cell of an inexpensive kitchen scale and a simple signal-conditioning circuit. (From Prutchi [1999]. Reprinted with permission from *Popular Electronics*, June 1999 © Gernsback Publications Inc.)

222

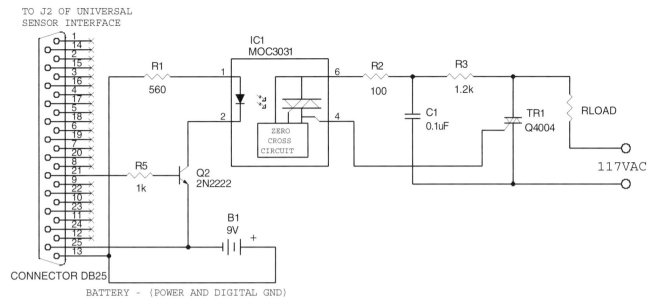

Figure 5.10   The digital-output lines of the universal sensor interface can be used to control ac loads in response to sensor inputs. This line-control interface is suitable for turning on or off relatively light ac-powered loads. (From Prutchi [1999]. Reprinted with permission from *Popular Electronics*, June 1999 © Gernsback Publications Inc.)

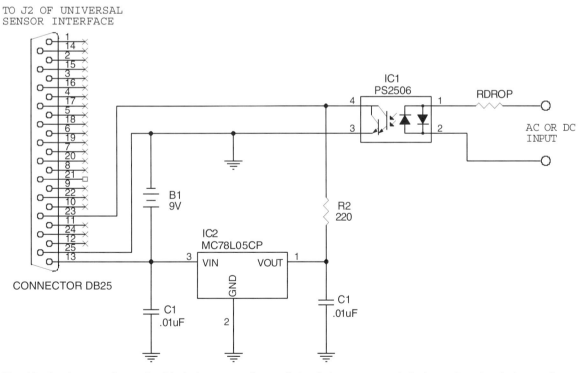

Figure 5.11   Use this circuit to translate and safely isolate external ac or dc levels (e.g., power switch closures) to signals that can be read by the digital input lines of the universal sensor interface. (From Prutchi [1999]. Reprinted with permission from *Popular Electronics*, June 1999 © Gernsback Publications Inc.)

SAMPLING RATE AND THE NYQUIST THEOREM

Without doubt you have heard that according to the Nyquist theorem, a signal should be sampled at twice its frequency. Yes, and no. There is no data acquisition concept that is more quoted and less well understood than the Nyquist theorem. Let's explore what this theorem actually implies for proper data acquisition. Nyquist stated that any *bandwidth-limited* signal can be reconstructed from its digitized equivalent if the sample rate is at least twice the highest-frequency component.

Signal components with a frequency above half the sampling rate are *aliased* and show up in a reconstruction as a component with a frequency at the difference between its real frequency and the sampling rate. This effect is commonly seen in films of moving cars, where the wheels seem to be rotating impossibly slow, or even going backward. The aliased rotation frequency is caused by the slow shutter rate of the camera relative to the fast rotational speed of the wheels' rims. For a dramatic demonstration, pay attention to the apparent behavior of the wheels of a speeding stagecoach in virtually any old western movie.

The way of preventing aliasing is to ensure that there are absolutely no signal components at frequencies above half the sampling rate. Assuming that to sample a signal of approximately x hertz you simply need to select an A/D rate of $2x$ just doesn't cut it. The only ways of making sure are to supersample[2] the signal and to apply antialiasing filtering. Perfectionists insist that all signals to be sampled must be low-pass filtered to prevent aliasing. In reality, however, many signals can be sampled fast enough that they are naturally low-pass filtered by the response of the sensor or by the process being measured. For example, temperature changes in the body occur so slowly that sampling a temperature sensor even once per minute suffices to eliminate aliases by supersampling. Despite this, care must be taken that power line noise or other high-frequency interference does not contaminate the sensor signal by using appropriate shielding, differential amplification, and/or a simple RC low-pass filter. A good rule of thumb to avoid aliasing when an antialiasing filter is not used is to supersample at a sampling rate of at least 10 times the highest expected (unfiltered) signal component.

Ten times supersampling can be unachievable when your application involves the acquisition of high-frequency signals. Here, the use of antialiasing filters is unavoidable. The ideal antialias filter would be a sharp low-pass filter that passes all frequencies below its cutoff at half the sampling frequency and totally eliminates any components above that frequency. As we saw in Chapter 2, however, real-world filters do not yield a perfect step in the frequency domain, and they will always allow through some components above their corner frequency. This means that, in practice, sampling must happen at a rate higher than twice the filter's cutoff frequency. Please note that the antialiasing filter must be an analog implementation—it is too late to use digital filtering once you have done the sampling.

The other common misunderstanding about the Nyquist theorem is that although it states that all the information needed to *reconstruct* the signal is provided by sampling at least at twice the highest signal frequency, it does not say that the samples will look like the signal. Figure 5.12 shows a 48-Hz signal that is sampled at 100 Hz—fast enough according to Nyquist's theorem—barely more than twice per cycle. It is clear from Figure 5.12c, though, that if straight lines are drawn between the samples, the signal looks amplitude modulated (although the signal's frequency is correctly reproduced). This effect arises because each cycle is taken at a slightly different part of the original signal's cycle. Many engineers would take the modulated signal as an indication that it was sampled improperly. On the contrary, there is enough information to reproduce the original

---

[2]Most engineers have heard the term *oversampling* applied to data acquisition. Although it is intuitive that sampling and playing back something at a higher rate looks better than a lower rate—more points in the waveform for increased accuracy—that's not what oversampling usually means. In fact, oversampling usually refers to output oversampling and it means generating more samples from a waveform that has *already* been digitally recorded.

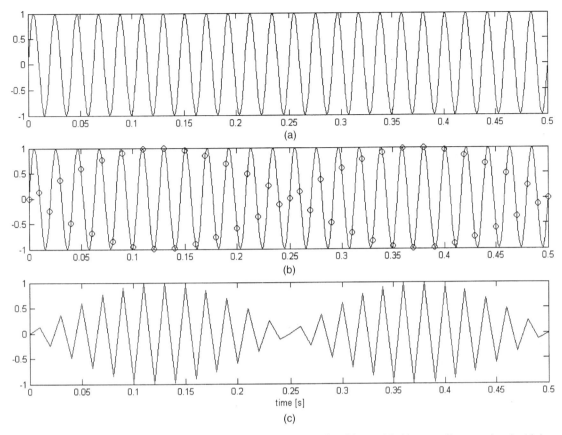

Figure 5.12 Nyquist stated that all the information needed to reconstruct a signal is provided by sampling at twice the highest signal frequency at least, which does not imply that the samples will look like the signal. (*a*) A 48-Hz signal (*b*) is sampled at 100 Hz (circles). (*c*) Simply drawing lines between the samples does not reconstruct the original signal.

signal, but the correct reproduction is not achieved just by drawing straight lines between the samples.

The correct way of reconstructing a sampled signal is by using an interpolating filter. The low-pass reconstruction filter interpolates between the samples to make a smoothly varying output signal. Let's assume for a moment that the reconstruction filter is an ideal low-pass filter which has an infinitely steep cutoff. It eliminates all frequencies above the cutoff and has no effect on either the amplitude of phase of frequencies below the cutoff. The impulse response of this low-pass filter is the sin function (sin x)/x. For the reconstruction, the stimuli fed to this filter are the series of discrete impulses that are the original digitized samples. Every time an impulse hits the filter, it "rings"—and it is the superposition of all these peaky rings that reconstructs the proper signal. If the signal contains frequency components that are close to one-half the sampling rate, the reconstruction filter has to be very sharp indeed. This means that it will have a very long impulse response—long enough to fill in the signal even in region of the low-amplitude samples.

Channel skew is another issue that must be considered when establishing how to sample data. Channel skew is the undesirable effect of time-shifting sampled data points that otherwise should be time aligned. Let's assume that you are sampling multiple biopotential signals acquired from the same source through an electrode array. Take, for example, the signals displayed in Figure 1.33, which were recorded differentially using 32 surface electrodes placed 2.54 mm apart over the biceps brachii muscle. A proper sampling rate for these signals is about 10 kHz. Sequential sampling of the 32 channels would result in a time skew of 3.1 ms between the first and last channels, a situation that is intolerable for many applications. The way of eliminating time skew between channels is to sample-and-hold each channel individually at the same time. Digitization can then occur at the A/D's leisure. Most often, however, the impact of channel skew is minimized by a technique called *burst sampling*. Here, all EMG channels would be sampled at the maximum rate of the converter once every 1/10,000 of a second. Assuming the use of a 1-MHz A/D, the maximum time skew between the first and last channels drops to barely 31 µs.

Channel skewing has become a very critical consideration, since multichannel recordings enjoy wide popularity in research and diagnostic uses of electrophysiological activity. For example, 32-, 64-, and even 128-channel systems are commonly used in the acquisition of EEG and evoked potentials; and 64-channel systems are now commercially available for body potential mapping (BPM) ECG.

FREE DATA ACQUISITION CARD IN YOUR COMPUTER

If you need to measure one or two audio-frequency signals that do not include a dc component, don't overlook the data-acquisition card that your PC probably has already—your PC's sound card. Audio-range signals can be applied to a sound card via stereo $\frac{1}{8}$-in. mini jacks. Inside the sound card, these signals are ac-coupled, with one pair of inputs being sent through preamplifiers for boosting low-level signals (intended to amplify microphone signals), and another pair feeding signals within a $\pm 0.5\,V_{p-p}$ range directly into the card's analog multiplexer/mixer. The input impedance of a sound card is typically above $10\,k\Omega$, and channels typically respond within the frequency band 20 Hz to 20 kHz. Signals at the output of the mixer/multiplexer are digitized by a delta-sigma A/D that is part of the sound card's codec IC.

Consumer-grade sound cards offer 11.025-kHz ("telephone quality"), 22.05-kHz ("music quality"), and 44.1-kHz ("CD quality") sampling rates. Low-cost sound cards often use a single codec chip that is multiplexed between the two input channels, which means that these sound cards would sample both inputs only at 22.05 kHz or lower. Better cards feature two A/D converters or at least two sample-and-hold devices so they can sample both inputs simultaneously at rates up to 44 kHz. There are some "professional-quality" sound cards that feature more input channels, higher resolution, and dc input response, but they require proprietary software to operate, and their prices are similar to those of a full-function data acquisition card.

Of course, the most common use for a sound card is to generate the various annoying sounds of explosions, phasers, and kicks that make video games so realistic. The output jack of a typical sound card carries an amplified ac-coupled signal capable of driving 8-Ω speakers directly with some 2 W of power. Sounds are generated by using an FM synthesizer within the sound card to combine harmonically related sine waves or by selecting data from a wave table that contains digitized samples from various musical instruments. A third option available on higher-end sound cards is to use an onboard musical instrument digital (MIDI) synthesizer to produce an even wider variety of sounds.

There is quite a bit of free software available on the Web to turn a PC equipped with a sound card into all sorts of audio-range virtual instruments, such as two-channel oscilloscopes, spectrum analyzers, signal generators, frequency counters, and noise generators. The following are some of the freeware packages that deserve special attention.

Signal Generator

- BIP Electronics Labs Sine Wave Generator v3.0 (sine30.zip, freeware for Windows 3.1, but works well in most cases under Windows 9x) by Marcel Veldhuijzen
- Sweep Sine Wave Generator v2.0 (swpgen20.zip, freeware for Windows 9x) by David Taylor

Audio Oscilloscopes

- BIP Electronics Labs Digital Scope v3.0 (scope30.zip, freeware for Windows 3.1, but works well in most cases under Windows 9x) by Marcel Veldhuijzen
- Oscilloscope for Windows v2.51 (osc2511.zip, freeware for Windows 9x) by Konstantin Zeldovich

Audio Spectrum Analyzers

- Spectrogram v5.0.5, Dual Channel Audio Spectrum Analyzer (gram501.zip freeware for Windows 9x) by Richard Horne
- Audio Wavelet Analyzer v1.0 (audiowaveletanalyse.zip freeware for Windows 9x) by Christoph Lauer

Another very neat thing that can be done with a sound card is to use it as the basis for a transfer-function analyzer. In essence, a full-duplex sound card can be used to generate a test signal (e.g., a sweeping tone) that is fed to the input of a system under test. The output of the system is recorded by the sound card and analyzed in the context of the excitation waveform. For example, you could play back a sweeping tone through your stereo set and acquire the sound using a good microphone placed at a favorite listening spot in the room. A plot of the frequency response of the complete system (amplifier, speakers, room acoustics, and microphone) would show distortions, room resonances, and other effects that alter the music that you listen to. One such software package is the RightMark Audio Analyzer v2.5 (rmaa25.zip freeware for Windows 9x, 2000 and NT) by Alexey Lukin and Max Liadov.

The flexibility and potential of a PC sound card as a simple data acquisition system has been recognized by commercial software vendors, and some major data processing packages (e.g., Matlab by The MathWorks Inc.) include commands to acquire data directly into its environment from the sound card (e.g., wavrecord.m) and output data streams as sound (e.g., wavplay.m). Interfacing a signal line to a sound card requires some experimentation. First, you will need a way of delivering the signals to the sound card's stereo $\frac{1}{8}$-in. jack. Avoid using premade cable splitters. Instead, build your own cables using RG-174 miniature coaxial cable and metal-shell $\frac{1}{8}$-in. plugs.

Next, use a good signal generator to feed a signal to the sound card. Find settings for the generator's output (the range is usually limited to $\pm 0.5\,V_{p-p}$) and the sound card's mixer (accessed through the speaker icon in your Windows tray) that give you a full-scale deflection without clipping. You could use the microphone input for better sensitivity, but the noise floor is higher, and many cards output a voltage for the microphone supply. If that input is used, a capacitor for blocking the microphone bias voltage is necessary.

Remember that a sound card's signal return connects to the ac power ground through the PC's chassis, which opens up the possibility of introducing interference into measurements by forming unwanted ground loops. If the input signal does not include a dc component, you can use a 1 : 1 audio transformer for signal isolation or to convert a differential input into the single-ended input expected by the sound card.

Consider the fact that sound cards are designed to be very inexpensive, so that sound card manufacturers do not usually spend the few extra cents necessary to include overvoltage protection components, making sound cards vulnerable to destructive overloads. In addition, older sound cards sometimes include permanently enabled automatic gain control (AGC), which can produce misleading measurements because it compresses signal peaks. Finally, since our ears don't care much about a compressive versus an expansive waveform, signals may be inverted by the sound card when acquired. The best advice is that you should always use a real oscilloscope to compare the input signal against the signal acquired by the sound card before you trust the acquisition setup.

Converting the Sound Card into a Precision DC-Coupled A/D

Sound cards suffer serious limitations when used for acquiring physiological signals; for one, typical PC sound cards are ac coupled through series capacitors on the signal path. Typical high-pass cutoff frequencies are above 20 Hz, making it impossible to record waveforms containing dc or low-frequency components, a characteristic shared by most biopotential and physiological signals. Furthermore, consumer-grade sound cards often exhibit poor sound-recording quality characteristics, especially a lack of passband flatness and harmonic generation, which results in signal distortion.

However, a voltage-controlled oscillator (VCO) and some software can turn a sound card into a precision dc-coupled A/D. The VCO of Figure 5.13 generates an audio tone that varies in frequency as a function of a control signal input. The VCO's output is a whistling sound that is easily recorded even with low-end sound cards. The original signal is then recovered by software FM demodulation of the VCO audio recorded. IC3 is an Exar XR-2206 function generator that implements the VCO. The carrier frequency (in hertz) is given by $1/(R13 + R14)C4$. Good performance is achieved with most sound cards by setting the carrier frequency somewhere in the range 2 to 10 kHz. The frequency of oscillation is modulated by applying a control current in the ± 3 mA range to pin 7, which is biased within the XR-2206 at $+3$ V. R16 sets the offset voltage of IC2A such that zero control voltage applied to IC2B results in zero current across R15 and R18. R18 adjusts the modulation level (frequency deviation per volt). A good range for most sound cards is around $\pm 80\%$ of the carrier for the full-control voltage input range. The sound card sampling rate should be selected to be at least five times higher than the highest VCO frequency expected.

R6 and R7 are used to trim the harmonic distortion of IC3's sinusoidal output. The unadjusted distortion is specified to be $<2.5\%$, so R4, R6, R7, and R9 are optional. R3 should be adjusted together with the sound card's slider volume control to produce a clean-sounding tone. Please note that the stability of the VCO circuit depends on the stability of the frequency-setting components. Proper performance requires the use of low-temperature-coefficient, high-tolerance components. Resistors should be precision 1% tolerance type of the RN60D variety. Capacitors should be Mylar, polyester film, or other types that remain stable with age and which are not sensitive to temperature variations.

The following Matlab code shows how easy it is to obtain digitized data (vector x) from the VCO audio (vector y):

```
Fc = 2144; % Select VCO carrier frequency
vcok = 0.176; % Select VCO constant
Fs = 40000; % Select sound card sampling frequency
samptime = 3; % Select sampling time (in seconds)
y = wavrecord(samptime*Fs, Fs, 'double'); % Sample sound into y
wavplay(y, Fs); % Play the recorded sound
x = (demod(y,Fc,Fs,'fm',vcok)); % FM demodulate the sound
cutoff=100/(Fs/2); % Cutoff frequency for post-demodulator low-pass filter
in radians/s is desired cutoff/(1/2 sampling frequency)
[b,a] = butter(4,cutoff); % Design Butterworth low-pass filter
xfilt = filter(b,a,x); % Filter demodulated signal
time = 1/Fs:(1/Fs):samptime; % Generate time vector
plot(time,xfilt) % Plot filtered data vs time
grid % Overlay grid on plot
xlabel('time [s]') % Label x axis
ylabel('Input [Volts]') % Label y axis
```

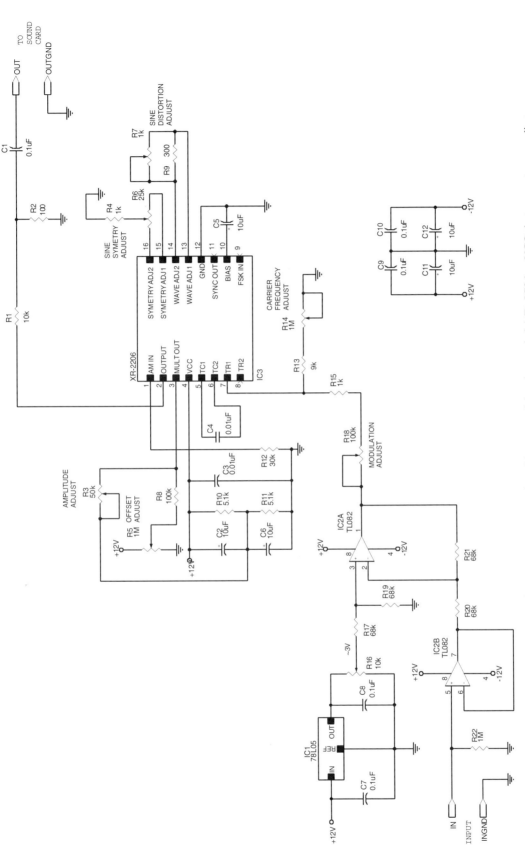

Figure 5.13 A VCO and some software can turn a sound card into a precision dc-coupled A/D. IC3 is a VCO that generates an audio tone that varies in frequency as a function of a control signal input. The original signal is then recovered by software FM demodulation of the VCO audio recorded.

You can use the following Matlab command to look at the actual frequency shift and distortion of the VCO signal sampled through the sound card:

```
specgram(y,512,Fs,kaiser(256,5),220)
```

If you are not a Matlab user, here is how you can write a program to determine the instantaneous frequency (or phase) of the sampled VCO output:

1. Transform the incoming signal to a complex data stream using a Hilbert transform.[3] This gives I (in-phase) and Q (quadrature-phase) data. A true Hilbert transform is unrealizable, but you can get excellent approximations with FIR or IIR filters. A thorough description of how to implement the discrete Hilbert transform is available in P. A. Regalia, Special Filter Designs, in S. K. Mitra and J. F. Kaiser (eds.), *Handbook for Digital Signal Processing*, Wiley, New York, 1993.

2. Pass these complex samples through a polar discriminator (i.e., multiply the new complex sample by the conjugate of the old sample).

3. The output is the phase difference vector from the two samples.

4. Compute $\arctan(Q/I)$ to get the phase angle of the vector. You can use an easy polynomial fit, Taylor series, or lookup table to estimate the $\arctan(\cdot)$ function in order to extract the phase angle.

Figure 5.14 displays a signal acquired through this method. The artificial ECG test signal was generated by an Agilent 33120A ARB. Note that the demodulated signal faithfully reproduces the dc offset and low-frequency components of the ECG.

Since the VCO output is in the audible range, the modulated signal can be transmitted to the sound card via a voice radio or telephonic link for remote data acquisition. This is the principle of ECG transtelephonic monitoring, a common technique used to follow up pacemaker patients using a transmitter such as that of Figure 5.15. To do so, however, the tone frequencies produced by the VCO for a full-scale input must be limited within the bandpass of the communications channel. For a plain telephone line, this range is 400 Hz to 3 kHz, while a commercial FM audio link is specified to cover the audio bandwidth 30 Hz to 15 kHz. Another interesting possibility is to use a small 1 : 1 audio isolation transformer and a floating power supply to turn the VCO into an isolated data acquisition front end.

It should be noted that the full bandwidth of a single sound card channel can be shared by multiple VCOs occupying separate audio bands to convey various simultaneous low-frequency signals to an array of software bandpass filters and demodulators. This is exactly what FM telemetry systems do, and the U.S. Army's Inter-Range Instrumentation Group of the Range Commanders Council (IRIG) has established a standard (IRIG-106-96) that covers all aspects of frequency modulation (FM) and pulse code modulation (PCM) telemetry, including transmitters, receivers, and tape recorders. Owing to its success as a proven standard and its wide support by telemetry equipment manufacturers, most commercial data acquisition systems use the same IRIG standard channels.

The IRIG standard specifies ways of performing frequency-division multiplexing (FDM) over a telemetry channel, that is, how to generate a composite signal consisting of a group of subcarriers arranged so that their frequencies do not overlap or interfere with each other. Various FM subcarrier and deviation schemes are available to accommodate different

---

[3]The *Hilbert transform* is a mathematical operation that decomposes a waveform into an instantaneous phase and an instantaneous amplitude waveform. If the input is a pure sine wave, the output instantaneous amplitude waveform will have a constant value while the phase will increase linearly over time. If the sine wave were amplitude modulated, the instantaneous amplitude waveform would show this modulation. Similarly, frequency modulation will affect the instantaneous phase waveform.

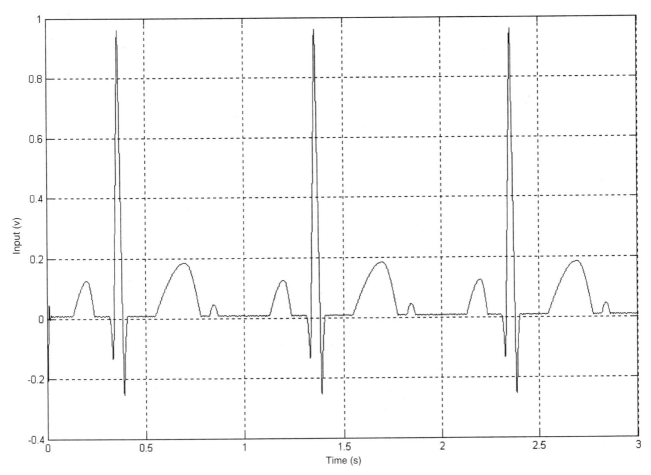

Figure 5.14 An artificial ECG test signal generated by an Agilent 33120A ARB was acquired using the circuit of Figure 5.13 and a software FM demodulator. Note that the demodulated signal faithfully reproduces the dc offset and low-frequency components of the ECG.

channel needs. For example, Table 5.4 shows some of the IRIG subcarrier frequencies and FM deviations used to yield channels whose bandwidths are constant and independent of their carrier frequency. On the other hand, Table 5.5 shows some choices for subcarrier frequencies and deviations which result in channels where their bandwidth is proportional to their carrier frequency. For a consumer-grade sound card, the maximum sampling rate is 44.1 kHz, which imposes an absolute maximum tone frequency limit of around 18 kHz, which ultimately constrains the number of channels that can be sampled simultaneously with the FM carrier technique.

IRIG publications can be ordered from:

Defense Technical Information Center
Attn: DTIC-OCP
John J. Kingman Road, Suite 0944
Fort Belvoir, VA 22060-6218
(703) 767-8019
(703) 767-8032 (fax)

The complete VCO circuit of Figure 5.13 (one channel) can be built for under $10. However, if you have a cost-sensitive application that can tolerate somewhat lower precision and

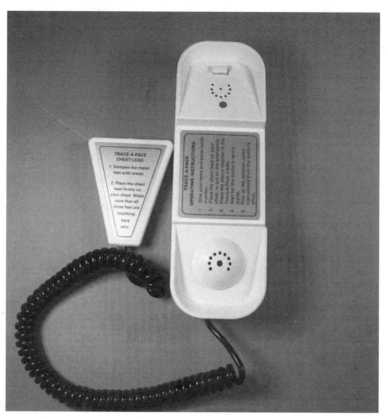

Figure 5.15 A transtelephonic ECG monitor provides a simple, effective solution for monitoring pacemaker patients from the comfort of their home. To transmit real-time ECGs, the patient simply dials the receiving center and places the telephone handset on the transmitter's cradle. These monitors typically use a center frequency of 1965 Hz with a deviation of 100 Hz/mV to convey a 0.5 to 100 Hz ECG signal detected through three chest electrodes that are dampened with water.

linearity, you may design a simpler VCO using two op-amps or a 555 timer IC to produce a square or pulsed output. Straightforward FM demodulation should not be used for these waveforms. Rather, a pulse-width or pulse-position demodulator should be used. The Matlab demodulation command (demod.m) can be told to use these demodulation methods simply by replacing 'fm' for 'pwm' or 'ppm' in its argument line.

If your application requires remote acquisition with transmission of the modulated tone via voice link, don't do it with a pulsed waveform, since these are distorted significantly by telephone lines and do not properly modulate conventional voice transmitters. If using a VCO with sine-wave output (e.g., the XR-2206, Intersil's ICL8038, or Maxim's MAX038) is still out of the question, you can convert a frequency-modulated pulse train into a quasi-sinusoidal tone using the circuit [Allen, 1981] of Figure 5.16. A 4018 Johnson counter is wired to sequence resistors R1–R5 to produce a 10-step staircase waveform that approximates the shape of a sine wave. The sharp edges of the staircase are removed using a simple *RC* low-pass filter formed by R6 and C4 before they are attenuated by R7 and ac-coupled by C5 so they can be fed to the microphone input of a telephone or voice transmitter. Since the square-to-sine converter divides the input frequency by 10, the VCO should be made to output at a frequency range that is 10 times higher than the sinusoidal tone audio range required.

TABLE 5.4   Some Center Frequencies and Frequency Deviations Specified by IRIG to Yield Constant-Bandwidth Channels for Acquisition through FM Subcarriers

Frequency Deviation and Nominal Frequency Response per Channel				
±2 kHz	±4 kHz	±8 kHz	±16 kHz	±32 kHz
400 Hz	800 Hz	1.6 kHz	3.2 kHz	6.4 kHz
Channel Center Frequency (kHz)				
8	16	32	64	128
16	32	64	128	256
24	48	96	192	384
32	64	128	256	512

TABLE 5.5   Many Channels of Differing Bandwidth Can Be Accommodated by Selecting Channel Bandwidths Proportional to Their Carrier Frequencies[a]

IRIG Proportional Bandwidth Channel	Center Frequency	Lower Deviation Limit	Upper Deviation Limit	Bandwidth	Channel's Nominal Frequency Response
±7.5% Deviation Channels					
1	400	370	430	60	6
2	560	518	602	84	8
3	730	675	785	110	11
4	960	888	1,032	144	14
5	1,300	1,202	1,398	196	20
6	1,700	1,572	1,828	256	25
7	2,300	2,127	2,473	346	35
8	3,000	2,775	3,225	450	45
9	3,900	3,607	4,193	586	59
10	5,400	4,995	5,805	810	81
11	7,350	6,799	7,901	1,102	110
12	10,500	9,712	11,288	1,576	160
13	14,500	13,412	15,588	2,176	220

[a]These IRIG channels are useful for acquiring multiple physiological signals with different bandwidth requirements. For example, the subject's skin conductance signal fits well within the 11-Hz bandwidth of channel 3, the 45-Hz bandwidth of channel 8 could accommodate an EEG channel, and a three-lead wideband ECG can be acquired through channels 11, 12, and 13.

SPECTRAL ANALYSIS

The analysis of a signal based on its frequency content is commonly referred to as *spectral analysis*. Because of the rhythmic nature of practically every phenomenon in a living body, frequency-domain analysis is one of the most powerful tools in the examination of physiological signals. The mathematical basis for this operation, the Fourier transform, has been known for many years, but it was the introduction of the fast Fourier transform (FFT) algorithm that made spectral analysis a practical reality. Implementing the FFT in personal computers and embedded DSP systems has allowed efficient and economical application of Fourier techniques to a wide variety of measurement and analysis tasks. Moreover, because the FFT has been found to be so valuable in applications such as medical signal processing, radar, and telecommunications, DSP chips are often designed to implement it with the greatest efficiency.

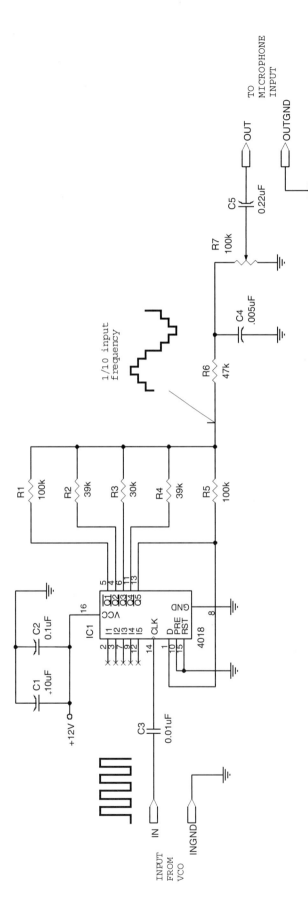

Figure 5.16 A 4018 Johnson counter can be made to convert a frequency-modulated pulse train into a quasi-sinusoidal tone for remote data acquisition. IC1 is wired to sequence resistors R1–R5 to produce a 10-step staircase waveform that approximates the shape of a sine wave. Since the square-to-sine converter divides the input frequency by 10, the input frequency range should be 10 times higher than the required sinusoidal tone audio range.

In most instances, the powerful Fourier techniques used in EEG systems and CT or ultrasound scanners are hidden from the user, who does not have to worry about their mathematical implications. In other cases, however, human interpreters must make diagnostic decisions based on frequency-domain representations of data processed through Fourier transforms. For example, many digital storage oscilloscopes offer the user the option to convert time-domain signals into the frequency domain through the use of an FFT running on an embedded DSP that displays the results directly on screen. It is also common for scientists and engineers to write short FFT-based routines to display a spectral representation of experimental data acquired into a personal computer. It is in these cases where the unwary may fall into one of the many traps that FFTs conceal.

FFT users often forget that real-world signals are seldom periodic, free of noise, and distortion, and that signal and noise statistics play an important role in their analysis. Because of this, FFTs and other methods can only provide estimates of the actual spectrum of signals, which require competent interpretation by the user for their correct analysis. Moreover, the FFT has certain inherent problems that make it unsuitable for high-resolution applications.

FFTs and the Power Spectral Density

Using a typical data acquisition setup, a signal is sampled at a fixed rate of f_S(samples/s), which yields discrete data samples $x_0, x_1, \ldots, x_{N-1}$. These N samples are then spaced equally by the discrete sampling period $\Delta t(\text{s}) = 1/f_S$. The discrete Fourier transform (DFT) represents the time-domain data with N spaced samples in the frequency domain $X_0, X_1, \ldots, X_{N-1}$ through

$$X(f) = \Delta t \sum_{n=0}^{N-1} x_n e^{-2\pi j f_n \Delta t}$$

where the frequency $f(\text{Hz})$ is defined over the interval $-\Delta t/2 \leq f \leq \Delta t/2$. The FFT will efficiently evaluate this expression at a discrete set of N frequencies spaced equally by $\Delta f(\text{Hz}) = 1/N\Delta t$.

In its simplest form, the energy spectral density estimate of the time-domain data is given by the squared modulus of the FFT of these data, and the power spectral density (PSD) estimate $P(f)$ (or simply, the spectrum) at every discrete frequency f is obtained by dividing the latter by the time interval $N\Delta t$:

$$P(f_m) = \frac{|X_m|^2}{N\Delta t} \qquad m = 0, 1, \ldots, N-1$$

where $f_m(\text{Hz}) = m\Delta f$. When real data are used (usually, the case when sampling from real-world signals), the PSD for negative frequencies will be symmetrical to the PSD for positive frequencies, making only half of the PSD useful. However, at times it may be necessary to compute the PSD for complex data, and relevant results will be obtained for both positive and negative frequencies.

Although obtaining the PSD seems to be as simple as computing the FFT and obtaining the square modulus of the results, it must be noted that because the data set employed to obtain the Fourier transform is only a limited record of the actual data series, the PSD obtained is only an estimate of the true PSD. Moreover, as will be seen later, meaningless spectral estimates may be obtained by using the estimate of $P(f_m)$ without performing some type of statistical averaging of the PSD.

Pitfalls of the FFT

When sampling a continuous signal, information may be lost because no data are available between the sample points. As the sampling rate is increased, a larger portion of the information is made available. We explained above that according to Nyquist's theorem the bare minimum sampling rate to avoid aliasing must be at least twice that of the highest-frequency component of the waveform. Because aliased components cannot be distinguished from real signals after sampling, aliasing is not just a minor source of error. It is therefore of extreme importance that antialiasing filters with very high roll-off be used for all serious spectral analysis.

Beyond appropriate sampling practices, the FFT still exposes other inherent traps which can potentially make impossible the analysis of a signal. The most important of these problems are leakage and the picket-fence effect. The first problem is caused by the fact that the FFT works on a short portion of the signal. This is called *windowing* because the FFT can only see the portion of the signal that falls within its sampling "window," after which the FFT assumes that windowed data repeat themselves indefinitely. However, as shown in Figure 5.17, this assumption is seldom correct, and in most cases the FFT analyzes a distorted version of the signal that contains discontinuities resulting from appending windowed data to their duplicates. In the PSD, these discontinuities appear as leakage of the energy of the real frequency components into sidelobes that show up at either side of a peak.

The second problem, called the *picket-fence effect* or *scalloping*, is related inherently to the discrete nature of the DFT. That is, the DFT will calculate the frequency content of a signal at very well defined discrete points in the frequency domain rather than producing a continuous spectrum. Now, assuming a perfect system, if a certain component of the signal would have a frequency that falls between the discrete frequencies computed by the DFT, this component would not appear in the estimated PSD.

To visualize this problem, suppose that an ideal signal is sampled at a rate of 2048 Hz and processed through a 256-point FFT. There will then be a spectral channel every 4 Hz: at dc, 4 Hz, 8 Hz, 12 Hz, and so on. Now suppose that the signal being analyzed is a pure

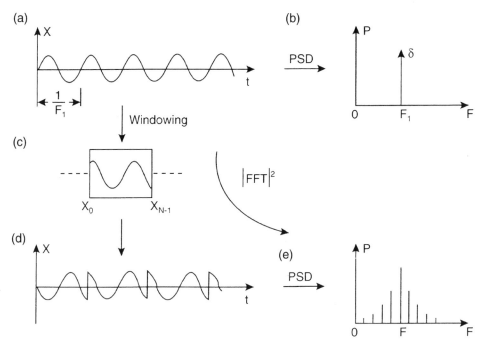

Figure 5.17 A purely sinusoidal signal (*a*) has a single impulse as its true spectrum (*b*). However, the signal is viewed by the FFT through a finite window (*c*), and it is assumed that this record is repeated beyond the FFT's window (*d*). This leads to leakage of the main lobe to sidelobes in the spectral estimate (*e*).

TABLE 5.6   Typical Window Functions for Use with the FFT in Spectral Estimation

Window[a]		3-dB Bandwidth (bins)	Scallop Loss (dB)	Highest Sidelobe (dB)
Rectangular				
$w(n) = \begin{cases} 1 & \text{for } -\dfrac{N}{2} \le n \le \dfrac{N}{2} - 1 \\ 0 & \text{for all other} \end{cases}$		0.89	3.92	-13
Triangular				
$w(n) = \begin{cases} 1 - \dfrac{\lvert 2n + 1 \rvert}{N} & \text{for } -\dfrac{N}{2} \le n \le \dfrac{N}{2} - 1 \\ 0 & \text{for all other} \end{cases}$		1.28	1.82	-27
Hamming				
$w(n) = \begin{cases} 0.54 + 0.46 \cos \dfrac{2\pi(2n + 1)}{2(N - 1)} & \text{for } -\dfrac{N}{2} \le n \le \dfrac{N}{2} - 1 \\ 0 & \text{for all other} \end{cases}$		1.30	1.78	-43
Hanning				
$w(n) = \begin{cases} 0.5 + 0.5 \cos \dfrac{2\pi(2n + 1)}{2(N - 1)} & \text{for } -\dfrac{N}{2} \le n \le \dfrac{N}{2} + 1 \\ 0 & \text{for all other} \end{cases}$		1.44	1.42	-32

[a]Windows are N-point long and are assumed here to be symmetric around $n = 0$.

sinusoidal with a frequency of 10 Hz. In a perfect system, this signal would not appear in the PSD because it falls between two discrete frequency channels, much as a picket fence allows us to see details in the scene behind it only if they happen to fall within a slot between the boards. In reality, however, because the FFT produces slightly overlapping "bins" of finite bandwidth, components with frequencies that fall between the theoretical discrete lines are distributed among adjacent bins, but at reduced magnitudes. This attenuation is the actual picket-fence or scalloping error.

Both of these problems are somewhat corrected by the use of an appropriate window. So far, all samples presented to the FFT have been considered equal, which means that a weight of 1 has been applied implicitly to all samples. The samples outside the FFT's scope are not considered, and thus their effective weight is zero, resulting in a "rectangular"-shaped window. This ultimately leads to discontinuities that cause leakage. A number of windows have been devised that reduce the amplitude of samples at the edges of the window, while increase the relevance of samples toward its center. By doing so, these windows reduce the discontinuity to zero, thus lowering the amplitude of the sidelobes that surround a peak in the PSD. In addition, use of a nonrectangular window increases the bandwidth of each bin, which results in a decreased scalloping error.

Some typical window functions and their characteristics are presented in Table 5.6. In essence, these functions produce N weights $w_0, w_1, \ldots, w_{N-1}$ which are weighted (multiplied) one to one with their corresponding data samples $x_0, x_1, \ldots, x_{N-1}$ before subjecting them to an FFT:

$$X(f) = \Delta t \sum_{n=0}^{N-1} w_n x_n e^{-2\pi j f_n \Delta t}$$

The price paid for a reduction in leakage and scalloping through the use of a nonrectangular window is reduced resolution. In fact, if it is necessary to view two closely spaced peaks, the rectangular window's narrow main lobe will allow the user to obtain analysis results that report the existence of these closely spaced components, whereas any of the other windows would end up fusing these two peaks into a single smooth crest.

Use of a rectangular window is also appropriate for the analysis of transients. In these cases, zero signal usually precedes and succeeds the transient. Thus, if the FFT is made to look at the complete data record for the transient, no artificial discontinuities are introduced, and full resolution can be obtained without leakage. As you may well see, there is no single window that outperforms all others in every respect, and it is safe to say that selecting the appropriate window for a specific application is more of an art than an exact science.

When the signal rides on a relatively high dc level or on a strong sinusoidal signal, it is advisable to remove these components from the data before the PSD is estimated, because otherwise the biasing and strong sidelobes produced by them could easily obscure weaker components. Whenever expected physically, the dc component of a signal can usually be removed by subtracting the sampled data mean $\bar{x} = (1/N)\sum_{n=0}^{N-1} x_n$ from each data sample to produce a "purely ac" data sequence $x_0 - \bar{x}, x_1 - \bar{x}, \ldots, x_{N-1} - \bar{x}$.

Zero-Padding an FFT

An interesting property of the FFT is that simply adding zeros after a windowed data sample sequence $x_0, x_1, \ldots, x_{N-1}$ in order to create a longer record $x_0, x_1, \ldots, x_{N-1}, 0, 0, \ldots, 0$ before performing an FFT will cause the FFT to interpolate transform values between the N original transform values. This process, called *zero padding*, is often mistakenly thought of as a trick to improve the inherent resolution of an FFT. Zero padding will also provide a much smoother PSD and will help annul ambiguities regarding the power and location of peaks that may be scalloped by the non-zero-padded FFT.

Classical Methods

As mentioned before, a common mistake is to assume that the solution to $P(f_m) = |X_m|^2/N\Delta t$, $m = 0, 1, \ldots, N-1$, the *periodogram*, is a reliable estimate of the PSD. Actual proof of this is beyond the scope of the book, but it has been demonstrated that regardless of how large N (the number of available data samples) is, the statistical variance of the estimated periodogram spectrum does not tend to zero. This statistical inconsistency is responsible for the lack of reliability of the periodogram as a spectral estimator.

The solution to this problem is simple, however. If a number of periodograms are computed for different segments of a data record, their average results in a PSD estimate with good statistical consistency. Based on this, Welch [1967] proposed a simple method to determine the average of a number of periodograms computed from overlapping segments of the data record available. Welch's PSD estimate $\hat{P}(f)$ of M data samples is the average of K periodograms $P(f)$ of N points each:

$$\hat{P}(f) = \frac{1}{K}\sum_{i=1}^{K} P(f)$$

where the $P(f)$'s are obtained by applying $P(f_m) = |X_m|^2/N\Delta t$, $m = 0, 1, \ldots, N-1$ to appropriately weighted data. It is obvious that if the original M-point data record is divided into segments of N points each, with a shift of S samples between adjacent segments, the number of periodograms that can be averaged is

$$K = \text{integer}\left(\frac{M - N}{S + 1}\right)$$

High-Resolution Methods

The main limitation of FFT-based methods is restricted spectral resolution. The highest inherent spectral resolution (in hertz) possible with the FFT is approximately equal to the reciprocal of the time interval (in seconds) over which data for the FFT are acquired. This limitation, which is further complicated by leakage and the picket-fence effect, is most noticeable when analyzing short data records.

It is important to note that short data records not only result because of the lack of data, such as when sampling a short transient at a rate barely enough to satisfy Nyquist's criterion, but also from data sampled from a process that varies slowly with time. Although there are many applications in the medical field, the best example comes from the oil field. By analyzing the vibrations picked up from an oil-well drill, the operator can monitor the buildup of resonance in the long pipe that carries torque to the drill bit, avoiding costly damages to the equipment [Jangi and Jain, 1991]. Although a continuous signal from the vibration transducers is available for sampling, the vibrations on the drill assembly change rapidly, resulting in a limited number of data samples which represent each state of the drill bit. It is here that high-resolution estimates would be desirable, even though the data available are limited.

A number of *high-resolution spectral estimators* have been proposed. These alternative methods do not assume, as the FFT does, that the signal outside the observation window is merely a periodic replica of what is observed through the window. For example, one of these methods, the parametric estimator, relies on the selection of a model that suitably represents the process that generates the signal in order to capture the true characteristics of data outside the window. By determining the model's parameters, the theoretical PSD implied by the model can be calculated and should represent the signal's PSD.

Many signals encountered in real-world applications are well approximated by a rational transfer function model. For example, human speech can be characterized by the resonances of the vocal tract that generate it. These resonances, in turn, are well represented by the poles of a digital filter. Parameters for the filter can then be estimated, such that the filter would turn a white noise input into the signal of interest, and from the filter's transfer function we could easily estimate the PSD of the signal.

Various kinds of filter structures exist, and they are often classified according to the type of transfer function that they implement. An all-pole filter is called an *autoregressive* (AR) *model*, an all-zero filter is a *moving-average* (MA) *model*, and the general case of a pole–zero filter is called an *autoregressive-moving-average* (ARMA) *model*. Using the past example, the model best suited for speech is then an AR model. Although high-resolution estimators have been implemented for all these models, AR model-based estimators are the most popular because many computationally efficient algorithms are available. A well-behaved set of equations to determine the AR parameters with a computationally efficient algorithm has been introduced by Marple [1987].

In the model of Figure 5.18, the AR filter coefficients $a_0, a_1, \ldots, a_p$ are estimated by Marple's algorithm based on the input data samples $x_0, x_1, \ldots, x_{N-1}$. The model assumes that a white-noise source drives the filter, in which the output is regressed (thus the name *autoregressive*) through a chain of delay elements z^{-1}, from which p taps feed the AR coefficients. The system's transfer function can then be computed efficiently through the FFT, resulting in an estimate of the signal's PSD. The performance of Marple's estimator is startling. Figure 5.19b presents three spectral estimates obtained from a short 64-point complex test data set suggested by Marple. Estimates obtained through the zero-padded FFT periodogram, Welch's averaged periodogram, and Marple's method can be compared to the theoretical spectrum of Figure 5.19a. Only positive-frequency PSD estimates are shown for clarity.

Notice that the closely spaced components cannot be resolved by either of the classical methods, but they appear clearly separated in the estimate produced by Marple's method.

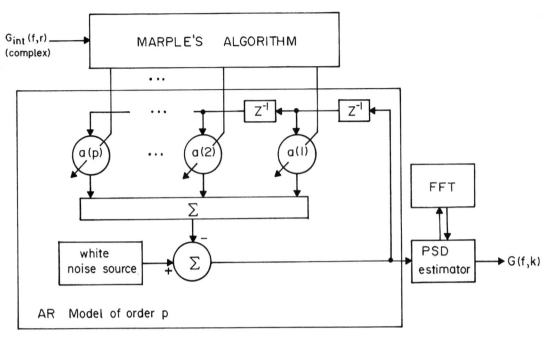

Figure 5.18   In one implementation of a parametric high-resolution spectral estimator, the coefficients $a_0, a_1, \ldots, a_p$ of an AR filter are determined from input data through Marple's algorithm. The transfer function of the filter is then evaluated efficiently by the FFT, resulting in a high-resolution estimate of the input data's PSD.

You may also notice that Marple's estimate is "peaky" even for the smooth continuous spectral components at the far right and far left of the PSD. The reason for this is that a purely autoregressive filter will generate a spectrum based on pure resonances, and only through use of a moving average could these resonances be damped to produce a perfectly smooth spectrum in regions where this is necessary. Although this limitation of AR-based estimators would lead to errors in the actual amplitudes of the PSD components, it is very well suited for high-resolution detection of periodicities in the signal.

A price must be paid for the increase in resolution, and just as you might suspect, the computational burden of these high-resolution methods far exceeds that for a simple FFT. In addition, like the selection of an appropriate window for the classical estimators, the rules for selecting an appropriate model, parameter estimation method, and model order are essentially inflexible.

Implementing Spectral Analysis Algorithms

Program SPECTRUM.BAS at the Wiley ftp site for this book demonstrates implementation of the spectral estimation methods discussed above. The program was written in QuickBasic 4.5, but should run with little trouble under any other Basic compiler on any PC. The FFT, as well as Welch's averaged periodogram and the AR spectral estimation routines, are based on the Fortran programs that accompany Marple's book. However, Basic does not support complex-number arithmetic, so explicit operations have been used in which variable names with the suffix "r" represent the real portion of that variable, and those with the suffix "i" represent the imaginary portion of the same. The ftp site also includes a version of this program (LPT8SPEC8.BAS) that acquires evenly sampled data through the Universal Sensor Interface described above and then displays the spectrum estimated via the zero-padded FFT, an averaged-periodogram

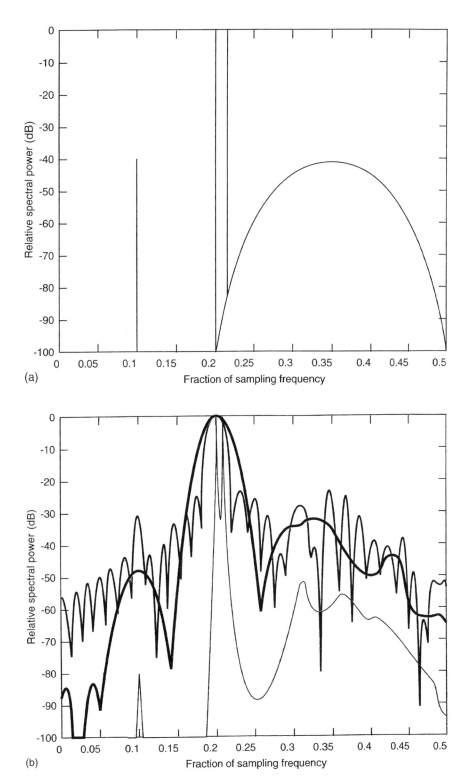

Figure 5.19 Positive-frequency spectral estimates for Marple's 64-point complex data test set: (*a*) theoretical spectrum; (*b*) spectrum estimated using three different methods: (i) zero-padded FFT periodogram (medium line); (ii) Welch's method (bold line, $N = 32$, $S = 16$); (iii) Marple's method (thin line, $p = 15$).

method (Welch's estimator), and a parametric estimator (Marple's autoregressive method).

After being defined by the user, the program will read a file containing the N-data-point sequence to be analyzed. These data can be either a single column of (plain ASCII) samples, or two columns, one containing the real and the other the imaginary parts of complex data samples. The program will estimate the spectrum of the input data using three methods:

1. A single periodogram of the data record is obtained by zero-padding the data up to npsd data points (npsd = 512 for complex data, 1024 for real input data), from which the squared modulus of the FFT is computed. A rectangular window is assumed.
2. Welch's method with a Hamming window is applied using the number of samples per periodogram and the shift specified by the user.
3. Marple's method is used to estimate the PSD of the data using an AR model with model order given by the user.

Prior to its display in the output screen, the PSD is normalized relative to its maximum, and transformed to decibels. For complex input data, both the positive- and negative-frequency sides of the spectrum are plotted. Otherwise, only the positive-frequency spectrum is presented. Because of screen resolution limitations, the number of PSD points computed for display has been limited to 512. If a larger PSD record is required, however, npsd can be increased to any desired power of 2, and a file can be opened to receive the estimated PSD results.

A few simple demonstrations can be set up to compare the performance of the methods. First, you may generate a data file for a signal consisting of a single sinusoid at $f_s/4$ with white noise added to it through

```
pi = 3.14159262
OPEN "noise.dat" FOR OUTPUT AS #1
FOR i = 1 TO 256
      x = 2 * (RND - .5) + (SIN(2 * pi * i / 4))
      PRINT #1, x
NEXT i
CLOSE #1
```

You may vary the signal-to-noise ratio by changing the value of the coefficient of the noise component. You may also vary the frequency of the sinusoidal component by changing the denominator of the sine's argument. Of course, from Nyquist's theorem, a denominator smaller than 2 will produce an aliased signal, and you may want to experiment with the effect that this has on the PSD estimate. In addition, the resolving power of the estimators may be compared by using a signal containing two closely separated sinusoidal components. This can be accomplished by adding the second component to the program line that computes x: for example,

```
x = 2*(RND - .5) + (SIN(2*pi*i/4)) + (SIN(2*pi*i/4.1))
```

Regarding the AR model order that you should use, a rule of thumb that often helps is to keep it smaller that one-third of the number of data samples available and to allow for at least twice the number of spectral components expected. SPECTRUM.BAS will announce an error whenever mathematical ill-conditioning is encountered due to too-high a model order, but an unreasonably "peaky" spectrum is often obtained before ill-conditioning can be detected.

Array Signal Processing

The greatest interest in high-resolution spectral estimators has been generated in the field of array signal processing. Here, a number of sensors are placed at various locations in space to detect traveling waves. For example, in seismology, a number of sensors capable of detecting the shock waves of a tremor or earthquake are spread over a certain area. As the shock waves travel under the sensor array, signals from each sensor can be sampled along time, producing a data record which also contains information regarding the spatial characteristics of the waves (because the sensor locations are known). The processing of resulting spatiotemporal data is meant to estimate the number, vector velocity (speed and direction), and waveshape of the overlapping traveling waves in the presence of interference and noise. Array signal processing has been applied successfully to biomedical diagnosis and has been used to track weak electrical potentials from the brain, nerves, and muscles. Other applications involve image reconstruction from projections, such as MRI and medical tomography.

The most common form of traveling wave is the plane wave. In its simplest form, a plane wave is a sinusoidal wave that not only propagates through time t but also through space. In the direction of propagation r, this wave can be represented by

$$g(t, r) = A \sin \left[2\pi \left(ft - \frac{r}{v} \right) \right]$$

where A is the amplitude of the wave, f its temporal frequency (in hertz = 1/s), and v the velocity (in m/s or any other suitable velocity units) at which the wave propagates through space.

If one such simple plane wave is sampled discretely along time and space, we would obtain a record similar to that presented on the left side of Figure 5.20a. As you may well see, at any given time the spatial sampling of the wave will also form into a sinusoid with frequency k_1. The spatial frequency (in 1/m) of such a simple plane wave, called the *wavenumber*, is given by $k = f/v$. Its physical meaning indicates that at a distance r from the origin, the phase of the wave accumulates by $2\pi kr$ radians.

The two-dimensional spectrum of the plane wave in our example would be an impulse δ (the spectrum of a sinusoid) located in the frequency–wavenumber (f–k) plane at f_1, k_1. Through this type of spectral analysis we can infer not only the components of the waveform but also their velocity, because the slope at which the components are found is equal to their propagation velocity. In this case, $v_1 (\text{m/s}) = f_1 (1/\text{s})/k_1 (1/\text{m})$.

By adding a second component with a different frequency and propagation velocity (Figure 5.20b) to the original component, we obtain a plane wave (Figure 5.20c) that regardless of its simplicity, can hardly be recognized in the space-time domain. However, the two-dimensional frequency–wavenumber spectrum of the signal clearly resolves the components and their propagation velocities.

The two-dimensional spectrum can be computed with ease knowing that the two-dimensional DFT is computable as a sequence of one-dimensional DFTs of the columns of the data array, followed by a sequence of one-dimensional DFTs of the rows of this new array, or vice versa. As such, the simplest two-dimensional PSD estimator is implemented through the FFT. In practice, however, due to the limited number of spatial samples (because only a few sensors are normally used), high-resolution estimators must be used.

Since enough samples $x_0, x_1, \ldots, x_{N-1}$ can usually be obtained from each of the R sensors through time, a hybrid two-dimensional spectral estimator can be implemented by combining

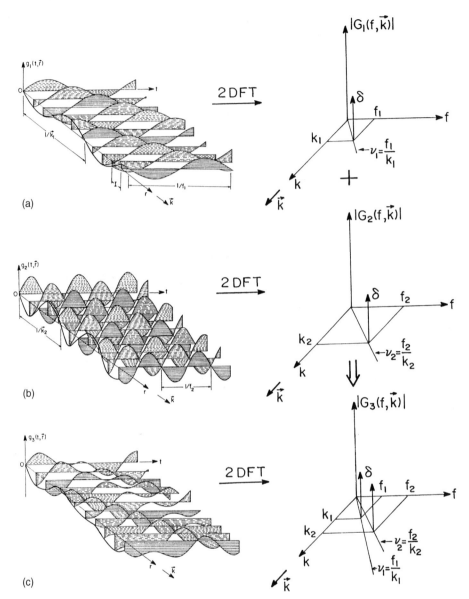

Figure 5.20 Frequency–wavenumber spectra of plane waves: (*a*) spatiotemporal (left) and frequency–wavenumber (right) representations of a sinusoidal wave of frequency f_1 traveling at propagation velocity v_1; (*b*) sinusoidal wave of frequency f_2 traveling at propagation velocity v_2; (*c*) sum of the above. The two-dimensional PSDs clearly show the component waveform spectra and propagation velocity.

the classical and high-resolution spectral estimation approaches. As shown in Figure 5.21, using spatiotemporal data $g(t,r)$,

$$g(t, r) = \begin{bmatrix} x_{0,0} & x_{1,0} & \cdots & x_{N-1,0} \\ x_{0,1} & x_{1,1} & \cdots & x_{N-1,1} \\ \cdot & \cdot & & \cdot \\ \cdot & \cdot & & \cdot \\ \cdot & \cdot & & \cdot \\ x_{0,R-1} & x_{1,R-1} & \cdots & x_{N-1,R-1} \end{bmatrix}$$

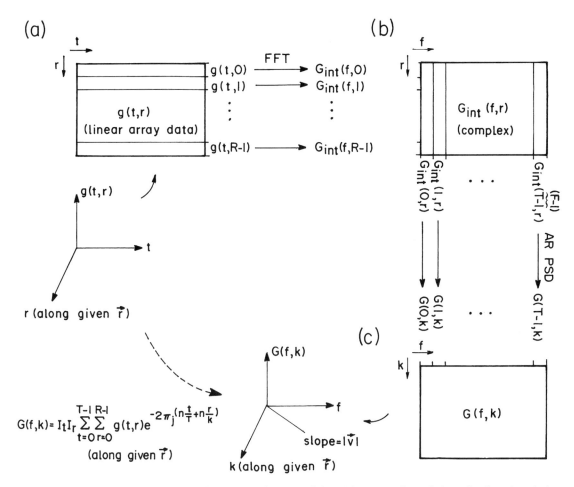

Figure 5.21 A hybrid two-dimensional spectral estimator: Spatiotemporal data (*a*) are transformed along the time domain into an intermediate array (*b*) through the application of a windowed FFT to every row of the original data. Applying an AR PSD estimator to every column of the intermediate array completes the two-dimensional PSD estimation process (c). [Reprinted from *Med. Eng. Phys.*, 17, D. Prutchi, A High-Resolution Large Array (HRLA) Surface EMG System, pages 442–454, 1995, with permission from Elsevier.]

an intermediate transform $G_{int}(f, r)$ is computed by applying the FFT along each row (time domain) of appropriately weighted data. The two-dimensional spectral estimate $G(f, k)$ is then completed by obtaining the AR-PSD of each column of complex numbers in the intermediate transform. In the more general case, using an array of sensors spread out over an area, and with a plane wave traveling in any direction under the array, a three-dimensional hybrid spectral estimator can determine not only the wave's components and their velocities but also each component's bearing.

A practical example of the use of this method is the analysis of biopotentials that can be picked up from skeletal muscle fibers using electrodes attached to the skin [Prutchi, 1994, 1995]. These biopotentials are caused by the sum of currents from action potentials that travel down individual muscle fibers responsible for the contraction of muscles. The conduction velocity, as well as the origin of these potentials, contains a wealth of information that can be used, for example, as an aid in the early diagnosis of nerve and muscle diseases. The large number of convoluted signals and the very small differences between their waveforms make it impossible to determine this information from spatiotemporal data (Figure 5.22*b*) recorded differentially using 32 surface electrodes placed 2.54 mm

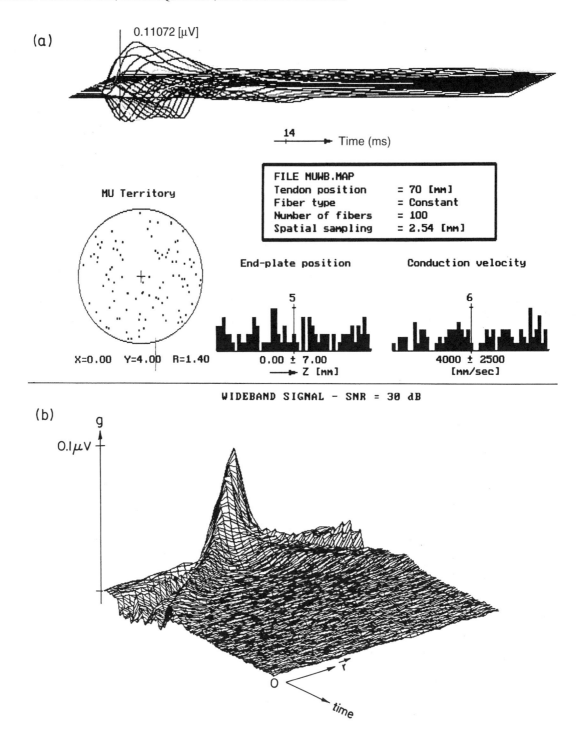

Figure 5.22 Frequency–wavenumber spectral estimation has been applied to analysis of the biopotentials recorded from a muscle twitch. (*a*) The complex spatiotemporal waveform has been analyzed to show information regarding the conduction velocity, origin, and location of the component potentials. Part (*b*) is a magnified view of the spatiotemporal data.

apart over the biceps brachii muscle. However, a complete analysis (Figure 5.22*a*) is possible through the use of multidimensional spectral estimates.

A Final Thought on Spectral Estimation

Of course, the Basic program in the book's ftp site may be too slow to cope with most real-time applications, but implementing both classical and high-resolution methods on a DSP is a relatively easy task. First, modern DSP chips are designed specifically to perform the convolution, vector arithmetic, and FFT operations in a minimal number of clock cycles. In addition, optimized subroutines to implement the most popular high-resolution algorithms are available, often in the public domain.

Multidimensional PSD estimation has a very high intrinsic parallelism because spectral estimates are taken independently for every dimension and thus can be solved efficiently within a parallel architecture. In other words, in array signal processing, where tasks require specific operations to be performed on innumerable data blocks, a parallel system exploits the full power of a number of processors that work concomitantly on different portions of the data toward solution of the larger problem.

As you can see, spectral analysis is a very convenient tool that can serve a number of engineering applications. Moreover, today's PCs have the power to implement modern PSD estimation algorithms with sufficient efficiency for experimenting and even for some real applications. With the enhanced capabilities of DSP chips, PCs with DSP coprocessors and laboratory spectrum analyzers with embedded DSPs become truly powerful and useful instruments. However, as you certainly understand by now, obtaining good spectral estimates is not only a matter of applying the algorithm blindly and watching the screen. Rather, knowledge about the spectral estimation methods, as well as empirical experience in their use, are of foremost importance in obtaining consistent results.

REFERENCES

Ackerman, B., High-Resolution Timing on a PC, *Circuit Cellar INK*, 24, 46–49, December 1991–January 1992.

Allen, G. R., Amateur Telemetry, *73 Magazine*, 72–76, July 1981.

Haykin, S., ed., *Array Signal Processing*, Prentice-Hall, Englewood Cliffs, NJ, 1985.

Jangi, S., and Y. Jain, Embedding Spectral Analysis in Equipment, *IEEE Spectrum*, 40–43, February 1991.

Marple, S. L., Jr., *Digital Spectral Analysis with Applications*, Prentice-Hall, Englewood Cliffs, NJ, 1987.

Prutchi, D., DSP Methods for Frequency–Wavenumber Analysis of the Array Surface Electromyogram, *Proceedings of the 5th International Conference on Signal Processing Applications and Technology*, 177–182, 1994.

Prutchi, D., A High-Resolution Large Array (HRLA) Surface EMG System, *Medical Engineering and Physics*, 17(6), 442–454, 1995.

Prutchi, D., Universal Sensor Interface, *Popular Electronics*, 39–44, June 1999.

Schulze, D. P., A PC Stopwatch, *Circuit Cellar INK*, 19, 22–23, February–March 1991.

Welch, P. D., The Use of a Fast Fourier Transform for the Estimation of Power Spectra: A Method Based on Time Averaging over Short Modified Periodograms, *IEEE Transactions on Audio and Electroacoustics*, AU-15(6), 70–73, 1967.

6

SIGNAL SOURCES FOR SIMULATION, TESTING, AND CALIBRATION

A common practice in evaluating the behavior of signal processing or control circuitry is to make use of an analog function generator to produce the necessary test input signals. The typical cookbook waveforms of the function generator are then used to investigate the behavior of the circuit when stimulated by sine, square, and triangle waves of different amplitude and frequency. Take, for example, the circuit of Figure 6.1. This module generates an extremely accurate sine-wave calibration signal. Its output signal is continuously variable within the range $10\,\mu V_{p\text{-}p}$ to $1\,mV_{p\text{-}p}$. The frequency is set by component selection in the range 2 MHz to 20 kHz. This circuit is meant for the precise calibration of biopotential amplifiers. Because of its high stability (sine-wave distortion of less than 0.2%), this circuit is ideal for measuring channel phase shift in topographic brain mappers and other biopotential array amplifiers.

As shown in Figure 6.2, the heart of the sine-wave generator circuit is Burr-Brown's Model 4423 precision quadrature oscillator. The frequency of the stable sinusoidal signal that appears at pin 1 is determined by the values of capacitors C6 and C7 as well as the values of resistors R1 and R2. A separate frequency-selection module should be assembled on an eight-pin DIP header for each frequency desired. The value and type of C6 and C7 (where C6 = C7) should be chosen according to Table 6.1.

After selecting the capacitor value required for a desired frequency range, the value of resistors R1 and R2 (where R1 = R2) in kilohms can be obtained from the expression

$$R1 = R2 = \frac{3.785f(C6 + 0.001)}{42.05 - 2f(C6 + 0.001)}$$

where f is the frequency desired in hertz and C6 = C7 is expressed in microfarads. It takes a certain amount of time to build up the amplitude of the sine-wave output to its full-scale value. By pressing the RESET pushbutton switch SW1 momentarily, the amplitude of the sinusoidal output is built up instantaneously. Degradation of signal distortion does not occur, as resistor R3 does not remain connected permanently within the oscillation circuit.

Resistors R4, R5, and R6 determine the gain of the op-amp (uncommited within the 4423 package). The op-amp, in which oscillatory behavior is prevented by C5, is used as a

Design and Development of Medical Electronic Instrumentation   By David Prutchi and Michael Norris
ISBN 0-471-67623-3   Copyright © 2005 John Wiley & Sons, Inc.

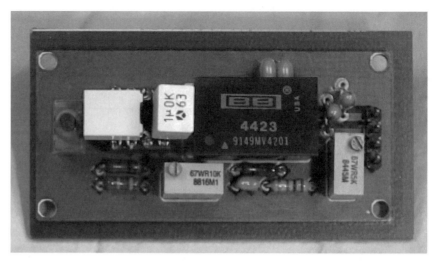

Figure 6.1   This module generates an extremely accurate, low-amplitude ($10\,\mu V_{p\text{-}p}$ to $1\,mV_{p\text{-}p}$) sine-wave signal for precise calibration of biopotential amplifiers. Because of its high stability (sine-wave distortion of less than 0.2%), this circuit is ideal for measuring channel phase shift in topographic brain mappers and other biopotential array amplifiers.

power output stage to decouple the precision quadrature oscillator from the output load. R5 is used to trim the amplitude of the signal at the op-amp output to exactly $1\,V_{p\text{-}p}$. R7 and R8 form an attenuator that divides the signal at the output of the op-amp to any amplitude desired between $10\,\mu V_{p\text{-}p}$ and $1\,mV_{p\text{-}p}$. To calibrate a single-ended biopotential amplifier, connect the ACTIVE input of the amplifier to J1-1 and the REFERENCE input to J1-2. For differential amplifiers, connect both REFERENCE and SUBJECT GROUND to J1-2.

ANALOG GENERATION OF ARBITRARY WAVEFORMS

In many applications, repetitive sine, square, and triangle waves are seldom representative of the signals that the equipment under test is designed to process. For example, the heart's electrical signal is a waveform consisting of a complex mixture of these basic waveshapes intertwined with intermittent baseline segments. Since a constant live feed of such signals may be impractical or even dangerous for testing biomedical equipment, dedicated signal sources had to be developed to be capable of synthesizing waveforms similar to those generated by their physiological counterparts. Similar requirements are evident for the generation of test signals representative of those produced by medical imaging sensors and other sources that cannot be simulated by plain sines, ramps, or square waves.

Figure 6.3 shows a simple circuit that can generate multiple synchronized repetitive waveforms of arbitrary shape. IC2 is a binary counter that causes the outputs of IC3 to go high in sequence. One at a time, each of these lines causes current to flow through R21 by way of its associated diode (D1–D15) and linear slider potentiometers (R1–R15). As the counter cycles, a stepped waveform appears across R21.

Think about a graphic equalizer—your stereo set probably has one, or you have seen them in pictures of recording studios. This type of equalizer is a multiband variable audio filter using slide controls as the amplitude-adjustable elements. It's named *graphic* for the positions of the sliders graphing the resulting frequency response of the equalizer. In a similar way, the basic shape of the stepped waveform across R21 tracks the shape described by the linear slider potentiometers as shown in Figure 6.4. Counter IC2 is clocked by the

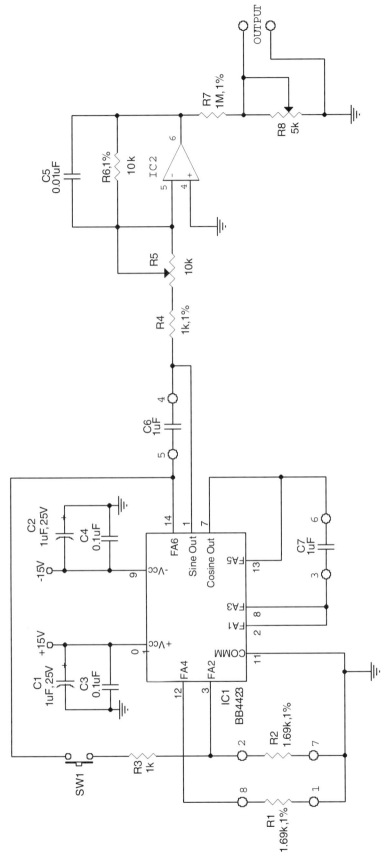

Figure 6.2 The sine-wave generator circuit is based on a Burr-Brown 4423 precision quadrature oscillator. The frequency of the stable sinusoidal signal is determined by the values of capacitors C6 and C7 as well as the values of resistors R1 and R2. A separate frequency-selection module should be assembled on an eight-pin DIP header for each frequency desired. With the component values shown, oscillation will occur at exactly 10 Hz.

251

TABLE 6.1   Capacitor Values and Types for the Precision
Sine-Wave Generator of Figure 6.2

Frequency Range	C6 = C7 (µF)	Capacitor Type
0.002–0.02 Hz	1000	Polycarbonate
0.02–0.2 Hz	100	Polystyrene
0.2–2 Hz	10	Teflon
2–20 Hz	1	Teflon
20–200 Hz	0.1	NPO ceramic
200 Hz–2 kHz	0.01	NPO ceramic
2–20 kHz	—	—

astable multivibrator formed by IC1A and IC1B. An approximate frequency for this oscillator is given by $f_s = 1/1.39(\text{R17} + \text{R18})\text{C1}$. Counting can happen only when IC2's enable line (pin 5) is set low. When trigger selector switch SW1 is in position 1, a trigger from the astable multivibrator formed by IC1C and IC1D clocks flip-flop IC4A, enabling IC2 to count. Once IC2 reaches a count of hex FF, output S15 of IC3 goes high. Flip-flop IC4B is clocked after the rising edge on output S15 propagates through Schmitt triggers IC1E and IC1F. The output of the flip-flop goes high, which resets IC4A and thus inhibits the counter. This action also pulls the counter's parallel load line (IC2 pin 1), which asynchronously forces the counter to hex 00. This state lasts for half a cycle of the sampling clock, when IC4B is reset by IC1A. It is possible to trigger the signal generator externally by supplying a 12-V clock signal to J1 and placing SW1 in position 2. Alternatively, placing SW1 in position 3 lets IC2 count freely, assigning equal time to each of its 16 cycles.

The setting of R1 defines a baseline level for the stepped signal voltage. The baseline level lasts as long as the counter is in the hex 00 state. As such, the stepped waveform developed across R21 is composed of a sequence of 14 levels (set by R2–R15), each presented for one cycle of the sample clock (output of IC1B), followed by a baseline level (set by R1) that lasts for one trigger clock cycle minus the time it takes the counter to clock through the 14 signal levels.

Although the step levels defined by the linear slider potentiometers are analog in nature, the time domain for the waveform is discrete. Of course, few real-world signals are appropriately represented by a stepped waveform with coarse jumps. Exceptions such as video-like streams from scanned sensor arrays do exist, but in most cases, the waveforms needed to develop or test a medical instrument should be representative of the smooth physiological signals that they are designed to process.

In Chapter 5 we discussed use of a low-pass filter to reconstruct a signal from samples acquired at a rate close to the Nyquist frequency. The same considerations apply here. After the signal across R21 is offset to zero by IC5A, it is low-pass-filtered by R24 and C13 to reconstruct a smooth waveform. Selecting the proper sampling rate and filter cutoff requires some consideration. Let's take, for example, the way in which you would set up the waveform generator to simulate an ECG signal. Figure 6.5 shows how the P–QRS–T complex could be represented by the position of the 14-step linear slider potentiometers (let's disregard the U-wave), while the isoelectric baseline level between the end of the T-wave and the start of the P-wave is set by R1. Each step would last 40 ms, which requires a sampling clock frequency of 25 Hz. If C1 = 0.1 µF and R17 = 100 kΩ, R18 needs to be set somewhere near 187 kΩ[1] for the multivibrator to oscillate at 25 Hz.

---

[1]The exact frequency of oscillation of a CMOS astable multivibrator also depends on the supply voltage and on the logic threshold voltage of the specific 40106 chip being used. The logic threshold can vary from 33 to 67% of the supply voltage from device to device.

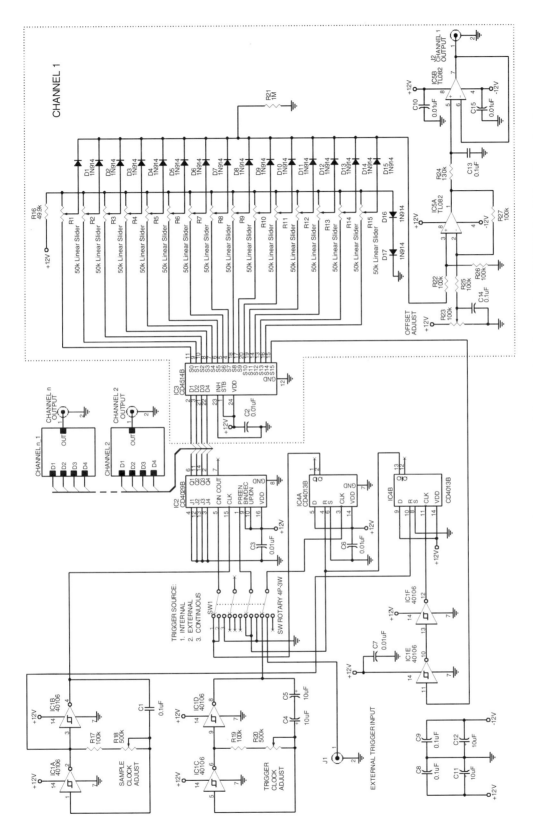

Figure 6.3 In this analog arbitrary signal generator, a counter sequentially scans through an analog ROM (IC3, D1–D16, and associated potentiometers) to generate a stepped waveform across R35 that is later smoothed through a low-pass filter (R38, C13). IC1A and IC1B generate the clock, which scans through the samples before resting at a baseline level. The waveform repeats once a new trigger is received by IC4A.

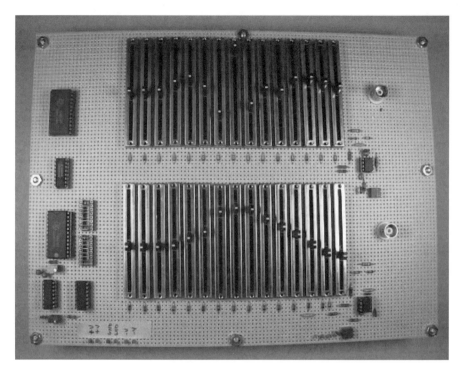

Figure 6.4   Just as in a graphic equalizer, the basic shape of the waveform generated by this arbitrary signal generator tracks the shape described by a bank of linear slider potentiometers.

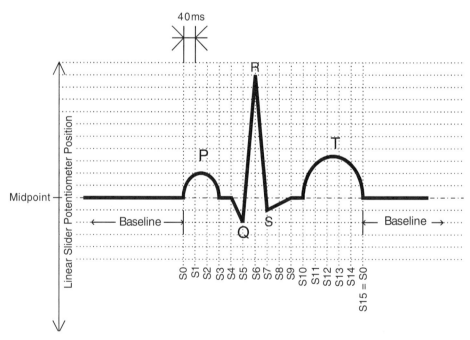

Figure 6.5   The P–QRS–T complex can be reconstructed from 14 analog samples. One more sample is needed to represent the isoelectric baseline level between the end of the T-wave and the start of the P-wave. Each step would last 40 ms, which requires a sampling clock frequency of 25 Hz. The trigger clock frequency needs to be 1 Hz to simulate a rate of 60 beats/min.

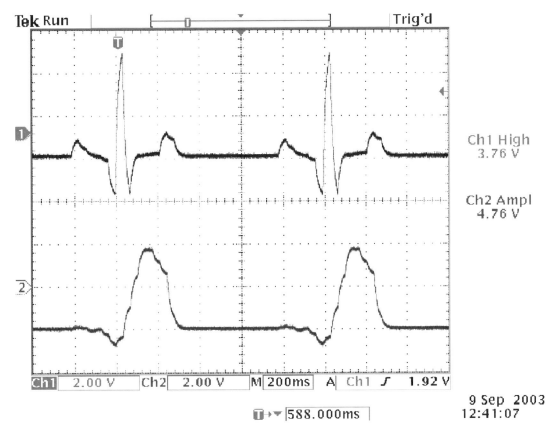

Figure 6.6   Synchronous signals can be added by connecting more channels to the clock data bus. This figure shows an oscilloscope print-out obtained from a two-channel cardiac signal generator set to simulate a patient's ECG and left-ventricular blood pressure signals.

The cutoff frequency of the output low-pass filter should be selected to be at least half of the sampling rate. Then, for $25\,Hz/2 = 1/2\pi(R24)(C13)$, and selecting $C13 = 0.1\,\mu F$, $R24 = 127,323\,\Omega$, which can be approximated by a 130-kΩ resistor. If a heart rate of 60 beats/min is chosen, R19, R20, C4, and C5 have to be selected to cause the trigger clock multivibrator to oscillate at 1 Hz. If $C4 = C5 = 10\,\mu F$ and $R19 = 100\,k\Omega$, R20 needs to set somewhere near 43 kΩ to yield a period of 1 s.

The capability of generating one or more synchronous signals can be added by connecting more channels to the clock data bus. This is especially useful when working with multiparameter monitors. In Figure 6.6, which shows the output of a two-channel cardiac signal generator, channel 1 simulates the surface ECG and channel 2 simulates a catheter-measured left-ventricular pressure waveform.

DIGITAL GENERATION OF ANALOG WAVEFORMS

More complex, nonstandard, real-world stimuli waveforms can easily be created as a numerical array and played back through a digital-to-analog converter (D/A or DAC) to yield analog waveforms of arbitrary complexity. This is the operating principle of a digital arbitrary waveform generator, or *arb*. Despite the simplicity of the concept, a PC program that would copy digital values stored in an array into a DAC would severely limit the maximum frequency of spectral components for the arbitrary signal. Even an assembly-language program copying the contents of sequential addresses on RAM to an I/O location would result in DAC writing rates of a few megapoints per second at the most.

Instead of having a DAC interfaced to memory through a microprocessor, arbs have dedicated RAM that is interfaced directly to the DAC. In this way, update rates are limited only by the access time of the RAM and by the speed of the DAC. As such, commercial arbs can currently be purchased with maximum writing rates in the vicinity of 1 gigapoint/s, yielding bandwidths of up to 500 MHz.

Direct Digital Synthesis

A generator capable of directly synthesizing an analog signal from digital data has at its core a memory that contains the full time-domain digital representation of the waveform desired. To generate the analog signal, the discretized point-by-point version of the waveform is played in a sequential manner through the generator's DAC. A simple form of such generator is a direct digital synthesizer (DDS). As shown in Figure 6.7, an address generator circuit controls the way in which samples stored in ROM are delivered to the DAC's input. On each clock pulse delivered to the address generator, a new address is issued to the ROM such that data for the next point in the sequence are presented to the DAC.

The ROM in a DDS generator most often contains data for generation of a complete single cycle of a sinusoidal waveform. The address generator is a simple counter, and the addresses it generates constitute the various phase angles ϕ for which samples of the sine wave $\sin \phi$ are available in ROM. The series of values coming out of this ROM lookup table as a function of incrementing phase angles is translated into an analog sine wave by a D/A converter (DAC).

Obviously, if the clock presented to the phase accumulator counter remains constant, the rate at which phases are generated remains constant, and the end result is a sine wave

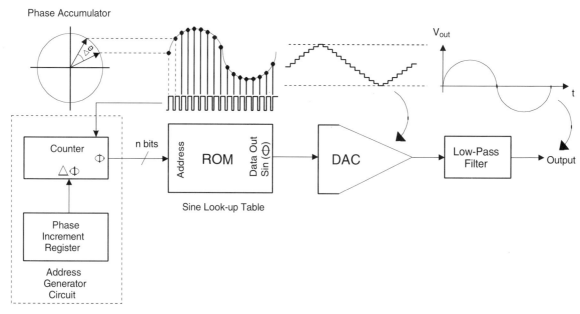

Figure 6.7 In a direct digital synthesizer (DDS), an address generator circuit or phase accumulator controls the way in which samples stored in a ROM lookup table are delivered to the DAC's input. Control over the output frequency is achieved by selecting an appropriate phase accumulator increment.

of a specific frequency. However, there is a very simple way in which DDS generators can be made to vary their sine output frequency without varying the clocking frequency. This is accomplished by allowing programmability of the phase increment value $\Delta\phi$. As such, if the output of the phase accumulator increments by $\Delta\phi$ on each incoming clock pulse, the frequency of the output sine wave is given by

$$f_{\sin e} = \frac{f_{\text{clock}} \Delta\phi}{360°}$$

The frequency resolution f_r of a DDS generator is thus defined by the number of bits n of the phase accumulator increment register and the clocking frequency:

$$f_r = \frac{f_{\text{clock}}}{2^n}$$

and the output frequency is set directly by the value W of the phase accumulator increment register:

$$f_{\sin e} = \frac{W f_{\text{clock}}}{2^n}$$

Since wide registers, large counters, and ample ROMs are easily integrated, IC DDS generators are becoming available capable of generating sine waves into the hundreds of MHz with incredibly high resolution.

Take, for example, the circuit presented in Figure 6.8. Here a Harris HSP45102 IC implements the phase accumulator and sine lookup table. The phase accumulator increment register in this IC is 32 bits wide and accepts a clock frequency of up to 40 MHz. In this way, the DDS IC is capable of providing data for the generation of sine waves as low as 0.009 Hz and as high as 20 MHz with a resolution of 0.009 Hz! The sinusoidal signal at the DAC's output is not infinitely pure, since at least some distortion is introduced by the fact that the digital samples presented to the DAC for translation are quantized in both time and amplitude. Time quantization results from the fact that the signal can change only at specific time intervals dictated by the clock. Amplitude quantization results from the discrete nature of the digital system itself. Samples of the infinitely continuous series of a sine are stored in ROM with finite resolution.

Obviously, time quantization errors are reduced by using as large a lookup table as possible. In the case of the Harris HSP45102, the lookup table is 8192 samples wide. It must be noted, however, that since the number of samples used to reconstruct the sinusoidal wave is equal to the ratio of the clock frequency (40 MHz) and the output frequency selected, time quantization errors get worse as the output frequency increases. Voltage quantization errors, on the other hand, are reduced by increasing the width of the data word presented to the DAC. Because price and complexity of a high-frequency DDS circuit increases with the DAC's resolution, a number of hobbyist projects have been presented using only 8-bit video DACs to gain simplicity at the expense of not taking full advantage of the HSP45102's 12-bit amplitude resolution [Craswell, 1995; Portugal, 1995].

In the DDS circuit of Figure 6.8, a 12-bit TTL-input-compatible ECL DAC takes full advantage of IC1's data word width. High-frequency harmonics generated by aliasing are low-passed by IC3. In more sophisticated systems, a very steep digitally tunable low-pass filter is used to pass the selected fundamental frequency and reject all of the sampling aliases. The use of an appropriate low-pass filter (usually, an elliptic filter) becomes critical for generating clean output at high frequencies since steps become increasingly large, and the DAC output resembles a sine wave less and less. For example, while a 40-kHz

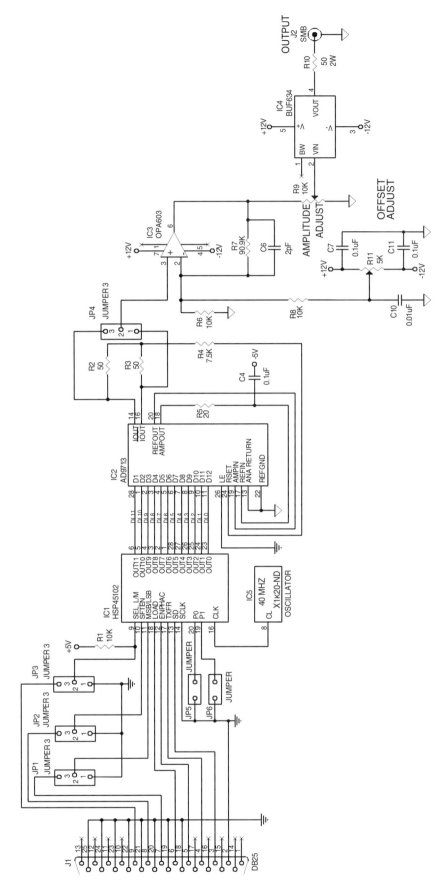

Figure 6.8 A simple and very versatile DDS generator can be built around a Harris HSP45102 IC, which implements the phase accumulator and sine lookup table. An ECL DAC converts 12-bit sine data into an analog output which is then filtered, buffered, and scaled. The DDS is programmed through the PC printer port.

TABLE 6.2   Input Lines P0 and P1 (Pins 19 and 20) of the HSP45102 Control the Introduction of a Phase Offset to the Output of the Phase Accumulator

P1	P0	Phase Shift
0	0	0
0	1	90
1	0	270
1	1	180

output signal is generated using 1000 samples per cycle, a 13.33-MHz signal is generated using barely 3 samples per cycle!

Two 32-bit phase accumulator increment registers are available onboard the HSP45102. A digital input on pin 9 selects which one of these registers is used at any given time for the generation of a sine wave. This allows direct frequency-shift keying (FSK) modulation of the output. In addition, the DDS generator allows us to change the phase on the fly by selecting the state of the P0 and P1 lines (pins 19 and 20), as shown in Table 6.2. These lines can then be used for selecting in-phase or quadrature data bits for QPSK modulation of the analog signal output.

Programming the HSP45102 is accomplished by loading 64 bits of data for the two-phase accumulator increment registers through the data input pin (SD) in serial format. While maintaining the shift-enable (*SFTEN) pin low, each data bit is fed by a rising edge on the clock input pin (SCLK) of IC11. Sine-wave generation is turned on and off by controlling the *ENPHAC pin. The TXFER* input line is used to control the transfer of the phase accumulator increment register selected by the SEL_L/M* line (pin 9) to the phase accumulator's input register. In this design we have retained printer port pin use compatibility with a DDS generator described in QST [Craswell, 1995], since the control software for that DDS is freely available over the Web as freeware, downloadable from `www2.arrl.org/files/qst-binaries/`.

ARB BASICS

As shown in Figure 6.9, an arbitrary waveform generator shares the basic building blocks of a DDS generator. Instead of a ROM sine lookup table, however, a full time-domain digital representation of the arbitrary waveform is downloaded into RAM. In addition, the counter is not thought of as a phase accumulator, and the means are provided to be able to define arbitrarily the last data point of the waveform cycle (end address). In this way, the waveform can be replayed continuously by looping back from the last point of the waveform's sequence to the address of the RAM location for the first point of the sequence (start address).

For some applications, it is necessary that the waveform be issued only once after a trigger event. To do so, additional circuitry within the address generator can receive a trigger signal that is used to allow addresses to be cycled once between the beginning and end of a waveform sequence for every triggering event. A typical application requiring this capability is, for example, the testing ultrasonic echo systems, where the simulated "echo" generated by the arb must be synchronized to excitation of the transmitting transducer.

One other difference with a DDS generator is that instead of maintaining the clock rate constant and jumping over sample points to change the period of a cycle, the clock frequency in an arb is programmable. In this way, the waveform can be compressed or expanded through time, resulting in a controlled shift in frequency of all spectral components of the

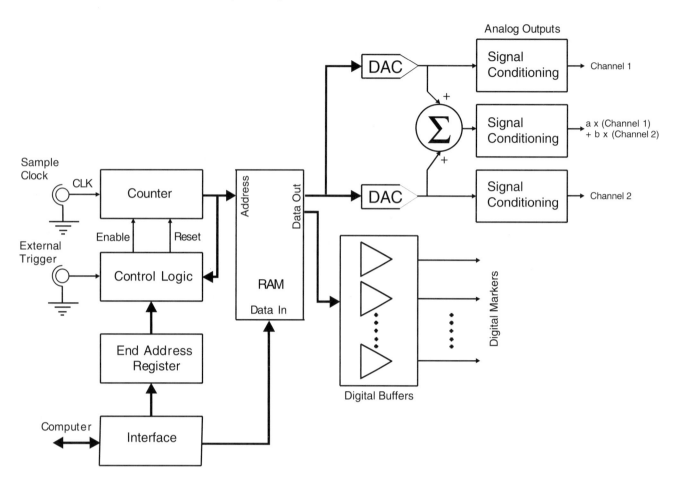

Figure 6.9   An arbitrary waveform generator has at its core a RAM that contains the full time-domain digital representation of the waveform desired. To generate the analog signal, the discretized point-by-point version of the waveform is played in a sequential manner through the generator's DAC. Waveforms can be replayed continuously by looping back from the last point of the waveform's sequence to the address of the RAM location for the first point of the sequence. Triggered operation is also possible by cycling through the memory contents only upon receiving a trigger signal.

waveform. Of course, reproducing a signal requires the stored waveform to have been sampled at a rate of at least twice its highest-frequency component, and an appropriate interpolating low-pass filter should be used. In turn, this means that the complexity and time duration of the reproduced waveform are limited by the arb's memory size, or *depth*. The time duration of the output waveform is given by

$$T_{\text{waveform}} = \frac{\text{number of waveform points}}{f_{\text{clock}}} = \frac{\text{end address} - \text{start address}}{f_{\text{clock}}}$$

Speech, for example, requires a sampling minimum speed of approximately 8 kilosamples/s. An arb with a depth of 32 kilosamples would then suffice for only 4.096 s of recording. When keeping the memory size fixed, longer waveform durations can be achieved only by limiting the bandwidth to allow lower sampling rates. Obviously, limiting the bandwidth results in limiting the complexity of the waveform by reducing the number of spectral components available to describe the details of waveform.

If more than sufficient memory is available in an arb to generate a waveform, additional memory can be used for a second waveform channel. Since both channels are generated using a single clock, the two output waveforms are precisely synchronized in time. This capability is essential for testing instruments that derive their measurements from the phase relationship between two signals. Additionally, purely digital lines or *marker channels* are sometimes offered to provide synchronization and position markers which are coincident with specified points of the arb waveform. These can be very useful for triggering oscilloscopes or other external instruments at specific times within the arb waveform cycle.

Beyond generating synchronized signals, however, the greatest advantage of having an additional channel is the possibility of summing both channels. In this way, two synchronized arbitrary components of a single waveform can be controlled independently, making it possible to test the effect of each component on the system. For example, in order to study the immunity of a circuit to an unwanted phenomenon, channel 1 could be loaded with the waveform that is normally seen by the system under test. Channel 2, however, could be loaded with the anomaly at the desired time within the normal waveform. Then, by varying the gain of channel 2, the amplitude of the anomaly can be adjusted without changing the amplitude of the normal signal.

Summing of two arb channels can also be used to extend the dynamic range of the combined signal beyond the maximum dynamic of each independent channel. Setting the gain of the summed channels to different values makes it possible to generate large signals that have very small features. Here, the macroscopic changes would occupy the full dynamic range of one of the channels. The smaller "details" of the waveform would then be programmed to occupy the full dynamic range of the other channel. By ratioing the gains between channels correctly, the summed signal can have a theoretical maximum resolution equal to the sum of the independent channels' resolutions.

PC-Programmable Arb

The instrument shown in Figure 6.10 is a simple arb built using standard SRAMs, a few counter ICs, some glue logic, and DAC ICs. In the arb project presented here, three 32k × 8 bit RAMs are used to store two 12-bit waveforms. An additional RAM IC provides seven marker channels, and the additional bit is used to encode the last valid data sample of a waveform sequence. As shown in Figure 6.11, the 15-bit address generator of the arb is formed by a chain of 74LS191 synchronous counters (IC1–IC4). The output of the counter chain is presented to 50-ns access-time SRAMs (IC5–IC8). From Figure 6.12 it can be appreciated that as long as IC1 is enabled, each clock pulse supplied in parallel to all counter ICs causes the address to advance. This process continues until the address points to a data element (D31) on IC8, in which bit 7 is low. This causes the asynchronous reset of the counter chain.

Notice how the arb's circuitry ensures that each sample of the waveform sequence has equal length. Data contents presented on the RAM data bus (D0–D30) are latched on the opposite edges of the clock than those which cause address transitions. In this way, since the data at the output of latches IC9–IC12 lag the data at the inputs of these latches by half of a clock cycle, the reset signal issued when the counters reach END_ADDRESS+1 (the location in which bit 7 of IC8 is low) causes the address to be reset to zero without upsetting the data corresponding to END_ADDRESS. The next falling edge on the clock line causes the data contents of the first RAM address to be presented at the output of the latches. Obviously, while the amount of time for which the first address is available is shorter than that of any other address, the data corresponding to it is presented at the output of the latches for exactly the same amount of time as for any other address.

Figure 6.10   This simple yet versatile arbitrary waveform generator can be programmed from the printer port of a PC. Once loaded with an array of digital data, this arb acts as a stand-alone instrument capable of delivering two simultaneous analog signals, each with an amplitude resolution of 12 bits, and variable temporal resolution down to 50 ns (20 megapoints/s).

When triggered rather than continuous cycling through the waveform is desired, flip-flop IC14 is allowed to control IC1's enable line by way of switch SW1. In the triggered mode, the flip-flop's *Q output goes low when enabled by the rising edge of a trigger pulse presented to its clock input line. This state is maintained until reset at the end of the waveform cycle by the same reset pulse that zeroes the counter chain. In this mode, trigger ambiguity is less than one clock cycle.

Downloading and uploading RAM waveform data from and to the PC is done through the printer port under a simple serial protocol implemented in the software available in the accompanying CD-ROM. On the Arb, the chain of 74LS323 ICs (IC16–IC19) of Figure 6.13 forms a 32-bit serial-to-parallel and parallel-to-serial converter. When the remote mode of operation is selected by the computer (digital low on bit 1 of the output port of the printer port), LOC/*REM goes low, causing IC20 to transfer control of the clock (CLK), address generator reset (*RESET), RAM output-enable (*OE), and RAM write (*WR) to the lines of the printer port. The mode control lines (pins 1 and 19) of the 74LS323s select between hold, shift left, shift right, and parallel load of the bits of the chain's 32-bit register. Data are clocked serially into IC16 and shifted down the chain toward IC19 by each rising edge of the serial clock line (SCLK). Once a complete 32-bit word is positioned in the register of the chain, IC16–IC19 drive the RAM data bus, and a write strobe is used to store the register's contents in the current address. The address generator is then advanced, and the cycle is repeated to store successive waveform data points. Data can be read from the RAM into the computer by reversing this process. Once an address is selected, data can be loaded from the RAM data bus into the register formed by

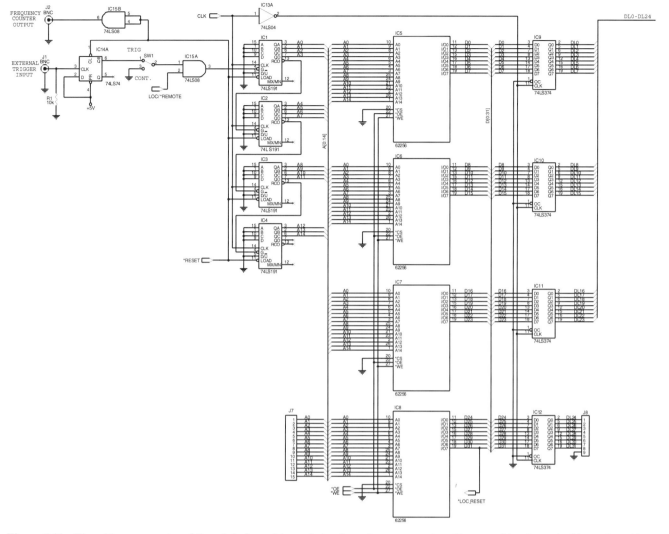

Figure 6.11 The address generator of the arb is formed by a chain of synchronous counters. Upon reaching the end address, the address generator resets, and the next data latched corresponds to that of the first RAM address. For nonvolatile operation, the RAMs should be mounted on Dallas Semiconductor's DS1213D SmartSockets.

the chain of IC16–IC19. The register's contents are then shifted out of IC16 into one of the status input lines of the printer port (pin 10 of J4).

As shown in Figure 6.14, two different DACs can be used with the arb. An Analog Devices AD9713 high-speed ECL DAC capable of updating its output at up to 100 megasamples(MS)/s can restore high-frequency signals with high resolution. Alternatively, the lower-cost AD667 provides more limited performance for applications which require DAC writing speeds of no more than 300 kilosamples/s. Unfortunately, it is difficult to take full advantage of the AD9713s, since when using these high-speed DACs, the arb's speed will be limited by the RAMs' access time. Under the direct addressing architecture used by this arb, RAMs with 50 ns access time will allow a maximum writing speed of $1/50 \times 10^{-9}\,\mathrm{s} = 20\,\mathrm{MS/s}$. Achieving 100-MS/s writing speeds would require 10-ns RAMs, which although available (e.g., cache RAM), are very costly, limited in size, and generally power-hungry.

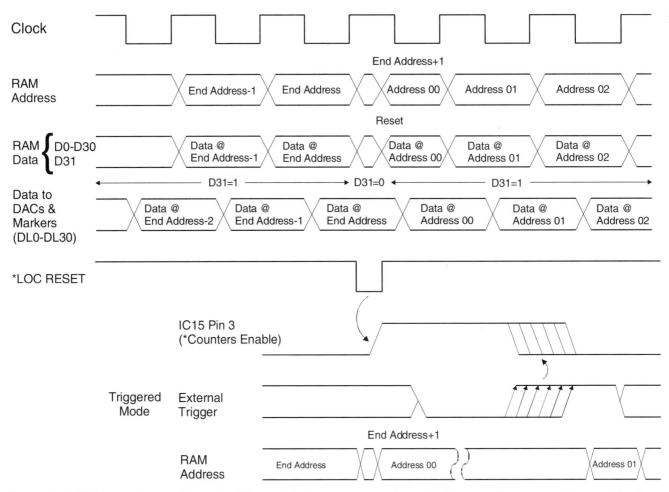

Figure 6.12   Under operating conditions, the arb's control logic ensures that each sample of the waveform sequence has equal length by presenting the data contents only on the opposite edges of the clock from those that cause address transitions. In the triggered mode, trigger ambiguity is less than one clock cycle.

Rather than using a direct addressing scheme, very high speed arbs overcome the RAMs' access time shortcomings by operating several RAM banks in parallel. In this multiplexed addressing scheme, one or more RAM banks are being accessed and allowed to settle while current data are taken from a different RAM. As the address is updated, data are taken from one of the RAMs that already has valid data available. A 4:1 multiplexed memory arb, for example, could use four low-cost 50-ns RAMs to achieve 80 MS/s. We chose not to use the more complex multiplexed approach, since 20 MS/s provided us with sufficient flexibility for our applications in the generation of relatively low frequency signals for evaluating biomedical instruments.

Once analog signals are available at the DAC outputs, the circuit provides for their offseting and scaling prior to being buffered for output. In addition, a summing channel is provided to expand the arb's versatility. The local sampling clock is generated by IC33, Maxim's MAX038 high-frequency waveform generator IC. Although this IC is typically used as a function generator, it is used within the circuit of Figure 6.15 as an oscillator whose frequency can be controlled over the range 20 Hz to 20 MHz. Alternatively, the

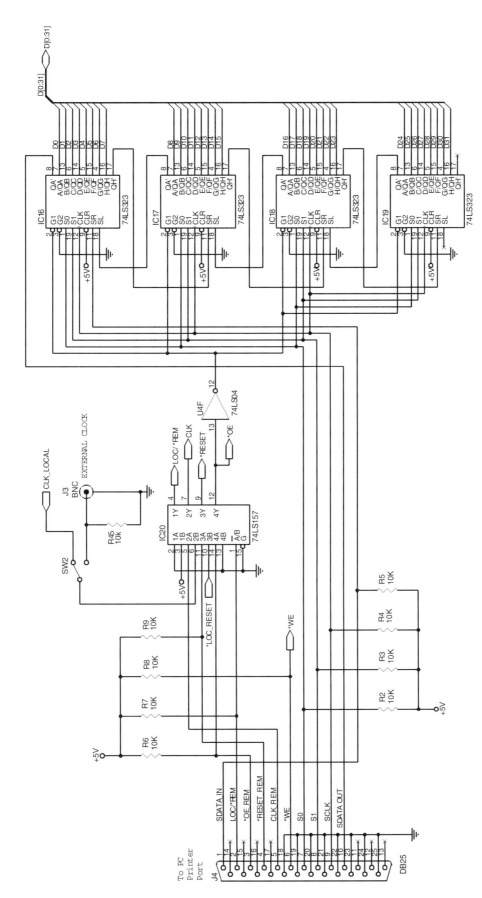

Figure 6.13 Downloading and uploading RAM waveform data from and to the PC is done through the printer port under a simple serial protocol. The chain of 74LS323 ICs forms a 32-bit serial-to-parallel and parallel-to-serial converter.

265

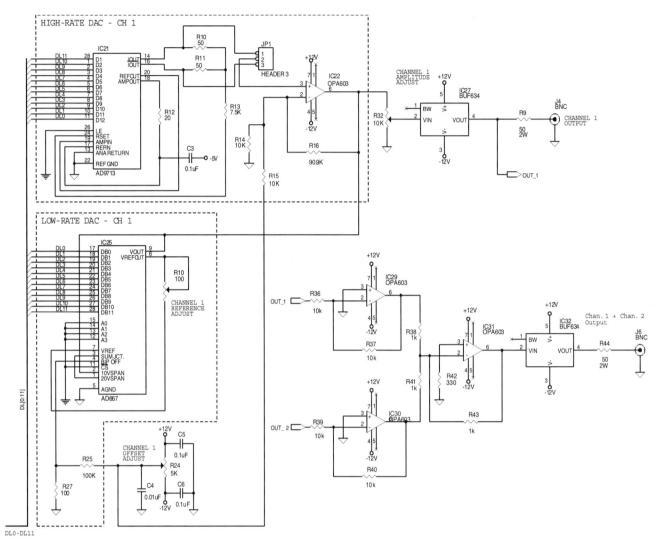

Figure 6.14 Two different DACs can be used with the arb: A high-speed ECL DAC capable of updating its output at up to 100 mega-samples/s can restore high-frequency signals with high resolution. Alternatively, a lower-cost DAC provides more limited performance for applications that require writing speeds of up to 300 kilosamples/s. The DAC analog outputs are then offset and scaled as needed. In addition, a summing channel is provided to expand the arb's versatility.

sampling clock can be supplied by an external TTL-level clock through connector J3 and switch SW2.

The arb's circuitry requires a supply of $+5$ V for the logic circuitry, -5 V for the ECL logic of the high-speed DACs, and ±12 V for the analog circuitry. The power supply of Figure 6.16 generates these voltages from a 12-V ac input. As shown in these figures, the arb, will lose the waveform data as soon as power is removed. For nonvolatile operation, however, the RAMs may be mounted on Dallas Semiconductor's SmartSocket DS1213D intelligent sockets. Remember that these sockets are designed to be compatible with RAMs of up to 128k $\times$ 8. For this reason, four more PCB pads than those required for each RAM IC are required when using the DS1213Ds. The μPD43256B RAMs are then mounted on pins 3–30 of the SmartSockets. Another option is to use Dallas Semiconductor DS1210 ICs to handle RAM power backup from a small battery [Bachiochi, 1990].

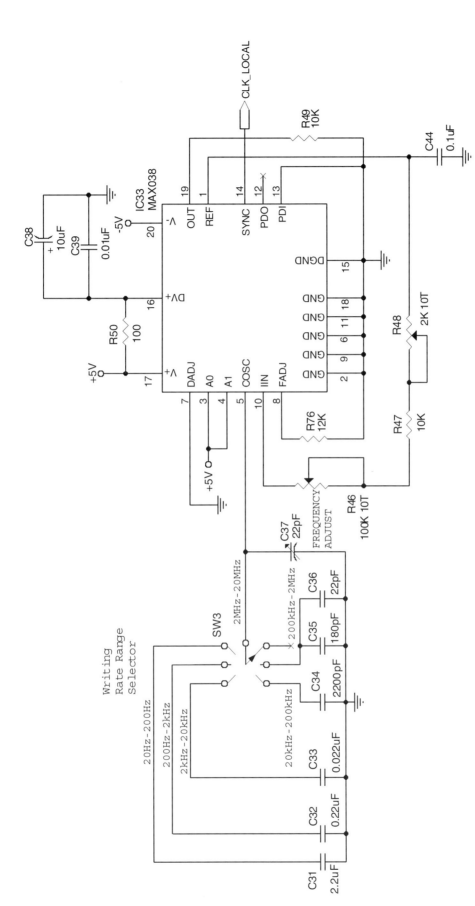

Figure 6.15 The local sampling clock is generated by IC33, Maxim's MAX038 high-frequency waveform generator IC. Although this IC is typically used as a function generator, it is used within the arb as a clock oscillator whose frequency can be controlled over the range 20 Hz to 20 MHz.

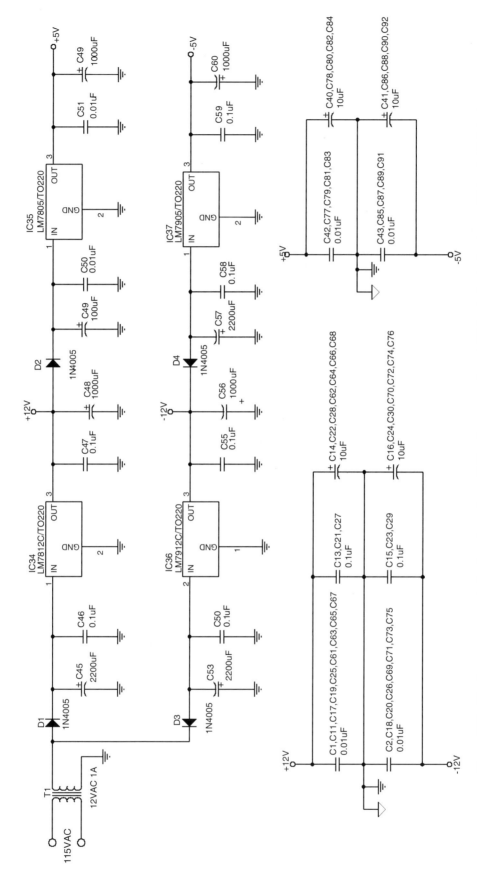

Figure 6.16 The linear power supply for the arb requires a 12 V ac input to produce +5 V for the logic circuitry, −5 V for the ECL logic of the high-speed DACs, and ±12 V for the analog circuitry. The same circuit can be used for powering the DDS generator.

Finally, a word of caution when building this arb. High-frequency clocks and signals demand proper layout techniques [Prutchi, 1994]. Preferably, use a multiple-layer printed circuit board. Separate the analog ground from the digital ground and join them at a single point at the power source. Maintain interconnection over the buses short and of equal length. In addition, use good-quality high-frequency capacitors to decouple the power rails in close proximity to the power input pins of each IC.

Creating Arbitrary Waveforms

Signal creation for reproduction by an arb is usually accomplished in one of two ways: (1) capturing an analog signal using a digital storage oscilloscope (DSO), or (2) creating the waveform on a PC by generating a numerical representation of the waveform's mathematical formulation. One of the most popular and useful packages used in the development of signal-processing algorithms is Matlab by The MathWorks Inc. Complex waveforms can be generated by this package's enormous library of mathematical functions. With an arb, test signals synthesized within Matlab or any other scientific programming environment can be used to excite the real hardware of a medical instrument, not just its algorithms.

The following Matlab function (mat2arb.m) is capable of saving two vector variables from its environment in a format that can be loaded into the PC-programmable arb described earlier.

```
function v = mat2arb(x, filename);
%MAT2ARB saves 2 vectors as a file that can be loaded into the
%2-channel Arbitrary Waveform Generator
% x = input matrix [vector 1 ; vector 2] sample value range = 0 to 4095
% filename = filename where arb file is to be saved entered as 'filename.ext'
%
%
%
if min(min(x))<0;
  disp('minimum sample value found')
  disp(min(min(x)))
  error('sample value range must be between 0 and 4095')
end
if max(max(x))>4095;
  disp('maximum sample value found')
  disp(max(max(x)))
  error('sample value range must be between 0 and 4095')
end
n = length(x);                % n is the length of the vectors
zer=zeros(1, n);              % generate vector of zeros for
                             % marker channels
y = 1:n;                     % generate vector of sample

                             % indexes
v=[x(1, y);x(2, y);zer(y);zer(y)];  % marker channels always low
fid = fopen(filename, 'w');   % open file for output
fprintf(fid,' %1g , %1g ,    % write file in format suitable
%1g , %1g \r', v);
                             % for ARB
fclose(fid)                  % close file
```

Although short Basic programs or numerical processing packages such as Matlab can be used to generate waveform data, truly flexible waveform creation is possible only through dedicated software. One of our favorite packages is Pragmatic Instruments' WaveWorks Pro since it offers a very intuitive environment for the creation of waveforms from a comprehensive menu of standard waveform templates, math operations, and transfer functions. Waveforms can also be imported from other programs or uploaded directly through GPIB or RS232 from popular DSOs. In addition, waveform synthesis and analysis can be performed in either the time or frequency domains.

Figure 6.17 presents an example of how easy it is to create waveforms with a waveform design package such as WaveWorks Pro. The software has 30 standard waveshapes with programmable parameters. These alone provide immediate solutions for the generation of test waveforms for general-purpose applications (e.g., sinusoidal, square, triangular waves), communications testing (e.g., AM, FM, BFSK, QPSK, NTSC waveforms), as well as other signals for advanced signal processing and control (e.g., $\sin(x)/x$, ECG waveform, digital and analog noise). After a waveform is defined, it can be modified using the 20 predefined transfer functions or 13 mathematical operators. Once the desired waveform is created, an FFT-based spectral estimator offers frequency analysis with the possibility of spectral editing and IFFT-based transformation back into time domain. Finally, a long and

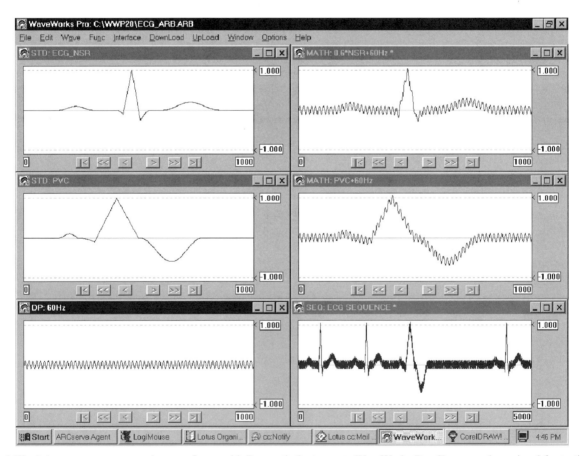

Figure 6.17   It is easy to create complex waveforms with Pragmatic Instruments WaveWorks Pro. For example, a signal for testing electrocardiography equipment can be created by first defining the basic components of the waveform from predefined templates. On the left: a beat corresponding to the heart's normal sinus rhythm (NSR), a premature ventricular contraction (PVC), and 60-Hz sinusoidal interference. A seamless link of the signals "contaminated" by 60 Hz (right) results in a realistic-looking ECG containing normal beats mixed with PVC episodes followed by a long compensatory pause before returning to NSR.

complex waveform can be created within WaveWorks Pro by looping and seamless linking of previously created waveforms.

As you can imagine, the introduction of cheaper, faster, high-resolution DACs, wider RAMs, and higher-performance microprocessors is making it possible for digital waveform generators to replace analog sources rapidly in many applications. High-performance integrated DDS generators have taken over in the spread-spectrum communications field and are the key elements that enable low-cost high-speed data links to be integrated into a wide variety of patient monitors and wirelessly networked medical instruments.

Arbs are also becoming very popular with design and test engineers. Arbs are more versatile sources than their analog counterparts. In fact, even when generating "standard" waveforms, arbs can compete with analog generators. For example, arbs can output ramps and triangle waves with higher linearity and sharper corners than an analog generator. Similarly, an arb-generated sine can have far better THD and frequency accuracy than the one generated by an analog circuit.

Of course, the neat control and waveform-design screens of commercial arbs, their powerful DSP microprocessors, and their exotic high-frequency mixed-mode circuitry makes them costly pieces of equipment. Most commercial arbs are priced in the range $3000 to $7000. On the other hand, an analog signal generator with similar bandwidth costs just a few hundred dollars. So don't feel that your reliable analog waveform generator no longer deserves its space on the workbench, but keep the arb in mind when your application demands ultimate flexibility without a compromise on performance.

PC Sound Card as an Arb

The PC sound card is a true audio-range arb (useful in the range of 20 Hz to 20 kHz). It takes the waveform definition stored in the computer's memory and plays it back as an analog signal. The simplest way of generating an arbitrary wave through the PC sound card is to store it as a .wav file and play it back using Window's Media Player utility. A .wav file is just a series of samples, preceded by a header that tells the player program important things such as the sampling rate and the number of bits in the sample. The player program reads the header, sets up the sound card, and then feeds the samples to the card's digital-to-analog converter.

PC multimedia data are often encoded in the RIFF file format. RIFF is based on chunks and subchunks. Each chunk has a type, represented by a four-character tag. This chunk type comes first in the file, followed by the size of the chunk, then the contents of the chunk. The .wav format is a subset of RIFF used for storing digital audio and requires two types of chunks: (1) the format (fmt) chunk, which describes the sample rate, sample width, and so on, and (2) the data chunk, which contains the actual samples. .wav can also contain any other chunk type allowed by RIFF, including LIST chunks, which are used to contain optional kinds of data, such as the copyright date and author's name. Chunks can appear in any order. In its simplest form, the .wav format starts with the RIFF header of Table 6.3.

TABLE 6.3   Contents of the RIFF Header of a .wav File

Offset	Length (bytes)	Contents
0	4	RIFF
4	4	<file length - 8> where the "8" is the length of the first two entries (i.e., the second entry is the number of bytes that follow in the file)
8	4	WAVE

The .wav specification supports a number of different compression algorithms. The format tag entry in the fmt chunk indicates the type of compression used. A value of 1 indicates linear pulse-code modulation (PCM), which is a straight, or uncompressed, encoding of the samples, which is just the exact amplitude of each sample. The fmt chunk describes the sample format (Table 6.4), and the data chunk contains the sample data (Table 6.5). All numeric data fields are in the Intel format of low–high byte (usually referred to as *little-endian*). Eight-bit samples are stored as unsigned bytes, ranging from 0 to 255; 16-bit samples are stored as 2's-complement signed integers, ranging from -32768 to 32767.

If you are a Matlab user, you can avoid the hassle of file formatting by playing a data stream directly from within the Matlab environment using the "sound" command. Matlab can also write .wav files from data variables or read the .wav file PCM-encoded signal into data that can be manipulated by Matlab. Another possibility is to use a professional waveform

TABLE 6.4   Contents of the Format Chunk of a .wav File

Offset	Length (bytes)	Contents
12	4	fmt
16	4	0x00000010
		which is the length of the fmt data (16 bytes)
20	2	0x0001
		which is the data-encoding format tag: 1 = PCM
22	2	<channels>
		which defines the number of channels: 1, mono; 2, stereo
24	4	<sample rate>
		in samples per second (e.g., 44,100)
28	4	<bytes/second>
		sample rate × block align
32	2	<block align>
		channels × bits/sample/8
34	2	<bits/sample>
		8 or 16

TABLE 6.5   The Data Chunk of a .wav File Contains the Actual Sample Data

Offset	Length (bytes)	Contents
36	4	data
40	4	<length of the data block>
44	As needed for data	<sample data>
		For multichannel data, samples are interleaved between channels:
		sample 0 for channel 0
		sample 0 for channel 1
		sample 1 for channel 0
		sample 1 for channel 1
		.
		.
		.
		where channel 0 is the left channel and channel 1 is the right channel; the sample data must end on an even byte boundary.

design package such as WaveWorks' Pro to create the desired signal, and then use software that is freely available on the Web, which can play data written in plain ASCII straight through the PC's sound card. For example, SoundArb version 1.02 (sasetup.exe freeware for Windows 9x, NT) by David Sherman Engineering Co. is a free PC sound card signal generator program that not only lets you select standard waveforms but also load and play arbitrary waveforms from a text wave table file with full control over frequency, amplitude, and trigger mode. Updates to SoundArb, as well as more sophisticated arb software, are available from David Sherman at `www.wavebuilder.com`.

The output jack of a typical sound card carries an amplified ac-coupled signal (20 Hz to 20 kHz) capable of direct driving of 8-Ω speakers with some 2 W of power. The actual output level is uncalibrated and will depend on the settings of the volume lever (which you can access by double-clicking the speaker icon in the Windows tray). The only way to set the amplitude to a known voltage is by observing the waveform on an oscilloscope. Since sound cards are meant to output sound, the volume control usually has a limited number of discrete steps (e.g., 16) that follow a two-part piecewise-logarithmic curve.

Converting the Sound Card into a Precision DC-Coupled Arb

Unfortunately, the typical 20-Hz high-pass cutoff frequency of consumer-grade sound cards makes them unsuitable for simulating most physiological signals. In addition, the output stage of sound cards does not usually have the output linearity or passband flatness required for accurate reproduction of low-frequency signals. However, a phase-locked loop (PLL) circuit and some software can turn a sound card into a precision dc-coupled arb. The idea is to use a software FM modulator to turn the arbitrary signal to be generated into an audio tone that is played through the PC sound card. The tone's frequency varies as a function of the arbitrary signal desired. The arbitrary signal is then recovered by hardware FM demodulation of the audio signal.

Matlab has a function (vco.m) that simulates operation of a voltage-controlled oscillator, essentially an FM modulator. The following code shows how easy it is to use this function to generate an FM signal by modulating a carrier (of frequency Fc) with an arbitrary signal contained in vector x (sampled at a rate Fs of more than twice Fc, with an amplitude range of ± 1):

```
Fs = 5000;        % Select arbitrary signal sampling frequency in Hz
Fc = 1687;        % Select VCO carrier frequency in Hz
moddev = 40;      % Percent FM frequency deviation
y = vco(x, [1-moddev/100 1+moddev/100]*Fc, Fs); % VCO simulation
sound(y, Fs)      % Play modulated signal through PC sound card
```

You can use the following Matlab command to look at the spectrum of the FM signal that is played through the sound card:

```
specgram(y, 512, Fs, kaiser(256, 5), 220)
```

If you are not a Matlab user, you can write a program to generate the FM signal from the arbitrary waveform by remembering that an FM signal $s(t)$ is expressed by

$$s(t) = A_c \cos\left[2\pi f_c t + 2\pi k_f \int_0^t m(\tau)\, \partial \tau\right]$$

where $m(t)$ is the modulating signal (the arbitrary waveform), f_c the carrier frequency, A_c the carrier amplitude, and k_f defines the frequency deviation caused by $m(t)$. The instantaneous frequency of the signal is larger than the carrier frequency when the signal $m(t)$ is positive and is smaller when $m(t)$ is negative.

The circuit of Figure 6.18 is used to demodulate the FM signal from the sound card. The signal coming from the sound card is ac-coupled by C17 and amplitude-limited by IC1D. Then a NE565C PLL IC demodulates the FM signal. The PLL tracks the incoming carrier signal and internally estimates the signal based on the frequency of its internal VCO set by R2, R4, and C2. The "error" between the actual carrier frequency and the estimate is the data signal when the PLL is locked. A suitable FM frequency deviation for this circuit is ±40%, which allows the bandwidth of the arbitrary signal to be reproduced to be approximately 18.5% of the carrier frequency. Table 6.6 gives the signal reproduction characteristics for some of the standard frequencies used in FM tape recorders, an application that uses the same FM modulation/demodulation schemes. The loop output of the PLL IC is fed into a unity-gain differential amplifier (IC1C). The common-mode rejection of this amplifier is used to eliminate DC and high-frequency carrier components present at the output of the PLL.

A Maxim MAX280 switched-capacitor filter IC is used to remove residual carrier-frequency signal components from the waveform desired. This IC is a fifth-order all-pole low-pass filter with no dc error, making it an excellent choice for processing low-frequency signals. The filter IC uses an external resistor (R9) and capacitor (C10) to isolate the fourth-order filter implemented within the IC from the dc signal path. The external resistor and capacitor are used as part of the filter's feedback loop and also form one pole for the overall filter circuit. The values of these components are chosen such that

$$f_{\text{cutoff}} = \frac{1.62}{2\pi(\text{R9})(\text{C10})}$$

where R9 should be around $20\,\text{k}\Omega$.

Now, for the Matlab code example given above, the demodulated signal bandwidth is expected to be $18.5\% \times f_c = 18.5\% \times 1687\,\text{Hz} = 312\,\text{Hz}$, which is where the -3-dB cutoff frequency for the low-pass filter should be placed. Selecting the closest standard-value components, R9 = $18.2\,\text{k}\Omega$ and C10 = $0.047\,\mu\text{F}$, the -3-dB cutoff will be 301 Hz.

The chip's internal four-pole switched capacitor filter is driven by an internal clock that determines the filter's cutoff frequency. For a maximally flat amplitude response, the clock should be 100 times the cutoff frequency desired. The filter has a cutoff frequency ratio of 100:1. The internal oscillator runs at a nominal frequency of 140 kHz that can be modified by connecting an external capacitor (C11) between pin 5 and ground. The clock frequency is given by

$$f_{\text{clock}} = 140\,\text{kHz}\left(\frac{33\,\text{pF}}{33\,\text{pF} + \text{C11}}\right)$$

TABLE 6.6   Signal Characteristics for Various Standard FM Tape Recorder Frequencies (kHz)[a]

Carrier Frequency	Carrier Deviation Limits (for 40% FM Deviation)		Modulating Frequency Bandwidth
	Plus Deviation	Minus Deviation	
1.687	2.362	1.012	dc–0.312
3.375	4.725	2.835	dc–0.625
6.750	9.450	4.050	dc–1.250
13.500	18.900	8.100	dc–2.500
27.000	37.800	16.200	dc–5.000

[a]The response bandlimits (dB) for a 100-Hz frequency response are ±1%.

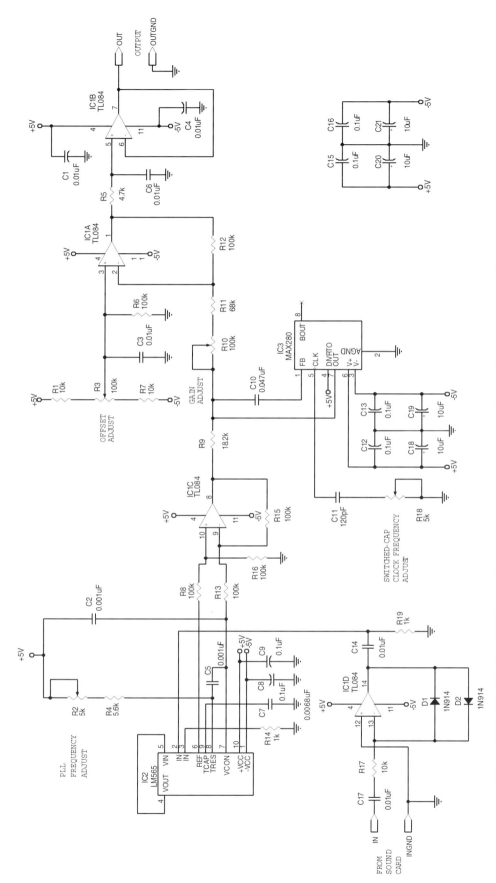

Figure 6.18 This PLL-based FM demodulator is used to generate dc-coupled signals from frequency-modulated signals generated through a PC sound card. IC1D is a limiter for the input signal. IC2 is the PLL. The common-mode rejection of IC1C eliminates dc and high-frequency carrier components present at the output of the PLL. IC3 is a fifth-order all-pole low-pass filter with no dc error used to remove residual carrier-frequency signal components from the waveform desired. IC1A adjusts the gain of the circuit and removes any offset introduced by the preceding stages.

For the example, when the cutoff should be around 300 Hz, which requires that $f_{\text{clock}} = 30$ kHz, obtained with C11 = 120 pF, a series resistor (R18) can be added to trim the oscillation frequency. In this case, the new clock frequency is given by

$$f_{\text{clock}} = \frac{f_{\text{clock}}^{\text{R18} = 0}}{1 - 4\,(\text{R18})(\text{C11})f_{\text{clock}}^{\text{R18} = 0}}$$

where $f_{\text{clock}}^{\text{R18} = 0}$ is the oscillator frequency when R18 is not present (obtained through the prior equation). After filtering, the gain is adjusted through R10 and all offset from the preceding stages is compensated with IC1A by setting R3. Finally, the *RC* low-pass filter formed by R5 and C6 removes any switching noise introduced by IC3.

The stability of the PLL circuit depends on the stability of the frequency-setting components. Proper performance requires the use of low-temperature-coefficient high-tolerance components. Resistors should be precision 1% tolerance type of the RN60D variety. Capacitors should be Mylar, polyester film, or other types that remain stable with age and which are not sensitive to temperature variations. Figure 6.19 demonstrates the performance of the technique. The test signal is a real ECG that was digitized at a sampling rate of 5 kHz with 12-bit resolution. Figure 6.19*b* shows the spectrum of the FM signal, and

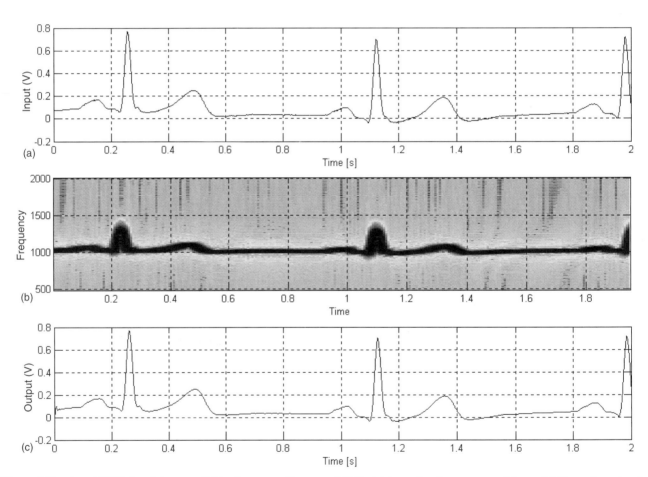

Figure 6.19   A real ECG signal that was digitized at a sampling rate of 5 kHz (a) is used to frequency-modulate a 2-kHz carrier using the Matlab vco.m function (b). The output of the PLL-based FM demodulator is shown in part (c), showing how the demodulated signal faithfully reproduces the dc offset and low-frequency components of the ECG.

Figure 6.19*c* shows how the demodulated signal faithfully reproduces the dc offset and low-frequency components of the ECG.

Since the sound card output is in the audible range, the modulated signal can be transmitted to the demodulator via a voice radio or telephonic link for remote signal generation. To do so, however, the tone frequencies produced by the sound card for a full-scale input must be limited to the bandpass of the communications channel. For a plain telephone line, this range is 400 Hz to 3 kHz, while a commercial FM audio link is specified to cover the audio bandwidth 30 Hz to 15 kHz. Another interesting possibility is to use a small 1 : 1 audio isolation transformer and a floating power supply to turn the demodulator into an isolated output stage. Finally, it should be noted that the full bandwidth of a single sound card channel can be shared by multiple software modulators occupying separate audio bands to convey various simultaneous low-frequency signals to an array of PLL demodulators

RESPONSIVE SIMULATORS

Signal generators are OK for testing medical instruments that only measure, process, analyze, or display physiological signals. However, many medical devices are used to deliver a therapy that dynamically changes the physiological signals that are measured. In this case, output-only signal generators are of only limited use. Take, for example, a DDD pacemaker, described in Chapter 8. This pacemaker can pace both the right atrium and the right ventricle separately at dynamically variable time delays to mimic the natural heartbeat whenever one or both chambers fail to contract on their own. To do so, the pacemaker can sense intrinsic electrical signals from both chambers. Whenever timely intrinsic activity is present in both atrium and ventricle, the device inhibits pacing. However, when ventricular intrinsic activity does not follow the atrial activity in a timely manner, the device triggers pacing on the ventricle in sequence after the atrium.

Testing a DDD pacemaker requires a simulator that is capable of emulating many of the heart's electrophysiological properties. Many subtleties about the heart's conduction system need to be designed into the simulator, including the way in which the atria and ventricles become refractory for some time after being excited intrinsically or artificially. In addition, a cardiac simulator suitable for interacting with a pacemaker should be able to exhibit pacing thresholds similar to those of a typical heart, and the cardiac signals generated by the simulator's "chambers" must have morphologies, amplitudes, and timings similar to real P- and R-waves detected with intracradiac electrodes.

We designed a responsive cardiac simulator as a test tool for three-chamber pacemakers (pacemakers that cannot only stimulate the right atrium and right ventricle, but can also synchronize the activity of the left ventricle to the pumping of the right heart). The circuit and timing characteristics for this responsive simulator are shown in Figures 6.20 to 6.28.

The heart's electrical activity as seen by intracardiac electrodes is simulated by the *CENELEC signal*.[2] Three signal generators, one corresponding to the heart's right atrium, one to the right ventricle, and the last to the left ventricle, provide programmable, 0- to 9-mV CENELEC signal outputs. The following explanation refers to the signal generator for the right atrium but is also applicable to the other two, since the right- and left-ventricle signal generator circuits are similar. The CENELEC waveforms are stored as 12-bit values

---

[2]*CENELEC* stands for "Comité Européen de Normalisation Electrotechnique" (European Committee for Electrotechnical Standardization). The CENELEC signal is specified in Figure FF.103 of the preliminary draft of the EN-45502-2-1 standard: *Active Implantable Medical Devices—Part 2-1: Particular Requirements for Active Implantable Medical Devices Intended to Treat Bradyarrhythmia (Cardiac Pacemakers)*, January 2001. This waveform is intended as a test signal used for the exact determination of sensitivity (sensing threshold) of pacemakers.

in IC37, the PIC16C77 microcontroller shown in Figure 6.20. The 12-bit values are output serially through the microcontroller's SPI port to IC1, an LTC1451 D/A converter. The output from each D/A converter is scaled by a programmable attenuator made up by IC3 and the resistor network, R2 and R5–R14. A thumbwheel switch connected to analog multiplexer IC3 selects the resistor used in the attenuator circuit. The signal from the attenuator is buffered by IC2A and changed from voltage to current by IC2B, IC2C, and IC2D. This current signal flows into R74, a 10-Ω resistor, and is seen by the device under test as a voltage signal. Figure 6.21 shows the schematic of the right-atrial and right-ventricular CENELEC signal generators.

A typical cardiac cycle would start with generation of the right-atrial CENELEC signal followed by the right-ventricular CENELEC signal. The delay from the start of the right-atrial signal to the beginning of the right-ventricular signal is fixed in this simulator at 100 ms. The left-ventricular signal would then start after the right-ventricular signal, with a time delay that

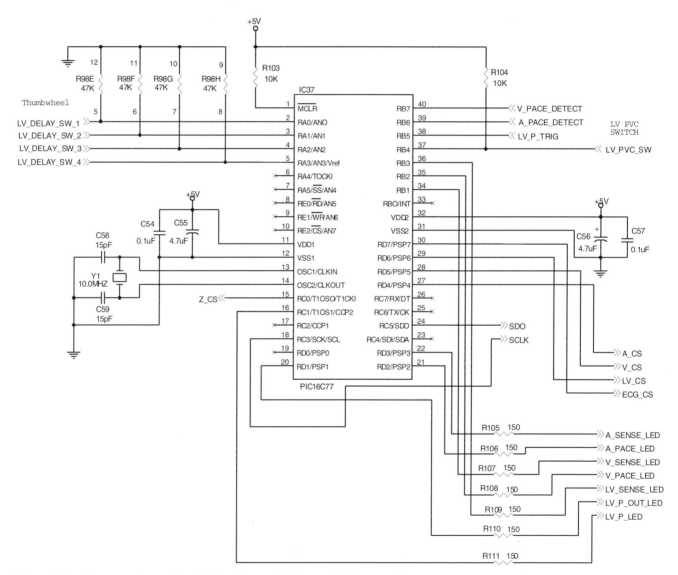

Figure 6.20   The responsive cardiac simulator is controlled by a PIC16C77. In addition, this PIC contains digitized CENELEC, ECG, and impedance waveforms that are played back via LTC1451 D/A converters to simulate the heart's electrical and mechanical activities.

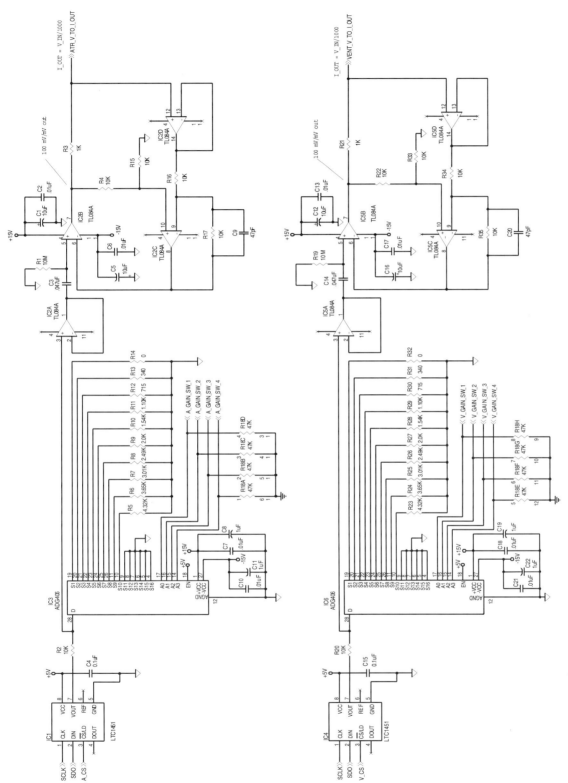

Figure 6.21 Right-atrial and right-ventricular intracardiac electrogram signal generators for the responsive cardiac simulator. The output from each D/A converter is scaled by a programmable attenuator. A thumbwheel switch connected to an analog multiplexer selects the resistor used in the attenuator circuit. The signal from the attenuator is changed from voltage to current and converted into a perfect differential voltage across to be seen by the pacemaker under test as a voltage signal.

can be selected from the following settings: 1, 10, 20, 30, 40, 50, 60, 70, 80, 90, 100, 110, 120, 130, 140, or 150 ms using the thumbwheel switch. Figure 6.22 show the timing relationship of right-atrium, right-ventricle, and left-ventricular CENELEC signal generation.

When a physician programs a pacemaker to the patient's needs, the patient's heart is usually monitored with a single-lead ECG. The cardiac simulator incorporates the simple ECG signal generator circuit of Figure 6.23 to provide an output that can be monitored with a standard single-lead ECG machine. An ECG signal is generated from digital values stored in IC37, a PIC16C77 microcontroller, and output through the microcontroller's SPI port to D/A converter IC34. R99 and R100 attenuate the output of IC34 to approximately 2 mV peak to peak. The P- and R-wave ECG signals are synchronized with the right-atrial and right-ventricular CENELEC output signals.

In Chapter 8 we explain how pacemakers can use impedance signals to derive control information based on the heart's contractile state. For the time being, suffice it to say that as the heart pumps, the volume of the blood pool around the electrodes changes. Since blood is more conductive than is muscle tissue, these volume changes result in a varying electrical impedance between the electrodes. The cardiac simulator incorporates a voltage-to-impedance converter to simulate cardiac impedance signals. This circuit is based on an idea by Belusov [1996] and was designed by Greg Martin, now a project manager at HyTronics. As shown in Figure 6.24, IC36, an LTC1451 D/A converter, generates a voltage waveform that is stored in the microcontroller's ROM in the same way as are the ECG and CENELEC signals. In this case, however, the voltage output of the D/A is converted into an impedance signal.

This impedance waveform is synchronized with the left-ventricular CENELEC output signal. R79 sets the baseline impedance value (typically, 500 Ω). IC35, a MAX038 sinusoidal waveform generator, simulates the impedance variations caused by respiration. The frequency of the respiratory component is set by R102. The intracardiac impedance and respiration waveforms are summed into the inverting input of IC40A. The output of IC40A, a summing, inverting amplifier circuit, is given by

$$V_z = -\left(\frac{V_Z}{1.6} + \frac{V_r}{2}\right)$$

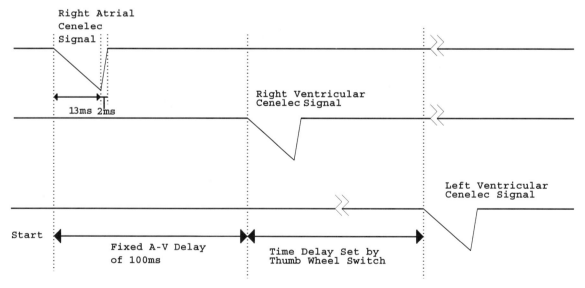

Figure 6.22   A typical "cardiac" cycle starts with generation of a right-atrial CENELEC signal followed by the right-ventricular CENELEC signal. The delay from the start of the right-atrial signal to the beginning of the right-ventricular signal is fixed in the cardiac simulator at 100 ms. The left-ventricular signal starts after the right-ventricular signal with a programmable time delay.

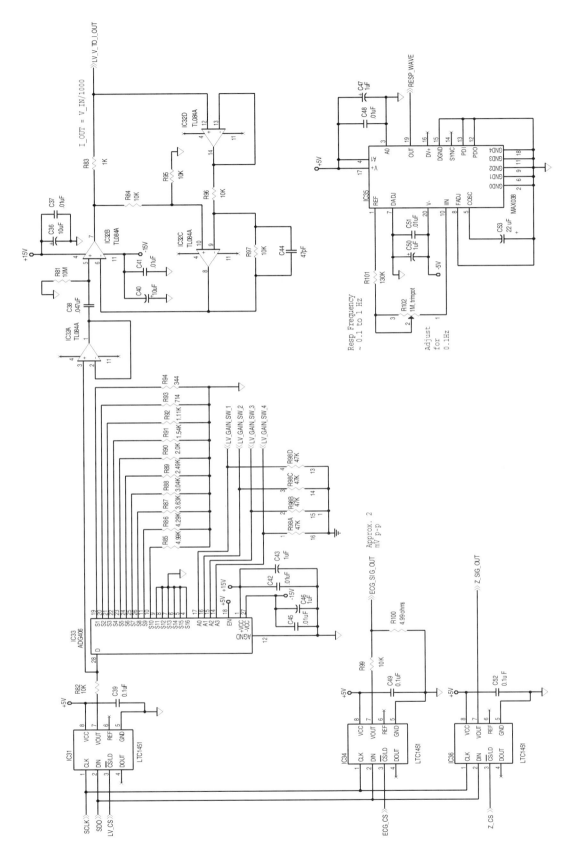

Figure 6.23 The responsive cardiac simulator incorporates a simple ECG signal generator circuit to provide an output that can be monitored with a standard single-lead ECG machine. An ECG signal is generated from digital values stored in the microcontroller and output through the D/A converter IC34. R99 and R100 attenuate the output of IC34 to approximately 2 mV peak to peak.

281

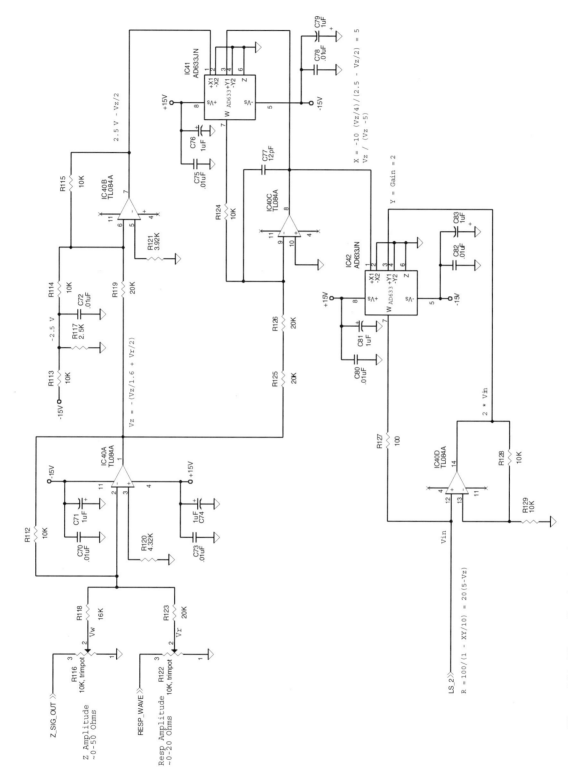

Figure 6.24 The cardiac simulator incorporates a voltage-to-impedance converter to simulate cardiac impedance signals. The voltage output of the D/A is converted into an impedance signal that is synchronized with the left-ventricular CENELEC output signal. R79 sets the baseline impedance value (typically, 500 Ω). Impedance variations caused by respiration are simulated using the sinusoidal signal generated by IC35 in Figure 6.23.

IC40B scales the summed signal and adds an offset of -2.5 V dc. IC40B's output is thus

$$V_{out} = 2.5 - \frac{V_Z}{2}$$

The change in impedance, which is seen between IC40D, pin 12, and ground, is given by

$$Z_{out} = \frac{100}{1 - XY/10} = 20(5 - V_Z)$$

The function of a pacemaker is to stimulate the heart when the heart's intrinsic pacemaker or conduction mechanisms fail to generate an action potential. This is where the responsive behavior of the simulator comes into play. Pacing pulse detectors are used to sense pacing pulses generated by a pacemaker connected to the simulator. This cardiac simulator responds to these stimuli by simulating the excitation of the heart chamber through which the pacing pulse was received. The atrial pacing detector circuit shown in Figure 6.25 is made up of IC7A, an inverting amplifier with a gain of 2, a comparator, IC8, and an inverter, IC9A. An external atrial pacing signal delivered by a pacemaker across R73 and R74 is input to IC7A, which inverts and amplifies the pacing signal by a factor of 2. IC8 compares the voltage on its noninverting terminal, which is set to approximately 1.0 V by trimmer R43. If the inverted pacing signal detected on pin 3 of IC8 is less than the voltage on pin 2, the output of the comparator is set high ($+5$ V). This digital high is inverted by IC9A to a low and input to the microcontroller's RB6 input. When the inverted pacing signal is greater than 1.0 V, the output of IC36 is low, which is inverted by IC35E to a high. This low-to-high transition generates a "change on port B" interrupt, which causes the microcontroller to execute the interrupt service routine code. R40 and C25 ensure that the pacing signal has sufficient amplitude and duration to trip the comparator. This guarantees that narrow transients do not pass through and trigger the microcontroller. The ventricular pacing detector circuit works in the same way as the atrial pacing detector described above, with the exception that the input pacing signal is input across R75 and R76.

The responsive cardiac simulator was designed to exercise three-chamber pacemakers and other devices intended to treat congestive heart failure. These sometimes deliver bipolar current pulses to the left ventricle, and it is important to monitor the approximate value of the positive and negative current pulses. The bar graph LED display of Figure 6.26 shows the amplitude of the left-ventricle pacing pulse in 2-mA steps. D1 and D2 steer each phase of the bipolar current pulse to the correct display circuitry. The current for the positive phase of the left-ventricular pacing pulse flows into the junction of R52–R60. A minimum of 2 mA is required to turn on optocoupler IC19 and the first segment of the bar graph display.

For discussion purposes let's suppose that the pacing pulse delivered to the left ventricle consists of a $+4$-mA current pulse followed by a -4-mA pulse. Then 2 mA would flow through R61 and turn on the optocoupler and the first section of the bar graph display. The voltage across R61 will forward bias the base–emitter junction of the PNP transistor in IC19 (across pins 5 and 6). Once the PNP transistor is forward biased, the input current can also flow through the base–emitter junction of the NPN transistor. This current flow switches on the NPN transistor in the IC and allows the additional 2 mA (what's left of our 4 mA) to flow through R52, the collector–emitter junction of the NPN transistor and the second LED of the bar graph display. Current will flow through the second LED of the bar graph display and through R53 and should be enough to forward bias the next PNP transistor (IC12). But there would not be enough remaining current flow to turn on the LED in the third section of the bar graph display. Each successive LED of the bar

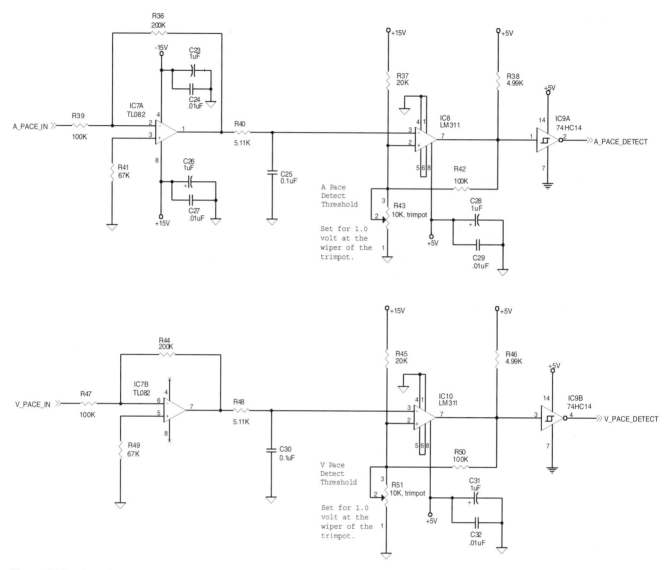

Figure 6.25   The atrial and ventricular pacing detector circuits sense the presence of external pacing signals delivered by a pacemaker under test. The comparator circuits generate interrupts for the microcontroller. An *RC* network (R40 and C25 for the atrial pacing detector) ensure that the pacing signals have sufficient amplitude and duration to "capture" the simulated heart.

graph display will be on if there is enough current to forward bias each PNP transistor in the IC, and each LED.

When enough current flows through the internal LED of the optocoupler (IC19) to turn the NPN transistor on, the input to IC9E is pulled to ground. This digital low is inverted by IC9E and signals the microcontroller of a left-ventricular pacing event. Either a positive or a negative pacing signal can trigger a left-ventricular pacing event as the outputs of the optocouplers are wire-ORed together. Figure 6.27 shows the power supply for the simulator. An external "brick" type of power supply produces ±15 V dc from 117 Vac. IC38 and IC39 produce ±5 V dc from the ±15-V power supply. Diodes D3–D10 drop approximately 0.6 V dc each and lower the input voltage to each voltage regulator. This reduces the power dissipation by each device.

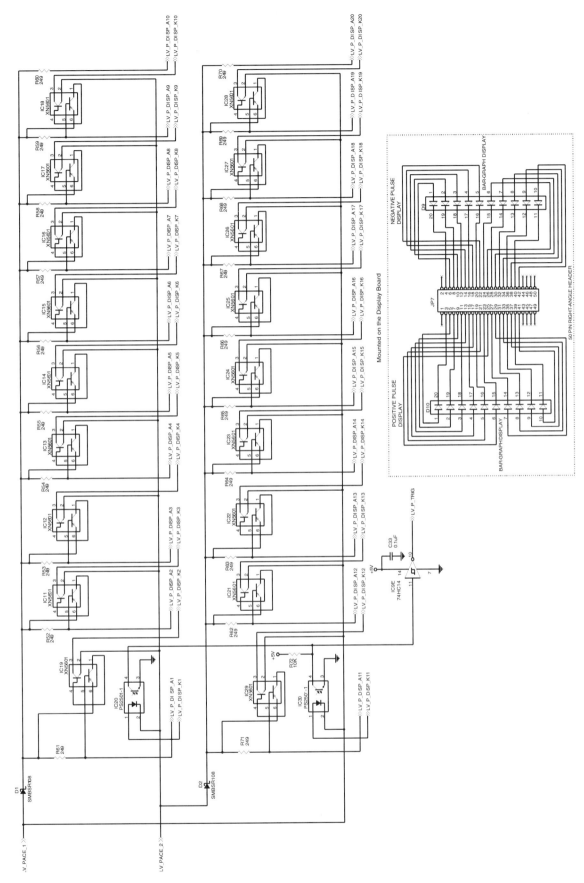

Figure 6.26 This bar graph display circuit monitors the intensity of bipolar current pulses delivered to the left ventricle.

285

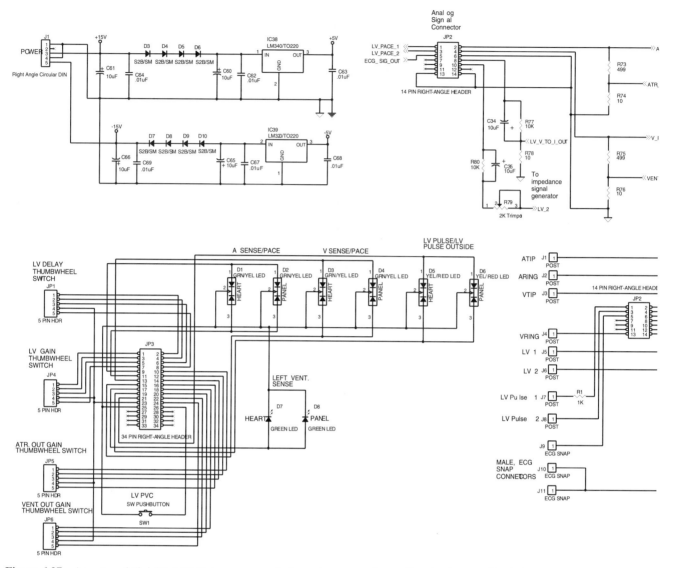

Figure 6.27   An external "brick" ±15-V dc power supply is used to power the cardiac simulator. ±5 V dc is generated onboard for the digital portions of the circuit.

Firmware for the Cardiac Simulator

Figure 6.28 shows the state machine for the main loop. The main routine performs the following tasks:

- Reads the LVS_Delay (left-ventricular sense time delay) hex switch setting and stores the value.
- Checks to see if the LV_P_IN signal was found outside the 180-ms window.
- Checks for an atrial capture event (A_CAP_EVENT flag set to 1). If the atrial capture event is true and the atrial capture action flag is enabled, the DO_A_CAP subroutine is run and the user returns to the main loop. The atrial capture action flag is used to disable atrial events in the main loop. If an atrial capture event is detected but the action flag is not set to 1 (enabled), the event flag is cleared.
- Checks for a ventricular capture event (V_CAP_EVENT flag set to 1). If the ventricular capture event is true and the ventricular capture action flag is enabled, the DO_V_CAP

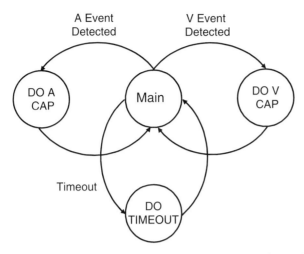

Figure 6.28   State machine diagram of the main loop of the responsive cardiac simulator.

subroutine is run and the user returns to the main loop. The ventricular capture action flag is used to disable ventricular events in the main loop. If a ventricular capture event is seen but the action flag is not set to 1 (enabled), the event flag is cleared.

- Checks to see if the PVC button was pressed (i.e., the LV_PVC_EVENT flag is set). If the event flag is set to 1 and the action flag (LV_PVC_ACTION_FLAG) is enabled, the DO_LV_PVC subroutine is executed. After execution the subroutine returns to the main loop. If the action flag is not set, the event flag is cleared.

- Controls the value stored in the delay timer of time spent in the main loop through main loop timeout. While inside the main loop the microcontroller checks continuously for capture events. If no capture events are detected, the delay timer times out, and the subroutine whose pointer is stored in TIMEOUT_INDX is executed. The delay timer is decremented once every millisecond in the TIMER1 interrupt service routine.

The interrupt service routine is comprised of two main sections. The first section is the TIMER1 interrupt routine, which is executed once every 1 ms.

- *Z delay timer.* A delay is used from the start of a ventricular event such as a pace detect event to the beginning of the impedance waveform output. This delay is set to 66 ms for a PVC event or 96 ms for a normal ventricular event. The Z delay timer is decremented and tested for a value of zero each time the TIMER1 interrupt occurs. The Z delay timer is first decremented and if found equal to zero, the impedance output flag is set to 1 (Z_FLAG = 1), which starts the output of the impedance waveform data from the microprocessor to the impedance D/A converter.

- *LVP window timer.* The left ventricular pacing pulse from the device under test must be completed within 180 ms after a right-ventricular event (either paced or intrinsic). The LV_P_WIN_TIMER is loaded with a value of 180 by the DO_V_CAP (capture event) or the DO_V_INTRIN (intrinsic event) routines. Each time through the TIMER1 interrupt service routine, the LV_P_WIN_TIMER is decremented by 1 and the LV pacing input (LV_P_TRIG) is checked to see if it has transitioned from a high to a low state (1 to a 0). If the LV_P_TRIG input does transition from a high state to a low state within the 180-ms period, the LV_P_ON subroutine is executed setting the LV_P_LED output to a 1. The purpose of this test is to verify that the LV pacing pulse output by the device under test does not exceed

180 ms beyond activation of the RV. This would be considered an intrusion into the "vulnerable period" of the heart.

- *A and V event timers.* The A and V event timers are incremented each time through the TIMER1 interrupt service routine. These timers are used to calculate the time since the last A and V events.
- *Right-ventricle event to left-ventricle sense delay timer.* The right-ventricle event to left-ventricle sense output time delay, if not equal to zero, is decremented each time through the TIMER1 interrupt service routine. If after decrementing the timer, the timer is equal to zero, the left-ventricular sense waveform output is started. The delay is read from the front-panel hex thumbwheel switch.
- *LED timers.* Each LED on the front panel is on for a fixed length of time. The on time for each LED is loaded into a timer when the LED is switched to the on state by one of the subroutines. Each of the LED timers, if not equal to zero, is decremented; after decrementing, if the timer is found to be equal to zero, the LED is switched off.
- *Delay timer.* A general-purpose delay timer, if not equal to zero, is decremented each time through the TIMER 1 interrupt service routine, and if after decrementing is equal to zero, the DELAY_FLAG is cleared.
- *Test ECG out.* Check to see if the ECG waveform output flag is enabled (ECG_FLAG = 1). If the flag is enabled, write the next word in the ECG waveform table to the ECG D/A converter.
- *Test Z out.* Check to see if the Z waveform output flag is enabled (Z_FLAG = 1). If the flag is enabled, output the next word from the impedance waveform table to the impedance D/A converter.

The second section is the TIMER0 interrupt service routine, which is executed once every 200 μs:

- *Test port B inputs.* Check if any of the inputs on the upper 4 bits of port B went from a low to a high state. If the enable flags for each of these inputs is set to a 1, the event flag for each input is then set to a 1. The event flags are used to signal that one or more external events have been detected.
- *Test the PVC switch input.* Check to see if the PVC switch was pressed and that a LV_PVC_ENABLE event is not already pending. If not, signal an LV PVC event by setting the LV_PVC_ENABLE flag.
- *Test A out.* Check to see if the right-atrial CENELEC waveform output flag is enabled (A_FLAG = 1). If the flag is enabled, write the next word from the CENELEC waveform table (CEN_TABLE) to the A sense D/A converter.
- *Test V out.* Check to see if the right-ventricular CENELEC waveform output flag is enabled (V_FLAG = 1). If the flag is enabled, write the next word from the CENELEC waveform table (CEN_TABLE) to the V sense D/A converter.
- *Test LVS out.* Check to see if the left-ventricular CENELEC waveform output flag is enabled (LVS_FLAG = 1). If the flag is enabled, write the next word from the CENELEC waveform table (CEN_TABLE) to the LV sense D/A converter.

When Radio Shack Parts Are Not the Answer

Responsive simulators such as the one described earlier can be designed to emulate the deterministic behavior of a biological system. There are occasions, however, when the relevant physical properties of the biological system are too complex for simulation with relatively simple electronic circuits. Here comes a story: Circa 1995, there were various reports about implantable defibrillator patients being shocked when passing through electronic article surveillance (EAS) systems, such as those used in stores against shoplifting. There were also reports about pacemaker patients who fainted in the proximity of these systems, presumably because their pacemakers were inappropriately inhibited by EMI from the EAS.

EAS systems consist of electromagnetic field emitters which illuminate and interrogate a uniquely identifiable "tag" which is affixed to an object. When the tagged object is in the field produced by the EAS equipment, the EAS sensors detect its presence and activate a response. Back in our Intermedics[3] days, we conducted much of our research together with Dr. Thomas Fåhræus in Lund, Sweden. He became very interested in the potential problem of pacemaker interference by EAS systems. After all, EAS systems are widely used to track and monitor merchandise, for inventory control and theft prevention, and in some cases, to track living specimens. Because of their widespread use in retail stores and commercial establishments, wearers of implantable medical devices have a high likelihood of entering the fields produced by such systems.

Thomas terminated a ventricular bipolar pacemaker lead with a LED to be able to tell when the pacemaker paced normally, and built a simple setup simulating an implant configuration (Figure 6.29). He programmed the pacemaker to the VVI mode and a low pacing frequency (e.g., 40 ppm). He then toured the shops in Lund looking for signs of EAS interference on his pacemaker implant simulation. If the LED stopped blinking, that meant that the EAS system was inhibiting the pacemaker. If the LED started to blink rapidly (e.g., 70 ppm), on the other hand, that meant that the pacemaker had entered its noise reversion mode.

Thomas called us in alarm when he found that most of the EAS systems in town interfered with his pacemaker setup. However, that observation was not consistent with what we were seeing with real implants. Patients were not falling dead like flies when walking through EAS systems. In fact, most of the cell phone and EAS interference reports were highly anecdotal, to the point where they reeked of urban myth. It wasn't difficult to explain the difference between the test setup and a real implant. The LED was acting as the detector diode in a crystal radio. The interfering signals were really generated by the LED's nonlinearity, an element that is negligible in the real implant situation.

Thomas was very ingenious in replacing his pacing detector by a more realistic kind. He figured out that the tongue can detect the pacing pulses while making the electrode–tissue interface similar to that of the real implant. This time he made us all accompany him with different pacemaker models on his interference hunting trip. It was not even April 1, and there we were, in the best stores of downtown Lund, with wires stuck in our mouths, moving bizarrely next to the shoplifting gates. Good that everyone in town knew that Fåhræus was a respectable physician. End of the story: Our not-too-scientific study found no evidence of pacemaker interference by EAS systems. No one really wanted to check whether or not a defibrillator would react differently.

This is why when determining the effects of electromagnetic environments on implantable medical devices, it is essential that the measurements be performed while operating the test devices in an environment that simulates the absorption and shielding characteristics of the human body. A common physical test environment consists of a $0.027\,M$ saline solution in a rectangular tank that is essentially transparent to the incident radiated field. The $0.027\,M$ NaCl concentration has a resistivity of approximately $375\,\Omega\cdot\text{cm}$, which is used to simulate the electrical characteristics of the human body tissue and fluid.

---

[3]Intermedics was the world's third-largest implantable cardiac pacemaker/defibrillator company. It was acquired by Guidant Corporation in 1998.

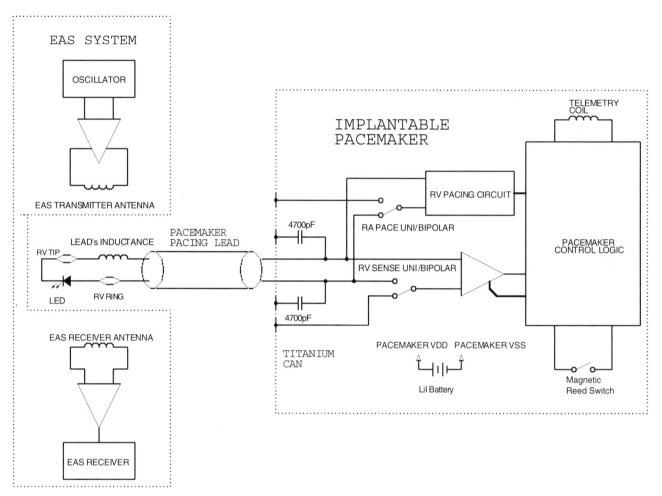

Figure 6.29   A ventricular bipolar pacemaker lead with a LED does *not* make a good simulation setup to evaluate how electromagnetic environments such as those generated by electronic article surveillance systems (EAS) interfere with implantable pacemakers and defibrillators. The LED acts as the detector diode in a crystal radio. Interfering signals are then demodulated by the LED, giving a false indication that the device is susceptible to electromagnetic interference.

A torso simulator model standard is now used by the pacemaker industry. The setup is described in ANSI/AAMI standard PC69:2000 (*EMC Test Protocols for Implantable Cardiac Pacemakers and Implantable Cardioverter Defibrillators*), Annex B, and was based on research by Ruggera et al. [1997]. Figure 6.30 shows a torso simulator that our colleague Paul Spehr constructed based on the standard to test the immunity of pacemakers against cellular telephone interference. The materials used were:

- White louver light grid (nonconducting), Home Depot part 074567432008
- Under-bed box 28-qt/27-L Sterilite No.1856 white, $23\frac{1}{4}$ in. $\times$ 17 in. $\times$ 6 in. (59 cm $\times$ 43 cm $\times$ 15 cm)
- Stainless steel nuts and bolts, 10-24 $\times \frac{1}{2}$ in. Phillips head and nut, four each
- Titanium sheet
- Silicone adhesive sealant
- Four-foot nylon threaded rod, $\frac{1}{2}$-13

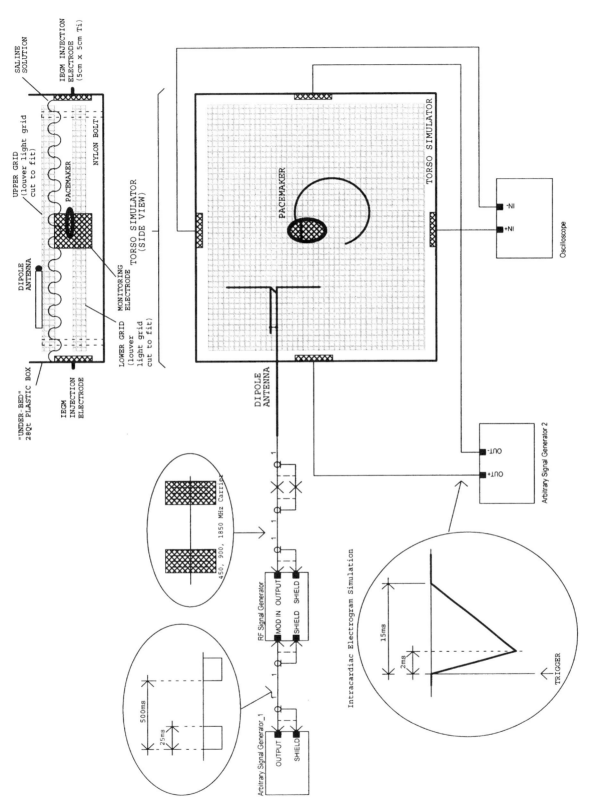

Figure 6.30 A torso simulator model for electromagnetic interference studies can be constructed using some white louver light grid (used in fluorescent lamp assemblies) and a 28-qt under-bed plastic box. Tissue properties are simulated using 0.027 M saline solution. Four electrodes on the sides of the box are used to monitor operation of the "implanted" device as well as to inject simulated intracardiac electrogram signals.

291

- Nylon $\frac{1}{2}$-13 hex nut
- Monofilament, 20-lb test, Shakespeare OMNIFLEXR fishing line
- Five gallons of supermarket distilled water
- Sodium chloride, NaCl USP grade (Kosher salt works equally well)

The titanium sheet stock was cut to four pieces $5\,\text{cm} \times 5\,\text{cm}$. Holes were drilled in the center of each plate to allow for attachment with screws. Holes were drilled in the center of each side of the 28-qt plastic box. Stainless steel screws were coated with silicone adhesive and then used to attach the titanium pieces to the inside of the box, providing electrical connection to the outside of the box. The louver grid was cut to give two pieces, one fitting the top ridges and one on the bottom. Six nylon nuts were attached to the bottom grid with medical adhesive. The nylon threaded rod was cut to 4-in. lengths, six each. A slot was cut in the top of each rod.

The saline solution should be prepared to the proportions recommended by ANSI/AAMI PC69:2000, Table 2: $0.027\,M = 1.8\,\text{g/L}$ or 0.18% NaCl concentration at 21°C. To do so, the salt is first dried in an oven set at 200°C for 30 minutes. Then 30.6 g of dry salt can be added to 17 L of distilled water to make enough solution with a concentration of 1.8 g/L. Submersion in a conducting fluid facilitates monitoring of the implantable medical device's operation while minimizing the electromagnetic field distortion and detection effects of directly attached probes. Monitoring of the implantable medical device's responses to the applied electromagnetic environment is accomplished by sampling the test device's output pacing pulses via the square electrodes submerged in the saline solution. The potentially interfering signals are applied and their magnitude is gradually changed while the response of the medical device is monitored.

Pacemakers and defibrillators are usually tested in each of their normal operating modes, and test equipment is set up to determine if, and when, the test device is inhibited or operating in its noise mode. To do so, these tests are conducted both with and without external simulated intracardiac electrogram signals injected via the saline solution into the sensing inputs of the test device.

Shocking Water

Whereas simulating a single biopotential signal channel is not too complex, generating accurate signals to validate array processing methods is very problematic. Take, for example, the case of validating the performance of *inverse solutions*. In inverse electrocardiography, researchers attach a large array (e.g., 128 or 256 electrodes) of electrodes to the chest instead of the usual 12-lead ECG. The idea is to process the array ECG signals, taking into consideration the specific geometry of the chest and body organs to create three-dimensional images of the potentials in the heart muscle itself. The potential of such a technique is tremendous. It would give physicians a noninvasive method to identify patients at risk of sudden death, for specific diagnosis of rhythm disorders, and for localization of disturbances in the heart in order to guide intervention.

Several computational approaches that attempt to solve the electrocardiographic inverse problem have been developed to estimate heart surface potential distributions in terms of torso potentials, but to date their suitability for in vivo and clinical situations has not been firmly established. The same can be said about the solution of other inverse problems in biomedical engineering, such as inverse electroencephalography and electrical impedance computed tomography. Before any inverse electrical imaging procedure can be used as a noninvasive diagnostic tool with confidence, it must first be validated so that recorded experimental observations can be faithfully reproduced.

Experimental validation studies for inverse imaging solutions can involve animal preparations, completely synthetic physical materials, or even a combination of the two in order to simulate the ideal conditions of biopotential sources inside a human body. Because of the technical challenges of measuring source parameters and geometry from animal models, most validation studies for inverse problems use synthetic electrical sources embedded in conducting media as a way to obtain controlled physical models of the source organ and body.

Early modeling of the heart for inverse electrocardiography experiments used a current bipole to simulate the source because it is a direct equivalent of the single heart dipole vector that still serves as the basis of much of clinical electrocardiography. A *potentiostat–galvanostat* is a general-purpose instrument that may be used for controlling either the potential difference between the electrodes of the bipole source (potentiostatic operation) or the current between them (galvanostatic operation).

Figure 6.31 shows the circuit for a potentiostat–galvanostat. When the potentiostat–galvanostat switch (SW1) is set for potentiostatic (constant-voltage) operation, the potential difference between the "active" and "reference" electrodes is fed into the error amplifier IC1, where it is compared against the control voltage. Any error existing between these two potentials will be amplified by IC1 and buffered by IC2, causing a current to flow between the active and reference electrodes. This imposed current will force the potential difference between the electrodes to move in a direction so as to reduce the error between it and the control voltage. In a very short time (~1 ms) a steady state will be achieved in which the current is just sufficient to maintain a very small error voltage.

With the potentiostat–galvanostat select switch set for galvanostatic operation, the voltage from current-sensing resistor R1 is amplified by IC4 and fed into the error amplifier IC1, where it is compared against the control voltage. As described above, the error amplifier will amplify any error existing between the two voltages, causing some additional current to flow between the electrodes so as to reduce the error voltage. A steady state is achieved when the error becomes very small. However, in this case it is the current, not the electrode potential, which is maintained at the desired value.

The control voltage can be dc or ac. With the component values shown in Figure 6.31, the electrode potential difference or electrode current will track the control voltage as long as the input signal is limited to the range dc to 300 Hz. IC5A and IC6A buffer the electrode voltage and current signals from the outputs of IC3 and IC4. These can be monitored through an oscilloscope to evaluate the dc or ac impedance of the electrodes.

Probably the simplest physical models used to shed light on the inverse problem in electrocardiography were two-dimensional models. Grayzel and Lizzi [1967] used conductive paper (Teledeltos) to create two-dimensional inhomogeneous models of the human thorax, to which they attached current source–sink pairs (bipoles) to represent the heart. Teledeltos paper is a resistive paper that has uniform resistance. As shown in Figure 6.32, Teledeltos paper can simply be cut, after suitable scaling, to the shape of the body region to be investigated. Then electrodes are painted on the paper with conductive silver ink so that sources (voltage and current sources, and sinks or loads) can be attached to the paper at the appropriate places to set up an analog of the boundary conditions desired. The extent and value of inhomogeneities are controlled by means of perforations or silver spots applied to the conductive paper. The ratio of hole diameter to hole spacing determines the relative increase in resistivity of the area punched. The ratio of silver-dot diameter to center spacing determines the relative decrease in resistivity.

To find solutions, one simply reads out the field intensity values by using a sharp-tipped voltmeter probe applied to the paper at any point that field intensity is desired. Thus, a very simple laboratory setup can be used to solve sets of complicated partial differential equations empirically without the user even knowing what partial differential equations he or she is actually solving!

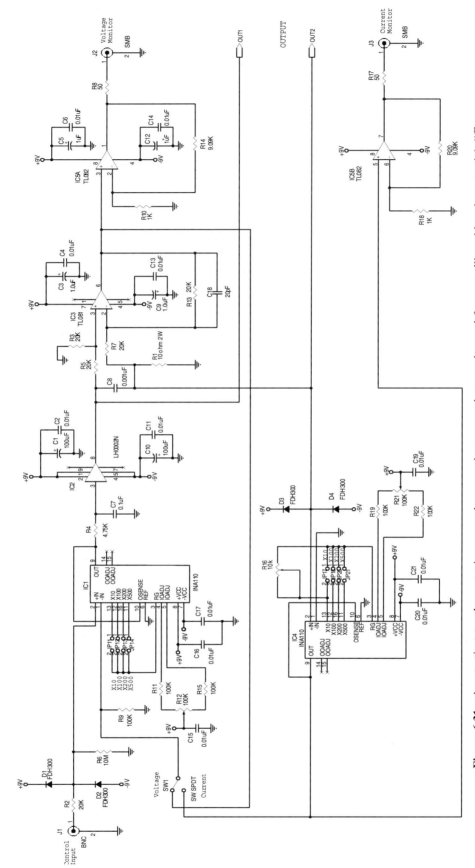

Figure 6.31 A potentiostat–galvanostat is a general-purpose instrument that may be used for controlling either the potential difference between the electrodes of the bipole source (potentiostatic operation) or the current between them (galvanostatic operation). The control voltage can be dc or ac, and the electrode potential difference or electrode current will track the control voltage as long as the input signal is limited to the range dc to 300 Hz.

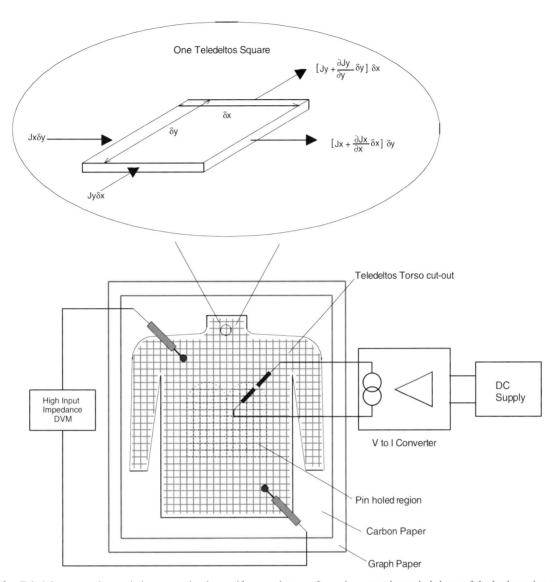

Figure 6.32 Teledeltos paper is a resistive paper that has uniform resistance. It can be cut to the scaled shape of the body region to be investigated. Electrodes painted with conductive ink allow connection to current–voltage sources. Inhomogeneities are introduced by means of perforations or silver spots (e.g., pinhole region simulating lungs). Field intensity values are read with sharp-tipped voltmeter probes. A sheet of carbon copy paper and graph paper make it easy to trace equipotential lines and gradients.

Consider how Teledeltos can be used to solve electrical field problems. The current flow is essentially in two dimensions (i.e., across the surface of the paper but not through its thickness from one face to the other). Suppose that the vector $\mathbf{J}$(A/m) represents the current density per unit width at some point on the Teledeltos sheet. Then the electric field (or voltage gradient) at the same point in the sheet is

$$E = J\rho (\text{V/m})$$

where ρ is the specific resistance of the sheet. Teledeltos paper has a uniform grid printed on it, and ρ represents the resistance between opposite sides of a grid square.

Now consider a square element of the surface of the Teledeltos paper. The net outflow of current from the element is

$$\delta y\left(J_x + \frac{\partial J_x}{\partial x}\delta x\right) + \delta x\left(J_y + \frac{\partial J_y}{\partial y}\delta y\right) - \left(\delta y J_x + \delta x J_y\right) = \left(\delta y \frac{\partial J_x}{\partial x}\delta x\right) + \left(\delta x \frac{\partial J_y}{\partial y}\delta y\right)$$

neglecting terms in δy^2 and δy^2. Since we are considering a steady state, there must be no net rate of loss or accumulation of charge in the element. Thus, the expression given above must be equal to zero. Dividing by the area of the element yields

$$\frac{\partial J_x}{\partial x} + \frac{\partial J_y}{\partial y} = 0$$

The relationship, called the *equation of continuity*, applies to the flow of any fluidlike material whose volume or quantity does not change when flowing across a region in steady conditions. It applies to the flow of heat, fluids, electricity, and mass particles. In this case it shows that the current flows completely through the element without shedding any charge. In the days before digital computers were able to run finite-element models, Teledeltos paper was used to quickly obtain practical solutions to tough electromagnetic, thermodynamic, and hydrodynamics problems.

Back to our discussion on modeling with Teledeltos, the electric field **E** across the surface of the sheet may be expressed in terms of the potential V by

$$\mathbf{E} = -\left[\frac{\partial V}{\partial x}\mathbf{x} + \frac{\partial V}{\partial y}\mathbf{y}\right] = -\boldsymbol{\nabla}V$$

where **x** and **y** are unit vectors along the x and y axes. Then the continuity equation can be rewritten as

$$\frac{\partial^2 V}{\partial x^2} + \frac{\partial^2 V}{\partial y^2} = \boldsymbol{\nabla}^2 V = 0$$

This equation, commonly referred to as *Laplace's equation*, at first glance looks difficult to solve. Although there are many known solutions, the problem is to find a solution that fits the given boundary conditions. Obviously, the solution given by Teledeltos paper is the one that completely satisfies the analog boundary conditions modeled using cutouts and/or painting conductive silver ink. Don't disregard Teledeltos just because it is "old-fashioned." By applying just a bit of common sense, you can use this simple, elegant method to solve very complicated problems.

PASCO Scientific sells black conductive paper to go along with its E-field mapping kit. One hundred sheets of conductive paper marked with a centimeter grid sells for $30 (part PK-9025). You can make electrodes using the type of conductive silver ink pen that is used for PCB repairs (which you can also buy from PASCO as part PK-9031B) or by attaching 3M's 1181 electrical tape, which is copper foil with a conductive acrylic adhesive.

Using the simple Teledeltos model, Grayzel and Lizzi [1969] were able to find that the relationship between source location and body surface, as expressed by the standard lead field, was much more variable and complex in the inhomogeneous than in the homogeneous torso. This led them to the conclusion that several "standard" lead systems of their day showed a sharp deterioration in performance after adding inhomogeneities to the torso model.

Burger and van Milaan [1948] opted for a more realistic three-dimensional model and constructed an electrolytic tank out of a michaplast shell molded on the statue of a supine human to create a torso model. The tank split horizontally to provide access to the interior, which was filled with copper sulfate and outfitted with copper foil electrodes fixed to the inner surface. Their heart source model was a set of copper disks oriented along one of the body axes and adjustable from outside the tank by means of a rod. The tank included inhomogeneous regions constructed from cork and sandbags to simulate the electrical distortions caused by the spine, ribs, and lungs. Other investigators have used inflated dog lungs and agar gel models of human lungs inserted into the tank.

If your research calls for an electrolytic body-shaped tank, a visit to your local hobby and crafts store will prove invaluable. Describe what part of your anatomy you would like to make a cast of, and the store's attendant's will help you select the best materials and techniques. (Hint: don't ask for "michaplast.") For the electrolyte solution you can use iso-conductive saline made by mixing 1.8 g of noniodized salt (use Kosher salt if you don't want to overpay for analytical-grade NaCl) per liter of distilled water (supermarket distilled water is okay). Electrodes should be Ag/AgCl, and inhomogeneous regions to simulate the lungs and other organs can be cast of high-density agarose gel (e.g., any standard agarose gel with about $1200 \, g/cm^2$ strength, melting range about 87 to 89°C, gelling range about 36 to 39°C, sold by VWR Scientific) or synthetic sponge foam.

As shown in Figure 6.33, a bipolar electrode to model cardiac potentials can be made from two Ag/AgCl disk electrodes, some plastic tubing, and epoxy adhesive. This bipolar source is fashioned after the one described by Brandon et al. [1971]. It yields a source with uniform distributions of current, 3 mm in diameter, separated by 12.7 mm. The potentiostat–galvanostat of Figure 6.31 makes an excellent current source to drive this bipolar electrode.

Electrolyte tanks have also been used to investigate the way in which currents applied through skin electrodes distribute within the body to cause tissue stimulation. This is an important issue in external pacing, where high-voltage pacing pulses applied through skin-surface electrodes not only cause the heart to contract, but cause quite a bit of pain to the patient. Another area of special interest in the solution of the same problem of stimulation overflow to nontarget tissue is functional neuromuscular stimulation (FNS; see Chapter 7). We [Sagi-Dolev et al., 1995] used a phantom model in which a saline tank model was improved by adding a layer to simulate skin impedance properties. The purpose of the study was to look for electrode array geometries required to achieve target muscle activation with minimal overflow and to avoid pain or burning.

The tank was a small 40 cm × 30 cm × 30 cm glass aquarium filled with isoconductive saline solution to simulate the volume conductor of the human forearm. To simulate the way in which the skin's epidermis and panniculus adiposus distort the spatiotemporal characteristics of FNS stimulation pulses, we made a carbon-loaded silicone elastomer that had similar electrical properties to skin.

Rosell's method for the measurement of skin impedance [Rosell et al., 1988] was adapted to measure the electrode–skin impedance for the specific conductive-silicone electrode materials used for FNS and for the specific spectral content of the FNS waveform (1 to 9 kHz). A stimulating ring with inside diameter 12.5 mm and outside diameter 15 mm was constructed from carbon-loaded silicone electrode material (Ag/AgCl electrodes of similar dimensions were used by Rosell). Rectangular 2 mm × 5 mm voltage-sensing electrodes were positioned, one within the center of the stimulating ring and the second parallel to the first at a distance of 2 mm from the ring as shown in Figure 6.34. The electrodes were embedded at a depth of 1.5 mm within an isolating cast. The 1.5-mm cavities allowed for a consistent conductive gel volume. A large self-adhesive Ag/AgCl ECG electrode was used as a reference. The isolated amplifier input impedances had an equivalent input impedance of 1 MΩ in parallel with 20 pF.

The average impedance measurements were 7.7 kΩ at 1 kHz and 4.8 kΩ at 10 kHz for a 1-cm² electrode surface area. By assuming that a parallel resistor–capacitor suitably models the skin impedance, these impedance values result in an equivalent of a 7.64-kΩ resistor in parallel to a 2479-pF capacitor for a 1-cm² electrode surface area. The artificial "skin" was fabricated to match these values. Varying amounts of colloidal graphite (MacDermid PTF4150 flexible graphite conductive paste) and silicone rubber (RTV No. 159) were

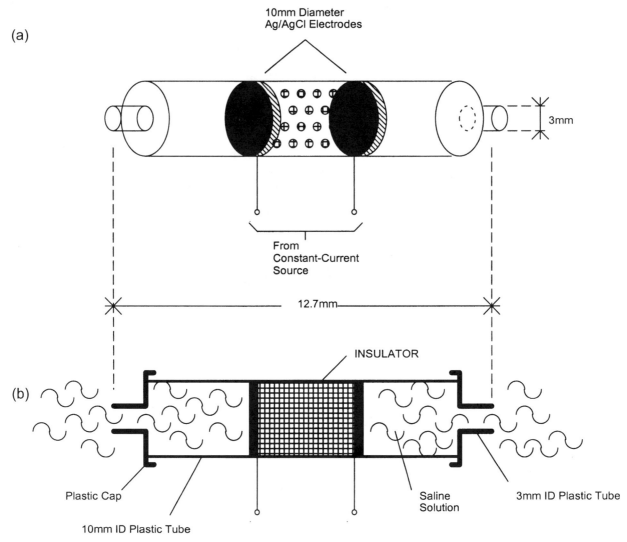

Figure 6.33   A bipolar electrode to model cardiac potentials can be made from two Ag/AgCl disk electrodes, some plastic tubing, and epoxy adhesive: (a) perspective view; (b) section view. This bipolar configuration yields a source with uniform distributions of current, 3 mm in diameter, separated by 12.7 mm.

thoroughly blended to form homogeneous pastes, which were then formed and cured into thin layers. Impedances of each sample were then measured using the same setup as before (Figure 6.34) but using a saline-soaked sponge to simulate the arm's underlying tissues. The correct graphite–silicone rubber ratio was found to be 20% (by weight).

Signals within the physical model were measured as shown in Figure 6.35. A differential probe was constructed with two gold wires 1 mm in diameter, each exposed 1 mm to the saline solution and placed 1 mm apart parallel to the y axis representing the long axis of the forearm. This distance was chosen because of its physiological significance, since the threshold potential difference for stimulation of myelinated axons must be reached at approximately the internodal distance.

As shown in Figure 6.36, the computer triggered the delivery of one FNS stimulation pulse by strobing the TRIGGER input. Probe signals were amplified differentially through an AD521K instrumentation amplifier. The signal reference was established through a large plate electrode placed external to the glass tank. The instrumentation amplifier output was integrated

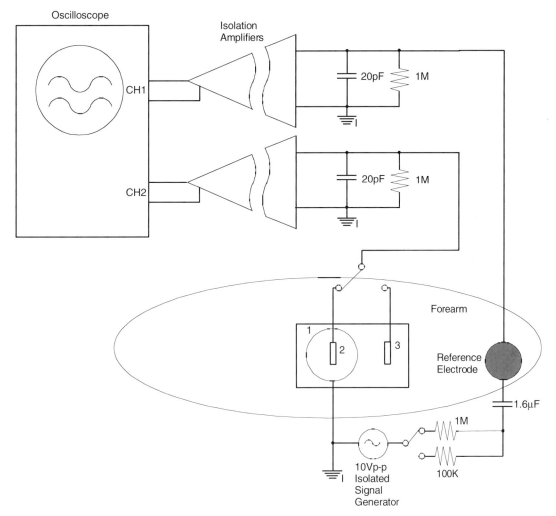

Figure 6.34 Rosell's method applied to measure electrode–skin impedances using carbon-loaded silicone electrodes: 1, stimulating ring electrode; 2 and 3, voltage-sensing electrodes. The electrodes are embedded at a depth of 1.5 mm within an isolating silicone rubber cast.

(with a time constant given by R3 and C2) throughout the duration of the stimulation phase of the FNS pulse to emulate the integration process carried out by the excitable tissue membrane. After the stimulation phase, the integrated voltage was read through the computer's A/D card. The actual position of the probe was measured by digitizing the voltage output from the x-, y-, and z-axis potentiometers linked to the three-dimensional manipulator. The computer reset the integrator by strobing the RESET line prior to taking a new measurement.

VERY REALISTIC PHYSIOLOGICAL SIGNAL SOURCES

No physical model can really replace the biological realism that can be achieved with an intact animal model with implanted instrumentation. However, the intact animal presents the problem that body geometry and exact organ configuration are very difficult to integrate as part of an experimental study. A compromise solution is to use an isolated preparation of the organ generating the biopotentials placed in a synthetic volume conductor that simulates the human cavity holding the organ.

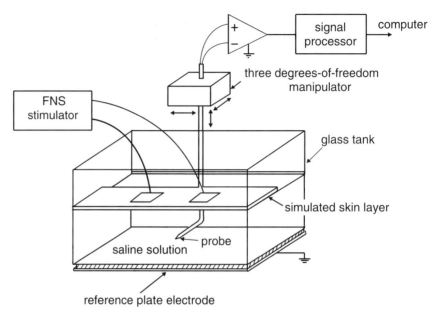

Figure 6.35 An electrolyte tank can be used to investigate how FNS currents applied through skin electrodes distribute within the body to cause tissue stimulation. The setup consists of a 40 cm × 30 cm × 30 cm glass tank (a small aquarium) filled with isoconductive saline. A carbon-loaded elastomer layer simulates the patient's skin. A differential probe made with two gold wires exposed 1 mm to the saline solution and placed parallel 1 mm apart simulate the stimulation spatial constant of myelinated axons. (From Sagi-Dolev et al. [1995]. Reprinted with permission from the International Federation for Medical & Biological Engineering.)

A common trend in recent studies for the validation of inverse electrocardiography has been to replace the synthetic signal source with a perfused dog heart, thus achieving a much higher degree of realism by eliminating the limitations of representing the heart as a single dipole rather than as a distributed source of bioelectric current. Obviously, the main advantages of this type of preparation over instrumented whole-animal experiments are the relative ease of carrying out the experiments and the increased level of control they provide. The isolated heart is directly accessible when suspended in an electrolytic tank, which permits manipulations of its position, pacing site, coronary flow, and temperature. The simplified geometry of the tank also makes constructing customized geometric models simpler and faster than when a complete medical imaging scan is required for a whole animal.

Isolated heart systems are available from Hugo Sachs Electronik (now part of Harvard Apparatus). The isolated heart apparatus uses a synthetic bloodlike solution to provide oxygen and metabolic substrates for the heart to survive. Since the coronary system of the heart relies on the pressure created by the heart itself for flow, the heart must be assisted externally after explant from the donor animal. This is accomplished by pumping fluid directly into the coronary system and removing fluid from the heart's four chambers to ease the burden using a technique called the *Langendorff mode of perfusion*. Once the heart is capable of maintaining pressures and flows independently, it is weaned out of Langendorff mode and the native flow pattern is reestablished. At that point the heart is said to be in the *four-chamber working mode*, in which it generates pressures in all four chambers and is responsible for the work required to maintain flow through the coronary system, hence the name.

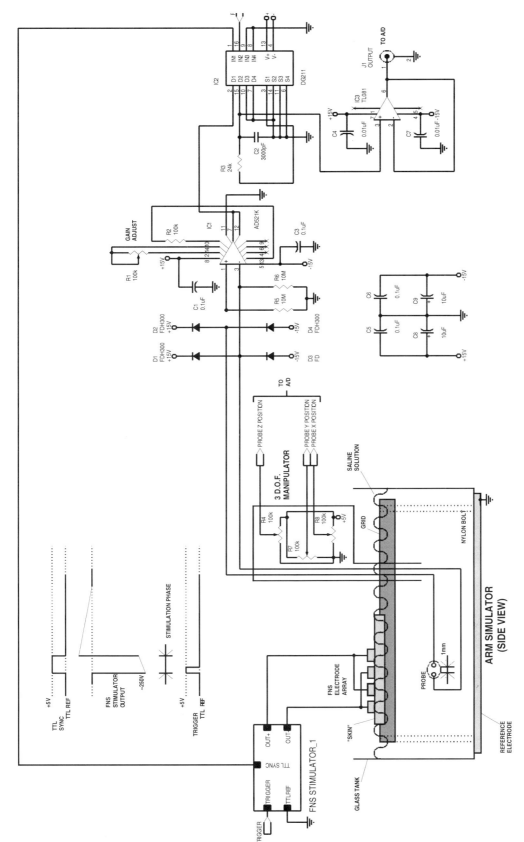

Figure 6.36 After the computer triggers the FNS stimulator, the energy deposited on the probe by the stimulation phase of the FNS signal is amplified through an instrumentation amplifier and integrated by R3 and C2. After the stimulation phase, the integrated voltage can be read through the computer's A/D card. The actual position of the probe can be measured by digitizing the voltage output from the *x*-, *y*-, and *z*-axis potentiometers linked to the three-dimensional manipulator.

Unfortunately, this preparation inevitably builds up metabolites and electrolytes which change its electrophysiological characteristics within a few hours of explant. To overcome this problem, many investigators [MacLeod et al., 1995] use a second dog to provide circulatory support for the isolated heart, which achieves very stable physiologic conditions over many hours (Figure 6.37). This preparation makes it possible to regulate coronary flow rate and blood temperature. In addition, it allows the infusion of cardioactive drugs to examine the effects of physiological change on forward and inverse solutions, since the support dog eliminates the drugs and their metabolites through its urine output.

This model has a high level of realism, yet maintains adequate control over the relevant parameters. The tank can be made to have a shape identical to the human torso and can be instrumented with an almost unlimited number of recording electrodes located both on the surface and within the volume of the tank. Agarose or conductive polymer "lungs" and "ribs" can be placed in the tank to simulate volume conductor inhomogenities. The isolated animal heart provides a very realistic and versatile bioelectric source which can be instrumented and manipulated to mimic many pathologies; for example, burns can be made on the myocardium to simulate myocardial infarcts.

Despite this flexibility, the model is not perfect. To start with, there is no autonomic nervous system present in the isolated heart, so that many responses to external physiological influences do not mimic the behavior of a real human heart. In addition, the mechanical behavior is altered significantly because it hangs freely in the electrolyte without a pericardium or the constraining influences of other organs.

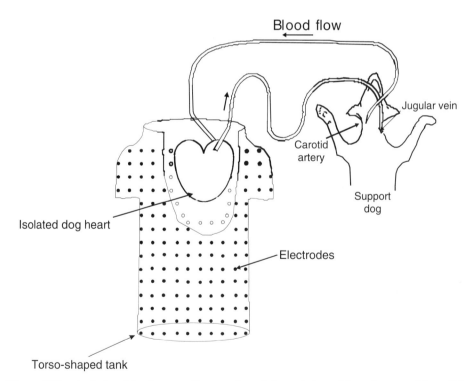

Figure 6.37   An isolated dog heart and human-shaped electrolytic tank have been used for validation of inverse electrocardiographic methods. This type of preparation uses a second dog to provide circulatory support for the isolated heart, which achieves very stable physiologic conditions over many hours.

We believe that experimental studies in the bioengineering field should be done, above all, to advance the development of methods to diagnose and cure disease. At the end of the day we must remember that men are not mice—simulation can go only so far. There is no model that can substitute *completely* for a human patient.

REFERENCES

Bachiochi, J., Creating a Nonvolatile RAM Module, *Circuit Cellar INK*, 65–71, August–September 1990.

Belusov, A., Programmable Impedance Has 12-bit Resolution, *EDN*, 105, May 23, 1996.

Brandon, C. W., D. A. Brody, C. P. Eddlemon, H. A. Phillips, and F. H. Terry, A Bipolar Electrode and Current Pump for Volume Conductor Experiments, *IEEE Transactions on Biomedical Engineering*, (1), 70–71, 1971.

Burger, H. C., and J. B. van Milaan, Heart-Vector and Leads—Part III: Geometrical Representation, *British Heart Journal*, 10, 229–233, 1948.

Craswell, J., Weekend DigiVFO, *QST*, 30–32, May 1995.

Grayzel, J., and F. Lizzi, The Combined Influence of Inhomogeneities and Dipole Location, *American Heart Journal*, 74, 503–512, 1967.

Grayzel, J., and F. Lizzi, The Performance of VCG Leads in Homogeneous and Heterogeneous Torsos, *Journal of Electrocardiology*, 2(1), 17–26, 1969.

MacLeod, R. S., B. Taccardi, and R. L. Lux, Electrocardiographic Mapping in a Realistic Torso Tank Preparation, *Proceedings of the IEEE Engineering in Medicine and Biology Society 17th Annual International Conference*, 245–246, IEEE Press, Piscataway, NJ, 1995.

Portugal, R. J., Programmable Sinewave Generator, *Electronics Now*, 43–66, January 1995.

Prutchi, D., Designing Printed Circuits for High-Speed Logic, *Circuit Cellar INK*, 38–43, January 1994.

Rosell, J. J., Colominas, P. Riu, R. Pallas-Areny, and J. G. Webster, Skin Impedance from 1 Hz to 1 MHz, *IEEE Transactions on Biomedical Engineering*, 35, 649–651, 1988.

Ruggera, P. S., et al., In Vitro Testing of Pacemakers for Digital Cellular Phone Electromagnetic Interference, *Biomedical Instrumentation and Technology*, 31(4), 358–371, 1997.

Sagi-Dolev, A. M., D. Prutchi, and R. H. Nathan, Three-Dimensional Current Density Distribution under Surface Stimulation Electrodes, *Medical and Biological Engineering and Computing*, 33(3), 403–408, 1995.

7

STIMULATION OF EXCITABLE TISSUES

An electrically excitable cell in its resting state is essentially a charged capacitor. The cell membrane is the dielectric, the ionic solutions on either side of the membrane constitute the plates, and differences in the concentrations of ions on each side generate a potential difference of about -70 to -90 mV (measured inside the cell against a reference in the extracellular fluid). To generate an action potential, the membrane capacitance must be discharged by about 15 mV in a small region. This results in a brief sequence of openings and closings of sodium and potassium channels in the membrane, which results in the flow of the action current. The action current depolarizes and then repolarizes adjacent regions of the cell membrane, giving rise to the action potential.

Excitable cells can be activated by a variety of stimuli, which include burning, mechanical trauma, electrical currents, and very intense variable magnetic fields. If sufficiently strong, any of these stimuli can depolarize the membrane of the excitable cells to a threshold voltage level at which the regenerative mechanisms of the action potential take over. However, the most common method of stimulating excitable tissue artificially is to pass an electrical current through the target tissue.

Hodgkin and Huxley's classical experiments on excitable cells were carried out by placing electrodes inside the cells under study. They wisely chose a huge cell membrane (at least as far as cells go), the giant squid axon, to make it easier to manipulate the electrodes without destroying cells. To analyze the nonlinear properties of ion conductances underlying action potentials, Hodgkin and Huxley [1952] used the voltage-clamp technique[1] developed by Kenneth Cole. As shown in Figure 7.1, space-clamp experiments usually involve inserting two electrode wires into the axon, one for recording the transmembrane voltage and the other for passing current into the axon. In the voltage clamp, the same

---

[1]Cell electrophysiology is outside the scope of this book. However, if you are interested in the subject, we would like to refer you to what we consider is the most no-nonsense source of information on cell electrophysiology and biophysics techniques: Axon Instruments Inc. publishes *The Axon Guide for Electrophysiology & Biophysics*, which is a practical laboratory guide covering a broad range of topics, from the biological basis of bioelectricity and a description of the basic experimental setup (including how to make pipette microelectrodes) to the principles of operation of advanced electrophysiology lab hardware and software. Best of all, you can download the complete guide free from Axon's Web site at www.axon.com/MR_Axon_Guide.html.

Design and Development of Medical Electronic Instrumentation   By David Prutchi and Michael Norris
ISBN 0-471-67623-3   Copyright © 2005 John Wiley & Sons, Inc.

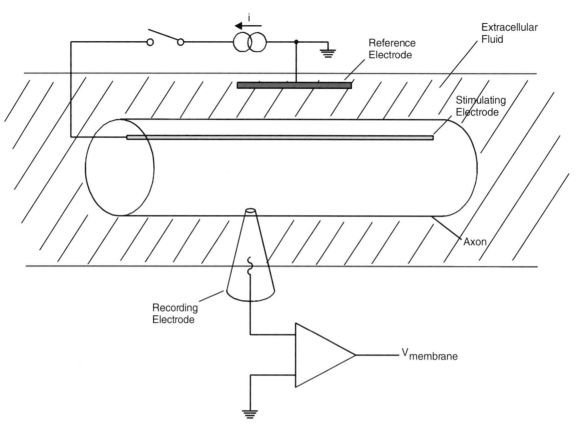

Figure 7.1 The classical space-clamped current-pulse stimulation of the giant squid axon is simulated by the Matlab program HODKIN-HUXLEY.M using Hodgkin and Huxley's membrane model.

setup is used as in the space clamp, but the current source is driven by a voltage-feedback circuit to keep the transmembrane voltage constant (or "clamped") at any value desired, providing a reading of the amount of current required to do a task. Since current is proportional to conductance at a constant voltage, this gives a measurement of the membrane conductance. In contrast, a measurement of the membrane potential without the voltage clamp shows little about the membrane conductance.

Using a voltage clamp, Hodgkin and Huxley forced open selective ion channels that reside in the membrane and are normally closed at rest and reveal themselves by the current that suddenly flows through them. The potassium current was separated experimentally by either bathing the axon in a sodium-free solution or by adding a selective sodium-channel blocker, the puffer fish poison tetrodotoxin (TTX). To determine the sodium current, they then subtracted the potassium current from the total current. [Today, the selective blocking agent tetraethylammonium or (TEA) may be used to isolate the sodium current.] Based on their experiments, Hodkin and Huxley came up with an empirical model of an excitable membrane. A Matlab program (HODKINHUX-LEY.M, available in the book's ftp site) can simulate the behavior of the axon membrane based on their model and can be used to play neurophysiologist without acquiring the squid-slicing skills of a sushi chef.

Figure 7.2 shows what happens when the simulated axon setup of Figure 7.1 is stimulated with weak ($1\,\mu A/cm^2$) and strong ($10\,\mu A/cm^2$) depolarizing ($+$ electrode inside the cell) and hyperpolarizing ($-$ electrode inside the cell) stimuli. The change in membrane voltage evoked by the weak stimuli is related primarily to the change in charge across the membrane's

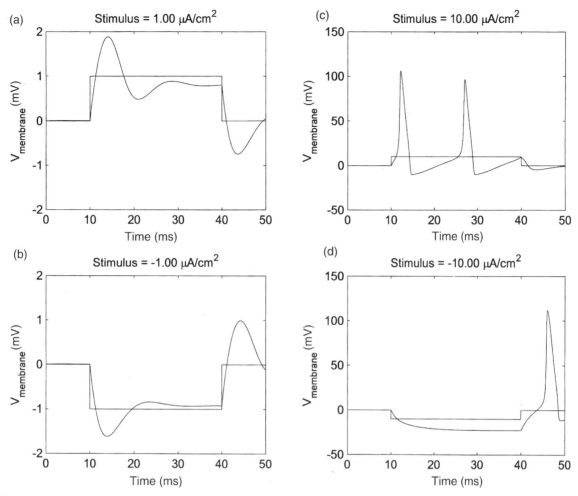

Figure 7.2 Simulation results from the stimulation of a giant squid axon with weak (1 μA/cm^2) and strong (10 μA/cm^2) depolarizing (positive electrode inside the cell) and hyperpolarizing (negative electrode inside the cell) stimuli. (*a*) and (*b*) The change in membrane voltage evoked by the weak stimuli is related primarily to the change in charge across the membrane's capacitance. (*c*) A strong depolarizing stimulus (+10 μA/cm^2) takes the membrane voltage over the threshold, causing action potentials for the duration of the stimulus. (*d*) A strong hyperpolarizing stimulus (−10 μA/cm^2) yields an action potential at the trailing edge of the pulse through rebound excitation.

capacitance. However, the strong depolarizing stimulus (+10 μA/cm^2) takes the membrane voltage over threshold, causing action potentials for the duration of the stimulus.

The strong hyperpolarizing stimulus (−10 μA/cm^2) also yields an action potential, but only at the trailing edge of the hyperpolarizing pulse. This is what Hodgkin and Huxley referred to as anode break excitation or *rebound excitation*. The hyperpolarizing pulse decreases the potassium conductance and removes sodium inactivation. The former leads to less hyperpolarizing current and the latter to more depolarizing current. Since the kinetics of the potassium channels are slower than that for the sodium channel gate, a transient depolarization takes place after a prolonged hyperpolarizing voltage, which if large enough can generate an action potential.

What we would like you to remember as we move to discuss the clinical uses of electrical stimulation is that anodic currents are usually responsible for the activation of excitable tissue *when the current is delivered through an intracellular electrode*. In addition, we would ask you to remember that hyperpolarizing currents can also lead to activation of excitable tissue via the rebound excitation mechanism.

EXTRACELLULAR STIMULATION

In clinical practice, the stimulation methods used to elucidate the electrical properties of excitable cells are simply not suitable. To start with, clinically useful stimulation requires stimulating much more than a single nerve or muscle cell, making it impractical to build electrode arrays that can impale a large number of cells simultaneously. Next, even if a sufficient number of cells could be impaled with microelectrodes, it is next to impossible to keep them in place in a living, moving being. Because of this, in vivo stimulation almost always involves delivering the stimulating currents between a pair of electrodes placed near (but not inside) the target cells. Consider the simple model of Figure 7.3. Here the vector current flux is indicated by arrows and transmembrane current is assumed to flow only at the anode and cathode (in reality it flows at all parts of the cell) when the switch closes. As shown in Figure 7.3b, the current through the membrane will hyperpolarize the intracellular membrane region under the anode and depolarize the intracellular region under the cathode. Stimulation will occur when the transmembrane potential at the cathode crosses the

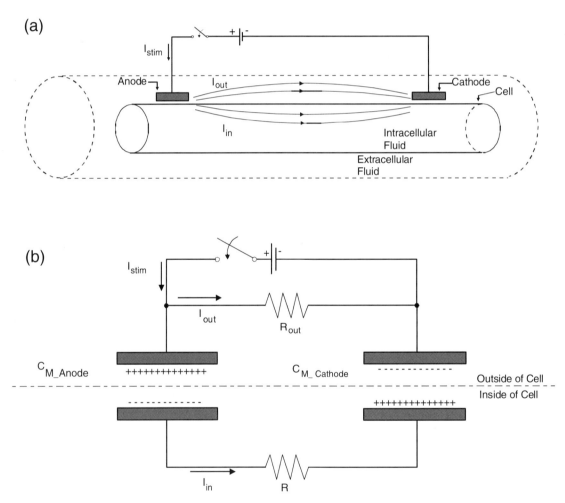

Figure 7.3 Simplified model of electrical stimulation of a cell by a current applied through extracellular electrodes. (*a*) A transmembrane current is assumed to flow only at the anode and cathode. The vector current flux is indicated by arrows. (*b*) The current through the membrane hyperpolarizes the intracellular membrane region under the anode and depolarizes the intracellular region under the cathode. Stimulation occurs when the transmembrane potential at the cathode crosses the membrane's threshold voltage.

membrane's threshold voltage. Note that this is the reverse of what happens with intracellular stimulation, where excitation occurs at the anode.

Depending on the arrangement of the electrodes, three stimulation modes can be distinguished:

1. *Bipolar*. Both electrodes are close to the target tissue.
2. *Monopolar* (also called *unipolar*). One electrode, normally the cathode, is close to the target tissue, and the other (anode) is remote from the target tissue, making its size and exact placement irrelevant.
3. *Field stimulation*. Both electrodes are remote from the target tissue.

The efficiency of bipolar and monopolar stimulation is similar. However, the current delivered in the monopolar mode often crosses through nontarget tissue on its way to the anode (yes, the conventional direction for current is in the opposite direction, but you know what we mean) and is sometimes capable of stimulating these nontarget excitable cells undesirably. Field stimulation is the most inefficient method but is very commonly the preferred mode of current delivery in nonchronic applications since it allows tissues to be stimulated using noninvasive skin-surface electrodes.

A stimulus must be of adequate intensity and duration to evoke a response. If it is too short, even a strong pulse will not be effective. The stimulation threshold is defined as the minimum strength of stimulus (expressed either in volts or in milliamperes) required for activation of a target tissue for a given stimulus duration. When thresholds for several durations are put together on the same graph, a strength–duration curve is formed. The nice thing about the strength–duration curve is that with one quick look one can determine whether or not a stimulus will be effective. Any stimulus that falls above the curve will excite the target tissue.

As shown in the stylized strength–duration curve of Figure 7.4, stimulus current and duration can be mutually traded off over a certain range. For a short pulse, the effectiveness of a stimulus is characterized by the product of current I and duration t, where delivered charge $Q = It$. Hence if the amount of charge required to activate the target tissue is $Q_{\text{threshold}}$ and the stimulus duration is t, the current $I_{\text{threshold}}$ required to achieve activation will be $I_{\text{threshold}} = Q_{\text{threshold}}/t$.

It would seem from this relationship that the strength–duration curve should show a decline to near zero as stimulus duration is increased. However, the strength–duration curve of real excitable tissue flattens out with long stimulus durations, reaching an asymptote called the *rheobase*. The root *rheo* means current and *base* means foundation; thus, the rheobase is the foundation, or minimum, current (stimulus strength) that will produce a response. When the stimulus strength is below the rheobase, stimulation is ineffective even when stimulus duration is very long.

The reason for the difference between the actual behavior and that predicted by $I_{\text{threshold}} = Q_{\text{threshold}}/t$ is that the latter assumes that the membrane is an ideal capacitor. This is not the case, and the leakage resistance shows its effect during prolonged stimulation (large values of t). The equation fails to predict the charge transfer across the cell membrane because under these conditions, more membrane current is carried by the leakage resistance and less is used to charge the membrane capacitance. Membrane potential thus rises exponentially to a plateau during prolonged stimulation instead of increasing linearly with time.

The strength–duration curve was characterized by Lapicque [1909] by the value of the rheobase (in volts or milliamperes) and a second number called the *chronaxie*. The root *chron* means time and *axie* means axis. The chronaxie is measured along the time axis and is defined as the stimulus duration (in milliseconds) that yields excitation of the tissue when stimulated at twice the rheobase strength. In the strength–duration curve of Figure 7.4, the

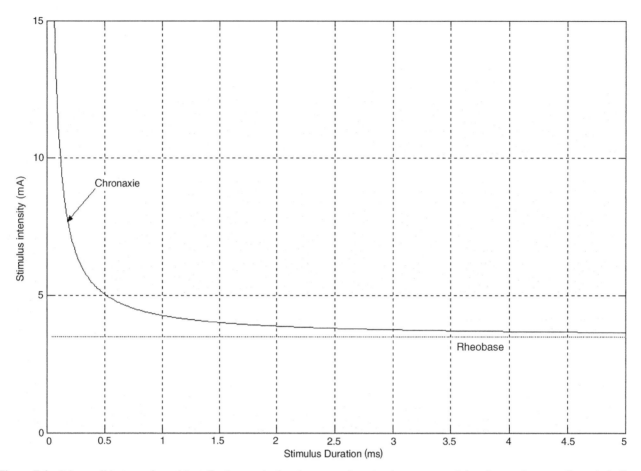

Figure 7.4   It is possible to see from this stylized strength–duration curve that stimulus current and duration can be mutually traded off over a certain range. The strength–duration curve was characterized by Lapicque by the value of the rheobase (in volts or milliamperes) and the chronaxie, which is measured along the time axis and defined as the stimulus duration (in milliseconds) that yields excitation of the tissue when stimulated at twice the rheobase strength. In this example, rheobase = 3.5 mA and chronaxie = 0.22 ms.

rheobase is the minimum stimulus strength that will produce a response, which is the point at which the curve asymptotes, about 3.5 mA. To determine the chronaxie, simply look for the stimulus duration that yields a response when the stimulus strength is set to exactly twice rheobase, or 7 mA. In this example, the chronaxie is 0.22 ms.

The strength–duration curve is highly dependent on the type of tissue being stimulated. For example, the chronaxie of human motor nerve is approximately 0.01 ms, about 0.25 ms for pain receptors, and approximately 2 ms for mammalian cardiac muscle. That is why there is rarely a need for pulses longer than 2 ms in nerve stimulation, whereas a pulse width as long as 10 ms is often necessary for direct stimulation of certain smooth muscles.

The empirical equations for the threshold current, charge, and energy for a rectangular stimulation pulse are

$$I_{\text{threshold}} = \text{rheobase} \times \left(1 + \frac{\text{chronaxie}}{t} \right)$$

$$Q_{\text{threshold}} = \text{rheobase} \times t \times \left(1 + \frac{\text{chronaxie}}{t} \right)$$

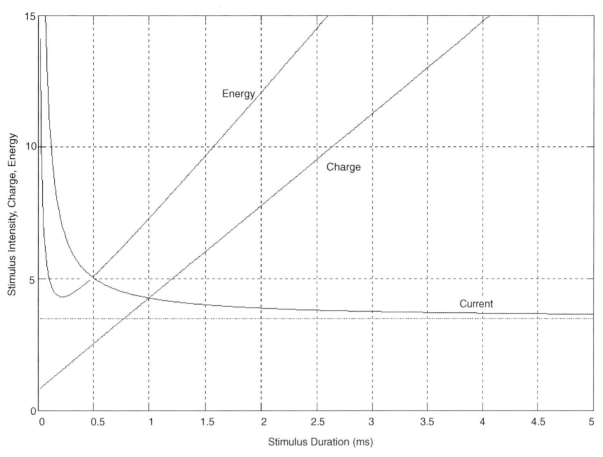

Figure 7.5   The strength–duration relationship can be expressed in terms of threshold current, threshold charge, or threshold energy needed to excite a tissue. For a rectangular current stimulus of duration t delivered to a load of resistance r, these relationships are given by $I_{\text{threshold}} = $ rheobase $\times (1 + \text{chronaxie}/t)$, $Q_{\text{threshold}} = $ rheobase $\times t \times (1 + \text{chronaxie}/t)$, and $E_{\text{threshold}} = $ rheobase$^2 \times r \times t \times (1 + \text{chronaxie}/t)^2$.

$$E_{\text{threshold}} = (\text{rheobase})^2 \times r \times t \times \left(1 + \frac{\text{chronaxie}}{t}\right)^2$$

where t is the stimulus duration and r is the resistance of the path through which the current flows (resistance of the wires, ionic resistance of the medium, and electrode–tissue interface impedance). Figure 7.5 presents the three strength–duration curves in a single graph.

In practice, stimulation parameters close to those of the strength–duration curve are seldom used. Owing to small fluctuations of excitability, the target tissue may not always be excited if the stimulus is only slightly above threshold. For this reason, the stimulus parameters are usually set to at least twice the threshold in applications that require reliable stimulation.

CLINICAL USES OF ELECTRICAL STIMULATION

Clinical electrical stimulation is simply the application of electrical currents to a body, be it for function or therapy. As we just discussed, the current of electrons passing through the wires is converted into a current of ions moved within the tissue, which are in turn capable of transporting electrical charge across the membranes of excitable tissues. The purpose of these applied currents is to cause the targeted depolarization of nerve and/or muscle to threshold.

The most common clinical applications of electrical stimulation are:

1. Cardiac pacing. Electrical stimulation of the heart's chambers relieves or eliminates the symptoms of bradycardia (a heart rate that is too slow). Rhythmic stimulation (pacing) increases the heart rate to meet the oxygen needs of the body. Cardiac pacing is discussed in detail in Chapter 8.

2. Cardiac defibrillation. High-energy stimulation of the heart (in the form of an electrical shock) interrupts a rapid heart rhythm (tachycardia) so that a more normal rhythm can be restored. Cardiac defibrillation is discussed in detail in Chapter 8.

3. Cardiomyoplasty. A skeletal muscle (e.g., the latissimus dorsi, which attaches at one end to the upper part of the upper arm bone and spreads out like a fan to attach to the spine and ribs) is dissected free from its normal attachments and then wrapped around the heart. The muscle is then stimulated to contract in synchronism with the heart. Since skeletal muscle is prone to fatigue, it must first be trained by converting its fibers to fatigue-resistant type 1 muscle fibers. Training is done with a low stimulation rate with only one pulse in a burst and over a period of six weeks increasing the repetition rate and the number of pulses in the burst.

4. Electroventilation. Electrical stimulation of the phrenic nerve or the diaphragmatic muscles is used to support ventilation. Candidates for breathing pacing include patients who require chronic ventilatory support because of spinal cord injury, decreased day or night ventilatory drive (e.g., sleep apnea), intractable hiccups (chronic hiccups often lead to severe weight loss and fatigue and can have fatal consequences), and damaged phrenic nerve(s). The physiological respiratory function provided by these devices is far superior to that provided by mechanical ventilators since the air inhaled is drawn into the lungs by the musculature rather than being forced into the chest under mechanical pressure.

5. Diagnostic stimulation of nerves and muscles. Nerve conduction studies are performed routinely to assess peripheral nerve function. Electrical stimulation is applied to a nerve and the nearby EMG signal is measured. This is done to determine the speed of transmission along the nerve. It also helps to determine if there is a blockage in the nerve or where the nerve connects to the muscle. In a similar way, electrical stimuli delivered at the wrist or behind the knee are used to evoke brain responses to sensory inputs. The somatosensory-evoked potentials are detected by coherent averaging of the EEG. From this information, the evaluator may determine whether there is a delay in conduction to the brain, a blockage at any point, or abnormally low or high activity in the brain. Another common diagnostic use of nerve stimulation is monitoring the depth of neurological blocks present in a patient following the administration of muscle relaxant drugs (e.g., prior to surgery, and after surgery following the administration of antagonist drugs).

6. Diagnostic stimulation of the brain. Very brief high-voltage pulses or pulse bursts to stimulate percutaneously human motor cortex, visual cortex, or spinal cord are used for intraoperative monitoring as well as for diagnosis of neurological diseases.

7. Pain relief. The technique of applying electric currents to the spinal cord or a peripheral nerve to relieve pain is known as *electroanalgesia*. Its use with both permanently implanted and nonsurgically applied devices is common practice in the treatment of patients suffering from chronic pain.

8. Control of epileptic seizures. Electrical stimulation of the vagus nerve [also known as vagus nerve stimulation (VNS)] involves periodic mild electrical stimulation of the vagus nerve in the neck by a surgically implanted device. VNS has been found effective in controlling some epilepsies when antiepileptic drugs have been inadequate, their side effects intolerable, or neurosurgery has not been an option. In some cases VNS has also been effective in stopping seizures. It carries minimal side effects (e.g., mild tingling sensations and voice hoarseness during stimulation), but unlike many medications, there seem to be no significant intellectual, cognitive, behavioral, or emotional side effects to VNS therapy. VNS is now the second most common treatment for epilepsy in the United States, and the improvement in seizure control is comparable to that of new antiepileptic drugs.

9. Control of Parkinsonian tremor. Electrical stimulation of neuron clusters deep inside the brain [also known as deep brain stimulation (DBS)] is now used to inactivate the subthalamic nucleus, which is overactive in Parkinson's disease. A multielectrode lead is implanted into the ventrointermediate nucleus of the thalamus. The lead is connected to a pulse generator that is surgically implanted under the skin in the upper chest. When the patient passes a magnet over the pulse generator, the device delivers high-frequency pulse trains to the subthalamic nucleus to block the tremor.

10. Gastric "pacing." Electrical stimulation of the stomach is currently being used to reduce symptoms of nausea and vomiting for patients suffering from gastroparesis (a stomach disorder in which food moves through the stomach more slowly than normal).

11. Restoration of lost sight. Electrical stimulation of the retina, the optical nerve, and the visual cortex is now developed to the point at which implants for functionally restoring sights to blind patients will soon be available commercially. Functional sight may be given to patients blinded by retinitis pigmentosa by using integrated circuits embedded in contact with the retina. The ICs contain an array of photovoltaic cells that directly power an array of microstimulators and electrodes to convert the image into a directly mapped electrical image, bypassing degenerated photoreceptors and directly stimulating the remaining nerve cells in the retina. For patients with blindness caused farther down the optical nerve, the possibility exists of stimulating the visual cortex directly using microelectrode arrays to generate coherent images from phosphenes (sensation of a spot of light) elicited by the electrical stimulation.

12. Restoration of lost hearing. Cochlear implants stimulate spinal ganglion cells of the auditory nerves, bypassing nonfunctional hair cells to restore limited hearing in some types of deafness. The cochlear implant system really consists of an implanted stimulator connected to an electrode array inserted in the cochlea and an external speech processor that codes the speech into stimulation patterns that can be translated back into sounds by the brain. The external speech processor also powers the implant via an inductive energy transfer link. Cochlear implants are now common and provide substantial benefits to many profoundly deafened children and adults. Benefits vary by person and range from increased perception of environmental sounds to the ability to use a telephone.

13. Restoration of lost or impaired neuromuscular function. Functional electrical stimulation (FES), also known as functional neuromuscular stimulation (FNS), is a rehabilitation strategy that applies electrical currents to the nerves that control paralyzed muscles in order to stimulate functional movements such as standing or stepping. FNS systems include either skin-surface or implanted electrodes, a control unit which often also receives motion information back from sensors, and a stimulus generator. A number of FNS units are now either available commercially or under clinical investigation. Typical applications of FNS include controlling foot drop, enabling lower-limb paraplegics to stand or sit, and restoring hand function to the paralyzed upper limb.

14. Maintenance or increase in range of movement. Electrical muscle stimulation (EMS) is used to strengthen muscle and facilitate voluntary motor function. Although EMS devices are often advertised for muscle toning and weight reduction, they are authorized by the FDA only as prescription devices for maintaining or increasing range of motion, relaxation of muscle spasm, prevention or retardation of disuse atrophy, muscle reeducation, increasing local blood circulation, and postsurgical stimulation of calf muscles to prevent the formation of blood clots.

15. Electroconvulsive therapy (ECT). This is a relatively painless procedure that is effective in treating major depression. A short, controlled set of electrical pulses is given for about a minute through scalp electrodes to produce generalized seizures. Biological changes that result from the seizure are believed to result in a change in brain chemistry which is believed to be the key to restoring normal function. Because patients are under anesthesia and have taken muscle relaxants, they neither convulse nor feel the current.

TABLE 7.1   Typical Parameters Used in Various Clinical Applications Involving the Stimulation of Tissues with Electrical Currents

Clinical Application	Typical Method of Current Delivery	Typical Waveform	Typical Current or Voltage
Cardiac pacing	Implanted electrodes in contact with heart; electrode impedance 250 Ω to 1 kΩ	0.1- to 2-ms capacitor-discharge pulse with charge-balancing phase	0.1 to 8 V peak delivered from 5- to 10-μF capacitor
	Gelled skin-surface electrodes placed on chest; electrode impedance ~50 Ω	Balanced biphasic current pulse 20 to 40 ms in duration	50 to 200 mA with 30-V compliance
Cardiac defibrillation	Implanted electrodes in contact with heart; electrode impedance 30 to 60 Ω	Biphasic capacitor discharge 5 to 10 ms in duration	2 to 10 A with capacitor bank charged to <1 kV
	Gelled skin surface electrodes placed on chest; electrode impedance 50 to 100 Ω	Monophasic or biphasic capacitor-discharge pulse 5 to 10 ms in duration	30 to 40 A with capacitor bank charged to <3 kV
Cardiomyoplasty	Platinum–iridium wire electrodes sewn across skeletal muscle a few centimeters apart, looped under the nerve branches that run along the surface of the muscle; electrode impedance 50 to 100 Ω	Burst of capacitive discharge pulses with charge-balancing phase, 0.06 to 1.0 ms in duration; 1 to 16 pulses per burst, at a pulse repetition rate of ~10 to 60 Hz; burst delivered in synchrony with cardiac activity at a burst-to-beat ratio of 1:1 to 1:16	0.1 to 8 V peak delivered from 5- to 10-μF capacitor
Electroventilation	Temporary electroventilation with gelled anterior axillary skin-surface electrodes; impedance 250 Ω to 1 kΩ	0.8-s bursts of balanced monophasic pulses 10 μs in duration at 35 Hz	200 mA to 1.5 A with up to 1500-V compliance
	Implanted electrodes in contact with phrenic nerve or innervation point of diaphragmatic muscles	0.8-s bursts of balanced biphasic current pulses 1 to 10 ms in duration or (phrenic nerve) 25 to 100 ms in duration (muscles); repetition rate 30 Hz	1 to 10 mA with 12-V compliance
Diagnostic stimulation of peripheral nerves	Bipolar pair of 2- to 5-mm-diameter spherical dry electrodes with interelectrode distance of 2 to 5 mm applied to skin over target nerve	Monophasic current pulses 50 μs to 2 ms in duration	0 to 100 mA with up to 400-V compliance
Diagnostic stimulation of brain (cortex) and spinal cord	Bipolar pair gelled electrodes (or corkscrew electrodes for intraoperative monitoring) with interelectrode distance of ~7 cm applied to skin	50-μs-wide transformer-isolated square wave	100 to 1000 V with a maximum current of 1.5 A (at a rate of current rise of 0.1 A/μs)
Pain relief	Implanted electrodes in contact with spinal cord or targeted peripheral nerve to block the sensation of pain	Monophasic or biphasic pulses ~210 μs in duration delivered at 30 to 80 Hz	0.1 to 12 V peak
	Gelled skin-surface electrodes (impedance 200 Ω to 1 kΩ) placed on painful region; often known as transcutaneous electrical nerve stimulation (TENS)	Monophasic or biphasic pulses 50 to 150 μs in duration delivered at 10 to 150 Hz	10 to 150 mA with <150-V compliance
Vagus nerve stimulation (VNS)	Implanted electrodes in contact with vagus nerve; electrode impedance 1 to 7 kΩ	Monophasic current pulses (with charge-balancing phase) 130 to 1000 μs in duration delivered at ~30 Hz for 30 s every 5 minutes	0.25 to 35 mA with 12-V compliance

TABLE 7.1   (*Continued*)

Clinical Application	Typical Method of Current Delivery	Typical Waveform	Typical Current or Voltage
Deep brain stimulation (DBS)	Thin electrode implanted deep into parts of the brain that are involved in control of movement; electrode impedance 600 Ω to 2 kΩ	60- to 450-μs charge-balanced capacitor-discharge pulses delivered at 2 to 185 Hz; burst on/off times depend on patient needs	0.1 to 10.5 V peak
Gastric pacing	Implanted electrodes stitched to the stomach muscle wall of the antrum 10 cm proximal to the pylorus; electrode impedance 200 Ω to 1 kΩ	Monophasic or biphasic pulses $\sim$210 μs in duration delivered at 30 to 80 Hz	0.1 to 12 V peak
Restoration of lost sight	Implanted electrode array in contact with retina; typical electrode impedance 1 to 10 kΩ	Balanced biphasic current pulse 100 μs to 5 ms in duration; repetition rate 60 to 500 Hz	10 to 600 μA with 6-V compliance
	Implanted electrode array in contact with brain's visual cortex; typical electrode impedance 10 to 100 kΩ	Balanced biphasic current pulse 100 μs to 2 ms in duration; repetition rate 10 to 250 Hz	1 to 60 μA with 6-V compliance
Cochlear stimulation	Implanted electrode array in contact with cochlea; typical electrode impedance 1 to 10 kΩ	Balanced biphasic current pulse 20 μs to 1.2 ms in duration; repetition rate up to 2 kHz	30 μA to 2 mA with 12-V compliance
Functional neuromuscular stimulation (FNS)	Implanted electrodes in contact with muscle; electrode impedance 200 Ω to 2 kΩ	Balanced biphasic current pulse 25 to 500 ms in duration; repetition rate up to 100 Hz	1 to 10 mA with 20-V compliance
	Gelled skin-surface electrodes placed over target muscle	<1-ms pulse (typically around 300 μs) with a frequency <100 Hz (due to the absolute refractory period of normal muscle) Biphasic 10- to 15-ms waveforms $\sim$10 Hz for denervated muscle	10 to 150 mA with <150-V compliance
Electrical muscle stimulation (EMS)	Gelled skin-surface electrodes placed over target muscles	<1-ms pulse (typically around 300 μs) with a frequency of <100 Hz (due to the absolute refractory period of normal muscle) Biphasic 10- to 15-ms waveforms $\sim$10 Hz for denervated muscle	10 to 150 mA with <150-V compliance
	Interferential mode: at least two pairs of skin surface electrodes delivering high-frequency signals that interfere at the target muscles; gelled skin-surface electrodes with impedance 100 Ω to 1.5 kΩ at 4 kHz	Sinusoidal current; one channel at a frequency of 4 kHz, second channel at 4 kHz $\pm$ selectable beat frequency	0 to 100 mA RMS with 150-V_{p-p} compliance
Electroconvulsive therapy (ECT)	Gelled skin-surface electrodes applied to the forehead; impedance 250 Ω to 1.5 kΩ	10-s burst of 0.25-ms pulses delivered at 10 to 100 Hz	Up to 1 A with 2.5-kV voltage compliance

The primary factors determining whether sufficient current flows to yield a desired clinical effect are impedance of the body tissues in the path of the current, electrode size and position, stimulation parameters, and the electrical characteristics of the tissue to be excited. These parameters are usually interrelated, as shown in Table 7.1 for the various clinical areas in which electrical stimulation is used.

The most commonly used stimulation signal waveshapes are those shown in Figure 7.6. The charge-balanced pulses of Figure 7.6a ensure that no net charge is introduced to the

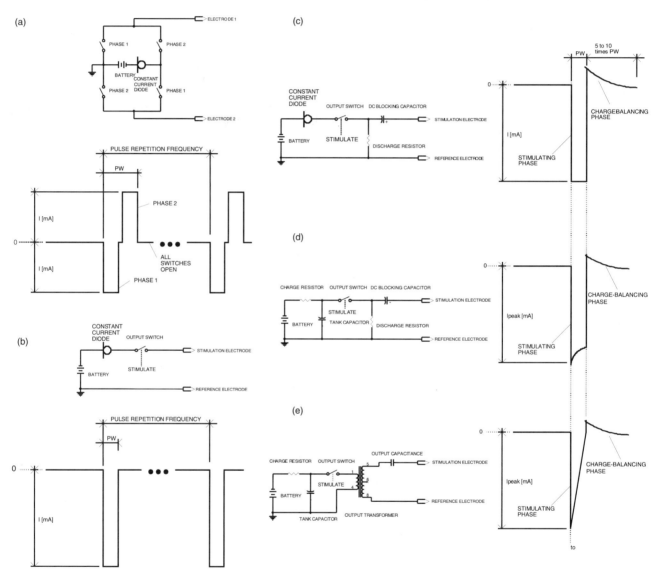

Figure 7.6   These are the most common stimulation signal waveshapes and the generic circuits used to produce them. (*a*) Charge-balanced pulses ensure that no net charge is introduced to the body. In the balanced bidirectional pulse pair, each stimulus pulse has two phases of identical duration and identical current magnitude, but of opposite polarity. (*b*) True monophasic waveforms are seldom, if ever, used to stimulate tissue because they introduce net charge through the tissue that can cause tissue damage. They are produced only when the energy source is switched along the way to the tissue and there is no way for the electrode–tissue interface capacitance to discharge. (*c*) Monophasic waveforms really tend to be asymmetric biphasic, as the net charge built up in the electrode–tissue interface or in a dedicated dc-blocking capacitor discharges. Not all stimulators deliver a constant-current stimulus. Some generate the stimulus current by discharging a capacitor into the tissue (*d*) or by using an impulse transformer to step-up the voltage (*e*).

body. Current is driven into the tissue first in one direction and then, after a brief interval, in the other. Thus, ions moving in the tissue would first be pushed one way and then quickly the other way, stimulating the tissue and leaving the ions in their former positions within the electrodes, interstitial fluids, and cells. This waveform is known as a *balanced bidirectional pulse pair*. Neurons are especially sensitive to poisoning by metallic ions released from the electrodes as well as by the products of electrolytic decomposition of salts and water. Microscopic studies of brains stimulated with this balanced pulse pair at low current densities showed that it causes no electrolytic damage to the neurons.

As shown in Figure 7.6*b*, true monophasic pulses are produced only when the energy source is switched along the way to the tissue and there is no way for the electrode–tissue interface capacitance to discharge. True monophasic waveforms are seldom, if ever, used to stimulate tissue because they introduce net charge through the tissue that can cause electrolysis and tissue damage. Monophasic waveforms really tend to be asymmetric biphasic, as the net charge built up in the electrode–tissue interface or in a dedicated dc-blocking capacitor discharges as shown in Figure 7.6*c–e*.

DIRECT STIMULATION OF NERVE AND MUSCLE

Implantable cardiac pacemakers have been around since the late 1950s. More recently, the same basic techniques have been applied to stimulate the vagus nerve for the control of epilepsy, to stimulate the sacral roots to control the bladder and correct erectile dysfunction, and to stimulate nerves in the spine for the control of pain and angina. In addition, interest in functional electrical stimulation (FES) has grown rapidly during recent years, due primarily to progress made in miniaturized hardware that makes multichannel stimulators possible. New surgical techniques enable the use of chronically implanted stimulators to stimulate specific nerves and brain sections directly within the body, making it possible to restore function lost due to disease or trauma. Advances are being made rapidly in the development of implants for restoring limbs, sight (e.g., through artificial retinas or by direct stimulation of the visual cortex), and hearing (e.g., through cochlear implants) [Loeb, 1989].

As shown in Table 7.1, and with the exception of cardiac defibrillation, all other applications in which the electrodes are placed in close contact with target tissue require the delivery of relatively narrow pulses ($<2\,\mathrm{ms}$) of low voltage ($<12\,\mathrm{V}$) at low current ($<35\,\mathrm{mA}$). These can easily be produced with miniature circuits that use standard bipolar or MOSFET transistors (discrete transistors or as part of an IC), tantalum capacitors, and implantable-grade lithium batteries. Implantable stimulators typically use either a constant-current source or a capacitor discharge circuit as output stages to generate stimulation pulses.

Capacitor-Discharge Stimulators

In a capacitor-discharge output stage, an energy-storage capacitor, usually called a *tank capacitor*, is charged to the desired peak voltage and then delivered to the target tissue. Figure 7.7 shows the circuit of a simple capacitor-discharge pulse generator circuit that can generate stimulation pulses with an amplitude of either 3 or 6 V from a 3-V source (e.g., a single lithium battery). When inactive, the control logic for the stimulator sets the HIGH AMPLITUDE STIMULUS line low, which charges tank capacitor C2 to VDD. The STIMULUS signal is maintained low to keep transistor Q2 open, and line ACTIVE DISCHARGE is maintained low to keep switch Q1 open. Coupling capacitor C1 slowly discharges by way of resistor R1 ($100\,\mathrm{k}\Omega$) through the tissue and electrodes connected to terminals V+ and V−.

When a stimulus is to be generated, and if the amplitude selected is 6 V, the HIGH AMPLITUDE STIMULUS line is set high, which closes Q4 and opens Q3. This causes

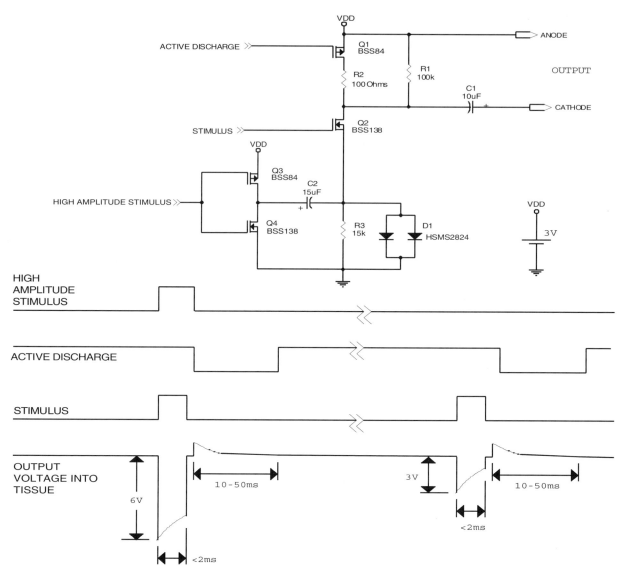

Figure 7.7 This capacitor-discharge output stage suitable for portable or implantable stimulators is capable of delivering 3- or 6-V stimuli using a single lithium battery as the energy source. The circuit can sustain sufficient stimulation currents for up to 2 ms when delivered into implanted electrodes that present a load impedance of approximately 500 Ω.

the positive terminal of capacitor C2 to be connected with the battery's negative terminal. When a 3-V stimulation is desired, HIGH AMPLITUDE STIMULUS is set low, which connects C2's positive terminal to the battery's positive terminal (VDD). In the first case, the potential difference between the negative terminal of C2 and VDD is 6 V, while in the second case the potential difference is 3 V.

The STIMULUS line is set high to deliver the stimulus to the tissue, which closes Q2 and connects the negative terminal of C2 to C1 (which is discharged). As such, the leading edge voltage of the pulse appearing across electrode terminals V+ and V− is equal to the voltage selected (3 or 6 V). This voltage decays throughout the stimulus pulse as C2 discharges and C1 charges. To terminate current delivery to the tissue, the control logic places all stimulus-related lines back to their rest condition.

Once the pulse has been delivered, coupling capacitor C1 remains charged. The delivery of a new stimulation pulse will require this capacitor to be discharged. This is accomplished

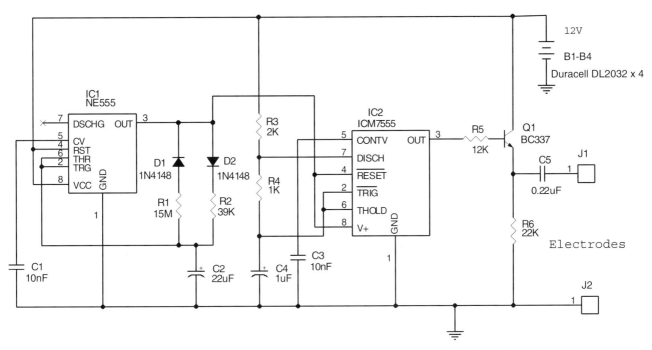

Figure 7.8 A variation on the capacitor-discharge circuit is the capacitor-coupled stimulator. In this circuit for the stimulation of denervated muscles, C5 differentiates the rectangular square waveform at the output of Q1 to yield a net-zero charge transfer across the tissue.

by delivering the energy stored in this capacitor through the tissue. A zero net current flow through the tissue results from passing the same amount of charge (albeit not within the same amount of time) through the tissue during the discharge phase as was delivered during the stimulus pulse, but in the opposite direction. Not doing so would cause electrochemical imbalance, which can result in electrode corrosion and tissue damage. Taking the ACTIVE DISCHARGE line low closes Q1, allowing the charge in coupling capacitor C1 to flow through the tissue via resistors R2 and R1 ($100\,\Omega$ in parallel with $100\,k\Omega$). Any remaining charge after the fast discharge time is delivered at a slower rate through R1 ($100\,k\Omega$). With the component values shown, this circuit is suitable for delivering stimulus pulses with durations of up to 2 ms into implanted electrodes that present a load impedance of approximately $500\,\Omega$.

A variation on the capacitor-discharge circuit is the capacitor-coupled stimulator. Figure 7.8 shows a stimulator based on the circuit configuration proposed by Sebille et al. [1988] for the stimulation of denervated muscles. In this circuit, timer IC2 generates a square wave that drives transistor Q1. The common-collector connection allows the output voltage to remain independent of the impedance presented by the tissue being stimulated. C5 differentiates the rectangular square wave to yield a zero net charge transfer across the tissue. Timer IC1 cycles IC2's power on and off, making it possible to vary the stimulation duty cycle by changing the values of R1 and R2.

Current-Source Stimulators

Hochmair [1980] described a CMOS low-power current source suitable for implantable devices. It uses a standard CMOS 4007 integrated circuit, which contains six enhancement MOSFETs, three n-channel and three p-channel. The n-channel bodies (p-silicon) are connected to pin 7 and must be kept at the most negative voltage used in the circuit. The p-channel bodies (n-silicon) are connected to pin 14 and must be kept at the most positive voltage used in the IC. The transistor elements are accessible through the package terminals.

The CD4007 is usually characterized as a CMOS digital IC, but it is perfectly useful as a FET for analog circuits.

In the current-source circuit of Figure 7.9, the control voltage V_{in} is delivered to the gates of the CMOS pair formed by transistors Q1 and Q2. The output capacitor C2 is charged by the n-channel transistor Q4. The thresholds of Q3 and Q4 are adjusted to approximately the same level of V_{in}, about 4 V. Q4 conducts above this level and Q3 conducts below it. When the current source is not active, either Q3 or Q4 is off completely, and the power consumption is under 400 μW.

Q1 and Q2 together with zener diode D1 act as source followers to increase the thresholds of the current-source output transistors. D1 is selected to compensate for

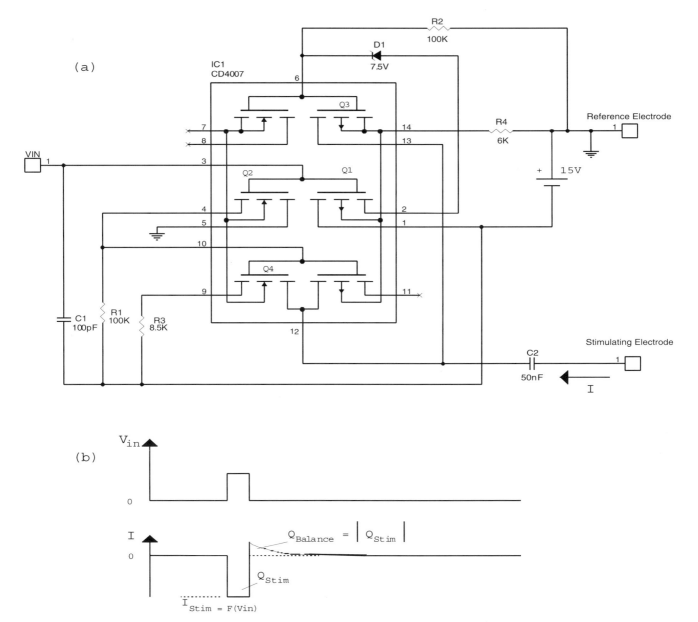

Figure 7.9   A MOS circuit based on a standard 4007 can be used as a low-power current-source stimulator capable of delivering charge-balanced pulses of up to 700 μA with a compliance of up to 10 V for a maximum delivered charge per pulse of 0.5 μC.

parameter variations of the 4007 IC such that the thresholds of Q3 and Q4 are adjusted to be almost identical. For a 15-V power supply, a 7.5-V zener is required to bring the threshold of Q4 up to 4 V. Figure 7.10 shows the transfer characteristics of the current source. The circuit can be implemented as a thin-film hybrid using bare-die 4007, thin-film resistors, and surface-mounted diodes and capacitors. Hochmair reported using this circuit as part of an eight-channel auditory nerve stimulator capable of delivering charge-balanced pulses of up to $700\,\mu A$ with a compliance of up to 10 V for a maximum delivered charge per pulse of $0.5\,\mu C$.

True symmetrical biphasic current waveforms are often generated through a voltage-to-current converter driven by the voltage representation of the current waveform desired. Figure 7.11a and b show the basic configurations for floating-load op-amp voltage-to-current converters. Floating-load circuits provide the best possible performance of any of the current output circuits, but require the load to float, which makes them unsuitable for many applications. In these circuits, the load current I_{out} develops a proportional voltage in R_{sense}, which is fed back for comparison to applied input. As long as voltage across R_{sense} is lower than the input voltage, the magnitude of the output voltage increases. The inverting circuit has the advantage of being able scale the transfer function up or down by selecting the proper ratio R_F/R_{in}.

The improved Howland current pump provides a topology for voltage-to-current converters that require the load to have one end ground referenced. The circuits of Figure 7.11c and d act as differential amplifiers with a differential input and a differential output. V_{in} is gained up by the ratio of R_F/R_{in} and impressed differentially across R_{sense}. I_{out} is thus the voltage across R_{sense} divided by the value of R_{sense}. Since the input is also differential, moving V_{in} to the opposite input simply reverses the relationship of I_{out} to V_{in}. The dominant error source in the Howland pump topology is ratio matching of the R_F/R_{in} resistors. The ratio of R_F/R_{in} for the negative feedback path should closely match the ratio of R_F/R_{in} in the negative feedback path.

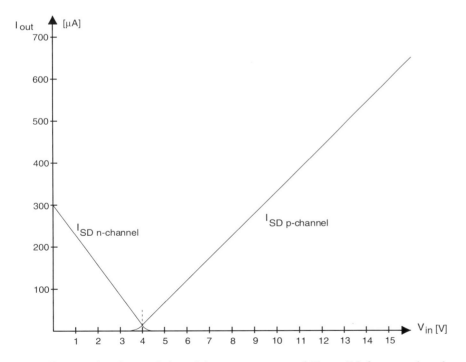

Figure 7.10 Transfer characteristics of the current source of Figure 7.9 for a supply voltage of 15 V.

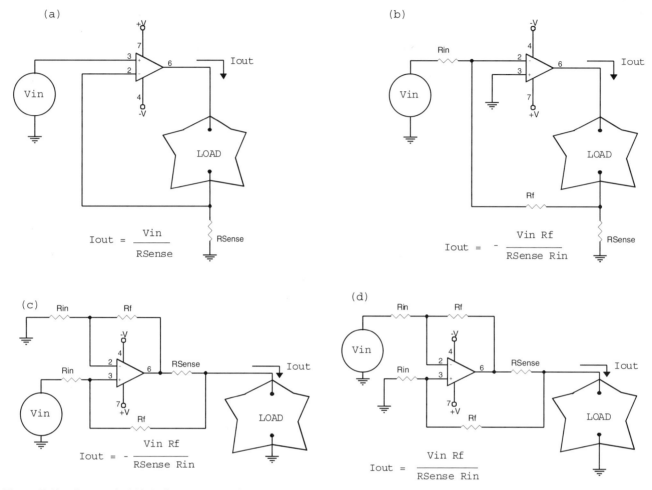

Figure 7.11 Symmetrical biphasic current waveforms are often generated through voltage-to-current converters: (*a*) floating-load noninverting configuration; (*b*) floating-load inverting configuration; (*c*) grounded-load noninverting improved Howland current pump; (*d*) inverting configuration of the Howland pump.

As an example, Figure 7.12 presents the output stage of a backpack stimulator based on the one developed by Livnat et al. [1981] for animal experiments involving chronic stimulation of the central nervous system with implanted electrodes. Pulses are generated by IC1 at a frequency set by R3 and R4 (e.g., 160 Hz). IC2A, IC3A, IC4A, IC4B, and IC5 reshape the output of timer IC1 into a symmetric bipolar square-pulse train. The voltage waveform at the output of IC5 is converted into a constant-current waveform by the inverting Howland pump built around IC6. The stimulation current can be selected by the setting of R1. In Livnat's stimulator, power for the output stage was turned on and off through a set of CMOS switches driven by a low-frequency sequencer that selected the burst duration and frequency at which bursts were delivered to the animal. The stimulator output was designed to deliver up to 1 mA into electrodes with impedances of up to 10 kΩ.

Another way of producing biphasic current pulses is to use a unipolar current source that is switched through an H-bridge so that the polarity presented to the tissue can be reversed. Take, for example, the optically isolated bipolar current source of Figure 7.13. This very neat circuit configuration was proposed by Woodward [1998] as a laboratory stimulator for neurophysiology research. In this circuit, the four optocoupler transistors of the H-bridge (IC2) are used not only to switch the polarity of the current source, they are themselves the

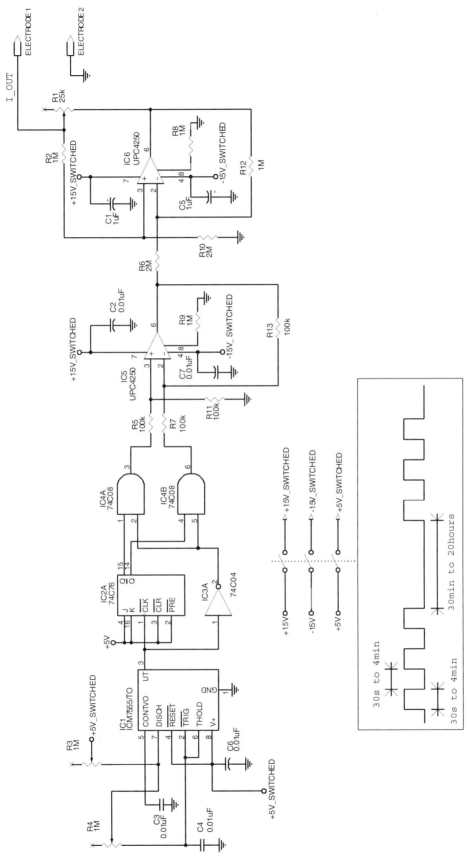

Figure 7.12 In this symmetric biphasic current stimulator, pulses generated by timer IC1 are reshaped by IC2A, IC3A, IC4A, IC4B, and IC5 into a symmetric bipolar square-pulse train. The resulting voltage waveform is converted into a constant-current waveform by the inverting Howland current pump built around IC6. Livnat et al. [1981] used implanted electrodes and this output stage for chronic stimulation of the central nervous system.

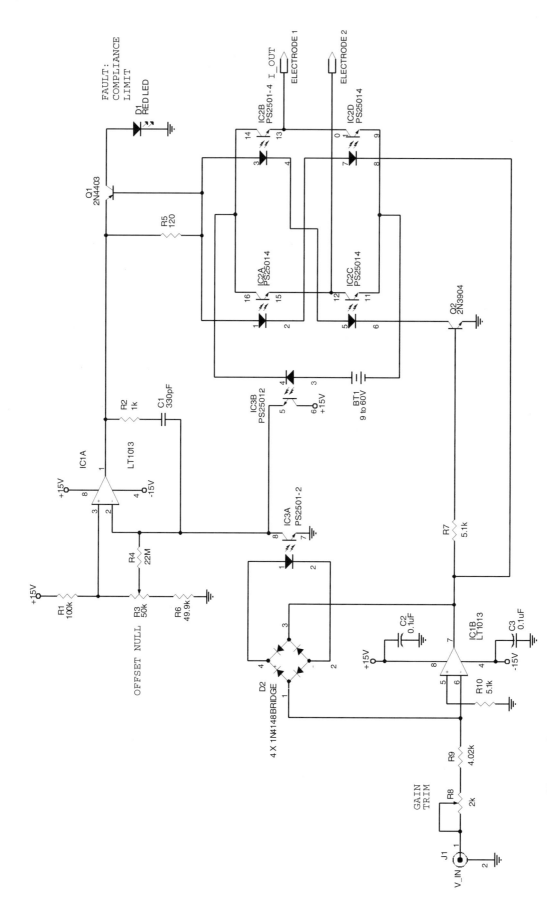

Figure 7.13   In this circuit for neurophysiology research, the four optocoupler transistors of the H-bridge are used to switch the polarity of the current source and to regulate the current delivered to the electrodes. The current delivered to the bridge is sensed by optocoupler IC3B. The current-control loop is closed via IC1A, which receives setpoint and feedback currents through IC3A and IC3B, respectively.

current-regulating elements of the current source. In a nutshell, the circuit works as follows: IC1B and Q2 will drive current through either the LEDs of IC2A and IC2D or those of IC2B and IC2C, depending on the polarity of the input signal. This will cause current from the battery to be driven into the load connected to the electrodes by the H-bridge transistors of multichannel optocoupler IC2. The current delivered to the bridge is sensed by optocoupler IC3B. The current-control loop is closed via IC1A, which receives setpoint and feedback currents through IC3A and IC3B, respectively. Q1 and LED D1 turn on when IC1A drives excessive current through the bridge, providing an indication that the voltage compliance of the source has been reached.

Op-Amp Bridge Stimulators

Another stimulator output stage configuration that is worth exploring is the op-amp bridge. The bridge connection of two op-amps provides output voltage swings twice that of one op-amp. This makes it possible to reach higher stimulation or compliance voltages in applications with low supply voltages or applications that operate amplifiers near their maximum voltage ratings in which a single amplifier could not provide sufficient drive.[2] In addition, the bridge configuration is the only way to obtain bipolar dc-coupled drive in single-supply applications.

Figure 7.14a shows a generic voltage-output bridge connection of two op-amps. Amplifier A1 is commonly referred to as the *master* and A2 as the *slave*. The master amplifier accepts the input signal and provides the gain necessary to develop full output swing from the input signal. The total gain across the load is thus twice the gain of the master amplifier. The master amplifier can be set up as any op-amp circuit: inverting or noninverting, differential amplifier, or as a current source such as a Howland current pump. In the latter, only the master amplifier is configured as a current source. As shown in Figure 7.14b, the slave remains as an inverting voltage amplifier.

This is, however, not the only current-source circuit that can be built in bridge configuration. In fact, U.S. patent 4,856,525 to van den Honert [1989] shows how to use two current sources per channel of a multichannel stimulator to minimize the cross-coupling of stimulating currents between channels. As shown in Figure 7.15a, each channel of a multichannel stimulator may have a current source that drives current between one electrode and a common ground electrode. However, a single current source for each channel requires only that the current in the loop containing the electrode and electrical ground be equal to the value of the current source and not necessarily that the current passing to ground go through any particular one of electrodes coupled to electrical ground. If the electrical power supplies of the channels of the electrical stimulator are not isolated electrically, a cross current from one electrode pair to another electrode pair could occur while maintaining the loop current requirements of each channel's single current source. Van den Honert invented a multichannel current source configuration with improved channel isolation that does not require electrically isolated power supplies.

As shown in Figure 7.15b, each channel of the electrical stimulator has a pair of current sources that work in concert. One current source is coupled to each electrode of each channel's electrode pair. The current sources operate harmoniously. When one current source is sourcing (or sinking) a certain current, its complementary current source is sinking (or sourcing) an almost identical current. As such, the current passing between the

---

[2]Op-amp bridge current sources are not only limited to low-voltage applications in direct stimulation of tissues. There are a number of op-amps that will operate with very high supply voltages, making it possible to use the same techniques in the design of high-voltage external stimulators. For example, the PA89 by Apex Microtechnology is rated for a total supply voltage of 1200 V.

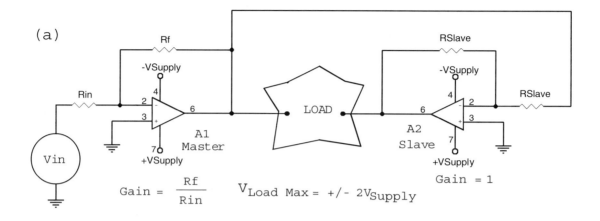

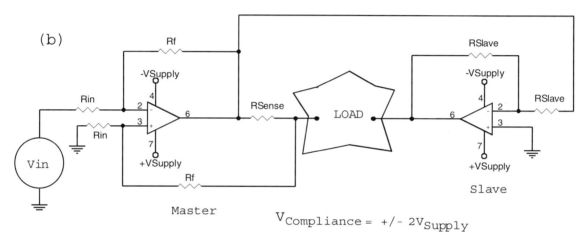

Figure 7.14   The bridge connection of two op-amps provides output voltage swings twice that of one op-amp, making it possible to reach higher stimulation or compliance voltages in applications that operate amplifiers near their maximum voltage ratings. (*a*) Generic voltage-output bridge connection of two op-amps. (*b*) The master amplifier can be set up as any op-amp circuit: for example, as a Howland current pump.

electrode pair is equal to the amount of current desired. Current leak between electrode pairs is minimized and channel isolation is improved significantly. Figure 7.15*c* shows how the circuit could be built with Howland current pumps.

Charge Injection through Implanted Electrodes

Electrodes used for stimulation behave radically different than electrodes that are used only for sensing biopotentials. This is because much stronger currents need to be produced through the tissue by converting electron flow in the leads to ion movement in the aqueous solution. Safe electrical stimulation of the nerves and muscle through direct-contact electrodes requires this conversion to happen only through reversible charge-injection processes. Irreversible reactions such as electrolysis result in the evolution of gases and acid or alkali solutions that can destroy the tissues. Other harmful irreversible faradic reactions include saline oxidation, metal dissolution, and oxidation of organic molecules.

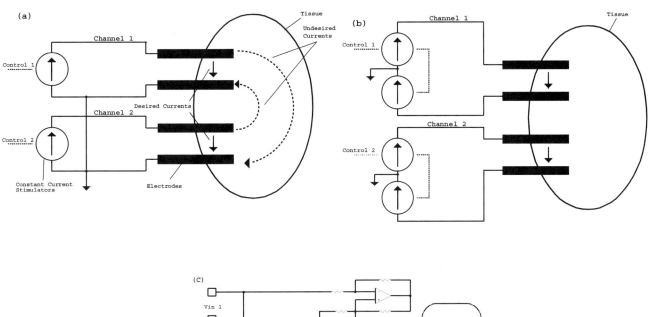

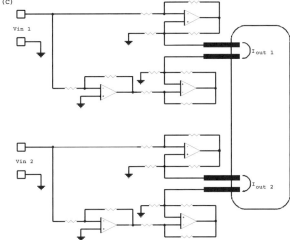

Figure 7.15   Van den Honert [1989] showed how to use two current sources per channel of a multichannel stimulator to minimize the cross-coupling of stimulating currents between channels. (*a*) A cross current from one electrode pair to another electrode pair can occur while maintaining the loop current requirements of each channel's single current source when isolated power supplies are not used. (*b*) Using two current sources per channel improves channel isolation without requiring electrically isolated power supplies. (*c*) The improved circuit could be built with Howland current pumps.

A simple fully reversible reaction is the charging and discharging of the capacitance between the conductive electrode and body fluids. However, the limit of reversibility is reached when the charge accumulated on the electrode–tissue interface capacitance yields a potential difference above the working voltage for the electrolysis of water (approximately $\pm 0.8\,\text{V}$). As such, the amount of charge that can be passed by this double-layer interface depends on the equivalent capacitance of the electrode–tissue interface. Just as in a conventional capacitor, the capacitance of the electrode–tissue interface depends on the surface area of the electrode and the thickness of the effective dielectric boundary of the interface.

Two safety demands are then obvious for implantable stimulating electrodes. The first is that the electrode material must be compatible with the body so that the formation of connective tissue layers whose thickness is greater than $100\,\mu\text{m}$ is suppressed such that the stimulation threshold remains largely constant. Second, a high double-layer capacitance should develop at the phase boundary between the electrode and the body fluids so that the polarization rise during the stimulation pulses remains under the limit of irreversible electrochemical reactions.

The classical materials for invasive stimulating electrodes have been stainless steel and platinum. These materials have a limited range for reversible charge injection by surface faradic processes in the in vivo saline environment before the onset of water electrolysis (approximately $0.5\,mC/cm^2$ for platinum and platinum alloys) or catastrophic corrosion (approximately $0.04\,mC/cm^2$ for stainless steel). These charge injection limits restrict their usefulness as stimulation electrodes to applications requiring only low charge-injection densities. However, even low charge-injection densities are known to produce corrosion of the metal, thereby releasing trace quantities of dissolved metal into the surrounding environment. In the case of platinum, dissolution products may be toxic to the tissue in which the electrodes are implanted. In the case of stainless steel electrodes, dissolution or corrosion of the electrode may result in electrode failure due to corrosion-induced fracture. In fact, dissolution often results in disappearance of the entire electrode.

Another metal in use today for stimulating electrodes is iridium. Pure iridium is extremely stiff and has a much lower impedance than that of any other noble metal. It is extremely inert and very resistant to corrosion. The de facto standard, however, is an alloy of platinum–iridium that has a lower concomitant impedance value than that of either tungsten or stainless steel for the same exposure, making electrodes less likely to erode during intense stimulation protocols. This alloy is excellent for chronic implants because it is very biocompatible. Table 7.2 shows the material properties of some of the most common metals used for the fabrication of electrodes. You can buy high-purity materials to make chronically implantable electrodes from a number of suppliers, including Alfa Aesar (high-purity raw materials), Noble-Met (drawn wire as thin as 0.001 in. in diameter), and Xylem Company (wires and metallic parts).

In the 1970s, the promise of electrical stimulation as a cure for pain, paralysis, deafness, and blindness seemed close at hand. Development in these areas prompted the search for new electrode materials to increase safe charge-injection densities. Significant developments were made in the 1980s with the introduction of metallic oxide layers. Electrodes with a large microscopic surface area were fabricated by anodizing titanium and tantalum electrodes. The anodized electrodes operate by charging and discharging the double-layer capacitance at the electrode–electrolyte interface. They provide an intrinsically safe means of charge injection because they form a dc-blocking capacitor right at the electrode–tissue interface that ensures charge balancing. Unfortunately, these anodized electrodes have limited charge densities (lower than those of platinum) and can be used for injecting anodic charge only, unless appropriately biased, whereas the physiological preference is for cathodic charge.

Later, coatings such as titanium nitride (TiN) [Konrad et al., 1984] and iridium oxide (IROX) [Robblee et al., 1986] were developed to overcome the shortcomings of anodized electrodes. IROX works by delivering charge to tissues through reversible reduction–

TABLE 7.2   Desirable Material Properties for Some Metals Commonly Used in Stimulating Electrodes

	Platinum	Iridium	Gold	Silver	Tantalum	Titanium	Stainless Steel	MP35N Biocompatible Superalloy
Corrosion resistance	✓	✓	✓		✓	✓	✓	✓
Biocompatibility	✓	✓	✓		✓	✓	✓	✓
Electrical conductivity			✓	✓				

oxidation reactions occurring within the oxide film. The iridium oxide layer becomes the charge transfer interface and enables charge injection densities up to $10\,mC/cm^2$ for either cathodic or anodic polarities without water electrolysis or other faradic reactions involved in corrosion of the underlying metallic electrode.

Thin films of hydrated iridium oxide have been used as low-impedance coatings for neural stimulation and recording electrodes. The iridium oxide provides a way of injecting charge into tissue while minimizing electrochemically irreversible processes at the electrode–tissue interface, where reduction and oxidation reactions occur to mediate between electron flow in the external circuit and ion flow in the tissue. Electrodes coated with iridium oxide are very good for long-term stimulation of nerves in the spinal cord [Woodford et al., 1996], in the ear's cochlea [Anderson et al., 1989], and in the brain cortex [Bak et al., 1990; Hambrecht, 1995].

The idea behind using IROX as an electrode material is that iridium can store charge by going through valence changes that cause reversible redox reactions. The fact that these reactions are reversible is important for biocompatibility. Reversibility means that no new substance is formed and hence no reactants are released into tissue. The state of the iridium will depend on the potential applied across the metal–electrolyte junction.

Since tissue-contact stimulation is usually accomplished with a constant-current source, the voltage applied is dependent on the charge storage of the oxide. As the oxide absorbs more charge (positive current), the potential across the interface will increase, which will result in the oxide reacting with the electrolyte. Initially, with no applied voltage, the iridium oxide is in the $Ir(OH)_3$ state. As potential is increased, the iridium oxide increases valence by ejecting protons into the solution. The following reaction summarizes the sequential change in the oxide as potential increases:

$$IrO_{x-1}(OH)_{4-x} \rightleftharpoons IrO_x(OH)_{3-x} + (x-1)(H^+ + e^-)$$

where $x = 1, 2, 3$ and increases with potential. The IrO_3 state ($x = 3$) is unstable and its degradation will result in oxygen evolution. This reaction defines the water window on the positive voltage side. Subsequently, as charge is removed from the oxide (negative current), the reactions reverse. Do you see where the name IrOx comes from?

Iridium oxide coatings are usually formed on electrodes by three different processes:

1. *Activated iridium oxide film* (AIROF): formed from iridium metal in an aqueous electrolyte by an activation process in which the electrochemical potential of the metal is cycled or pulsed between negative and positive potential limits close to those for electrolysis of water

2. *Sputtered iridium oxide film* (SIROF): formed by reactive sputtering of iridium medium in an oxidizing atmosphere

3. *Thermal iridium oxide film* (TIROF): formed by the decomposition of iridium salts to form an iridium oxide film on top of a metallic substrate electrode

The AIROF method is the simplest to use to home-brew IROX-coated electrodes by starting with electrode substrates made of pure iridium metal. Iridium grows a hydrous oxide layer on its surface when it is activated electrochemically in an electrolyte ($0.3\,M$ sodium phosphate dibasic, Na_2HPO_4). A standard three-electrode scheme is used to perform cyclic voltammetry using the setup shown in Figure 7.16. Cyclic voltammetry is an analytical technique that involves application of a time-varying potential to an electrochemical cell and simultaneous measurement of the resulting current.[3] This measurement can be used to provide oxidation–reduction information about the system being studied. However, in this application, the technique is used as a manufacturing process.

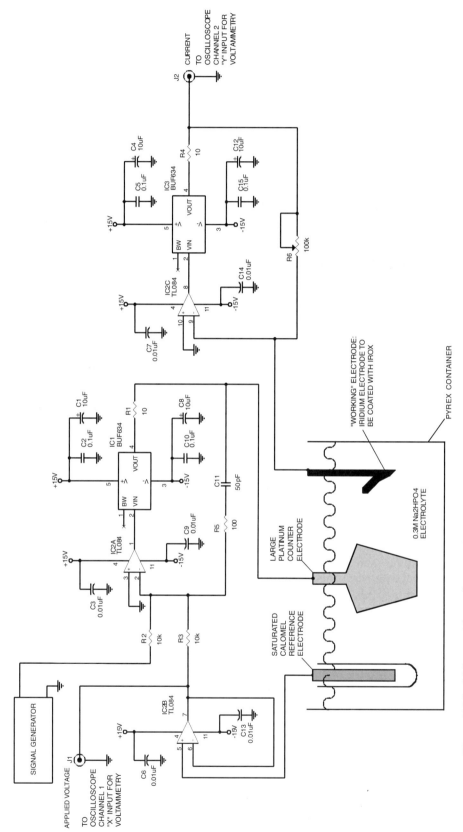

Figure 7.16 Iridium oxide coatings can be grown on iridium-metal electrodes by electrochemical activation in a Na_2HPO_4 solution. Electrochemical activation should be done using a standard three-electrode cyclic voltammetry setup. The potentiostat circuit (IC1, IC2A, and IC2B) tries to maintain the voltage applied between the working and reference electrodes with negligible current passing through the reference electrode. IC2C, IC3, R4, and R6 act as a current-to-voltage converter such that the counter electrode acts as a source or sink of electrons to balance the redox reaction occurring at the working electrode.

In this circuit, IC1, IC2A, and IC2B form a potentiostat, which tries to maintain the voltage applied between the working and reference electrodes and allows negligible current through the reference electrode (IC2B is configured as a unity-gain buffer, presenting virtually infinite input impedance to the reference electrode). In voltammetric measurements, the current is measured as a function of this applied potential. The counter electrode acts as a source or sink of electrons to balance the redox reaction occurring at the working electrode. IC2C, IC3, R4, and R6 act as a current-to-voltage converter. Therefore, the potential applied between the counter and working electrodes must be sufficient both to drive the appropriate electron transfer reaction at the counter electrode and to compensate for the potential drop due to the solution resistance between the counter and reference electrodes (this potential drop is given by Ohm's law, $V = iR_{solution}$, where i is the current generated by the electrochemical reaction and $R_{solution}$ is the solution resistance between the electrodes). The compliance voltage is then the maximum potential that can be supplied between the working and counter electrodes.

The voltage waveform is applied to a counter electrode (a large platinum electrode) in the electrolyte. The iridium electrode to be activated is the working electrode and provides a return current path. A saturated calomel electrode (SCE) provides a reference (Fisher Scientific 13-620-52). The voltage on the counter electrode is cycled between anodic and cathodic potentials while the iridium electrode is exposed to an electrolyte.

During the anodic sweep, an inner oxide (IrO_2) is formed from pure iridium. As the potential increases, the inner oxide changes to a hydrous outer layer [$Ir(OH)_3$]. The fact that this layer is hydrated (water molecules are attached) limits formation to a monolayer. The cathodic sweep causes reduction of the inner oxide back to iridium but does not go low enough to reduce the outer layer. The outer layer remains. Since the oxide is porous, the metal maintains contact with the electrolyte. Therefore, on the next potential sweep the process will repeat. In this way, a hydrous porous layer of iridium oxide is created. The potential limits depend on the electrolyte but should not exceed the potential that results in oxygen or hydrogen evolution.

The University of Michigan Center for Neural Communication Technology recommends the following method for the formation of high-quality IROX films:

1. Initially, hold sites at potentials of -3.0 and 2.5 V for approximately 3 minutes each to remove any oxide that has formed, essentially cleaning the metal.

2. Cycle the activation potential between -0.85 and 0.75 V. These limits are usually wide enough to grow an oxide but narrow enough to remain within the *water window* (the potential range that does not result in oxygen or hydrogen evolution). As a rule, the voltage limits should be set approximately 100 mV inside the water window.

3. Use a square wave (0.5 to 1 Hz) to activate. This allows the metal to remain at the critical levels for hydrous oxide formation (0.75 V) and inner oxide reduction (-0.85 V) for a longer time than a ramp wave would. Holding the potential at these levels allows more complete oxide formation and reduction and also reduces the number of potential cycles (500 to 1000 cycles) needed to grow the oxide.

4. Activate to a limit of 30 mC/cm^2. Although studies have shown 100 mC/cm^2 to be the maximum usable storage capacity for activated iridium, 30 mC/cm^2 allows the oxide to better maintain electrical characteristics and should suffice for most neural stim-

---

[3]You can still use this circuit as a classical cyclic voltammeter. Just change the output of the function generator to a triangular wave and connect the voltage- and current-monitoring outputs to a two-channel oscilloscope capable of working in the *x–y* mode. For a detailed discussion of the cyclic voltammetry technique, we recommend D. K. Grosser, *Cyclic Voltammetry Simulation and Analysis of Reaction Mechanisms*, Wiley-VCH, Weinheim, Germany, 1993. In addition, the accompanying CD-ROM includes a cyclic voltammetry simulator (VirtualCV v1.0 freeware for Windows 9x by Andre Laouenan) to help you understand how the technique works and how to analyze results.

ulation applications. In vitro pulse tests of the oxide grown using this method show that coating maintains its electrical characteristics over at least 100 million pulses.

The TIROF method is more difficult for home-based coating, but it can be used to deposit IROX on any common metallic substrate. Robblee's patent, "Iridium Oxide Coated Electrodes for Neural Stimulation" [1987], describes exactly how to coat platinum, platinum–iridium, stainless steel, titanium, or tantalum with a TIROF[4]:

1. An acid–alcohol solution containing dissolved iridium complexes is prepared by heating $IrCl_3 \cdot 3H_2O$ [Ir(III)trichloride] in 5.5 M HCl (4% wt/vol) until 75 to 80% of the solution is evaporated. Heating the Ir(III) trichloride in the acid results in conversion of the Ir(III)trichloride to a hexachloroiridate ion, $(IrCl_6)^{2-}$. The solution is restored to its initial volume with the addition of alcohol. Either isopropyl alcohol or ethyl alcohol can be used. This acid–alcohol solution is then aged for a period of time (e.g., 1 to 2 weeks), during which time the hexachloroiridate ion is slowly converted to a chloroiridate–alcohol complex of Ir(IV). A way of avoiding boiling iridium chloride in HCl is simply to buy chloroiridic acid [Hydrogen hexachloroiridate(IV) hydrate] from Alfa-Aesar (item 11031) and stir it with isopropyl alcohol to achieve the correct acid–alcohol ratio. The solution must then be left to age for 1 to 2 weeks in a tightly sealed jar.

2. The surface of the metallic electrode is pretreated prior to deposition of the iridium solution to enhance adhesion of the iridium oxide film formed. Such pretreatments include chemical or electrochemical etching and vary depending on the metallic electrode being coated. Common ways of preparing surfaces for coating include etching with HCl and sandblasting using a fine abrasive powder.

3. The metallic electrode is soaked in the aged acid–alcohol solution for a prolonged period of time, typically 16 hours, after which it is dried for 1 hour at room temperature (22°C) and annealed in air at 320°C for 80 to 90 minutes. Prolonged soaking of the metallic electrode allows intimate association of chloroiridate–alcohol complexes in solution with the surface of the metallic electrode so that the metallic surface becomes completely covered with iridium complexes. Shorter soaking times lead to incomplete coverage of the metallic surface so that areas of uncoated metal remain exposed.

4. Successive layers are added by soaking the electrode in the acid–alcohol solution of iridium again and again for 16 to 24 hours, followed by 3 to 6 hours annealing at 320°C after each period of soaking. Typically, two to four layers are applied. The high-temperature annealing converts the chloroiridate–alcohol complexes deposited on the surface of the electrode to an oxide of iridium [probably a combination of IrO_2, $IrO_2 \cdot H_2O$, and $Ir(OH)_4$]. The annealing temperature of 320°C was reported by Robblee to be optimal for obtaining electrodes of the highest charge-injection capacity. The long annealing times are required for complete conversion of the chloroiridate–alcohol complexes to iridium oxide and the elimination of chloride from the film.

5. Chloro complexes of iridium that may remain on the surface due to insufficient annealing are susceptible to passive leaching and dissolution from the film. Moreover, insufficiently annealed films that contain a high chloride content have a very low charge injection capacity, due to the smaller proportion of Ir(IV) oxide species, which are responsible for the surface faradic charge injection reactions desired. As such,

---

[4]Patents issued before June 5, 1995 had a 17-year term from the date of issue. Patents filed after that date have a 20-year term measured from the filing date. Patents filed before that date, and pending or issued on that date, have the longer of the two terms. Once a patent has expired the invention is in the public domain, and you are free to use it.

repeated neutralization with deionized water in an ultrasonic cleaner and thorough cleaning (e.g., with alcohol) are needed to ensure that the tissues are not poisoned by the chemical precursors of the IROX film.

IROX-coated electrodes acquire the electrochemical properties of the IROX and lose those of the underlying metallic electrode. For example, the voltage change across the IROX interface in response to a constant-current pulse is significantly less than the voltage change across an uncoated metallic electrode interface in response to a current pulse of the same magnitude. The corrosion resistance conferred upon the metallic electrode by an IROX coating is also enormous. In agreement with our own experience, Robblee found that no dissolution of the underlying metallic electrode occurred when using IROX-coated platinum–iridium electrodes stimulated with biphasic regulated current pulses, whereas the same electrodes with no coating lost between 2 and 8 µg of platinum during only 24 hours of in vitro stimulation under the same conditions.

The coating process is dependent on the purity and concentration of the chloriridic acid, the exact ratio of chloroiridate–alcohol complexes, and the preparation and cleanliness of the base metal. You'll end up going through a *lot* of chloroiridic acid (and sweat) before you achieve a repeatable coating process.

As for titanium nitride, the preferred process is to apply the porous nitride layer to the carrier material serving as substrate by means of reactive ion plating. If you are part of a university, you can probably ask around the material sciences department to see if they have the setup necessary for physical vapor deposition. If you really want to home-brew titanium nitride–coated electrodes, we recommend that you read the very practical articles on amateur sputtering by Steve Hansen that appeared in *The Bell Jar* (volume 8, numbers 3–4, pages 14–16, 1999; volume 9, number 1, pages 2–4, 2000; and volume 9, number 2, pages 10–12, 2000).[5]

An alternative to doing your own coating is to contact one of the specialized vendors that can coat electrodes for you. The two best known in the implantable devices industry are Hittman Materials (now Wilson Greatbatch Technologies, Inc.) and W.C. Heraus. Both companies are providers of precious metals and special material processing for implantable devices. If you approach vendors other than Hittman or Heraeus, be aware that "desirable characteristics" vary depending on the application. For example, one form of titanium nitride coating (a shiny golden finish) generates a very hard, smooth surface and is used on metal mold surfaces and surgical tools. But the form of titanium nitride preferred for

---

Warning! Chloroiridic acid, iridium trichloride, and hydrochloric acid are relatively dangerous materials. Do not breathe dust or mist and do not get in eyes, on skin, or on clothing. Sodium phosphate dibasic is an eye and skin irritant. When working with these materials, approved safety goggles or glasses must be worn. Contact lenses are not protective devices: Appropriate eye and face protection must be worn instead of, or in conjunction with, contact lenses. Wear disposable protective clothing to prevent exposure. Protective clothing to prevent skin contact includes a lab coat and apron, flame- and chemical-resistant coveralls, gloves, and boots. Follow good hygiene and housekeeping practices when working with these materials. Do not eat, drink, or smoke while working with them. Wash hands before eating, drinking, smoking, or applying cosmetics.

---

[5]*The Bell Jar* (ISSN 1071-4219) is a short magazine on vacuum technique and related topics for the amateur investigator. It is edited and published by Steve Hansen, who can be contacted through the *Bell Jar*'s Web site at www.belljar.net, or by mail at 35 Windsor Drive, Amherst, NH 03031.

electrodes (a dull charcoal color) is a microporous structure that has a very rough texture rather than a smooth one.

Our experience with electrodes coated with titanium nitride and IROX is that they perform similarly. Typical capacitances are in the range 10 to 20 μF/mm^2. Platinum black and activated glassy carbon are other electrode materials that have high capacitance and exhibit excellent biocompatibility. Other interesting possibilities are to coat the electrode with layers of microspheres to increase the wetted surface or to create porous metallic electrodes and then "plug" the pores with carbon.

Neuromuscular Electrical Stimulation

Neuromuscular electrical stimulation (NMES) is the use of electrical stimulation of the intact peripheral nervous system to contract a muscle, either through direct activation of the motor neurons in the mixed peripheral nerve, or indirectly through reflex recruitment. When the stimuli are used to activate muscles directly, without activation of the peripheral nerve, the modality is known as electrical muscle stimulation (EMS).

NMES may be used for therapeutic or functional purposes. Therapeutic use is directed toward lessening impairments, prevention of secondary complications, or halting progression of a disabling condition. Therapeutic NMES includes strengthening muscles, lessening of spasticity, preventing muscle atrophy, and improving regional blood flow. Functional NMES is more commonly known as functional neuromuscular stimulation (FNS), which is a replacement for lost or impaired motor control. Today, transcutaneous FNS is widely used in the rehabilitation of paralyzed patients in whom natural nervous control of muscular contraction has been lost due to a spinal cord injury or a central nervous system disorder. In its best known applications, FNS has been used to restore function to affected limbs by providing artificial electrical stimulation patterns that enable the subject, for instance, to use upper extremity functions, to stand up, or to walk [Kralj and Bajd, 1989].

One or more pairs of surface electrodes along with conductive creams or gels are used to activate the excitable tissues. Surface electrical stimulation typically consists of a train of regular monophasic or biphasic pulses. The rate at which the nerve fibers fire depends on the frequency of pulse repetition. A single pulse produces a short-lived muscle twitch of not more than 250 ms. If pulses are repeated more frequently than this, the muscle does not have time to relax between stimuli, and at some point tetanic (continuous) contraction occurs.

Regardless of what mechanism is used to evoke muscle contractions, nerve and muscle stimulators are FDA class II prescription devices. The international standard that specifically covers the design, performance, and testing of these devices is IEC-60601-2-10, *Medical Electrical Equipment—Part 2: Particular Requirements for the Safety of Nerve and Muscle Stimulators.*

Transcutaneous NMES devices are relatively simple. Figures 7.17 and 7.18 show the circuit diagram of a battery-powered stimulator that can be used in NMES therapy and as an output channel for transcutaneous FNS. Whenever a stimulus is required, Q1 discharges C1 into step-up transformer T1, which yields a high-voltage pulse on its secondary. The transformers that we have used successfully are the type built for the small inverters designed for fluorescent lamp lanterns. The high-voltage pulse (<150 V peak) charges capacitor C2 to a voltage regulated by zener D1. The setting of R5 selects the current that flows through the electrodes. With switch SW1 open, the pulse output is monophasic (although the capacitance of the electrodes will generate a discharge phase through R7). With the switch open, the charge accumulated in the coupling capacitor (C2 and C3 back-to-back in series = 5μF nonpolar) flows through the electrodes after the stimulus to yield net-zero charge transfer through the tissue.

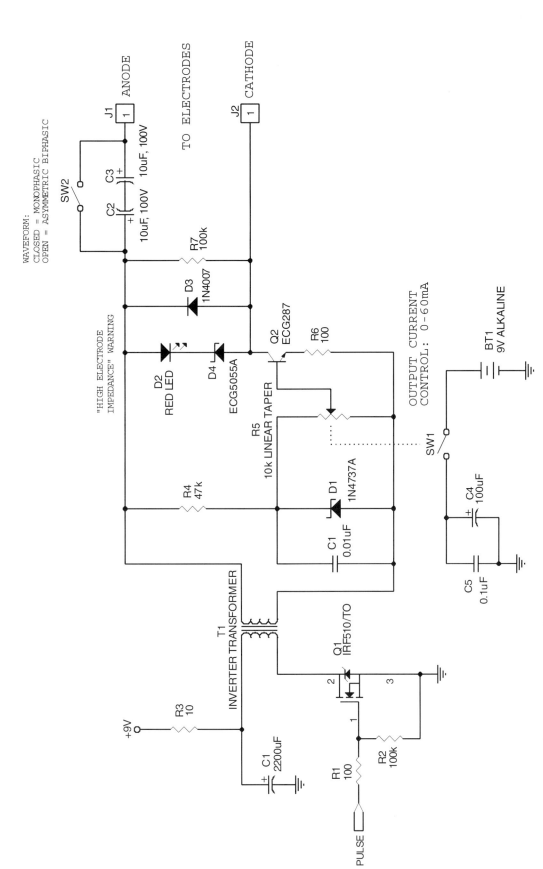

Figure 7.17 Output stage of a simple neuromuscular electrical stimulation (NMES) device that can be used for transcutaneous electrical stimulation of muscle contractions either through direct activation of the muscle fibers or through excitation of the motor neurons in the mixed peripheral nerve and/or through reflex recruitment.

335

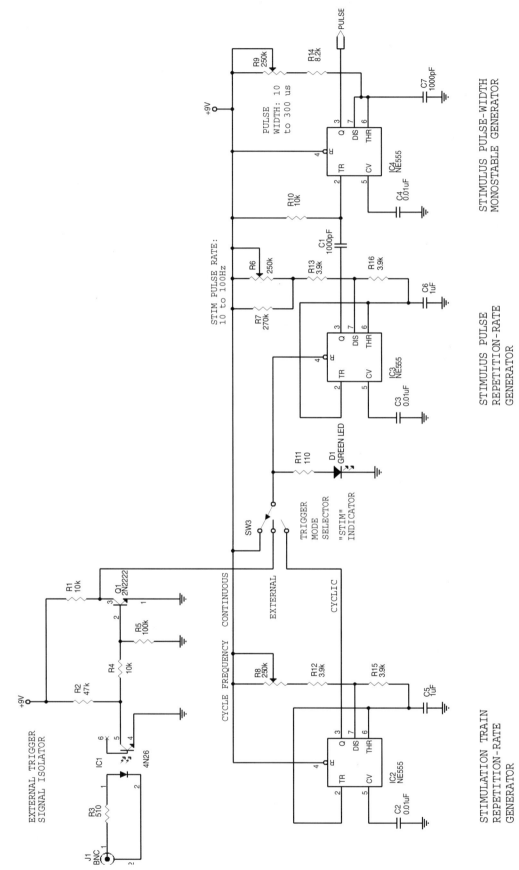

Figure 7.18 This circuit drives the output stage of Figure 7.17 to produce output pulses 10 to 300 μs in duration at a frequency of 10 to 100 Hz.

The circuit of Figure 7.18 drives the output stage to produce output pulses 10 to 300 μs in duration (based on the setting of potentiometer R9) at a frequency of 10 to 100 Hz (depending on the setting of potentiometer R6). Pulse-rate generator timer IC3 is either constantly enabled or turned on and off, depending on the setting of switch SW3. In the cyclic setting, the burst duration and frequency is defined by IC2 to the setting of R8. In the external mode, the trigger time and duration of the stimulus burst are set to the time that an external controller (e.g., a microcontroller in a FNS system) energizes the LED in optocoupler IC1.

There is clinical evidence showing that appropriate electrical stimulation can cause a denervated muscle to contract, which in turn may help limit edema and venous stasis within a muscle and therefore delay muscle fiber degeneration and fibrosis. Studies seem to indicate that proper use of EMS shortens recovery time following denervation.

EMS is not necessarily a benign therapy, however. Some studies indicate that contraction of a denervated muscle may disrupt regenerating neuromuscular junctions and subsequently, delay reinnervation. In addition, denervated muscle is more sensitive to trauma than is innervated muscle, and electrical stimulation may further traumatize the denervated muscle. Moreover, prolonged electrical stimulation until reinnervation occurs is often not worth the costs or time involved.

Television has popularized the cosmetic use of EMS devices. Late-night infomercials claim that muscle stimulators can help in losing weight and increasing muscle definition through "no-work" or "passive" exercising. Despite the "guaranteed results" offered by these advertisements, don't count on looking like Venus or Adonis by slapping on some gelled electrodes to twitch your beer belly. There is no clinical evidence that electrical muscle stimulation would provide any benefit toward weight loss and body toning.

In fact, the FDA has seen no evidence that these devices are safe and effective for home use or for applications touted in health spas and beauty salons. FDA considers electrical muscle stimulation to be misbranded and fraudulent when promoted for weight loss or body shaping. There are currently no acceptable uses for these devices when labeled for over-the-counter use. When labeled for medical use by a licensed practitioner, the following uses are generally recognized as acceptable:

- Relaxation of muscle spasm
- Prevention or retardation of disuse atrophy
- Muscle reeducation (e.g., rehabilitation of muscle function after a stroke)
- Increasing local blood circulation
- Immediate postsurgical stimulation of calf muscles to prevent venous thrombosis (formation of blood clots)
- Maintaining or increasing range of motion

The bottom line on the cosmetic use of electrical muscle stimulators: If you want to look like the models in muscle stimulation commercials, do not buy the stimulator, but do get on a diet and join a health club.

Transcutaneous Electric Nerve Stimulation

Transcutaneous electrical nerve stimulation (TENS) provides symptomatic relief from some forms of chronic or acute pain, including postsurgical and posttraumatic pain. TENS devices require FDA approval as class II medical devices and need physician prescription to be dispensed to or used on patients. Electrodes are placed at specific sites on the body for treatment with TENS. The current travels through electrodes and into the skin, stimulating nerve pathways to produce a tingling or massaging sensation that reduces the perception of pain. When used as directed, TENS is a safe, noninvasive, drug-free method of pain management. TENS has no curative value; it is a symptomatic treatment that suppresses pain sensation that would otherwise serve as a protective mechanism on the outcome of the clinical process. As such, it is used only to offer a better quality of life for people with pain when the source of the pain cannot be treated. It must be noted, however, that none of the studies have proven conclusive, and there is still quite a bit of debate about the degree to which TENS is more effective than placebo in reducing pain. Generally, TENS provides initial relief of pain in about 70% of patients, but the success rate decreases after a few months to around 25%.

The most accepted theories of how TENS may work are:

- *Gate control theory:* suggests that by electrically stimulating sensory nerve receptors, a gate mechanism is closed in a segment of the spinal cord, preventing pain-carrying messages from reaching the brain and blocking the perception of pain.
- *Endorphin release theory:* suggests that electrical impulses stimulate the production of endorphins and enkaphalins in the body. These natural, morphinelike substances block pain messages from reaching the brain, in a manner similar to conventional drug therapy but without the danger of dependence or other side effects.
- *Descending inhibitory pathway theory:* suggests that noxious stimuli excite the smaller pain fibers, leading to activation of the brainstem reticular formation. This releases serotonin (a neurotransmitter related to the feeling of well-being), which in turn inhibits pain at the spinal cord level. Putting it shortly, "pain inhibits pain."

The practitioners of quack treatments credit TENS with restoring "energy lines" and stimulating "acupuncture points." Don't waste your time considering such claims.

The circuit for a battery-powered two-channel TENS unit is shown in Figure 7.19. In this circuit, timer IC1 produces a pulse every time the stimulation channels need to be triggered. The rate at which the trigger pulses are generated is set via R1. With the component values shown in the circuit, trigger frequency can be varied between 3 and 100 Hz. Constant pulsing at a set frequency is known in TENS jargon as the *conventional mode*.

What is commonly known as *conventional TENS* is to stimulate at a relatively high frequency (40 to 150 Hz) and low current (10 to 30 mA), barely above the threshold of sensation. The reported onset of analgesia with these parameters is virtually immediate. Pain relief lasts while the stimulus is turned on, but it usually returns when the stimulation is turned off. Patients often leave the electrodes in place all day and turn stimulation on for 30 to 60 minutes whenever they need it. In the low-frequency conventional mode, the TENS unit is set to deliver pulses at 1 to 10 Hz, but the stimulus intensity is increased close to the patient's tolerance limit. The belief is that this will increase the time of analgesia after therapy because of the release of natural morphinelike substances. This stimulation mode is uncomfortable, and not many patients can tolerate it.

Burst TENS uses low-intensity stimuli firing in high-frequency bursts. Each burst delivers some 5 to 10 pulses at a rate of 100 Hz, with rests between bursts. Patients usually report that each burst is felt as a single pulse, making it more comfortable than single-pulse sensation. No particular advantage has otherwise been established for the burst method over the conventional TENS method. The TENS circuit of Figure 7.19 implements the low-frequency burst mode when SW2 connects the reset line of timer IC3. This IC3 periodically inhibits IC1 from oscillating. The fundamental burst frequency is approximately 2 Hz, which allows approximately seven trigger pulses to be issued by IC1 per burst cycle.

Ramped modulation is a feature that is often encountered in TENS units to cause a gradual rise and decline of amplitude and frequency, resulting in a comfortable stimulation sensation. It is very similar to the conventional mode, but a modulation in pulse rate and amplitude is thought to avoid nerve adaptation and accommodation, which diminishes the

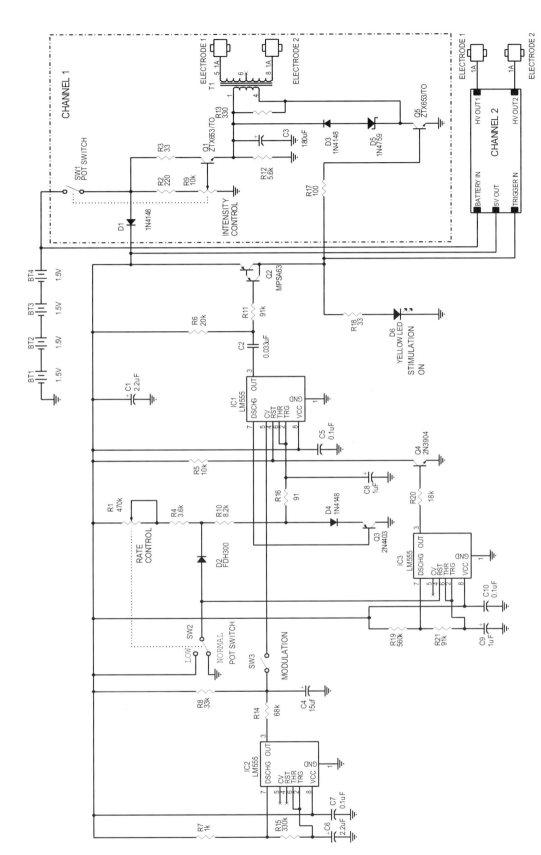

Figure 7.19 Circuit of a battery-powered, two-channel TENS unit. IC1 produces a pulse every time the stimulation channels need to be triggered. The trigger frequency can be varied between 3 and 100 Hz. Burst TENS is activated when SW2 connects the reset line of timer IC3 periodically to inhibit IC1 from oscillating. Modulated TENS is enabled by closing SW3, which allows a triangle wave generated by IC2 to modulate the frequency of IC1. The electrodes are driven by step-up transformer T1. Stimulation current is controlled via R9, which sets the current that Q1 allows across the primary of the step-up transformer. The peak current of a pulse into a purely resistive load of 500 Ω can be varied between 0 and 150 mA.

effectiveness of conventional TENS. When the circuit is operating in the conventional mode, modulation can be employed by closing SW3. This allows a triangle wave generated by timer IC2 to be fed to the control voltage input of IC1. The modulating signal frequency is approximately 1 Hz, which induces a -40% change in the selected conventional-mode frequency. Note that when modulation is enabled, the timing characteristics of the trigger pulses generated by IC1 change such that pulse amplitude is modulated to -25% on each modulation cycle.

The electrodes are driven by step-up transformer T1. A stimulation pulse is generated every time Q5 is driven into conduction by the Darlington pair Q2 connected to the output of IC1 via ac-coupling capacitor C2. LED D6 flashes each time that a stimulation pulse is delivered. The constant-current amplitude is set via R9, which controls the current that Q1 allows across the primary of the step-up transformer. The peak current of a pulse into a purely resistive load of 500 Ω can be varied between 0 and 150 mA. The shape of the pulse delivered to the skin electrodes is shown in Figure 7.20. The load used to simulate the body impedance is the one specified in the *American National Standard for Transcutaneous Electrical Nerve Stimulators* (ANSI/AAMI NS4-1985). The preferred waveform is biphasic, to avoid the electrolytic and iontophoretic (whereby ions and charged molecules can be driven through the skin by an electrical current) effects of a unidirectional current.

Power for the stimulator is controlled independently via the potentiometer switches SW1 of each channel. When either channel is on, current supplied by the four alkaline batteries in series is delivered to the timer ICs via diode D1 of the active stimulation channel. Each output channel is isolated from the other, which allows two distinct areas of pain to be stimulated independently.

TENS electrodes are usually placed initially on the skin over the painful area, but other locations, such as over cutaneous nerves, may give comparable or even better pain relief. TENS should not be applied over the carotid sinuses, due to the risk of acute hypotension (because of stimulation of the vagus nerve), over the anterior neck because of possible spasm of the larynx, or over an area of sensory impairment where the current could burn the skin without the patient becoming aware of it. Of course, TENS should not be used in patients with any active implantable medical device (e.g., pacemakers and implantable defibrillators) because of the risk of interfering with or damaging the implantable device. In addition, TENS should not be used during pregnancy because it may induce premature labor.

A relatively new TENS-like modality is *percutaneous electrical nerve stimulation* (PENS), which is often incorrectly called *electroacupuncture*. Rather than using surface electrodes, PENS uses needle probes as electrodes, placed just under the outer layers of the skin in the region where the patient feels pain. The only advantage of PENS over TENS is that it bypasses local skin resistance and delivers electrical stimuli at the precise level desired, in close proximity to the nerve endings.

Interferential Stimulation

Stimulating deep tissues using surface stimulation electrodes requires that very strong currents be delivered to the skin to yield sufficiently high currents to depolarize target tissue. Strong pulses are often painful, limiting their clinical applicability, especially when electrical stimulation is used for therapeutic purposes (e.g., TENS, or for the stimulation of deep muscles such as those of the pelvic floor). As shown in Figure 7.21, *interferential current therapy* (IFC) is based on the summation of two ac signals of slightly different frequency that are delivered using two pairs of electrodes. Each of the few-kilohertz "carriers" on their own do not cause skin sensations or stimulation of the underlying tissues. However, the tissue causes the signals to mix or interfere with each other, resulting in a low-frequency current that consists of cyclical modulation of amplitude, based on the

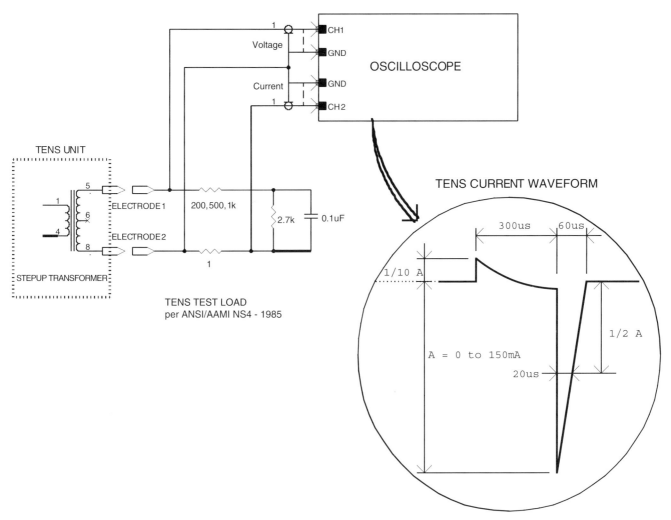

Figure 7.20 Stimulation pulse delivered to the skin electrodes by the TENS generator of Figure 7.19. The *American National Standard for Transcutaneous Electrical Nerve Stimulators* (ANSI/AAMI NS4-1985) specifies the load used to simulate the body impedance.

difference in frequency between the two carrier signals. When the signals are in phase, the low-frequency components sum to an amplitude that is sufficient to stimulate, but no stimulation occurs when they are out of phase. The beat frequency of IFC is equal to the difference in the frequencies of the two carrier signals. For example, the beat frequency, and hence the stimulation rate of an interferential unit with signals set at 4 and 4.1 kHz, is 100 Hz.

The interferential method is used most often for the therapeutic stimulation of nerves and muscles in the treatment of acute pain, edema reduction, and muscle rehabilitation. It is not a common modality for functional neuromotor stimulation because it requires greater energy consumption and a larger number of electrodes. Scientists in the former Soviet Union also use interferential currents delivered through scalp electrodes to produce narcosis (electronarcosis) and anesthesia (electroanesthesia). This last application, which does not involve causing convulsions as with ECT, is very controversial and seldom used in Western psychiatry.

IFC stimulators commonly use carrier frequencies around 4 to 5 kHz, with sinusoidal waveforms that can reach peak-to-peak voltages of 150 V and force currents of up to 100 mA

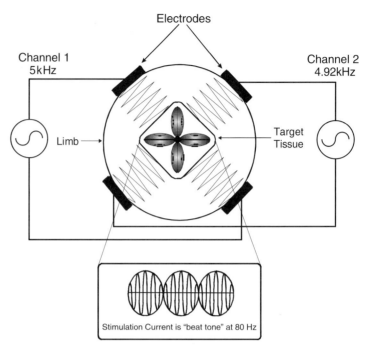

Figure 7.21 Interferential current therapy (IFC) is based on the summation of two ac signals of slightly different frequency that are delivered using two pairs of electrodes. Each of the few-kilohertz "carriers" on their own do not cause skin sensations or stimulation of the underlying tissues. However, the tissue causes the signals to mix or interfere with each other, resulting in a low-frequency beat current capable of stimulating the tissue.

RMS. The most common beat-frequency therapy ranges are 1 to 10 Hz for edema, 1 to 150 Hz for rehabilitation of muscle, and 80 to 150 Hz for the control of pain. Figure 7.22 shows a simple circuit that can generate interferential audio-frequency currents. The power amplifier sections are based around the ST Microelectronics TDA2005 10 W + 10 W audio amplifier IC. This IC is intended specifically for use in bridge amplifier designs in car audio systems. It is a class B dual audio power amplifier with a high output current capability of up to 3.5 A, ac and dc output short-circuit protection (one wire to ground only), and thermal shutdown protection, and is capable of driving very inductive loads. Although it can be used as a dual amplifier by operating each half of the device separately, in this application it is configured for operation as a bridge amplifier. With this mode, the current and voltage swings in and around the IC are twice that of a single amplifier, which results in a power output four times greater than that of a single amplifier with the same load connected to the output pins.

Two audio-output transformers are used in reverse to step up the amplifier's output voltage all the way up to the 150 V_{pp} necessary to push up to 100 mA RMS through the body. We used Hammond 1615 audio-output transformers in reverse; that is, the 8-Ω speaker outputs were connected to the outputs of the power amplifiers, and the 5-kΩ inputs were connected to the electrodes. As with all bridge amplifier designs powered from a single supply, both output terminals of the TDA2005 are held at half the supply voltage, which eliminates the need for the usually large-value dc blocking capacitor in series with the output transformers. The power amplifiers are driven through audio isolation transformers. The complete circuit should be powered from an IEC601-compliant power supply rated for 12 V at 5 A or more, for example using a Condor model MD12-6.8-A medical-grade linear supply.

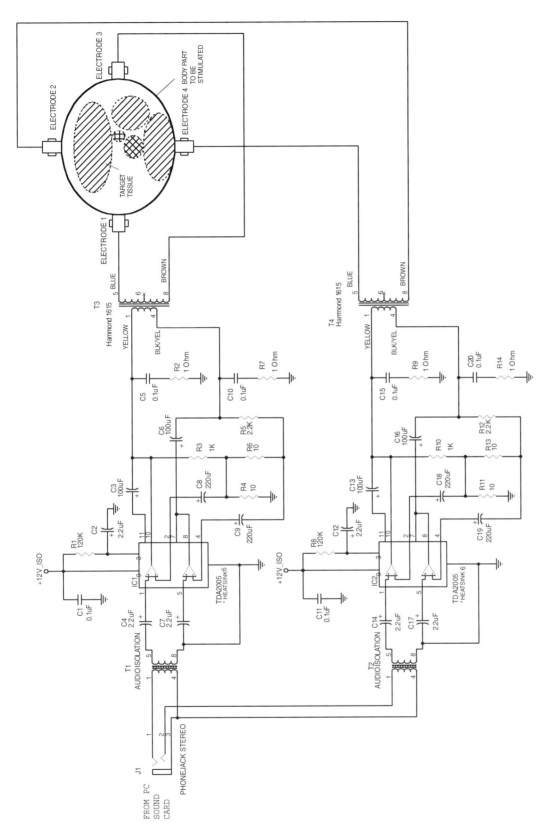

Figure 7.22 An experimental interferential-mode stimulator uses two audio-output transformers as step-up transformers. The primaries are driven by bridged audio amplifiers at carrier frequencies around 4 kHz with peak-to-peak voltages of 150 V and force currents of up to 100 mA RMS. Audio signals are generated by the PC sound card running a beat-tone generator program. The most common beat-frequency therapy ranges are 1 to 10 Hz for edema, 1 to 150 Hz for rehabilitation of muscle, and 80 to 150 Hz for the control of pain.

The exact frequency difference and phase relationship between the carrier signals has a tremendous effect on the site inside the body at which the differential stimulation will take place. For this reason, the majority of audio signal generators used in modern IFC stimulators are based on direct-digital synthesizers (see Chapter 6). The simplest way to drive the experimental IFC stimulator circuit of Figure 7.22 is to use the PC sound card and software that is freely available for download from the Web, such as the beat-tone generator (BEATINBRAIN.EXE, freeware for Windows 9x, 2000, NT) by the Physics Lab of Rutgers University, a copy of which is supplied for your convenience in the book's ftp site. Although this program is not designed specifically for IFC stimulation,[6] it generates two sine waves of different frequencies through the left and right sound card channels that can be used to drive the isolated high-voltage output stages. If you are a Matlab user, you can devise complex frequency and phase shifts between the carriers. The data streams can then be played using the `sound(y, Fs)` command, where y is an $N \times 2$ matrix that contains the sine-wave data to be played through each one of the sound card channels and Fs is the sampling frequency.

The risks associated with interferential stimulators is much larger than those for devices that deliver narrow stimulation pulses. The reason is that the carrier signals are constantly on and convey quite a bit of power through the tissue. IFC stimulators, especially those that do not have electrode impedance monitoring with automatic turn-off, require special attention to electrode selection, placement, and maintenance. If you ever use or build an IFC stimulator, first connect a 2-kΩ $\frac{1}{4}$-W resistor between one of the electrode pairs. Crank the power up until the resistor bursts into flames. This is exactly what will happen to skin if electrode contact is poor. As such, when placing electrodes for IFC, it is imperative that they are not touching and will not touch each other, since burns on the edges of the electrodes as well as on the skin will occur if electrodes are touching during stimulation. Even though the electrodes may appear to be far enough apart, remember that when a muscle comes into contraction, it can bring the electrodes together. In addition, if you use carbon-loaded silicone electrodes, always remember to apply and maintain a sufficient amount of an appropriate conductive medium, such as water-soaked sponges or especially formulated electrotherapy gel to prevent burns during stimulation. Finally, never even consider passing IFC currents through the brain. Remember the "this is your brain on drugs" advertising? Well, misapplication of IFC takes the fried-egg analogy one notch up on the realism scale.

General Safety Precautions and Contraindications for Transcutaneous Electrical Stimulation Therapies

All transcutaneous electrical stimulation devices (e.g., TENS, EMS, IFC, as well as iontophoresis units) are classified by the FDA as either class II or class III medical devices[7] and require physician prescription to be dispensed to or used on patients. The fact that some of these devices are sold without a prescription by unscrupulous online vendors may make a lot of people believe that they cannot do any damage. However, most electrical stimulation therapies carry significant risks with them. The following should be considered before using any form of transcutaneous electrical stimulator:

---

[6]The brain is able to combine two pure tones, each sent to a separate ear, to produce a beat tone at the difference frequency. Some researchers believe that this binaural beat effect can be exploited to affect brain states positively using difference frequencies related to those of the alpha, beta, and theta brainwaves. Even if it would be demonstrated that the brain can be trained to generate specific frequencies, whether or not altering the brainwaves has any effect on mind, body, or mood is subject to considerable debate.

[7]For the FDA's definition of class II and class III medical device, see the Epilogue.

- Cardiac demand pacemakers that detect a user's heart rate and turn on a pacemaker when the heart rate falls below a predetermined level can in certain circumstances be affected by stimulation. This is because the pulses from the stimulator may be confused with the heart's intrinsic signals and fool the pacemaker into thinking that the heart is beating faster than it is. A pacemaker or implantable defibrillator should be considered to be an absolute contraindication to upper limb and shoulder stimulation. Electrical stimulators can sometimes be used with caution for lower limb applications in these patients as long as the action of the pacemaker/defibrillator and heart is checked by a cardiologist while the stimulator is in use.

- Electrical stimulation should be used with extreme care in patients with any other type of active implantable medical device, such as a spinal cord stimulator or intrathecal pump. This is because electrical stimulation currents can interfere with the operation or even damage the circuitry of an implanted device. In addition, electrical stimulators should not be used over metal implants, whether active (pacemakers, spinal cord stimulators, etc.) or passive (orthopedic nails, metal plates, prosthetic joints, etc.) because these will distort the flow of current through the body and may cause internal hot spots with high current density that can destroy adjacent tissues.

- Electrical stimulation should not be applied over the eyes or the carotid sinuses (side of neck), due to the risk of acute hypotension through a vasovagal reflex.

- Electrical stimulation electrodes should not be placed over the anterior neck because of possible laryngospasm due to laryngeal muscle contraction.

- Electronic equipment such as ECG monitors, ECG alarms, sleep apnea monitors, and so on, may not operate properly when electrical stimulators are in use.

- Electrical stimulation should not be used during pregnancy because it may induce premature labor.

- With the exception of electrical stimulation that is specifically intended for the stimulation of the brain, electrodes should never be placed to cause current to flow transcerebrally (through the head), as this may induce seizures and have other undesirable neurological consequences.

- With the exception of electrical stimulation that is specifically intended for stimulation of the heart, electrodes should never be placed to cause current to flow across the chest (e.g., both arms simultaneously) because electrical stimulation currents may cause or lead to arrhythmic events. For the same reason, patients should be warned never to handle the electrodes while a stimulator is on since a current path through the heart could be created by accident.

- There are rare anecdotal reports that people who have poorly controlled epilepsy have had symptoms increased after using electrical stimulation. There is no known mechanism for this effect, but our advice is that electrical stimulators should not be used in patients with epilepsy that is not well controlled by drugs.

- Because electrical stimulation (especially EMS and IFC) will increase local blood circulation, it is possible that stimulation in the area of a malignant tumor might increase the rate of metastasizing and therefore the spread of the cancer. Electrodes should never be placed over the area of a known tumor.

- Long-term stimulation at the same electrode site may cause skin irritation through possible allergic reaction to tape or gel. Poor skin condition can be a problem when self-adhesive skin electrodes are used. This is because there is a greater chance of skin irritation. Electrodes should never be placed over broken skin or over rashes, blisters, spots, and so on.

- Electrical currents used in some modes of electrical stimulation (e.g., IFC) are large enough that they may cause skin burns under the electrodes. For this reason, the electrodes should not be placed in an area of sensory impairment (e.g., in cases of nerve lesions, neuropathies) where the possibility exists that the patient would not feel that the skin is being burned.

- People who have a spinal cord injury may be subject to episodes of autonomic dysreflexia. This is characterized by a rise in blood pressure elicited by a noxious stimulus such as electrical stimulation (e.g., FNS) applied below the level of the lesion.
- Electrical stimulators should not be used while driving or operating dangerous machinery.

MAGNETIC STIMULATION

Excitable tissue can be stimulated by strong, time-varying magnetic fields. As shown in Figure 7.23, a coil is placed over the tissue to be stimulated and a capacitor bank is discharged into the coil. Because of Faraday's law of induction, the time-varying current in the coil generates a time-varying magnetic field, which induces eddy currents in the tissue, causing stimulation. Note that we are not talking about some nonmedical therapy involving weak magnets alleged to promote health. We are talking about producing eddy currents that are strong enough to actually depolarize cell membranes and hence activate excitable tissue. This is electrodeless electrical stimulation, where the magnetic field is only the medium used to transfer electrical energy from the coil to the tissue. The magnetic field strengths need to reach a peak of several tesla (comparable to the static magnetic field of MRI machines, some 40,000 times the Earth's magnetic field), which is usually achieved by driving the stimulating coil with brief current pulses of several kiloamperes.

Plain electrical brain stimulation is possible noninvasively using scalp electrodes. However, transcranial electrical stimulation (TCES) is very painful because of activation of pain fibers in the scalp and hence is of limited clinical value. On the other hand, magnetic stimulation of the brain and peripheral nerves is painless. The reason for this difference is that in direct electrode stimulation, the stimulus current decays as a function of the impedance of the tissue between the electrodes and the target tissue. On its way to the brain, the current must pass through the highly resistive scalp and skull. Hence, to deliver sufficient electric current to neural tissue within the brain requires much higher currents to be delivered to the scalp. The narrow (e.g., 50 to 100 µs in duration) high-voltage pulses (<800 V) produce very high current densities close to the electrodes, which activate the pain receptors. In con-

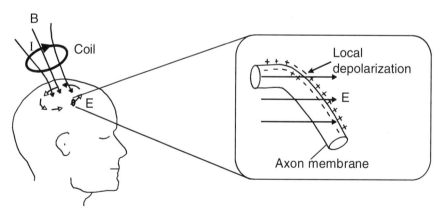

Figure 7.23   Excitable tissue can be stimulated by strong, time-varying magnetic fields. A coil is placed over the tissue to be stimulated and a capacitor bank is discharged into the coil. Because of Faraday's law of induction, the time-varying current $I(t)$ in the coil generates a time-varying magnetic field **B** which induces eddy currents in the tissue, causing an electric field **E** that leads to stimulation.

trast, a magnetic field penetrates all body tissues without alteration, falling off in magnitude only as the inverse square of the distance. As such, much lower currents are generated at scalp level to induce the same amount of current at the target brain structure.

The idea of magnetically stimulating excitable tissue is not new. The first experiments with magnetic stimulation were conducted by d'Arsonval in 1896. He reported phosphenes (visual sensations) and vertigo when placing a subject's head inside an induction coil. Magnetic stimulation of human peripheral nerves stimulation was accomplished by Bickford and Fremming [1965] using an oscillatory magnetic field that lasted 40 ms.

Modern magnetic stimulation is based on research by Polson et al. [1982], who constructed a pulsed magnetic stimulator for peripheral nerve stimulation. They used a pulse of 2 ms duration, which allowed them to record motor-evoked potentials (MEPs) by stimulating the median nerve.

Electric Field Induced by Magnetic Stimulation

The large time-varying magnetic field $\mathbf{B}$ required to induce an electric field $\mathbf{E}$ in the excitable tissue is obtained by passing a pulsed current $I(t)$ through a coil placed close to the target tissue. The relationship between the electric field induced and the magnetic field is obtained from *Faraday's law:*

$$\nabla \otimes \mathbf{E} = -\frac{\partial \mathbf{B}}{\partial t}$$

where the magnetic field produced by the coil is given by the *Biot–Savart law:*

$$\mathbf{B}(\mathbf{r}, t) = \frac{\mu_0 I(t)}{4\pi} \oint_C \frac{d\mathbf{l}(\mathbf{r}') \otimes (\mathbf{r} - \mathbf{r}')}{|\mathbf{r} - \mathbf{r}'|^3}$$

where the integration is performed with the vector $d\mathbf{l}$ along the coil windings C, $\mathbf{r}$ is the position in the tissue, and $\mathbf{r}'$ is a point in the coil. The permeability of free space $\mu_0 = 4\pi \times 10^{-7}$ H/m can be assumed, since the permeability of tissue is nearly equal to that of free space, and the conductivity is approximately $1/\Omega \cdot$m. The electric field $\mathbf{E}$ causes movement of charges in both the intra- and extracellular spaces. Any part of the cell membrane interrupting this charge flow becomes depolarized or hyperpolarized. Once the membrane threshold potential is exceeded, an action potential occurs.

The shape of the electric field induced in the tissue depends on the shape of the induction coil, the location and orientation of the coil with respect to the tissue, and the electrical conductivity structure of the tissue. We refer you to Ruohonen's Ph.D. thesis [1998] for a detailed model of the currents induced in the tissue.

Design of Magnetic Stimulators

From a device engineering point of view, these models are more than an academic exercise since they yield estimates of the threshold magnetic field needed to cause activation of excitable tissues. Since the cell membrane of excitable tissues behaves as a leaky integrator, the most effective induced current will have a duration in the same range as that of the membrane's time constant. The approximate value for the membrane of neurons is around 150 μs, and typical magnetic stimulators for use in neurology produce magnetic pulses that last anywhere between 100 and 600 μs. Coils usually consist of 10 to 20 turns of heavy copper tubing or wire. Typical peak currents driven through the coil are in the range 5 to 15 kA to generate peak magnetic field strengths in the range 1 to 5 T.

Figure 7.24 shows the circuit diagram for an experimental magnetic stimulator based on a design proposed by Merton and Morton [1980]. The high-voltage output of transformer T1 is rectified by D2–D4 and charges a 2100-μF capacitor bank through resistor

> **Warning!** This is an extremely dangerous device! It produces high voltages backed by sufficient energy to cause lethal electrical shocks. This is definitely a project that should not be attempted by anyone who is not experienced with high-voltage/high-energy devices. Just consider that TNT releases approximately 2 kJ/g. As such, a short across the capacitor bank would definitely blow up the weakest link in the chain with the force of a small explosive charge. Needless to say, any remains will turn into shrapnel that can cause further injury or death. Consider that 2 kJ is the equivalent energy of dropping a 100-kg anvil on your foot from a height of 2 m.

R2. The capacitor bank consists of nine Cornell Doubilier type 400X212U450BF8 2100-µF 450-V dc electrolytic capacitors in series–parallel arrangement. However, because of its photoflash rating, the PF212V500BF2B would be a better choice. Resistors R4–R6, R9–R11, and R14–R16 equalize the voltage drop among all capacitors. These resistors also bleed any remaining charge on the capacitors after the device is used. The capacitor bank can be charged to a maximum of 1350 V, resulting in the storage of up to $CV^2/2 =$ 1.93 kJ. A blinking LED and a Mallory Sonalert II alarm module warn the user when the capacitor bank is charged.

The desired capacitor charge voltage is selected by the user by pressing on the charge switch. A keylock switch and an interlock switch also need to be closed to complete the circuit to energize the charge relay K1, which in turn energizes the primary of high-voltage transformer T1. The complete high-voltage circuit of the stimulator is contained within a $\frac{1}{2}$-in.-thick polycarbonate enclosure that acts as a blast shield in case of component failure. Interlock switch SW2 automatically disables charging whenever this polycarbonate enclosure is opened to access the circuit.

The switching element is an International Rectifier type ST330S16P0 high-power thyristor (SCR). This specific device is rated for a maximum repetitive peak and off-state voltage of 1600 V. It can handle a 9-kA peak current with a maximum dI/dt value of 1 kA/µs. SCR triggering is done through a MOC3010 triac-output optocoupler. Whenever pushbutton switch SW5 is depressed, the optocoupler allows C4 to discharge between the main gate and the center amplifying gate of SCR1. Diodes D5–D8 prevent the voltage on the capacitor bank from reversing after the discharge.

Suitable coil inductances can be evaluated using the simplified PSpice model shown in Figure 7.25. The capacitors are each modeled as an ideal 2100-µF capacitor with a 40-mΩ resistor in series to simulate their high-frequency ESR. The heavy wire carrying the current to the coil and the coil's resistance are modeled by a 100-mΩ resistor (R_COIL). Since International Rectifier does not have a PSpice model available for the thyristor, we modified the instance model for a C228A SCR (which is totally different from a ST330S16P0) by setting the model parameters as follows:

- VDRM: max. nonrepetitive peak voltage   1700 V
- VRRM: max. off-state voltage   1600 V
- IH: max. holding current   600 mA
- VTM: max. on-state voltage   1.66 V
- ITM: max. one-peak nonrepetitive surge current   8380 A
- $dV\,dt$: max. critical rate voltage rise 400×10^6 V/s
- IGT: dc gate current required to trigger   200 mA
- VGT: dc gate voltage required to trigger   2.5 V
- TON: typical turn-on time   1 µs

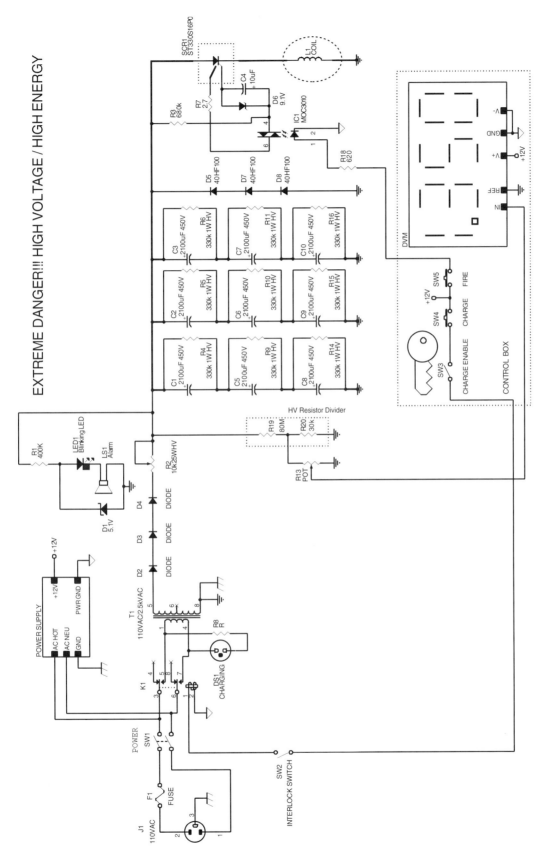

Figure 7.24 Schematic diagram of the experimental magnetic stimulator. A 2100-μF capacitor bank that can be charged to a maximum of 1350 V, resulting in the storage of up to 1.93 kJ. The switching element is an International Rectifier type ST330S16P0 high-power SCR rated for a maximum repetitive peak and off-state voltage of 1600 V. It can handle a 9-kA peak current with a maximum dI/dt of 1 kA/μs. SCR triggering is done through a triac-output optocoupler.

349

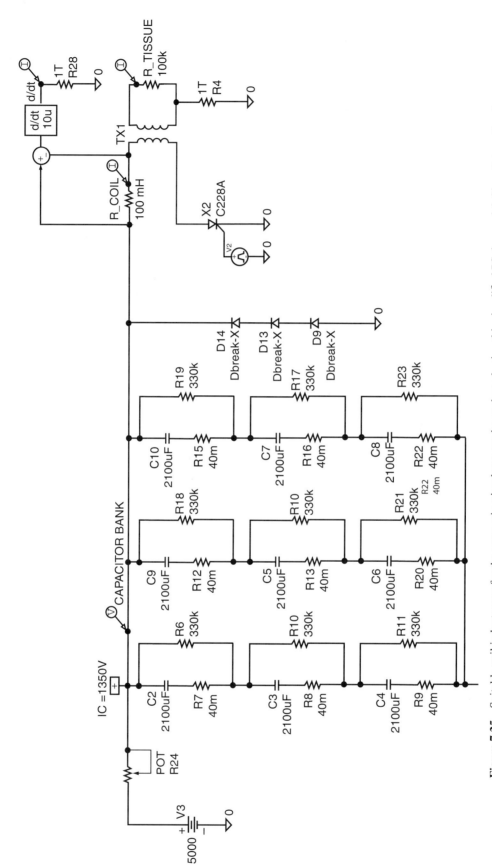

Figure 7.25 Suitable coil inductances for the magnetic stimulator can be evaluated using this simplified PSpice model. Since International Rectifier does not have a PSpice model available for the thyristor, we modified the instance model for a C228A SCR (which is totally different from a ST330S16P0) to model the parameters of the ST330S16P0. Similarly, a breakaway diode model (DBREAK) was modified to simulate the behavior of the 40HF100 diodes. The simulation is used to to find an inductance that keeps the peak current within the SCR's current-handling capability (~8.3 kA) while conforming to the SCR's switch-on capability (dII/dt <400 V/µs).

- TOFF: typical turn-off time   100 μs
- IDRM: max. off-state leakage current   50 mA

In a similar way, a breakaway diode model (DBREAK) is modified to simulate the behavior of the 40HF100 diodes:

- VRRM: max. peak reverse voltage   1000 V
- IRRM   9 mA
- IF(AV): max. average forward current   40 A
- IF(RMS): max. RMS forward current   62 A
- IFSM: max. peak surge current   570 A
- VF(TO)1: low-level value of threshold voltage   0.65 V
- VF(TO)2: high-level value of threshold voltage   0.70 V
- RF1: low-level value of forward slope resistance   4.29 mΩ
- RF2: high-level value of forward slope resistance   3.98 mΩ
- VFM: max. forward voltage drop   1.30 V

Now the trick is to find an inductance that keeps the peak current within the SCR's current-handling capability (~8.3 kA) while conforming to the SCR's switch-on capability ($dI/dt < 400$ V/μs). The PSpice simulation results of Figure 7.26 show that the peak current through the coil is approximately 8 kA and dI/dt is 270 A/μs, which are both within the ST330S16P0's ratings. The stimulating phase of the pulse lasts for approximately 100 μs. The other 900 μs of the current pulse duration doesn't do much as far as stimulating the tissue but causes considerable heat dissipation in the coil.

The energy delivered to the coil by the stimulator is really divided into Joulean energy and magnetic field energy. The Joulean energy is from the current flow through the system's resistance, and is lost as heat. However, some stimulator designs are capable of recovering the energy stored in the magnetic field (given by $E_{magnetic} = LI^2_{max}/2$). For example, Figure 7.27 shows a magnetic stimulator that uses a nonpolar capacitor bank. As shown in the PSpice simulation results of Figure 7.28, when the SCR is triggered, the capacitor discharges through the coil. Selecting L and C to yield an underdamped response, as the capacitor voltage reaches zero, the coil current and magnetic field are at their maximum. As the magnetic field collapses, the coil current continues to flow in the same direction until it charges the capacitor to the opposite polarity. A diode placed antiparallel to the SCR allows the charge to flow back through the coil in the opposite direction, charging the capacitor to its initial polarity. At this point the current stops because the SCR has turned off. If the loss to Joulean heating is low, most of the energy ends up back in the capacitor bank, and only a small amount of additional power is needed to restore the capacitor voltage.

Magnetic Stimulator Coil Design

In magnetic stimulation, the design of the coil is in many ways far more important than the design of the energy source. This is because the coil is placed in very close proximity to the subject, posing risks of electrocution, burning, and/or mechanical impact to both the subject and the operator. The following consideration must be made when designing stimulator coils:

1. The voltage across the coil is typically a few kilovolts. Depending on the coil's construction, the voltage across adjacent turns will typically be in the range 200 to 1000 V. For this reason, insulator materials must be selected such that they can pass

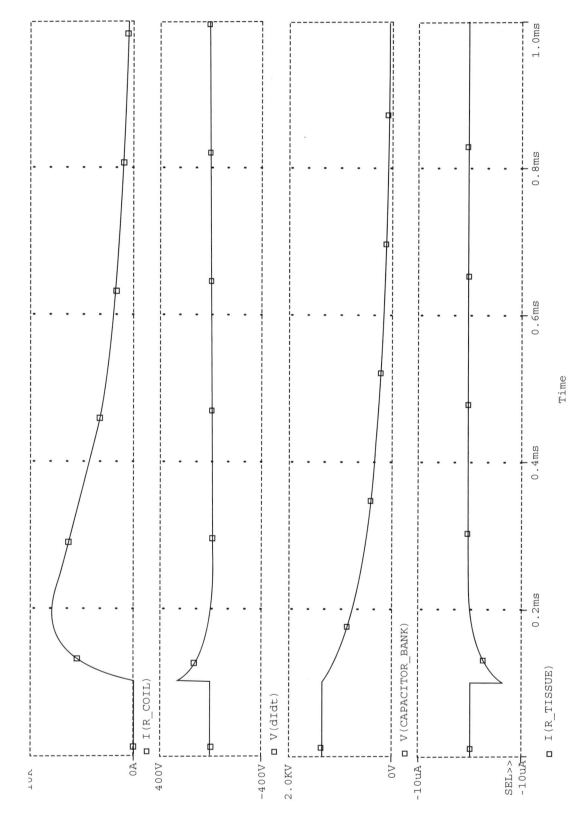

Figure 7.26 PSpice simulation results for the experimental magnetic simulator show that the peak current through the coil is approximately 8 kA and dI/dt is 270 A/μs, which are both within the ST330S16P0's ratings. The stimulating phase of the pulse lasts for approximately 100 μs.

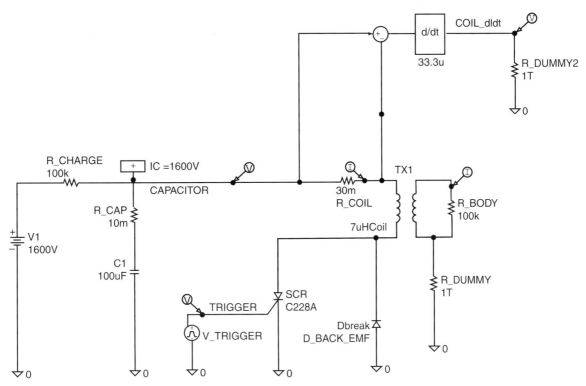

Figure 7.27 A magnetic stimulator that uses nonpolar capacitors is capable of recovering the energy stored in the magnetic field.

hiPot testing (see Chapter 3) for reinforced insulation. You may remember that parts carrying voltages U between 1 and 10 kV require their reinforced insulation to be tested at a voltage of $2(U + 2500)$ for 60 s.

2. The intense pulsed currents driven for stimulation cause very strong expanding and compressing forces in the coil. This is because when magnetic flux lines are packed together, they create magnetic pressure. The force of the expanding coil is proportional to the energy delivered to the coil and related inversely to the coil size. Typical pressures for circular coils are in the range 500 to 1500 lb/in², making it necessary to enclose the coil in a potting material (e.g., epoxy) to prevent the coil windings from flying apart when the stimulation pulse is delivered. A detailed analysis of magnetic stimulator coil forces was published by Mouchawar et al. [1991].

3. Joulean heating of the coil must also be considered, since according to the medical device standards, the surface temperature of the coil must not exceed 41°C. The main contributor to Joulean heating is the coil conductor's resistance, which is determined by the wire material, gauge, and coil geometry. Copper is used almost universally because of its relatively high conductivity and good mechanical properties. When the wire cross section exceeds 1 mm², the skin and proximity effects change the current distribution in the wire, increasing the dc resistance significantly. Striped, foil, or isolated multistrand (litz) wire is often used to reduce the skin and proximity effects.

The term *litz wire* is derived from the German word *litzendraht* meaning woven wire. This wire is constructed of individual film-insulated wires bunched or braided together in a uniform pattern. The multistrand configuration minimizes the power losses otherwise

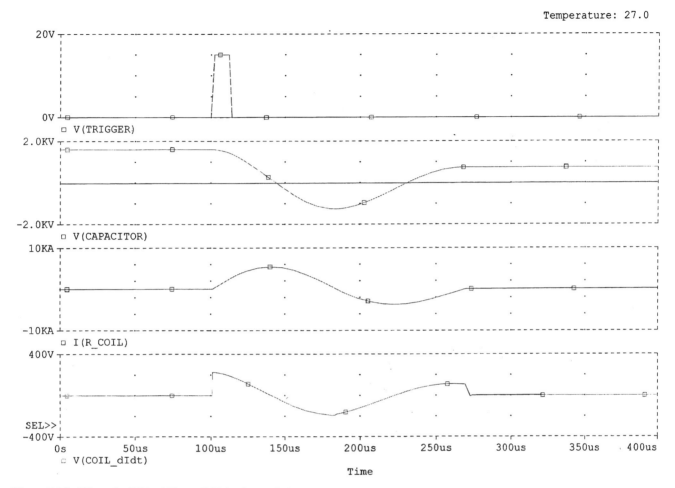

Figure 7.28   When the SCR of Figure 7.27 is triggered, the capacitor discharges through the coil. As the capacitor voltage reaches zero, the coil current and magnetic field are at their maximum. When the magnetic field collapses, the coil current continues to flow in the same direction until it charges the capacitor to the opposite polarity. A diode placed antiparallel to the SCR allows the charge to flow back through the coil in the opposite direction, charging the capacitor to its initial polarity. At this point, the current stops because the SCR has turned off.

encountered in a solid conductor due to the skin effect or tendency of radio-frequency current to be concentrated at the surface of a conductor. A good supplier of litz wire is WireTronic. You can download a free Windows utility from their Web site (WTSETUP. EXE is also provided for your convenience in the book's ftp site) that will give you dimensional data for a wide range of wire sizes, insulation specifications, current-handling capacities, resistance calculation, break strength calculation, and ordering information.

The coil for our experimental magnetic stimulator was constructed with WireTronic litz wire type 85/26HPN-DN in a flat pancake configuration. The inductance of an air-core pancake coil can be calculated using *Wheeler's formula:*

$$L(\mu H) = \frac{r^2 n^2}{8r + 11w}$$

where r is the radius to the center of the winding (in inches), n the number of turns, and w the width of the winding (in inches). Our prototype coil was wound on the inside of a PVC tube cap. A hole cut on the side of the cap was fitted with a PVC tube which acts as the coil handle and wire exit. 24-hour epoxy was used to pot the coil inside the cap. One must keep in mind that some glues can damage enamel on coil wire.

Coil geometry directly affects the depth and focality (precision of targeting) of the tissue to be stimulated. Standard stimulating coils are circular, with a diameter ranging from 50 to 150 mm. Coils designed for *focal stimulation*, that is, capable of stimulating one given area without stimulating a nearby area, usually consist of two adjacent coils with current flowing in opposite directions. Induced currents add at the center of these figure-eight coils, yielding twice the stimulus produced at the edges of the individual coils. Four-leaf coils have been constructed to produce a concentrated dot stimulus useful for peripheral nerve stimulation. Another design that is now being used is the half-toroid coil (as in a Slinky), which is wound with the turns in different angles while maintaining the tangency along one edge to drastically reduce the current induced away from the target stimulation site.

Mouchawar et al.'s paper [1991] provides a thorough analysis of the effect of coil geometry on stimulation. Despite this, designing coils for magnetic stimulators is more an art than an exact science. To play different "what if" scenarios, we recommend a free finite element analysis program called FEMM (Finite Element Methods Magnetics) by David Meeker (supplied for your convenience in the book's ftp site). FEMM is a suite of programs for solving static and low-frequency problems in magnetics. The programs currently address problems on two-dimensional planar and axisymmetric domains. FEMM comprises:

- *A preprocessor*: a CAD-like program for laying out the geometry of the problem to be solved and defining material properties and boundary conditions
- *A solver*: a device that takes a set of data files that describe a problem and solves the relevant Maxwell's equations to obtain values for the magnetic field through the solution domain
- *A postprocessor*: a graphical program that displays the resulting fields in the form of contour and density plots

Interestingly, many theoretical studies have demonstrated that today's coils are far from optimal. These studies also show that power consumption can be reduced drastically, opening many opportunities for individual innovators in coming up with new ideas that can become platforms for completely unexplored applications.

Commercial Magnetic Stimulators and Their Applications

Commercial magnetic stimulators are not much different than the simple prototype device discussed above. The three main commercial players in the magnetic stimulation field are Cadwell Laboratories, Inc. (Kennewick, Washington), Magstim Company, Ltd. (Whitland, UK) and Medtronic Dantec NeuroMuscular (Skovlunde, Denmark). One of the newest magnetic stimulation modalities is repetitive transcranial magnetic stimulation (rTMS), capable of delivering trains of stimuli at 1 to 50 Hz. rTMS was first produced by Cadwell Laboratories in 1988. Dantec and Magstim have add-on modules for their single-pulse devices that can be used to drive one coil with two to four pulses separated by 1 ms to 1 s. As you can imagine, the duration of sustained operation is limited by coil heating, and Cadwell solves this problem with continuous water cooling.

Single-stimulus magnetic stimulators are used routinely by neurologists as diagnostic tools to measure nerve-conduction time. In addition, transcranial magnetic stimulation (TMS) over the primary motor cortex evokes movement in the contralateral limb and has provided information on the anatomical organization and functional characteristics of the motor system. TMS can be used to map motor cortex representations precisely. Magnetic stimulation is now used for operating-room monitoring to directly assess the central motor pathways.

Recently, rTMS has shown therapeutic potential in the control of depression [George et al., 1996; Pascual-Leone et al., 1996]. Many groups of investigators have demonstrated that mood can be altered by rTMS in healthy subjects and improve mood even in patients with medication-resistant depression [George et al., 1995, 1997]. Much more clinical work is still required to turn rTMS into a clinical alternative to current antidepressant treatments. However, its promise is significant. An open trial reported comparable antidepressant efficacy for TMS and the most powerful of all antidepressant treatments, electroconvulsive therapy (ECT). However, unlike ECT, rTMS does not require sedation or cause cognitive impairment and can be administered in an outpatient setting.

At the time of this writing, commercially available magnetic stimulators are approved by the FDA for peripheral nerve and spinal cord stimulation only, but the FDA is allowing low-repetition-rate devices (<1 Hz) to be used by investigators for human cortical stimulation without the need for an Investigational Device Exemption (IDE; see the Epilogue). However, the FDA believes that TMS at frequencies of ≥1 Hz carries significant risk and thus requires an IDE for studies involving rTMS of the human cortex.

A different emerging area for the therapeutic use of repetitive magnetic stimulation is the excitation of peripheral nerves for painless treatment of neurological and neuromuscular disorders. For example, Neotonus Inc. (Marietta, Georgia) has focused its attention on the treatment of urinary incontinence. In June 1998, the FDA approved use of their magnetic stimulator to enhance peripheral nerve innervation in treating urinary incontinence in women.

Some groups, such as that of Loughborough University [Young et al., 2001], are pursuing a much more ambitious goal. They are working on pulsed power energized from a pair of parallel-connected 200-μF at 22 kV capacitors. The 118-μH double coil induces pulses of 500 μs in duration and 400 V/m at 5 cm from the coil. This has the potential to activate structures deep inside the body, such as the heart, bowel, bladder, spinal cord, and kidney. The aim of this group's program is to be able to stimulate the kidneys to restart or enhance peristaltic pumping to evacuate stone fragments left behind after extracorporeal shock wave lithotripsy. Threshold for peristalsis with a "low-power" (by Loughborough's standards) 4.2-kJ magnetic stimulator applied directly to the kidney during surgery was reported to be between 470 and 720 V/m, with an induced pulse width of 270 μs. They expect that the 500-μs pulse delivered by their 80-kJ stimulator will activate the unmyelinated nerve fibers in the kidney from outside the body. This is certainly a new field with plenty of potential. It's easy to see that as in many other areas of biomedical engineering, progress will be made with magnetic stimulation as ingenious solutions tackle today's engineering challenges.

Safety in Magnetic Stimulation

Guidelines for safe use of magnetic stimulation have not been established conclusively. It is obvious that magnetic stimulation should not be applied to subjects with implanted metallic or magnetic objects. The magnetic field of the stimulator attracts ferromagnetic objects and repels nonmagnetic conductors with a force that can harm surrounding tissues. It should also be obvious to practitioners of magnetic stimulation that an enormous magnetic pulse can destroy any implanted electronic devices (e.g., pacemakers, implantable defibrillators).

On the less obvious side of possible side effects, the experimenter should take into account possible induction of seizures by transcranial magnetic stimulation. Also, magnetic stimulator coils usually produce loud clicks when their windings try to expand during a pulse. The peak sound pressure is often in the range 120 to 130 dB at a distance of 10 cm from the coil. Most sound energy is in the frequency range 2 to 7 kHz, where the human ear is the most sensitive, and this noise may exceed criteria limits for sensorineural hearing loss. For this reason, hearing protection aids are recommended for both the experimenter and the subject.

OTHER CLINICAL APPLICATIONS OF ELECTRICAL CURRENT DELIVERY TO TISSUES

There are many clinical applications for the delivery of electrical currents to the body that do not involve the activation of excitable tissue. Unfortunately, the marketplace is also littered with all sorts of "energy medicine" contraptions and procedures that claim to "balance the flow of vital energy in the body" using currents delivered to the body. Needless to say, the latter are all either extremely controversial or totally unproven. Some, such as electroacupuncture, are simply old "alternative" treatments with an electrical twist. Others are without doubt the electrical equivalents of snake oil. Let's take a look at what is real and what is quack.

Electrosurgery and RF Ablation

The use of alternating currents for surgical techniques was attempted by D'Arsonval in 1891. However, the technique did not become practical until 1928, when physicist William T. Bovie developed an RF electrosurgical unit (ESU) for use in the operative environment. Developments in solid-state circuitry and advances in the methodology of RF signal modulation as well as inventions related to the precise control of tip temperature have enabled all sorts of interventional applications for RF-based surgical tools (Table 7.3).

Depending on how RF is applied, as well as the crest factor (the ratio of peak voltage to root mean square voltage), the waveform (damped or unmodulated sinusoid), and the power output, three different electrosurgical effects can be achieved:

1. *Electrosurgical cutting*: electric sparking to tissue with a cutting effect
2. *Electrosurgical fulguration*: electric sparking to tissue without significant cutting, also known as *noncontact coagulation* or *spray mode*; used to control diffuse bleeding
3. *Electrosurgical desiccation*[8]: low-power coagulation without sparking, also known as *contact coagulation* mode; used to control local bleeding and enables point coagulation of tissues

The RF current flows from a tip electrode into the tissue, producing ionic agitation in the tissue about the electrode tip as the ions attempt to follow the changes in direction of the alternating current. This agitation drives water from the cells, leading to desiccation and coagulation (which occurs at a temperature of approximately 48°C). Further increases in temperature due to ionic frictional heating in the tissue surrounding the electrode are responsible for cutting and fulguration.

Today, RF energy delivered via catheters to the heart is being used to treat cardiac arrhythmias. RF ablation (selective destruction) has replaced traditional surgical procedures in the treatment of refractory supraventricular arrhythmias and ventricular tachycardia by delivering energy to selectively destroy the accessory conduction pathways responsible for the arrhythmia.

Iontophoresis

Iontophoresis is an accepted method of drug delivery similar to the passive transdermal medication patches now on the market for smoking cessation and hormone therapy. However, iontophoresis uses low-level electrical current to speed up delivery of the drug ions into the skin and surrounding tissues (Table 7.4). Iontophoretic drug delivery systems

---

[8]Not to be confused with *electrocautery*, which is the coagulation of blood or tissue by means of an electrically heated wire, where the current heats only the wire and does not pass through the patient's body.

TABLE 7.3 Typical Parameters Used in Various Electrosurgery and RF Ablation Instruments

Clinical Application	Method of Current Delivery	Typical Waveform	Typical Current or Voltage
Electrosurgical cutting	Contact electrodes with various surface areas, depending on desired coarseness of cut; either unipolar (referenced against large surface-dissipative electrode) or bipolar pair on "scalpel"; electrode impedance 100 to 300 Ω	Continuous-wave sine wave 300 to 800 kHz causes quick, clean cutting effects with little or no hemostasis; this pure-cut waveform causes cells to swell and explode into steam and smoke	Up to 1 kV peak to yield up to 350 W
Electrosurgical fulguration	Noncontact electrode allowed to spark against tissue; reference electrode is usually a large dissipative plate	Modulated 300 to 800 kHz; coagulation occurs when the cell fluid is allowed to cool between heating. The waveform consists of a dampened cut wave with an ON time (heating), then an OFF time (cooling); most ESUs offer a "blend" mode that combines cut and coagulation currents	Up to 6 kV peak to yield up to 50 W
Electrosurgical desiccation[a]	Contact electrodes with various surface areas, depending on desired width of dessication; either unipolar (referenced against large surface dissipative electrode) or bipolar pair on dessication tool; electrode impedance 100 to 500 Ω		Up to 500 V peak to yield up to 60 W
Selective ablation of myocardiac tissue to terminate arrhythmic foci	Electrode inserted into the heart via a temporary catheter; electrode impedance 100 to 300 Ω	Sine wave 300 to 800 kHz	Up to 50 W
Perforation of soft tissue for the creation of atrial septal defects, treatment of pulmonary atresia, hypoplastic left heart syndrome, SVC occlusion and esophageal atresia	Fine-tip electrode inserted to desired perforation site via a temporary catheter; electrode impedance 2 to 4 kΩ	Sine wave 300 to 800 kHz	Up to 25 W

[a]However, not all forms of coagulation have to be performed with a "coagulation current". Endoscopic procedures, particularly coagulation, should be performed using non-modulated current ("cutting current") delivered through an electrode with larger surface area so that contact coagulation can be done at a lower voltage with reduced arcing.

use a low-level dc current of under 4 mA to push water-soluble ionized medication through the skin's outermost layer (stratum corneum), which is the main barrier to drug transport. Once the drug passes through the skin barrier, natural diffusion and circulation are required to shuttle the drug to its proper location.

The drug delivery electrode that contains the medication is placed over the area to be treated or where the procedure is to be done. For example, the drug delivery electrode loaded with a suitable local anesthetic (e.g., lidocaine) is placed over a vein in the bend of the elbow for a routine blood draw or intravenous start. The other electrode, the grounding (dispersive) electrode, is placed approximately 4 to 6 in. away from the drug delivery electrode.

Topical anesthetic agents are often delivered through iontophoresis prior to cut-down for dialysis, insertion of tracheotomy tubes, and insertion of catheters. Iontophoresis has

TABLE 7.4   Typical Parameters Used in Iontophoresis Instruments

Clinical Application	Method of Current Delivery	Typical Waveform	Typical Current or Voltage
Delivery of water-soluble ionized medication through the skin	Current delivered between drug-delivery and "dispersive" surface electrodes; typical electrode impedance at beginning of treatment in the range 20 to 100 kΩ; impedance at the end of treatment in the range 1 to 20 kΩ	Dc	0 to 4 mA with limited ~60-V compliance; electrode current density maintained under 500 mA/cm^2
Drawing glucose through skin by electroosmosis	Current delivered between extraction surface electrodes	Dc applied for 3 to 5 minutes, then glucose is analyzed for 7 to 10 minutes, followed by another extraction using dc in opposite direction for 3 to 5 minutes	200 μA to 1 mA to yield current density of ~0.3 mA/cm^2 at the cathode

also been investigated with some success as a way of delivering anti-inflammatory agents (dexamethasone sodium phosphate + lidocaine) to treat musculoskeletal disorders, hyaluronidase to reduce edema, as well as medication for many localized skin disorders, such as nail diseases, herpes lesions (e.g., delivery of acyclovir), psoriasis, eczema, and cutaneous T-cell lymphoma.

An iontophoresis dose is expressed as

$$\text{charge delivered (mA/min)} = \text{current (mA)} \times \text{treatment time (minutes)}$$

A typical iontophoretic drug delivery dose is 40 mA/min but can vary from 0 to 80 mA/min. Most people feel little or no sensation at all during an iontophoretic treatment. Some people feel a tingling or warm sensation under one or both of the electrodes. This is caused by small blood vessels in the skin expanding due to the presence of direct current, or because mast cells are responding to the current by releasing histamine.

It has also been demonstrated that electroosmosis can be taken advantage of in *reverse iontophoresis*. Here, imposing an electric current across the skin extracts a substance of interest from within or beneath the skin to the surface. Sodium and chloride ions from beneath the skin migrate toward the electrodes. Uncharged molecules, including glucose, are also carried along with the ions by convective transport (electroosmosis). This technique is now being used to monitor the subdermal concentration variation of glucose, allowing diabetic patients to track their blood sugar without painful finger pricks.

Modulation of Cardiac Contractility

A major determinant of contractile strength of cardiac muscle cells is the amount of calcium reaching the contractile proteins during a beat. Reduced calcium transients are believed to contribute to contractile dysfunction in heart failure. Shlomo Ben-Haim, the founder of Impulse Dynamics (Haifa, Israel), discovered that extracellularly applied electric fields delivered during the absolute refractory period can modulate myocardial contractility [Ben-Haim et al., 2002] (Table 7.5). Experimental evidence in situ as well as in vivo (healthy dogs and pigs, heart failure dogs, and heart failure human patients) indicates that electrical signals do modulate cardiac contractility [Burkhoff et al., 2001].

TABLE 7.5   Typical Parameters Used in Cardiac Contractility Modulation Instruments

Clinical Application	Method of Current Delivery	Typical Waveform	Typical Current or Voltage
Enhancement of cardiac contractility	High-capacitance implantable electrodes in contact with cardiac muscle; typical electrode impedance 200 to 500 Ω	Charge-balanced biphasic controlled-current or capacitor-discharge pulse burst 5 to 10 ms in duration per phase; one to three pulses per burst delivered 30 to 60 ms into the ventricular absolute refractory period	±5 to ±15 mA with 20-V compliance or up to ±8 V from 660-μF capacitor

Impule Dynamics' implantable cardiac contractility modulation devices [Prutchi et al., 1999] are currently undergoing clinical investigations to determine their safety and effectiveness as tools in the treatment of heart failure. Inside the implantable pulse generator's titanium can, sense amplifiers detect the heart's electrical activity through standard pacing leads. Specialized circuitry is used to generate and deliver the cardiac contractility modulation (CCM) signals to the heart muscle during the ventricular absolute refractory period [Mika et al., 2001]. An implantable-grade battery powers the device.

In principle, the same techniques can be applied to controlling other tissues that use calcium as their main signaling mechanism. In gastroenterology, for example, application of electric impulses is being researched by Impulse Dynamics to treat morbidly obese patient. The hope is that application of these nonexcitatory signals to these organs will alter cellular function in a predictable and reproducible way while avoiding the systemic side effects of pharmacological agents.

Bone Growth Stimulators

Electrical bone growth stimulators have been proven to hasten the healing process for certain types of fractures and bone fusions. Noninvasive, semi-invasive, and invasive methods of electrical bone growth stimulation are available (Table 7.6). There are two noninvasive bone growth stimulation techniques. The first, *pulsed electromagnetic fields* (PEMF), involves the use of paired coils that are placed on either side of a fracture site. It is believed that this triggers calcification of the fibrous cartilage tissue within the fracture gap. Ten hours of treatment per day are usually necessary. Because of the relatively strong fields that need to be generated, these units run on mains power and are not portable, which forces the patient to spend a substantial part of the day next to a wall power outlet.

The second type of noninvasive bone growth stimulator, the *capacitively coupled stimulator*, involves the use of skin-surface gelled electrodes through which a constant current 60-Hz sine-wave signal is delivered at 5 to 10 mA to produce an electrical field strength at the desired fusion site of some 2.0 V/m with a current density of approximately 300 mA/cm^2. The probable mechanism of operation of this method is through translocation of calcium into the cells at the fracture site through voltage-gated calcium channels.

Semi-invasive and invasive bone growth stimulators utilize a dc source to generate a weak electrical current in the underlying tissue. Semi-invasive or percutaneous bone growth stimulators use an external power supply and electrodes (cathodes) that are inserted through the skin and into the bone segment where growth is desired, while a self-adhesive gelled anode electrode is placed directly on the skin. The electrodes are then connected to a power pack that delivers 20 μA dc and is embedded within the cast. These units are typically designed to be applied for 12 weeks.

TABLE 7.6   Typical Parameters Used in Bone Growth Stimulation Instruments

Clinical Application	Method of Current Delivery	Typical Waveform	Typical Current or Voltage
Accelerate bone fracture healing or bone fusion	*Noninvasive capacitive coupling:* ac signal delivered between gelled skin surface electrodes	Ac constant-current, 60-Hz sine-wave signal	5 to 10 mA to produce electrical field strength at the desired fusion site of ~2.0 V/m with a current density of ~300 μA/cm^2
	Noninvasive electromagnetic: ac magnetic field applied through coils over desired bone fusion site	Inductively coupled field 15-Hz burst, 20 pulses per burst, and a pulse frequency of 4.3 kHz	Maximum magnetic field amplitude at the fusion site ~25 G to induce an electrical field of ~0.2 V/m
	Semi-invasive: cathode electrode is implanted within the fragments of bone graft at the fusion site; anode is a skin surface electrode	Dc	20 μA
	Fully implanted: cathode electrode is implanted within the fragments of bone graft at the fusion site; anode is the enclosure of the implanted dc power source	Dc	20 μA

Invasive bone growth stimulators require surgical implantation of a current generator in a subcutaneous pocket. A cathode electrode is implanted within the fragments of bone graft at the fusion site. The generator is often no more than a battery and a 20 μA constant-current diode to the cathode electrode via a silicone-coated lead. The device's metallic enclosure acts as the anode electrode. The implant delivers dc continuously for six to nine months and is then removed by a simple procedure and is done most often under local anesthesia.

The precise mechanism of how dc speeds bone healing is not clearly understood. However, it is believed that the localized increase in pH and lowered oxygen tension at the cathode favor the activity of bone-forming cells and inhibit bone-absorbing cells. Bone growth takes place with a current between 5 and 20 μA. Currents below 5 μA do not enhance growth, and levels above 20 μA cause cell necrosis and bone death.

Electrical Stimulation for the Treatment of Chronic Wounds

Electrical stimulation has been studied as a possible therapy for accelerating wound healing (Table 7.7). In vitro as well as in vivo animal studies have shown that externally applied electrical currents can increase ATP concentrations in tissues, increase DNA synthesis, promote healing of soft tissue or ulcers, cause epithelial and fibroblasts to migrate into wound sites, accelerate the recovery of damaged neural tissue, reduce edema, and inhibit the growth of some pathogens.

Research has been done to evaluate the effectiveness of dc, pulsed, and ac currents in promoting the healing of wounds. However, a technology assessment study conducted by the Emergency Care Research Institute (ECRI) [1996] concluded that whereas electrical stimulation does facilitate the complete healing of chronic ulcers compared to the use of plain dressings, it is no better than conventional wound care involving debridement, cleaning agents, antibiotics (systemic or local), and bandages.

TABLE 7.7 Typical Parameters Used in Chronic Wound Treatment Instruments[a]

Clinical Application	Method of Current Delivery	Typical Waveform	Typical Current or Voltage
Promoting healing of wounds and skin ulcers	Cathode wrapped in saturated (saline) gauze and placed directly over the wound site, gelled skin surface anode placed near the wound	Dc applied two or three times per day for 2 hours at a time	20 to 100 µA with a compliance of ~8 V
		Low-voltage monophasic current pulses, 50% duty cycle delivered at up to 150 Hz	30 to 40 mA with up to 12-V compliance
		High-voltage monophasic pulses, 50% duty cycle delivered at up to 150 Hz	100 to 250 V
		Biphasic 40-Hz square-wave current	15 to 25 mA with <100-V compliance
		TENS device generating charge-balanced current-controlled pulses, 0.1 to 0.2 ms duration at 80 to 90 Hz	10 to 50 mA "stimulation" phase with <150-V compliance
	Ac magnetic field applied through coils over wound dressings	Inductively coupled field 10 to 150 Hz burst, 10 to 100 pulses per burst and a pulse frequency of 2 to 10 kHz	Maximum magnetic field amplitude at the wound site <50 G to induce an electrical field of ~0.1 to 0.5 V/m
	Nonthermal high-frequency RF applied with small capacitive "antenna" over wound dressings	60 to 100 µs of high-peak-power RF bursts of ~27 MHz, repeated at 80 to 600 Hz	Peak pulse power of 200 W to 1 kW

[a]It should be noted that in 1996 the ECRI found no evidence that dc stimulation improves the healing rate of chronic, decubitus, or diabetic ulcers. All other forms of stimulation seem to improve the normalized healing rate of decubitus ulcers but not of chronic venous or diabetic ulcers.

Electrochemotherapy or Electroporation Therapy

Brief, intense electric pulses have been in use since the 1970s to create temporary pores in cells without causing permanent damage. This process, known as *electroporation*, happens when the electrical pulse causes a transmembrane potential of 0.5 to 1.5 V. Under this field intensity, the lipid bilayer of cells is temporarily rearranged, forming aqueous channels in the cell membrane. These "pores" make the cell membrane permeable to a large variety of hydrophilic molecules that are otherwise unable to enter into the cell. Once formed, these pores remain open for a duration of seconds to minutes.

Genetronics, Inc. (San Diego, California) introduced the use of electroporation for in vivo delivery of high doses of chemotherapeutic drugs to cancerous tumors [Hofmann et al., 1996] (Table 7.8). Electroporation therapy (EPT) makes it possible to introduce into cells potent anticancer drugs such as bleomycin, which normally cannot penetrate the membranes of certain cancer cells. Treatment is carried out by injecting bleomycin directly into the tumor and applying electroporation pulses through an array of needle electrodes. The field strength must be adjusted reasonably accurately so that electroporation of the cells of the tumor occurs without damage, or at least minimal damage, to any normal or healthy cells. Treatment with this therapy avoids the toxic effects associated with the systemic administration of anticancer agents, making it possible to kill selectively the cancerous cells while avoiding surrounding healthy tissue. As such, a patient should require a much lower dose of chemotherapy than is usually necessary to kill the tumor, drastically reducing nasty side effects such as hair loss, nausea, and vomiting that are associated with conventional chemotherapy.

TABLE 7.8 Typical Parameters Used in Electrochemotherapy (Electroporation) Instruments

Clinical Application	Method of Current Delivery	Typical Waveform	Typical Current or Voltage
Deliver high doses of chemotherapeutic drugs to cancerous tumors	*Genetronics:* six needle electrodes in a circle in contact with tumor driven in opposing pairs at any one time	*Genetronics:* voltage pulses, 100 μs in duration, delivered at a frequency of 1 to 100 Hz	Maximum pulse amplitude of 3 kV to yield field intensity of 600 to 1200 V/cm
	Daskalov et al. [1999]: pair of stainless-steel wires 0.8 mm in diameter and 14 mm in length spaced 5 to 30 mm	*Daskalov et al. [1999]:* burst of eight biphasic voltage pulses of 50 μs duration per phase at a frequency of 1 kHz	Maximum pulse amplitude of 1.25 kV to yield field intensity of 330 to 1250 V/cm

Electroporation therapy can be delivered to external tumors by placing electrodes on opposite sides of a tumor so that the electric field is between the electrodes. However, when large or internal tumors are to be treated, it is not easy to locate electrodes properly and measure the distance between them. For this type of tumors, needles are the preferred type of electrodes. Since the use of only two needles creates an inhomogeneous field, Genetronics uses an array of six needles to optimize the uniformity of the pore formation around the cell [Hofmann, 2001]. Genetronics drives the needle electrodes in opposing pairs because the resulting field is more homogeneous than that between opposing single needles. After each electroporation pulse the polarity of the needles is reversed immediately and the needle pair is pulsed again. After each of these paired pulsings, the sequence for the next electric field is rotated 60 degrees. A fairly uniform distribution, which maximizes pore formation in the cells over a circular section of the tissue, is generated by rotating the field three times.

Electrochemical Therapy

Electrochemical therapy (EChT), also known as *electrochemical tumor therapy* or *cancer galvanotherapy* consists of placing a platinum wire anode electrode into a tumor and a number of similar cathode electrodes in the tumor's periphery. <100 mA dc is passed between the electrodes until the delivered charge reaches some 50 to 100 C/mL of tumor (Table 7.9). The flow of direct current through tumor tissue triggers electrolytic processes at the electrodes. Positively charged ions (e.g., H^+, Na^+) migrate to the cathode, resulting in the formation of an extremely alkaline environment (pH $\approx$ 12.9). At the same time, negatively charged ions (e.g., Cl^-) migrate to the anode, creating an extremely acidic environment (pH $\approx$ 2.1). These local pH levels are well outside the physiologic range and have a destructive effect on tissue.

The hypothesis is that the cancerous tissue is more sensitive to than normal tissue extreme pH levels and is thus selectively killed. It is also thought that the change in ion concentrations permanently depolarizes cancer cell membranes, causing their further destruction. Last, proponents of the technique suggest that the EChT process may also generate heat shock proteins around the cancer cells, inducing the body's own killer cells to better target the tumor.

The technique has reportedly been applied with success, mostly in China and Japan, for the treatment of lung, breast, and bladder cancers. The technique has been used primarily in conjunction with systemic or topical chemotherapy agents (e.g., adriamycin), which because of their polar nature may be attracted by the electrodes and concentrate at the tumor site. Yuling [2000] reviewed the experience at 108 hospitals of treating 7642 patients of malignant tumors with EChT between 1978 and 1998. Yuling reported complete remission in 33.2% of the cases and partial remission in 42.8% of the cases, concluding that the

TABLE 7.9 Typical Parameters Used in Electrochemical Therapy Instruments

Clinical Application	Method of Current Delivery	Typical Waveform	Typical Current or Voltage
Electrochemical destruction of tumors	Two or more platinum needle electrodes positioned under local anesthesia in or at the site of the tumor with an interelectrode distance of <3 cm	Dc	5 to 100 mA with a voltage compliance of up to 20 V to deliver a total charge of approximately 30 to 100 C/mL of tumor volume
Attract lymphocytes to solid tumors or infections	12-μm-diameter IROX-coated wire electrodes in the tumor or infection and large platinum reference electrode at remote site	Charge-balanced biphasic current pulses 400 μs in duration delivered at 200 Hz for 10 minutes to 8 hours per session	$\pm 20\,\mu A$ to yield ~500 to 2000 $\mu C/cm^2$

technique was effective in 76% of cases. The effectiveness rate reported for superficial tumors only was 80.2%. However, because most of the trials were conducted at centers that are not bound by the strict research methodologies prescribed by the FDA to prove safety and effectiveness, we must consider these results to be mere anecdotal evidence that EChT may become a useful tool in the fight against cancer.

Unfortunately, EChT is not being given the attention that it deserves to rigorously prove or disprove its clinical efficacy. The problem is that the pioneer of the technique, Swedish physician Björn E. W. Nordenström, believes that the body has a second circulatory system of continuous energy circulation that he calls a *biologically closed electrical circuit* (BCEC). He believes that these currents participate in maintaining homeostasis and in controlling the healing process in living organisms, and presents EChT as a direct application of his BCEC concepts. This not only attracts the attention of a lot of quacks, but also makes the principle of EChT unpalatable to most mainstream oncologists.

In a somewhat related application, U.S. patent 6,038,478 to Yuen et al. [2000] describes the use of low-current biphasic pulses to attract lymphocytes to sites that can be accessed through surgery to place an array of electrodes. The technique is said to be useful in the treatment of solid tumors or in the treatment of certain infections, especially in poorly vascularized (e.g., brain) or inaccessible areas where surgical intervention is unadvised and electrode placement is feasible and less destructive.

Induction of Apoptosis via Nanosecond Pulsed Electric Fields

Apoptosis is the process by which a cell actively commits suicide, which is essential for maintaining tissue homeostasis. Cancer is believed to result from the failure to regulate apoptosis properly. High-intensity (>300 kV/cm), nanosecond (10 to 300 ns) pulsed electric fields (nsPEFs) are now being investigated as a tool to trigger apoptosis in cancerous tissue (Table 7.10). The hypothesized working principle behind nsPEF is cytochrome *c* release into the cytoplasm, suggesting that nsPEF targets the mitochondria, which are the initiators of apoptosis. This technique, pioneered by researchers at Old Dominion University in Norfolk, Virginia [Schoenbach et al., 2001], is still in its infancy, and its clinical efficacy, if any, is yet to be demonstrated.

Embolic Therapy

The traditional treatment for a ruptured intracranial aneurysm to prevent rebleeding is through microsurgery. The method comprises a step of clipping the neck of the aneurysm,

TABLE 7.10 Typical Parameters Used in Nanosecond Pulse Field Therapy Instruments

Clinical Application	Method of Current Delivery	Typical Waveform	Typical Current or Voltage
Induction of apoptosis via nanosecond pulsed electric fields (nsPEFs)	RF applicator in contact with tumor?	1 to 10 narrow high-voltage nanosecond-rise-time pulses, 10 to 300 ns in duration	Sufficient to yield field intensities of 50 to >300 kV/cm

TABLE 7.11 Typical Parameters Used in Embolic Therapy Instruments

Clinical Application	Method of Current Delivery	Typical Waveform	Typical Current or Voltage
Detachment of embolic device and initiation of platelet and RBC aggregation	Between guide wire (anode) and gelled surface electrode (cathode); typical impedance 1 to 4 kΩ	Dc delivered until embolic device separates from guide wire	0.1 to 1 mA with compliance of 10 V

performing a suture ligation of the neck, or wrapping the entire aneurysm. General anesthesia, craniotomy, brain retraction, and placement of a clip around the neck of the aneurysm are required in these surgical procedures. The surgical procedure is often delayed while waiting for the patient to stabilize, leading to the death of many patients from the underlying disease or defect prior to surgery.

A minimally invasive alternative to surgery involves reaching the interior of the aneurysm through a catheter introduced through a remote artery. Once the catheter is positioned, a long platinum microcoil that is fused to a guide wire is fed into the aneurism. Detachment of the coil from the guide wire is then achieved by the passage of low-voltage dc through the guide wire to hydrolyze a sacrificial link between the guide wire and coil (Table 7.11). The current along the coil also initiates platelet and RBC aggregation, promoting thrombosis. Within a short period of time after the filling of the aneurysm with the embolic coil, a thrombus forms in the aneurysm and is shortly thereafter complemented with a collagenous material that significantly lessens the potential for aneurysm rupture.

Microcurrent Stimulation and Other Energy Therapies

Microcurrent stimulators have been around since the 1960s. Proponents of the technique believe that delivering charge-balanced square waves of some 10 to 600 μA at frequencies between 0.5 and 100 Hz across the head "normalizes the activity of the nervous system" and is thus claimed to treat depression, anxiety, and insomnia. In 1976, when Congress chartered the FDA with enforcing the Federal Food, Drug, and Cosmetic (FD&C) Act, these devices were classified as cranial electrotherapy stimulation and their commercial approval was obtained through a grandfather provision of the FD&C Act. This is why these devices, although far from being proven effective by today's standards, are promoted by their manufacturers as "FDA Approved."

The alleged mechanism of operation of these devices certainly groups them along with the quackery promoted by homeopaths and chiropractors. Just look at what Electromedical Products International, probably the largest microcurrent stimulator manufacturer, claims about their devices: [The Alpha-Stim devices are] "based on the concept that the biophysics underlying the body's biochemistry also plays a significant role in regulating all of life's processes. . . . Alpha-Stim's proprietary waveform works by moving electrons through the body and brain at a variety of frequencies, collectively known as harmonic resonance. This normalizes the electrical activity of the nervous system and brain as measured by an electroencephalogram (EEG)." Controlled studies reported in reputable journals [Tan et al.,

2000] found that microcurrent stimulators do no better than placebo in treating depression, anxiety, and insomnia. When confronted, microcurrent stimulator manufacturers quickly tell you all about conspiracies against them and cite their own studies [Kirsch, 1999]. The debate for their effectiveness continues. We remain among the skeptics.

All other forms of energy therapy are either untested or unverified. Electroacupuncture and auricular acupuncture (acupuncture limited to the ear) claim to deliver currents to various "acupuncture points" to relieve pain and cure disease by correcting "imbalances" in the flow of "life energy." MORA units purportedly select, amplify, and reintroduce into the body the patient's own "good electromagnetic oscillations" filtering out the "pathological oscillations." Other unproven contraptions (e.g., Bioresonance, Rife, Zapper, and Radionic devices) claim to use sound and radio waves supposedly "tuned to the resonant frequency" of various pathogens to kill cancer cells, viruses, bacteria, and fungi. Again, we are skeptics.

REFERENCES

Anderson, D. J., K. Najafi, S. J. Tanghe, D. A. Evans, K. L. Levy, J. F. Hetke, X. Xue, J. J. Zappia, and K. D. Wise, Batch-Fabricated Thin-Film Electrodes for Stimulation of the Central Auditory System, *IEEE Transactions on Biomedical Engineering*, 36, 693–704, 1989.

Bak, M., J. P. Girvin, F. T. Hambrecht, C. V. Kufts, G. E. Loeb, and E. M. Schmidt, Visual Sensations Produced by Intracortical Microstimulation of the Human Occipital Cortex, *Medical and Biological Engineering*, 257–259, 1990.

Ben-Haim, S., N. Darvish, M. Fenster, and Y. Mika, Electrical Muscle Controller, U.S. patent 6,363,279, 2002.

Bickford, R. G., and B. D. Fremming, Neural Stimulation by Pulsed Magnetic Fields in Animals and Man, *Digest of the 6th International Conference on Medical Electronics and Biological Engineering* (Tokyo), Abstract 7-6, 1965.

Brighton, C. T., W. Wang, R. Seldes, G. Zhang, and S. R. Pollack, Signal Transduction in Electrically Stimulated Bone Cells, *Journal of Bone and Joint Surgery*, 83-A, 1514–1523, 2001.

Burkhoff, D., I. Shemer, B. Felzen, J. Shimizu, Y. Mika, M. Dickstein, D. Prutchi, N. Darvish, and S. Ben-Haim, Electric Currents Applied during the Refractory Period Can Modulate Cardiac Contractility In Vitro and In Vivo, *Heart Failure Review*, 6(1), 27–34, 2001.

Daskalov, I., N. Mudrov, and E. Peycheva, Exploring New Instrumentation Parameters for Electrochemotherapy, *IEEE Engineering in Medicine and Biology*, 62–66, January–February 1999.

Emergency Care Research Institute (ECRI), Health Technology Assessment Information Service, *Electrical Stimulation for the Treatment of Chronic Wounds*, 1996.

George, M. S., E. M. Wassermann, W. A. Williams, A. Callahan, T. A. Ketter, and P. Basser, Daily Repetitive Transcranial Magnetic Stimulation (rTMS) Improves Mood in Depression, *Neuroreport*, 6, 1853–1856, 1995.

George, M. S., E. M. Wassermann, W. A. Williams, J. Steppel, A. Pascual-Leone, and P. Basser, Changes in Mood and Hormone Levels after Rapid-Rate Transcranial Magnetic Stimulation (rTMS) of the Prefrontal Cortex, *Journal of Neuropsychiatry and Clinical Neuroscience*, 8, 172–180, 1996.

George, M. S., E. Wassermann, T. Kimbrell, J. Little, W. Williams, and A. Danielson, Mood Improvement Following Daily Left Prefrontal Repetitive Transcranial Magnetic Stimulation in Patients with Depression: A Placebo-Controlled Crossover Trial, *American Journal of Psychiatry*, 154, 1752–1756, 1997.

Hambrecht, F. T., Visual Prostheses Based on Direct Interfaces to the Visual System, *Bullieres's Clinical Neurology*, 4, 147–165, 1995.

Hochmair, E. S., An Implantable Current Source for Electrical Nerve Stimulation, *IEEE Transactions on Biomedical Engineering*, 27, 278–280, 1980.

Hodgkin, A. L., and A. F. Huxley, A Quantative Description of Membrane Current and Its Application to Conduction and Excitation in Nerve, *Journal of Physiology*, 117, 500–544, 1952.

Hofmann, G. A., S. B. Dev, and G. S. Nanda, Electrochemotherapy: Transition from Laboratory to the Clinic, *IEEE Engineering in Medicine and Biology*, 124–132, November–December 1996.

Hofmann, G. A., Apparatus for Electroporation Mediated Delivery of Drugs and Genes, U.S. patent 6,241,701, 2001.

Kirsch, D. L., *The Science behind Cranial Electrotherapy Stimulation*, Medical Scope Publishing, Edmonton, Alberta, Canada, 1999.

Konrad, M., H. Freller, and H. Friedrich, Electrode for Medical Applications, U.S. patent 4,603,704, 1984.

Kralj, A. R., and T. Bajd, *Functional Electrical Stimulation: Standing and Walking after Spinal Cord Injury*, CRC Press, Boca Raton, FL, 1989.

Lapicque, L., Définition Expérimentale de l'Excitabilité, *Comptes Rendus de l'Academic des Sciences*, 67, 280–283, 1909.

Livnat, A., R. P. Johnson, and J. E. Zehr, Programmable Miniature Backpack Stimulator for Chronic Biomedical Studies, *IEEE Transactions on Biomedical Engineering*, 28, 359–362, 1981.

Loeb, G. E., Neural Prosthetic Interfaces with the Nervous System, *Trends in Neurosciences*, 12, 195–201, 1989.

Merton, P. A., and H. B. Morton, Stimulation of the Cerebral Cortex in the Intact Human Subject, *Nature*, 285, 227, 1980.

Mika, Y., D. Prutchi, and Z. Belsky, Apparatus and Method for Timing the Delivery of Non-excitatory ETC Signals to a Heart, U.S. patent 6,263,242, 2001.

Mouchawar, G. A., J. A. Nyenhuis, J. D. Bourland, and L. A. Geddes, Guidelines for Energy-Efficient Coils: Coils Designed for Magnetic Stimulation of the Heart, *Magnetic Motor Stimulation: Basic Principles and Clinical Experience* (EEG Supplement), 43, 255–267, 1991.

Pascual-Leone, A., M. D. Catala, and P. A. Pascual-Leone, Lateralized Effect of Rapid-Rate Transcranial Magnetic Stimulation of the Prefrontal Cortex on Mood, *Neurology*, 46, 499–502, 1996.

Polson, M. J. R., A. T. Barker, and I. L. Freeston, Stimulation of Nerve Trunks with Time-Varying Magnetic Fields, *Medical and Biological Engineering and Computing*, 20, 243–244, 1982.

Prutchi, D., Y. Mika, Y. Snir, J. Ben-Arie, N. Darvish, Y. Kimchy, and S. Ben-Haim, An Implantable Device to Enhance Cardiac Contractility through Non-excitatory Signals, *Circulation*, 100 (Supplement 1), 1, 122, 1999.

Robblee, L. S., Iridium Oxide Coated Electrodes for Neural Stimulation, U.S. patent 4,677,989, 1987.

Ruohonen, J., Transcranial Magnetic Stimulation: Modelling and New Techniques, Ph.D. dissertation, Helsinki University of Technology, Espoo, Finland, 1998.

Robblee, L. S., M. M. Mangaudis, E. D. Lasinsky, A. G. Kimball, and S. B. Brummer, Charge Injection Properties of Thermally-Prepared Iridium Oxide Films, *Materials Research Society Symposium Proceedings*, 55, 303–310, 1986.

Schoenbach, K. H., S. J. Beebe, and E. S. Buescher, Intracellular Effect of Ultrashort Electrical Pulses, *Bioelectromagnetics*, 22, 440–448, 2001.

Sebille, A., P. Fontagnes, J. Legagneus, J. C. Mira, and M. Pecot-Dechavassine, Portable Stimulator for Direct Electrical Stimulation of Denervated Muscles in Laboratory Animals, *Journal of Biomedical Engineering*, 10, 371–372, 1988.

Tan, G., T. Monga, and J. Thornby, Efficacy of Microcurrent Electrical Stimulation on Pain Severity, Psychological Distress, and Disability, *American Journal of Pain Management*, 10, 35–44, 2000.

Van den Honert, C., Multichannel Electrical Stimulator with Improved Channel Isolation, U.S. patent 4,856,525, 1989.

Woodford, B. J., R. R. Carter, D. McCreery, L. A. Bullara, and W. F. Agnew, Histopathologic and Physiologic Effects of Chronic Implantation of Microelectrodes in Sacral Spinal Cord in the Cat, *Journal of Neuropathology and Experimental Neurology*, 55, 982–991, 1996.

Woodward, W. S., Optically Isolated Precision Bipolar Current Source, *Electronic Design*, 130–132, April 20, 1998.

Young, A. J., B. M. Novac, I. R. Smith, B. Lynn, and R. A. Miller, Two-Dimensional Modelling of a Double-Spiral Coil System for High Electric Field Generation for Biological Applications, *IEE Symposium on Pulsed Power 2001*, London, 19/1–19/4, 2001.

Yuen, T. G. H., W. F. Agnew, D. B. McCreery, L. A. Bullara, and M. Ingram, Lymphocyte Attraction by Electrical Stimulation, U.S. patent 6,038,478, 2000.

Yuling, X., Clinical Results of 7642 Cases of Malignant Tumors Treated with Electrochemical Therapy (ECT), *Abstracts of HIGH CARE 2000*, Audimax Ruhr-University, Bochum, Germany, February 2000.

8

CARDIAC PACING AND DEFIBRILLATION

WITH CONTRIBUTIONS BY FERNANDO BRUM, JULIO ARZUAGA,
PEDRO ARZUAGA, AND OSCAR SANZ

The heart is a pump that normally beats approximately 72 times every minute. This adds up to an impressive 38 million beats every year. The walls of the heart are made of muscle tissue. When they contract, the blood is ejected from the heart into the arteries of the body. As shown in Figure 8.1, the heart has four chambers, two on the left side and two on the right side. Each side is divided further into a receiving chamber (atrium) and a pumping chamber (ventricle). The atria and ventricles are separated by one-way valves that keep the blood flowing in the proper direction. The right side of the heart pumps blood to the lungs (via the pulmonary artery) and the left side of the heart pumps blood to the rest of the organs (via the aorta). The amount of blood the left ventricle pumps into the aorta every minute, known as *cardiac output*, is expressed in liters per minute. If cardiac output decreases in a significant manner, the body's organs are starved for oxygen. In the case of the brain, a very low cardiac output can cause lightheadedness, weakness, loss of consciousness and even death.

The heart contains two specialized types of cardiac muscle cells. The majority (around 99%) are *contractile cells* responsible for the mechanical work of pumping the heart. The second type of cardiac cells are the *autorhythmic cells*. Their function is to initiate and conduct action potentials that are responsible for the contraction of the working cells. Autorhythmic cells have pacemaker activity as opposed to a nerve or skeletal muscle cell which maintains a constant membrane potential until stimulated. Cells that display pacemaker activity have membranes that slowly depolarize between action potentials until threshold is reached, at which time the cell undergoes active depolarization, initiating an action potential. These action potentials, generated by the autorhythmic cardiac muscle cells, will then spread throughout the heart, triggering rhythmic beating without any nervous stimulation.

The coordinated contraction of the various chambers of the heart is mediated through an organized electrical conduction system within the heart. Disturbances within the electrical conduction system are the cause of all arrhythmias (rhythm disturbances). The electrical signal that initiates each normal heartbeat arises from a small structure located at the top of the right atrium called the *sinus node* or *sinoatrial node*. In a normal heart, the sinus

Design and Development of Medical Electronic Instrumentation   By David Prutchi and Michael Norris
ISBN 0-471-67623-3   Copyright © 2005 John Wiley & Sons, Inc.

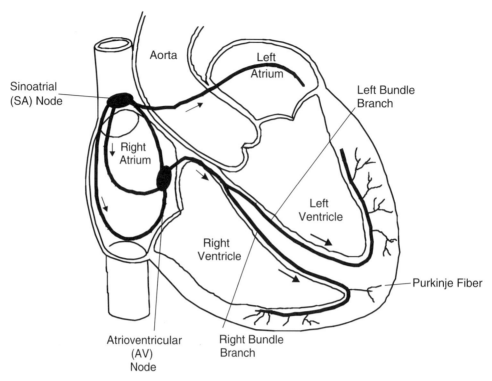

Figure 8.1 The human heart has four chambers, two on the left side and two on the right side. Each side is further divided into a receiving chamber (atrium) and a pumping chamber (ventricle). One-way valves separate the atria from the ventricles. The right side of the heart pumps blood to the lungs (via the pulmonary artery), and the left side pumps blood to the rest of the organs (via the aorta).

node acts as a natural pacemaker, setting the rate at which the heart beats. At rest, the sinus node fires 60 to 80 times per minute. When metabolic demand increases, for example due to physical activity or emotional stress, the sinus node accelerates to make the heart beat faster and increase cardiac output.

The depolarization of the sinus node creates a depolarization wave that spreads rapidly across the right atrium and the left atrium, causing them to contract. The atria are insulated electrically from the ventricles by the atrioventricular (AV) groove that runs around the outside of the heart. Electrical activity from the atria is transferred to the ventricles via a second electrical structure of the heart called the *atrioventricular node* or *AV node*, located deep in the center of the heart. The AV node is connected to the bundle of His, a bundle of specialized cells in the heart that can rapidly conduct the electrical signal to the muscle cells of the ventricles. The bundle of His branches downstream into a right bundle branch (to the right ventricle) and a left bundle branch (to the left ventricle). The fibers eventually branch out to the distant ventricular tissues and are at that point referred to as *Purkinje fibers*.

When the complete conduction system works properly, the atria contract about 200 to 300 ms ahead of ventricular contraction. This allows extra filling of the ventricles before they pump the blood through the lungs and peripheral circulation. Another important function of this system is that it allows all portions of the ventricles to contract almost simultaneously, which is essential for effective pressure generation in the ventricular chambers. As such, the heart must maintain a more or less steady rhythm in order to pump properly. Excessively slow heart rhythms make the cardiac output insufficient, causing symptoms of fatigue, weakness, lightheadedness, and loss of consciousness. On the other hand, if the heart beats too quickly, it does not get enough time in between beats to fill up with blood.

As a result, only a small amount of blood gets ejected with each beat, and cardiac output drops significantly.

Arrhythmias result from disturbances to the normal pacing and/or conduction systems of the heart. Sometimes the signal gets blocked along the way and the heart or a portion thereof does not beat when it is supposed to since the cells do not receive a stimulus to contract. This can cause slow heart rhythms, also known as *bradycardia* (from the Greek *brady* = slow + *cardia* = heart). Sometimes the electrophysiological parallel of a short circuit is formed, causing the depolarization wave to take an abnormal route or to get stuck in a reentrant circular pathway. This can cause the heart to beat rapidly, in a condition known as *tachycardia* (from the Greek, *tachy* = fast). Sometimes, a stimulus signal will arise from somewhere in the heart besides the sinus node and cause the heart to contract prematurely in what is known as a premature atrial contraction (PAC) or a premature ventricular contraction (PVC).

BRADYARRHYTHMIAS

The first obvious reason for bradycardia is that the natural pacing signal is generated by the sinus node at a low rate. Sinus bradycardia happens when the sinus node fires at less than 60 times per minute. However, sinus bradycardia is not necessarily abnormal. Healthy people often have heart rates below 60 beats/minute during rest or sleep. Athletes will usually have rates below 60 beats/minute at even moderate levels of activity because their training has made their hearts become very efficient at pumping. Sinus bradycardia is often seen in the elderly as a result of medications taken for high blood pressure, angina, or tachycardia. If there are no symptoms due to a sinus bradycardia, nothing needs to be done about it. However, a medical condition by the name of *sick sinus syndrome* arises when sinus bradycardia is associated with severe symptoms.

Sometimes the sinus node stops firing temporarily, and long pauses in the heart rhythm may result. This condition may occur because of aging, ischemia, neurological imbalances, and certain drugs. It causes symptoms of fatigue, lightheadedness, shortness of breath, and syncope. Sick sinus syndrome is usually treated with a permanent pacemaker which supplies a weak electrical signal to the atria that replaces the stimulus generated by the cyclic depolarizations of the sinus node.

Another common condition that causes bradycardia is the AV node failing to conduct atrial signals properly to the ventricles. This condition is known as *AV block* or *heart block*. The severity of AV block varies from mild to life-threatening:

- A first-degree AV block is an excessive delay (greater than 0.2 s) in the conduction of the depolarization signal from the atria to the ventricles, but all the action potentials are conducted. This benign condition is usually caused by disease in the AV node or by certain cardiac medications (Ca^+-channel blockers and β-blockers). No treatment is usually required at this stage.

- In second-degree AV block, some of the atrial signals fail to pass through the AV node. Depending on how often atrial impulses are blocked, second-degree AV block can cause symptoms that require a pacemaker implant.

- In third-degree AV block none of the atrial impulses can get through the AV node because of severe disease in cardiac conduction system. Under these conditions, ventricular contractions are initiated by autorhythmical (pacemaker) cells, whose pacemaker behavior is usually inhibited by the higher rate of the SA node, as shown in Table 8.1.

- Complete AV block usually results in severe bradycardia, and occasionally, the ventricles can stop beating, resulting in severe symptoms that may lead to death. The very first pacemakers were designed to treat patients with complete heart block.

TABLE 8.1   The Function of Autorhythmic (Pacemaker) Cells Is to Initiate and Conduct Action Potentials That Are Responsible for the Contraction of the Heart Muscle Cells[a]

Tissue	Action Potentials per Minute
SA node	60–80
AV node	40–60
Bundle of His	20–40
Purkinje fibers	20–40

[a]Since the sinoatrial node is capable of depolarizing at the highest rate, its activity usually controls the rate of the entire heart.

THE FIRST PACEMAKERS

In 1932, Alfred S. Hyman developed the first device used in the United States to stimulate the heart electronically. He called his creation an artificial *pacemaker*. His invention was used to resuscitate people who had suffered from shock or hypothermia by utilizing brief electrical currents. The Hyman device used mechanical induction of current to deliver stimuli of approximately 3 mA directly to the heart via a needle electrode at rates of 30, 60 or 120 impulses per minute. His device had to be rewound every 6 minutes. In 1952, Paul Zoll developed the first temporary external pacemaker. His device was large and limited the patient's mobility to areas where it could be wheeled and plugged to an electrical outlet. Electric shocks were coupled to the heart through electrodes placed on the skin, often causing pain and skin burns. Later, Earl Bakken modified the circuit of a transistorized metronome that he found in a back issue of *Popular Electronics* to operate as a portable pacemaker.

It was not until 1958, after low-leakage transistors became available, that an implantable pacemaker was constructed for human use. On October 8, 1958, Swedish physician Åke Senning implanted the first pacemaker in a human patient. Arne Larsson became the first recipient of the device developed by Rune Elmqvist of Elema-Schönander. The pacemaker was a hockey-puck-sized device that comprised a handful of electronic components, including two silicon transistors and a nickel–cadmium battery encapsulated in epoxy resin. The pulse generator delivered pulses of approximately 2 V in amplitude and 2 ms in duration. Batteries were charged noninvasively by RF induction of energy from a line-connected vacuum-tube radio-frequency generator driving a coil at 150 kHz. The first unit worked for only a few hours before it failed. On the next morning, a second unit was implanted and lasted for approximately six weeks.

The first successful long-term human implant of a pacemaker was achieved in Uruguay on February 2, 1960 by Orestes Fiandra and Roberto Rubio. The pacemaker was manufactured by Elmqvist and was implanted in Uruguay in a 34-year-old patient with AV block. Battery charging was done through the same RF link as that used in the first implant. One charging session, which was done overnight, was enough to power the pacemaker for about a month. This unit worked successfully for nine and a half months, until the patient died of sepsis from an infection. Other successful implants of the Swedish pacemaker followed in London. Soon after, William Chardack conducted the first human implant of a pacemaker in the United States. The device powered by 10 mercury–zinc cells was designed by Wilson Greatbatch.

In 1969, Fiandra began the Centro de Construcción de Cardioestimuladores del Uruguay (CCC) with the purpose of producing pacemakers for use in Latin America at prices well under those of U.S. devices. During the 1970s and 1980s CCC assembled pacemakers designed in the United States by Cordis. In the early 1990s, however, CCC formed an in-house design team. Today, CCC offers highly reliable pacemakers to markets that cannot afford the prices of devices manufactured in the United States or Europe.

In addition, CCC caters its design and manufacturing capabilities to companies interested in developing medical devices. Their field of expertise is in the design, prototyping, and manufacture of low-power circuitry for implantable and other critical-use medical devices.

In the sections that follow we describe the basic logic and circuitry of pacemakers. These were kindly contributed by CCC's engineering team: Fernando Brum designed the software architecture, hardware was developed by Pedro Arzuaga and Julio Arzuaga, and Oscar Sanz was responsible for the firmware.

Pacemaker State Machines

The first pacemakers were simple devices that generated a pacing pulse at a constant interval. Figure 8.2 shows a finite-state machine that represents the operation of such a pacemaker. This state machine has a single state [S]; [Time Out] is the event that causes the state machine to evolve; and [Pace] is the action that occurs as the state transition occurs. An arrowed line indicates the direction of a state transition and separates the event from actions taken during the transition.

In a simple pacemaker, the implementation of such state machine would consist of a timer with a fixed period. Every time that a length of time [Time Out] elapses, the state machine exits state [S], generates a pacing pulse (as described by the action [Pace]), and returns to state [S]. In early pacemakers, the timer's period, as well as the pacing pulse characteristics (amplitude, waveshape, and duration) were solely a function of the circuit. Take for example the circuit of Figure 8.3, which has been set up for PSpice simulation. The circuit is a replica of a 1960s design by Wilson Greatbatch. The story goes that around 1956, Greatbatch was designing a transistorized 1-kHz marker oscillator circuit to help record fast heart sounds. By mistake, he grabbed the wrong resistor from a box and plugged it into the circuit that he was making. Instead of producing the tone he expected, the circuit pulsed for 1.8 ms, stopped for 1 s, then repeated the cycle. Greatbatch recognized the "lub-dub" rhythm and the potential of the circuit for driving a sick human heart. On May 7, 1958, Greatbatch brought what would become the world's first implantable cardiac pacemaker to William Chardack and Andrew Gage. The three connected the oscillator circuit to the exposed heart of a dog. The device took control of the rate.

The blocking oscillator of the circuit is conceptually similar to that used in Greatbatch's first pacemaker. The circuit is self-starting and its output waveshape (pulse width and interval between pulses) remains almost constant despite drops in battery voltage. The circuit consumes almost no power between pulses. The original pacemaker used 10 zinc–mercury

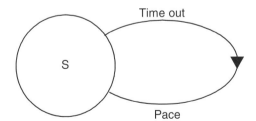

Figure 8.2   This finite-state machine represents the operation of an early pacemaker that generated pacing pulses at a constant interval. This state machine has a single state [S], [Time Out] is the event that causes the state machine to evolve, and [Pace] is the action that occurs as the state transition occurs. An arrowed line indicates the direction of a state transition and separates the event from actions taken during the transition.

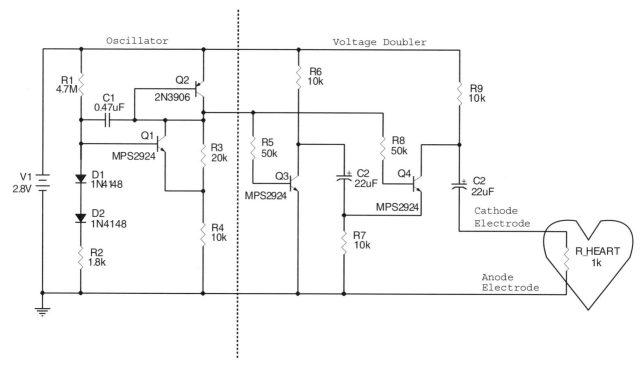

Figure 8.3 Early pacemakers had a period and pacing pulse characteristics (amplitude, waveshape, and duration) that were solely a function of their circuit. In this replica of a 1960s design by Wilson Greatbatch, a self-staring blocking oscillator drives a voltage doubler to produce pacing pulses with waveshape parameters (pulse width and interval between pulses) that remains almost constant despite drops in battery voltage.

batteries of 1.35 V and leads connecting the unit to the ventricle. The output was a 2-ms pulse of 5 to 8 V in amplitude every 1 s. Later, Greatbatch adopted the lithium–iodide battery chemistry, powering his circuits from a single 2.8-V cell. In the circuit of Figure 8.3, the output of the blocking oscillator drives a voltage doubler, making the pacing pulses delivered to the heart achieve sufficiently high amplitudes (approximately 5 V, as shown in the simulation results of Figure 8.4) for the pacing electrodes of the time to "capture" the heart.

Early pacemakers did not consider that the patient's heart could have spontaneous electrical activity. An important development in the field of cardiac pacing was the inclusion of circuitry that could detect the patient's intrinsic heart activity and pace only when the heart's rate fell below a predefined rate. Figure 8.5 shows that the logic needed to account for the patient's intrinsic activity simply requires the addition of a sense event to the state machine. When the pacemaker detects an intrinsic cardiac event, the timer in charge of issuing [Time Out] is retriggered.

In reality, however, the implementation of such a state machine is not all that simple, since it requires the inclusion of an amplifier and associated circuitry capable of detecting the heart's intrinsic activity. Since pacemaker sensing circuits usually limit their complexity to a low-power biopotential amplifier followed by a threshold detector, they detect intrinsic cardiac events based on the presence of a signal that surpasses the threshold voltage. This means that as the depolarization waveform sweeps past the pacing electrodes, the sense amplifier does not yield a single sharp transition that can be translated into a clear [Sense] event. Rather, it behaves as a very "bouncy" switch that generates a pulse train with unpredictable transitions and lasts as long as the cardiac signal remains within the range of the threshold comparator. In a similar way, the pacemaker's logic must be able to discriminate between an intrinsic beat and potentials resulting from pacing (the pacing

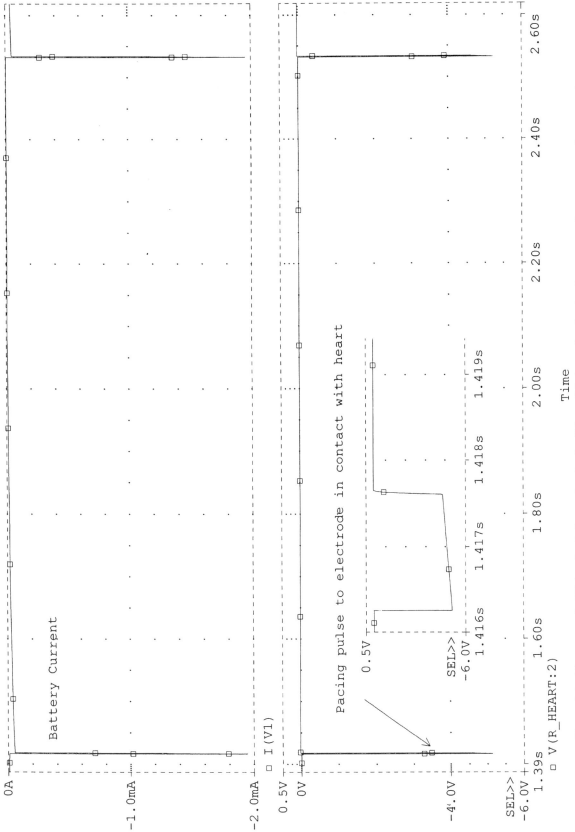

Figure 8.4 The output of a PSpice simulation on the circuit of Figure 8.3 shows that the circuit consumes almost no power between pulses. This simulation assumes use of a single lithium–iodide 2.8-V cell. The amplitude of the pacing pulses (approximately 5 V) was sufficient to "capture" the heart reliably with sub-1-V pacing pulses that last 0.5 ms.

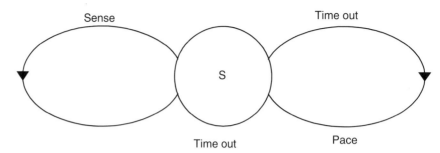

Figure 8.5   An important development was the inclusion of circuitry that could detect a patient's intrinsic heart activity and pace only when the heart's rate fell below a predefined rate. To do so requires the addition of a sense event to the state machine. In its simplest form, when the pacemaker detects an intrinsic cardiac event, the timer in charge of issuing [Time Out] is retriggered.

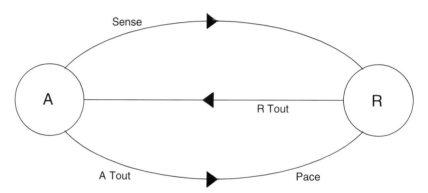

Figure 8.6   The practical implementation of the state machine of Figure 8.5 requires the inclusion of a refractory period to "debounce" the sense event detected by a simple threshold comparator that is generally used in the sense amplifier circuit. In this state machine, [A] is the "alert" state in which the pacemaker attempts to detect the heart's intrinsic electrical activity, while [R] is the refractory state in which the pacemaker ignores any external signal. The stimulus period is the sum of the time-out periods ([A Time Out] + [R Time Out]).

artifact and evoked potentials). To do so, the state machine is redesigned to incorporate the concept of a *refractory period*. In the state machine of Figure 8.6, [A] is the "alert" state in which the pacemaker attempts to detect the heart's intrinsic electrical activity, while [R] is the refractory state in which the pacemaker ignores any external signal. The stimulus period is now defined by the sum of two time-out periods [A Time Out] + [R Time Out].

An additional problem of using a simple signal detection scheme is the possibility that interfering signals, either from a source internal to the patient's body (e.g., potentials from the arm and chest muscles) or an external source (e.g., electromagnetic interference) will falsely trip the sense amplifier. One possible solution to this problem is considered later in the chapter.

Most pacemakers of the type described so far were used to stimulate the heart's ventricles in patients in whom intrinsic atrial signals would not propagate to the ventricles, most commonly due to a block in the atrioventricular conduction system. An obvious limitation of these pacemakers is their inability to make use of atrial function to enhance hemodynamic performance (the heart's capability to pump blood to the body) when the ventricles are paced. Improved physiological response of the pacemaker can be achieved by expanding the state machine to synchronize the activation of the ventricles to atrial activity. Figure 8.7 displays the state diagram of a dual-chamber pacemaker.

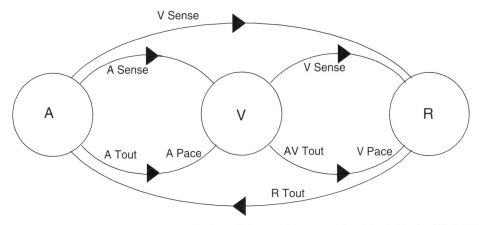

Figure 8.7   Dual-chamber pacemakers synchronize the activation of the ventricles to atrial activity. This simplified DDD pacemaker state machine can pace both atria ([A Pace]) and ventricles ([V Pace]). It has the following possible states: [A], atrial and ventricular alert, during which the pacemaker may sense signals from either or both chambers; [V], ventricular alert, during which sensing of intrinsic events is enabled only for the ventricle; and [R], refractory state. Possible events that cause transition between states are: [A Sense], sensing of an intrinsic atrial event; [V Sense], sensing of an intrinsic ventricular event; [A Time Out], maximum time that the state machine waits for an atrial sense event after exiting from the refractory period; [AV Time Out], maximum time that the state machine waits for a ventricular sense event after an atrial event; and [R Time Out], refractory period.

Possible events that cause transition between states are:

- [A Sense]   Sensing of an intrinsic atrial event
- [V Sense]   Sensing of an intrinsic ventricular event
- [A Time Out]   Maximum time that the state machine waits for an atrial sense event after exiting from the refractory period
- [AV Time Out]   Maximum time that the state machine waits for a ventricular sense event after an atrial event
- [R Time Out]   Refractory period

Possible actions taken by the state machine are:

- [A Pace]   Pacing pulse delivered to the atrium
- [V Pace]   Pacing pulse delivered to the ventricle

The states of this pacemaker state machine are:

- [A]   Atrial and ventricular alert, during which the pacemaker may sense signals from either or both chambers
- [V]   Ventricular alert, during which sensing of intrinsic events is enabled only for the ventricle
- [R]   Refractory state

If intrinsic events are not detected from the heart, the state machine's period is given by the following sequence:

1. Pacing pulse delivered to the atrium
2. AV delay ([AV Time Out])
3. Pacing pulse delivered to ventricle

4. Refractory period ([R Time Out])

5. Completely alert period ([A Time Out])

6. Go to 1.

This is an oversimplification of a real pacemaker's state machine because it does not consider the following situations that occur in clinical practice:

- Atrial tachycardia: an inappropriately high atrial rate caused by such a simplified state machine
- Noise detected by the sense amplifiers
- Need for atrial refractory of duration different from that of the ventricular refractory

As such, this state machine is presented for illustration purposes only. The description of the complete state machine of a modern pacemaker is outside the scope of this book.

PROGRAMMABLE PACEMAKERS

Modern pacemakers make it possible to program the various timeouts and pacing pulse parameters. This allows therapy to be tuned to the specific requirements of the patient. Possible programmable parameters for a microprocessor implementation of the simplified dual-chamber state machine described above are:

Timeouts

- [A Time Out]   Period during which the device is able to detect the heart's intrinsic activity
- [R Time Out]   Refractory period

Programmability of these parameters is accomplished by maintaining the programmed values in variables that can be handled by the device's firmware. Physicians are asked to enter desired values for the minimal heart rate (in beats per minute) and the refractory period (in milliseconds). The alert period is then calculated as

$$[\text{A Time Out}](ms) = \frac{60,000 \ (ms/min)}{\text{heart rate (beats/min)}} - \text{refractory period (ms)}$$

Pacing Stimulus Parameters

- *Pacing pulse width:* duration of the pacing pulse, can be implemented in the same way as timeouts
- *Pacing pulse amplitude:* initial voltage of the pacing pulse; requires the hardware to enable the firmware to adjust the pacing voltage to the desired level

Sensing Parameters

- *Atrial sensing sensitivity:* threshold voltage level (in millivolts) that the atrial electrogram signal must reach for the sense amplifier to report the occurrence of intrinsic atrial activity as an atrial sense event
- *Ventricular sensing sensitivity:* same as above, but for the ventricle

Pacing Mode This parameter selects the state machine to be used to deliver therapy. A three-letter code is used in medical nomenclature to specify a pacemaker's behavior. The first letter represents which chamber can be stimulated by the device (A, atrium; V, ventricle; D, both atrium and ventricle; O, no pacing enabled), the second represents the sensed chamber (A, atrium; V, ventricle; D, both atrium and ventricle; O, no sensing available or enabled), and the third the behavior of the device upon detecting a sensed event (I, inhibit pacing; T, trigger pacing; D, both inhibit or pace, depending on the current state of the state machine; O, disregards sensed events).

For example, the first pacemakers were asynchronous ventricular stimulators, also known as *VOO pacemakers*:

- V   Pacemaker can stimulate the ventricle.
- O   No sensing capabilities are present.
- O   There is no response to sensed events (since no sensing means are available).

The simplified state machine of Figure 8.6 corresponds to the VVI mode:

- V   Pacemaker can stimulate the ventricle.
- V   Pacemaker can sense the ventricle.
- I   Ventricular pacing is inhibited whenever the pacemaker detects a timely intrinsic ventricular event.

The simplified dual-chamber state machine of Figure 8.7 implements a simple DDD pacemaker:

- D   Both atrium and ventricle can be stimulated by the pacemaker.
- D   Both atrial and ventricular intrinsic signals can be detected by the pacemaker.
- D   Whenever timely intrinsic activity is present in both atrium and ventricle, the device inhibits pacing (I). However, when ventricular intrinsic activity does not follow the atrial activity in a timely manner, the device triggers (T) pacing on the ventricle in sequentially after the atrium.

Another mode which is commonly included in pacemakers is a *ventricular trigger* (VVT):

- V   Pacemaker can stimulate the ventricle.
- V   Pacemaker can sense intrinsic activity from the ventricle.
- T   Pacemaker stimulates the ventricle immediately following ventricular sensing.

COMMUNICATING WITH AN IMPLANTABLE DEVICE

Figure 8.8 shows how pacemakers have changed since their inception in the 1960s. The first pacemakers had a predetermined set of operating parameters. Later, the pacing rate of some early user-settable pacemakers could be changed using a fine needle screwdriver to change the setting of a potentiometer embedded within the material (e.g., epoxy and/or silicone rubber) encapsulating the pacemaker's circuitry. Today, pacemakers are programmable in a noninvasive manner using a bidirectional RF link that permits an external programmer to communicate with the implanted device's microprocessor. Operating parameters for external (temporary) pacemakers are usually selected through switches and dials.

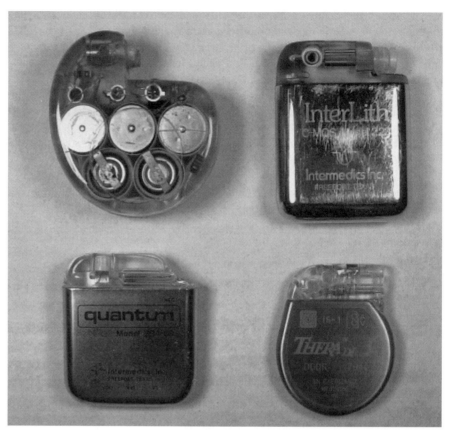

Figure 8.8 Evolution of implantable pacemakers. Top, left to right: Very early CCC VOO pacemaker powered by six mercury cells, Intermedics InterLith VVI with CMOS circuitry and powered by a lithium cell. Bottom: Intermedics "Quantum" programmable VVI/VVT pacemaker, a modern Medtronic Thera DDDR pacemaker.

Regardless of the programming method, the most convenient time for the microprocessor to read and update parameter settings is during the refractory period. Within that period, the microprocessor is free to modify timeouts and/or reconfigure the stimulation and sensing circuits. Obviously, RF communication with an implantable device is more demanding than polling a few switches in a temporary external pacemaker. The RF link requires the implementation of a hardware and software communications protocol between the implantable device's microprocessor and the microprocessor of a PC-like computer. In addition, modern programmers run graphical user interface (GUI) software to ease the programming of sophisticated devices by physicians and other medical personnel.

Since the circuitry of modern implantable devices is almost always encapsulated within a titanium case, communications is almost always achieved through a magnetically coupled link operating in the sub-100-kHz spectrum. This is a strong limiting factor to the data throughput through of the link. However, speed is not usually critical for programming operations, since the usual parameter set of an implantable pacemaker does not require the exchange of more than a few bytes per cardiac cycle, and at least 100 ms of the refractory period is available on each beat. Although communications speed is not a major constraint, the reliability of the protocol is of utmost importance. Communications protocols for implantable devices are usually designed with 100% redundance to ensure that the parameters selected by the physician are received and accepted correctly by the implantable device's microprocessor.

EXTERNAL VVI PACEMAKER

Figure 8.9 shows a simple implementation of a VVI pacemaker which will be presented as an example of the techniques involved in the design of pacemakers. The device is representative of a simple external pacemaker and is switch-programmable as follows:

- *Mode:* VVI, VOO
- *Rate:* 40, 50, 55, 60, 65, 70, 80, 90, 100, 110, 120, 140 beats/min
- *Refractory period:* 250, 350 ms
- *Pacing pulse amplitude:* battery voltage (e.g., 3 V) and two times battery voltage (e.g., 6 V)
- *Pacing pulse width:* 0.125 to 1.5 ms in 0.125-ms steps
- *Ventricular sensing sensitivity:* 1.0 to 6.0 mV in 1.0-mV steps

Although functional, the design is given only to illustrate the basic implementation of a pacemaker's circuits and state machines. The design does not include a number of important features found in modern temporary pacemakers, such as:

- Low battery detection
- "Runaway" detection (which provides protection against clock speedup which can result in pacing rates beyond 180 ppm)
- Regulated power supply
- Protection against EMI, electrosurgery, and defibrillation

As shown in Figure 8.10, the state machines that make up this pacemaker's logic run on a Microchip PIC16C76 microcontroller. This device spends the vast majority of time in the "sleep" mode and is awaken only after timeout of its counters or when a ventricular sense event is communicated via a high logic state on the circuit's input line, RB0. The timer is driven through a crystal-based oscillator running at 32.768 kHz. However, a 4-MHz *RC* clock is used to run the CPU (not the timer) once the microcontroller wakes up. The *RC* oscillator is preferable over a crystal oscillator because the former starts oscillation and stabilizes much more quickly once the microcontroller is awaken.

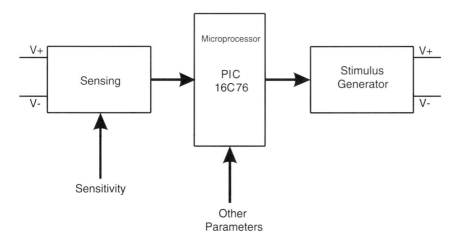

Figure 8.9 Block diagram of a simple external (temporary) VVI pacemaker circuit. Operating parameters are programmed through switches.

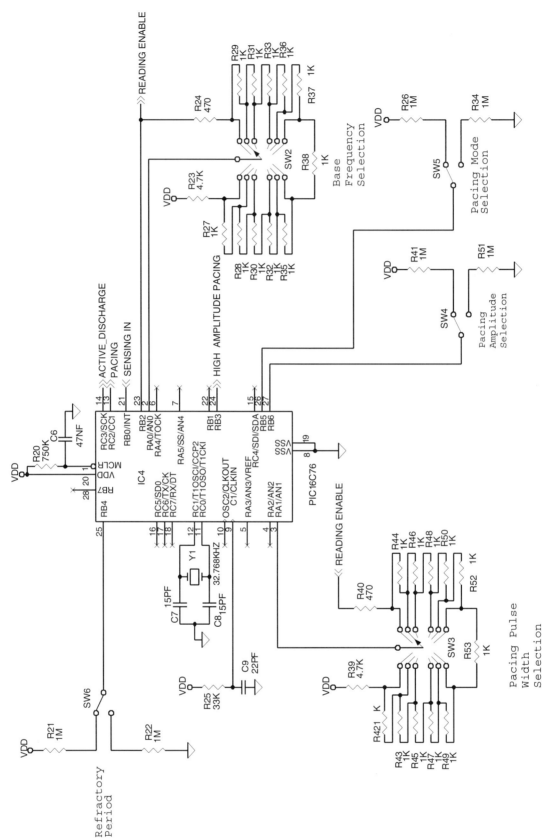

Figure 8.10 A Microchip PIC16C76 microcontroller runs the pacemaker's state machines. It spends the vast majority of time in the "sleep" mode and is awakened only after timeout of its counters or when a ventricular sense event is detected through RB0. The timer is driven through a crystal-based oscillator running at 32.768 kHz. A 4-MHz *RC* clock is used to run the CPU once the microcontroller wakes up.

Figure 8.11 presents the schematic diagram of the sense amplifier circuit. The ventricular electrogram signal detected through the electrodes attached to the ventricle is attenuated by the voltage divider formed by resistor R7 and one of the resistors among R15, R16, R17, R18, or R19. Selection is performed via five-position switch SW1, which sets the sensitivity of the sense amplifier. Past the variable attenuator, the electrogram signal is amplified by op-amp IC1. The bandpass of this amplifier is 88 to 100 Hz. The high-pass cutoff is set by the *RC* filter formed by R8 and C5. The low-pass characteristics are given by the bandwidth of the μPC4250 op-amp, which is a function of its bias set through resistor R12.

Comparators IC2 and IC3 detect whether or not the amplified and bandpassed signal has an amplitude with absolute value above 15 mV. The comparison range is defined by the divider formed by resistors R4, R6, R9, and R11. At rest, the outputs of both comparators remain at logic low. If the signal exceeds the positive threshold, the output of IC2 goes high. If the negative threshold is crossed, the output of IC3 goes high. The comparator outputs are ORed via the two diodes inside D1. A high signal at the output of the wired-OR is interpreted by microcontroller IC4 as a sensed event.

Figure 8.12 presents the schematic diagram of the of the pacing pulse generator circuit. This circuit is able to generate pacing pulses with an amplitude of 3 or 6 V, depending on the state of the pacing amplitude selector switch. When inactive, the microcontroller sets the HIGH AMPLITUDE PACING line low, which charges "tank" capacitor C2 to VDD. The PACING signal is maintained low to keep transistor Q2 open, and line ACTIVE_DISCHARGE is maintained high to keep switch Q1 open. Coupling capacitor C1 slowly discharges by way of resistor R1 (100 kΩ) through the heart's tissues and electrodes connected to terminals V+ and V−.

When a stimulus is to be generated, and if the amplitude selected is 6 V, IC4 sets HIGH AMPLITUDE PACING line high, which closes Q4 and opens Q3. This causes the positive terminal of capacitor C2 to be connected with the battery's negative terminal. When pacing at 3 V is desired, HIGH AMPLITUDE PACING is set low, which connects the C2's positive terminal to the battery's positive terminal (VDD). In the first case, the potential difference between the negative terminal of C2 and VDD is 6 V, while in the second case the potential difference is 3 V.

To deliver the stimulus to the tissue, microcontroller IC4 sets the PACING line high, which closes Q2 and connects the negative terminal of C2 to C1 (which is discharged). As such, the leading-edge voltage of the pulse appearing across electrode terminals V+ and V− is equal to the selected voltage (3 or 6 V). This voltage decays throughout the pacing pulse as C2 discharges and C1 charges. To terminate current delivery to the tissue, IC4 places all stimulus-related lines back to their rest condition.

Once the pulse has been delivered, coupling capacitor C1 remains charged. The delivery of a new pacing pulse will require this capacitor to be discharged, a procedure that is done by delivering the energy stored in this capacitor through the tissue. A net-zero current flow through the tissue is accomplished by passing the same amount of charge (albeit not within the same amount of time) through the tissue as was delivered during the stimulus pulse, but in the opposite direction. Not doing so would cause electrochemical imbalance, which can result in electrode corrosion and tissue damage. In this pacemaker, charge balancing is accomplished during the refractory period by taking line ACTIVE_DISCHARGE low, which closes Q1 allowing the charge in coupling capacitor C1 to flow through the tissue via resistors R2 and R1 (100 Ω in parallel with 100 kΩ). Any remaining charge after the fast discharge time is delivered at a slower rate through R1 (100 kΩ).

Pacing parameter selection in this demonstration external pacemaker is done through switches. As shown in Figure 8.10, parameters with only two possible values (i.e., mode, pacing pulse amplitude, and refractory period) use SPDT switches to deliver a logic high or a logic low directly to an input pin of the microcontroller. The rotary switch used for sensing sensitivity selection acts directly on the sense amplifier circuit. Rate and pacing

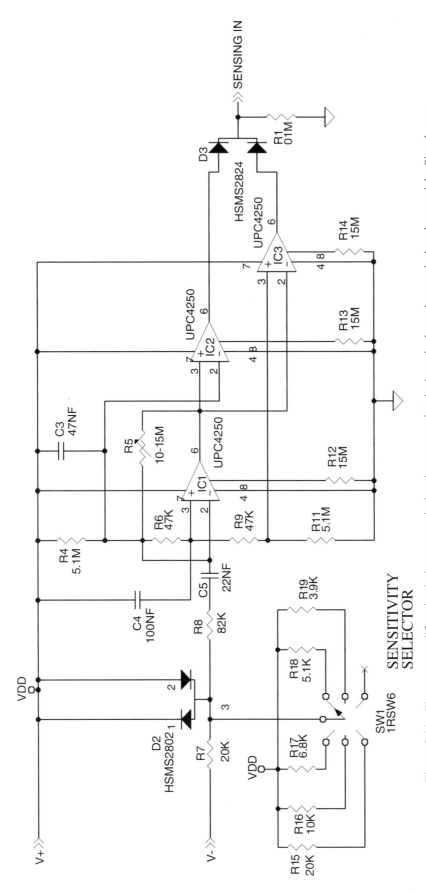

Figure 8.11 The sense amplifier circuit detects ventricular electrogram signals through electrodes attached to the ventricle. Signals are attenuated by the voltage divider formed by resistor R7 and one of the resistors among R15, R16, R17, R18, or R19. Selection of the sense amplifier sensitivity is through switch SW1. Signals within band 88 to 100 Hz are then amplified by IC1. Comparators IC2 and IC3 detect threshold crossings by the signal. Sense events are sent to microcontroller IC4 when either comparator is activated.

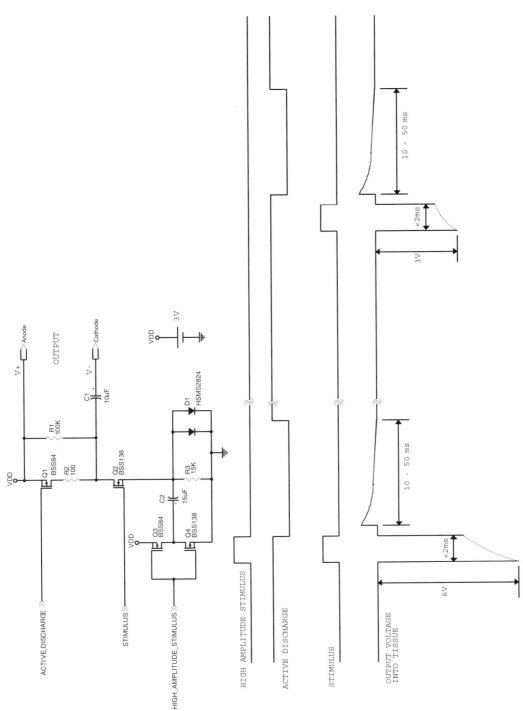

Figure 8.12 This circuit generates pacing pulses with an amplitude of 3 or 6 V, depending on the state of the pacing amplitude selector switch. When inactive, microcontroller sets the HIGH AMPLITUDE PACING line low, charging "tank" capacitor C2 to VDD. To generate a 6-V pacing pulse, HIGH AMPLITUDE PACING line is set low to connect the positive terminal C2 to the battery's negative terminal. For 3-V pacing, HIGH AMPLITUDE PACING is set high, to connect the positive terminal of C2 to VDD. A pacing pulse is delivered when the PACING line is set high, closing Q2, which connects the negative terminal of C2 to C1. Pacing current returns to VDD. Charge balancing is accomplished by closing Q1, allowing the charge in coupling capacitor C1 to flow through the tissue via resistor R2. Remaining charge is dissipated through R1 at a slower rate.

pulse width selection is accomplished through rotary switches that select a resistor used within a voltage-divider circuit fed from the battery voltage. The output of the resistive divider is measured by one of IC4's analog inputs. Different voltages are mapped by the microcontroller to the various parameter value selections.

Power for the circuit is obtained from a single nonrechargeable 3-V lithium battery (e.g., a Panasonic lithium carbon monofluoride battery). Please note that pacing pulse amplitude and sensing sensitivity vary as a function of battery voltage. Although two regular alkaline batteries in series could be used to power the circuit, the lithium carbon monofluoride chemistry has an almost flat discharge curve which minimizes the shift in the sensing threshold as battery capacity is used.

Almost all commercially available implantable pacemakers designed in the last 20 years use lithium–iodide cells (Li/I$_2$). These cells are designed to deliver current drains in the microampere range, reliably over long periods of time. They are available from Wilson Greatbatch Technologies, Inc. in a variety of sizes, shapes, and capacities. Lately, implantable-grade lithium carbon monofluoride (Li/CFx) are being used more and more in pacemakers and other implantable devices. The internal impedance of the CFx cell is much lower than that of the Li/I$_2$ cell throughout its entire life, allowing more flexibility in circuit design and performance. Wilson Greatbatch Technologies, Inc. now has Li/CFx batteries, which feature a titanium case, making it weigh half of a Li/I$_2$ cell of the same capacity.

Firmware for the VVI Pacemaker

The microcontroller runs algorithms that implement the state machine as well as stimulus routines. Firmware for pacemakers is usually coded in assembly language due to reliability concerns as well as real-time and power consumption issues. For clarity in this example, however, programming was done in C. Despite this, power consumption and real-time performance are reasonable, and use of a high-level language could be used to develop code for an implantable device.

The basic state machine for a VVI pacemaker was shown in Figure 8.6. However, enhancements are required to enable the logic to discriminate true intrinsic cardiac events from interference, such that pacing therapy is inhibited only when true ventricular activity occurs. A possible way of implementing a discrimination mechanism is to use dedicated hardware to prevent interfering signals from triggering a sense event at the microprocessor's input. For example, a retriggerable monostable together with edge-triggered sensing by the microprocessor would be able to cope with noise. However, this implementation requires additional circuitry and does not lend itself to real-time reporting of noise detection. Instead, this pacemaker design incorporates software mechanisms to detect noise and change the device's behavior to prevent noise from inappropriately inhibiting pacing therapy.

International standards that define the minimum requirements for pacemakers establish that devices must consider events detected repeating at more than 10 Hz to be noise. When such a condition is detected, a VVI pacemaker must automatically switch the mode to VOO. The device should remain in this asynchronous mode until normal sensing is resumed. Events detected at a rate below 10 Hz cannot be distinguished by simple circuitry from real cardiac events and may occasionally give rise to uncertain responses.

The state machine of Figure 8.13 is an enhanced version of the basic VVI state machine capable of detecting and responding to the presence of noise. Two new states [N] and [W] have been added. These states affect the sense condition, as well as the way in which the machine returns from the [R] state to the [A] state. The refractory period is now split in two: [R Time Out], which is an absolute refractory, which then proceeds to state [N]–a noise window within which events are sensed but not reported to the VVI state machine. Whenever a sense event occurs within state [N], the moment of occurrence is stored in time stamp variable [TS], but the machine remains in state [N] until a 100-ms timeout

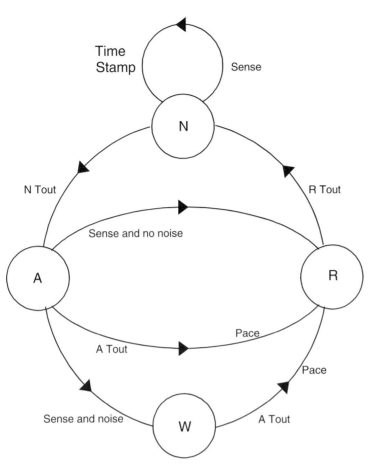

Figure 8.13 This enhanced version of the basic VVI state machine is capable of detecting and responding to the presence of noise. The refractory period is now composed of an absolute refractory [R Time Out] and noise window [N]. The presence of noise is assumed if the difference in time between sense events within [N] and [A] is equal or smaller than 100 ms, causing the machine to move to state [W] to complete the programmed period.

[Noise Time Out] elapses. By the time the machine enters the alert state [A], the complete refractory period will have elapsed, and variable [TS] will hold the time of the last sense event detected within the noise window. Under no-noise conditions, [TS] will be zero.

If no event is detected within the alert state [A], the device's behavior does not change. However, if an event is detected within [A], it is evaluated against the time of events that may have occurred within the noise window. If the difference between the time of occurrence of a sense event within the alert window and a time-stamped event is less than 100 ms, the presence of noise is assumed (since 1/100 ms = 10 Hz), and the state machine is moved to state [W] to complete the programmed period. If, on the other hand, the difference in time between the event sensed and the time-stamped event is larger than 100 ms (or no sensed events occurred within the noise window), the response to the sensed event will be to inhibit pacing and start a new cycle.

Despite the additional complexity introduced by this mechanism to the basic VVI state machine, its realization is relatively straightforward. An efficient and clear-cut method for implementing the finite-state machine is to encapsulate the event detection procedure within a routine [GetEvent()] and build an infinite loop that polls a variable that keeps track of the machine's current state.

The pacemaker state machine can then be implemented using the following code as the main software loop:

```
for( ; ; ) {
    switch( State ) {
        case ALERT:
            Event = GetEvent();
            DisableSense() ;
            switch( Event ) {
                case TOUT:
                    Pace();
                    Tout = RefractoryTout ;
                    State = REFRACTORY;
                    break;
                case SENSE:
                    if ( (TimeStamp == 0) || (SenseTime -
TimeStamp > T_100ms) ) {
                        Tout = RefractoryTout ;
                        State = REFRACTORY;
                    }
                    else {
                        Tout = AlertTout - Elapsed() ;
                        State = WAIT_END_ALERT ;
                    }
                    break;
                default:
                    State = FAILURE;
                    break;
            }
            break;
        case WAIT_END_ALERT:
            Event = GetEvent();
            switch( Event ) {
                case TOUT:
                    Pace();
                    Tout = RefractoryTout ;
                    State = REFRACTORY;
                    break;
                default:
                    State = FAILURE;
                    break;
            }
            break;
        case REFRACTORY:
            ReadParameters();
            Event = GetEvent();
            switch( Event ) {
                case TOUT:
                    EnableSense();
                    TimeStamp = 0 ;
                    Tout = T_100ms ;
                    State = NOISE_WINDOW;
                    break;
```

```
                default:
                        State = FAILURE;
                        break;
            }
            break;
    case NOISE_WINDOW:
            Event = GetEvent();
            switch( Event ) {
                case TOUT:
                        Tout = AlertTout ;
                        State = ALERT ;
                        break;
                case SENSE:
                        TimeStamp = SenseTime ;
                        Tout = Tout - Elapsed() ;
                        break;
                default:
                        State = FAILURE;
                        break;
            }
            break;
    case FAILURE:
    default:
    break;
    }
}
```

The variables for this implementation of the state machine are:

- *Unsigned char State:* keeps the current state of the machine. In addition to the various states of the machine, this variable can also have the value of FAILURE to terminate operation of the device in case of failure.
- *Unsigned char Event:* represents the event that occurred.
- *Unsigned int Tout:* global variable used to tell GetEvent() of a desired timeout.
- *Unsigned int RefractoryTout:* keeps the programmed refractory period minus 100 ms. These 100 ms correspond to the time that the machine spends in the noise window.
- *Unsigned int AlertTout:* keeps the programmed period minus the programmed refractory minus the programmed pacing pulse width.
- *Unsigned int TimeStamp:* keeps the time at which a sense event is detected within the noise window.
- *Unsigned int SenseTime:* keeps the time at which the last ventricular event is detected. It is used to calculate whether a sensed event that happens within the alert state should be classified as normal or as noise.

Auxiliary routines used by the main routine are:

- *Unsigned char GetEvent():* waits for the occurrence of an event. Its return value corresponds to the event that happened. Please note that the microcontroller will spend most of its time within this routine.
- *Void DisableSense():* prevents sense events from being reported by GetEvent(). This function could be used in an implantable device to turn off sensing circuitry when not needed in order to reduce power consumption.

- *Void EnableSense():* does the opposite of the previous routine.
- *Void Pace():* implements pacing pulse generation.
- *Unsigned int Elapsed():* returns the time elapsed while GetEvent() waited for an event to occur.
- *Void ReadParameters():* reads programmable parameters and updates them. Once this routine updates parameters, the device assumes the new parameter set at once.

The book's ftp site includes three source code file archives for firmware that can be used for this project:

1. VVI.zip implements a simple VVI state machine that disregards the VOO mode. This is the program that was discussed above.
2. VVI with embedded VOO.zip implements a VVI state machine that can be placed in VOO mode by disregarding sensed events. The state machine is identical to the one implemented in VVI.zip, but the EnableSense routine enables or disables sensing depending on the desired state.
3. VVI VOO with VOO state.zip is a VVI state machine in which the VOO state is implemented through a new state during which there is no sensing.

POWER CONSUMPTION

Together with safety and reliability, power consumption is a major design constraint in the development of implantable devices. One popular technique is to power-down circuits when they are not in use. For example, power to sense amplifiers can be shut down during the refractory period. In a similar manner, the telemetry and communications circuit can be kept off when out of range of a programmer.

Interestingly, software in an implantable device is often evaluated in connection with the power that it consumes. To keep power consumption down, clocks are kept to the lowest possible frequency, and software is usually designed to maximize the microcontroller's standby mode (using the SLEEP command). The [GetEvent()] routine implements this technique in the following manner:

```
unsigned char GetEvent()
{
      InitTout( Tout );
      Sleep();
      SenseTime = ReadTimer();
      if( INTF && INTE )  //sensing event
      {
          INTF = 0;
          return SENSE;
      }
      else if ( TMR1IF && TMR1IE )     // timeout
      {
          TMR1IF = 0;
          return TOUT;
      }
      else
          return UNKNOWN_EVENT;
      }
```

Of course, safety is by far the major concern in a medical device. Modern pacemakers incorporate fault-detection mechanisms such that if a fault condition arises, they will not cause harm to the patient. In some devices, redundancy is incorporated in the design of the pacemaker circuitry to ensure the maintenance of pacing therapy in the event of crystal oscillator, microprocessor, or other failures. When such a safety measure is implemented, the pacemaker may incorporate a backup pacing controller that runs on independent circuitry (either a separate microcontroller or dedicated logic) with conservative parameters that suffice to provide life-sustaining therapy to most patients.

The pacemaker presented in this section does not have redundant pacing capabilities. By way of example, however, the pacemaker's firmware incorporates some fault-detection mechanisms. Tripping any of the following fault detectors causes the device's pacing therapy to be deactivated:

- *RAM signature.*   Possible catastrophic RAM failures cause all volatile memory locations to be written with hex 00 or hex FF. These failures can be detected by initializing a specific volatile memory location with a test value (a "signature") and having the firmware check for the presence of this value on every cardiac cycle.

- *Parameter shadowing.*   Programmable devices keep parameters that reside in variables that should remain invariant at all times except when a programming operation takes place. For this reason, it is convenient to keep a duplicate copy of the parameter area (parameter "shadow") and check consistency of the main parameter area with its shadow on every cardiac cycle.

- *Program counter outside program area.*   Unexpected intrusions of the program counter into areas outside that used for the device's program can be detected by filling any free program memory with instructions to jump to a known safe program instruction.

SOFTWARE TESTING

Implantable pacemakers and defibrillators take care of conditions that could lead to the patient's death if left untreated. In addition, inappropriately delivered stimuli can cause dangerous and even lethal tachyarrhythmias. For this reason, testing of implantable cardiac stimulators has to be planned and performed with utmost care. Verifying that the hardware of the device works properly is relatively straightforward, since it usually involves injecting known test signals to the various subcircuits and verifying their operation. In addition, various kinds of hardware-related faults can be simulated readily using external hardware.

On the other hand, testing proper operation of the software is more involved and less of a common area for the uninitiated. Probably the most important tool for software testing starts with the design of the software itself. The key to simplifying testing is in writing clean, simple code which is as legible as possible and as close as possible to the formal language used to describe the operation of the device—in this case using finite-state machines and timer events.

In the example project presented above, testing would not only involve assessing proper operation of the various circuits and the ways in which the microprocessor controls them, but also the operation of the system as a whole, paying special attention to power consumption, operation of the [GetEvent()] routine and tests of the behavior of the device in the presence of noise. A reasonable approximation could be connecting the sensing input to a signal generator and verifying that the device enters the noise mode in the presence of signals sensed with a frequency over 10 Hz. Besides the clock, the device has a single external signal input, making it relatively easy to probe its behavior in response to different signal patterns.

RATE RESPONSIVENESS

The pacemakers described so far attempt to maintain the ventricular rate over a minimum, fixed programmed rate. However, most modern pacemakers have sensors that feed algorithms that attempt to determine what the heart rate should be from moment to moment. These sensors measure variables that are connected to the patient's activity level and/or emotional state, making it possible to calculate whether a heart rate higher than the programmed base rate should be maintained to supply the body with blood during times of physical and/or emotional stress. These pacemakers are said to be rate responsive and are identified by the letter "R" following the operational mode (e.g., DDD-R).

Rate-responsive pacemakers make use of different sensor technologies to determine the optimal heart rate. The goal is to control the heart rate of a pacemaker patient in a manner similar to the intrinsic heart rate of a healthy person with a normally functioning heart, under various conditions of rest and exercise, in a physiologically appropriate manner. The most common type of sensor used in pacemakers today is one that detects body movement. The more the patient is moving, the faster the heart rate should be. The level of activity or motion of the patient is picked up by an accelerometer or a piezoelectric crystal "microphone." Filtering and signal processing are applied to the signals detected by activity sensors to reduce the effects of disturbances unrelated to exercise which would otherwise affect the heart rate.

Another widely used method for rate responsiveness involves estimating a respiration parameter called *minute ventilation* (MV). Minute ventilation is the air volume being expired by the patient during 1 minute. Most minute ventilation sensors use the principle of impedance plethysmography, where the electrical impedance of the lung tissues is monitored and the changes in electrical impedance are interpreted as changes in the volume of air in the lungs. Higher impedance typically results from more air and less blood in the lungs following inhalation. Conversely, lower impedances result from less air and more blood in the lungs due to expiration.

Despite advances in the technology, activity sensors are inherently limited to detecting physical activity, leaving the pacemaker unresponsive to emotional stresses, which also require added hemodynamic support. In addition, activity sensors also have to make strong assumptions regarding what the patient's rate should be after exercise has stopped but when the patient still needs high levels of blood flow to let the tissues recover. For this reason, the very latest commercial pacemakers combine an activity sensor with an MV sensor to get the rate response closer to that of a healthy person.

There is still place for further developments in sensor technology for pacemakers. This is not only because better response to emotional stresses is desirable, but because the heart's own state should be considered in the calculation of rate. As such, it would be desirable to tune the parameters of a pacemaker to improve it hemodynamic state: namely, forces the heart has to develop to circulate blood through the cardiovascular system. The hemodynamic state of the heart is represented by the relationship between blood pressure and blood flow.

One promising hemodynamic sensor technology that we have worked on is the measurement of intracardiac impedance. In essence, impedance signals derived from electrodes attached to the heart contain information regarding the volume of blood held by the heart as a function of time. Intracardiac impedance measurements have been used to estimate the heart's stroke volume. The stroke volume of the heart is defined as the volume of blood expelled by the ventricle in a single beat. It is equal to the difference between the end diastolic volume (volume of blood to which the heart is filled when it relaxes) and the end systolic volume (volume of blood remaining in the heart when it reaches maximum contraction). In healthy humans, the stroke volume of the heart has been found to remain relatively constant over a wide range of exertion.

Now, cardiac output equals stroke volume multiplied by heart rate. Increases in cardiac output required to meet physiologic needs are provided primarily by increased heart rate, but some of the cardiac output during exertion is also provided by the heart, increasing its stroke volume. The stroke volume cannot increase, however, by a factor more than about 2 to $2\frac{1}{2}$. Beyond a certain rate, the heart does not fill up sufficiently, and increased heart rate beyond this point can actually result in decreases of stroke volume and ultimately, cardiac output. The point at which further increase in heart rate does not result in an increase in stroke volume is typically shifted markedly toward low rates for patients suffering from diseases of the heart muscle. The idea then is to use an impedance sensor to estimate stroke volume, and adjust the pacemaker's base rate and other parameters to optimize cardiac output.

IMPEDANCE TECHNIQUE

Impedance plethysmography (also known as *impedance rheography*) is one of the oldest applications of impedance measurement on living tissues. It is based on the fact that the impedance of body segments reflects the filling state of the blood vessels contained. This principle has been used in such diverse applications as monitoring cardiac hemodynamics (impedance cardiography), monitoring lung function and perfusion (rheopneumography), and monitoring cerebral blood flow (rheoencephalography). Since the conductivity of the body depends on the fluid content in various intracellular and extracellular compartments, body composition estimates can also be made using impedance measurements by assuming that bodily fluids subdivide the body mass into fat mass and lean body mass. In addition, impedance measurements are not limited to estimating the volume of bodily cavities. Bioimpedance techniques have also been used at the level of individual cells and small groups of cells to discriminate pathological states from changes in their equivalent resistive and capacitive parameters. For example, such changes have been detected in cancerous cells.

In the case of intracardiac impedance measurements, the impedance between two electrodes in the ventricular blood pool decreases as the ventricle is filled (since there are more conduction pathways for electrical currents), reaching a minimum at end diastole. At end systole, when the ventricle has expelled as much blood as possible, impedance measurements reach their highest values. Figure 8.14 depicts the most common methods for measuring the impedance of tissues. Here a constant-current source injects an ac current of constant amplitude into the tissue through two current-injection electrodes. This current causes a potential difference to be developed between any two points between the current-injection electrodes. This potential difference is related to the resistivity of the tissue between the voltage-sensing electrodes. The *equivalent resistance* is defined as the ratio of the voltage difference between the two voltage electrodes and the current flowing through the tissue.

Two-terminal measurements introduce some errors because the potential difference sensed between the two electrodes includes nonlinear voltages generated by the current flowing through the polarization impedance at the electrode–tissue interface. The four-electrode configuration yields a more precise measurements since the highly nonlinear effects of electrode–tissue contact impedance are reduced, as the sites of current injection and voltage measurement are physically separated. With a constant-current source, the injected boundary current becomes essentially independent of the contact impedance. Using voltage amplifiers with sufficiently high input impedance ensures that voltage measurements are virtually unaffected by the electrode–tissue contact impedance. For example,

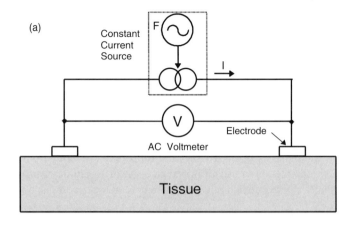

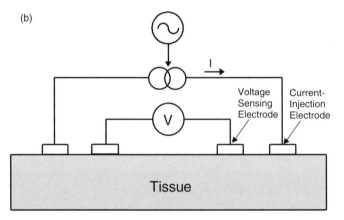

Figure 8.14   The most common method for measuring the impedance of tissues involves injecting an ac current of constant amplitude into the tissue and measuring the potential difference developed across two points between the current-injection electrodes. (*a*) Two-terminal measurement in which the impedance of the system at frequency *f* is given by Ohm's law $R = V/i$. (*b*) A four-terminal technique reduces errors of the two-terminal measurement caused by nonlinear potential differences generated at the current-injection electrodes by currents flowing through the electrode–tissue interface polarization impedance.

in an impedance cardiograph (Figure 8.15), one set of surface electrodes (usually, two pairs of gelled disk electrodes or two band electrodes) placed on the upper abdomen and upper neck are used to inject an ac current, providing more or less homogeneous coverage of the thorax with a high-frequency field. The voltage developed by the field is detected through a second set of electrodes located at the level of the root of the neck and the diaphragm.

The circuits of Figures 8.16 to 8.20 implement a general-purpose impedance plethysmograph. The circuit of Figure 8.16 is a 50-kHz sinusoidal oscillator that feeds the voltage-controlled current source of Figure 8.17 [AAMI, 1994]. The output of the *RC* oscillator formed around IC2A is feedback-stabilized by JFET transistor Q1. This transistor is controlled by the error signal between the amplitude setpoint given by R16 and the average oscillator output signal envelope calculated by the detector formed by D1 and C4.

The output of the oscillator is buffered by IC1A, scaled via potentiometer R30 and ac-coupled to modulate the current through Q3. IC3, Q2, and Q3 is a class B amplifier that is made to operate as a constant-current source. The 50-kHz input signal is adjusted via

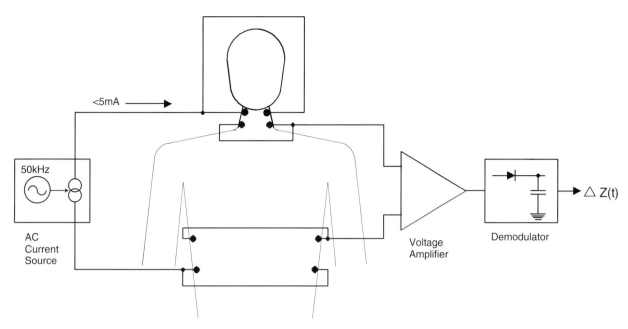

Figure 8.15   An impedance cardiograph uses two pairs of gelled disk electrodes placed on the upper abdomen and upper neck to inject an ac current, providing a more or less homogeneous coverage of the thorax with a high-frequency field. The voltage developed by the field is detected through a second set of electrodes located at the level of the root of the neck and the diaphragm. The demodulated voltage is inversely proportional to the content of fluids within the intrathoracic space. On a cardiac beat-by-beat basis, the majority of the impedance signal $\Delta Z(t)$ is caused by changes in aortic blood volume and blood velocity.

potentiometer R30 to cause the output current to have a sinusoidal waveform of sufficient amplitude (e.g., 100 µA to 1 mA) to allow detection of impedance changes associated with the phenomenon under study. Please note that this circuit is not designed to be connected to electrodes that can cause a significant portion of the current to cross the heart.

The 50-kHz voltage across the potential-measurement electrodes is demodulated using the *lock-in amplifier* circuit of Figure 8.18. A lock-in amplifier is a standard linear amplifier (in this case, IC5, an INA110 IC instrumentation amplifier that has its inputs protected by resistors R33 and R36 and by diodes D3–D6), followed by a synchronous detector (IC6). A synchronous detector has two inputs, one of which comes from the amplifier, the other being a signal of the same frequency in phase with the signal of interest. The synchronous detector is similar to an RF mixer in that the output includes the product of the two input signals. When the frequencies are the same, the output will contain a dc component whose value is proportional to the amplitude of the inputs and the sine of the phase angle between them. The reference signal is buffered through IC1B. Demodulation is maximized by correcting the reference signal phase through the phase-shifting circuit built around IC4B. The amplitude of the reference signal is controlled via potentiometer R48. The lock-in amplifier's output is scaled and low-pass filtered via the circuit built around IC4A.

Since the ac current delivered to the tissue is constant, the demodulated output voltage is related directly to the impedance of the tissue between the voltage-sensing electrodes. As shown in Figure 8.19, this signal is galvanically isolated from recording instruments via isolation amplifier IC7. A notch filter built around IC8 is available in case the impedance signals need to be filtered from power line interference. Note that a precision full-wave rectifier (Figure 8.20) built around IC14A is available for testing purposes. It provides the absolute value of the voltage generated by the current source across the current-injection electrodes.

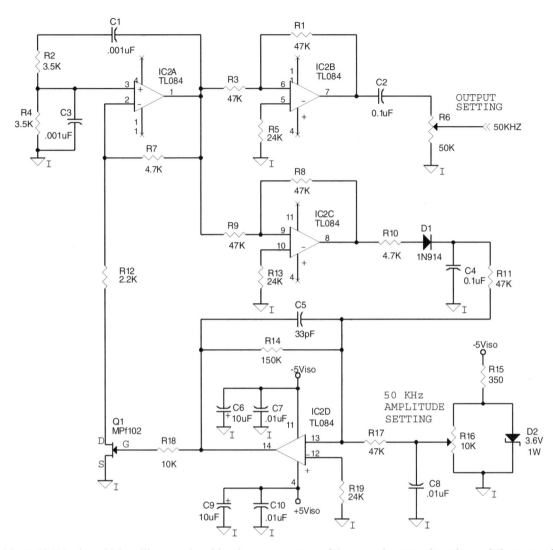

Figure 8.16 A 50-kHz sinusoidal oscillator used to drive the current source of the general-purpose impedance plethysmograph. The output of the *RC* oscillator formed around IC2A is feedback-stabilized by JFET transistor Q1, which is controlled by the error signal between the amplitude set point given by R16 and the average oscillator output signal envelope calculated by the detector formed by D1 and C4.

±15 should be supplied to the circuit from a medical-grade power supply. The isolation amplifier (IC7) generates isolated ±15 V, which are regulated via IC10 and IC13 to ±5 V used to power the 50-kHz oscillator, constant-current source, and synchronous demodulator. Alternatively, a 9-V battery can be used to power these circuits via the +5-V linear regulator IC10. If a battery is used, −5 V is generated using voltage pumps IC11 and IC12.

Measurements of ventricular volume using a "conductance catheter" are done in essentially the same way as for impedance cardiography, but using catheter-borne electrodes instead of surface electrodes. As shown in Figure 8.21, a multielectrode catheter is introduced in the left ventricle. A small (e.g., <10 μA), high-frequency (e.g., 50 kHz) constant current is injected via the two extreme electrodes, and voltages are measured by several intermediate pairs of electrodes. The idea is to divide the ventricular cavity into a number of cylindrical segments. The various voltage signals are each proportional to the volume held by each segment. The sum yields the total ventricular volume.

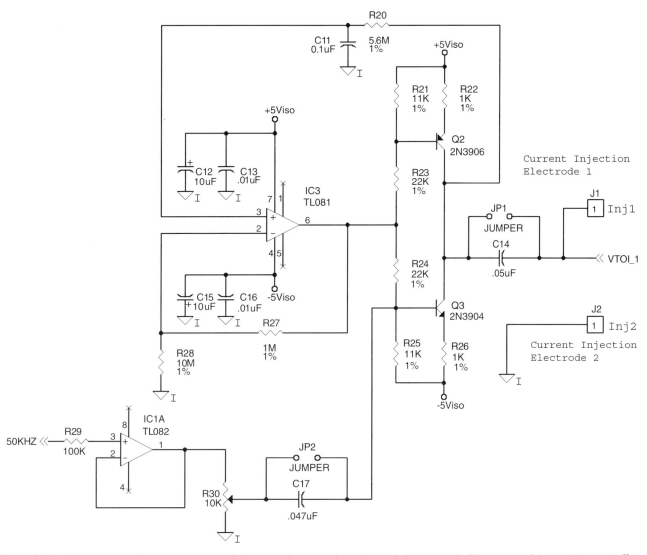

Figure 8.17 Voltage-controlled current source of the general-purpose impedance plethysmograph. The output of the oscillator is buffered by IC1A, scaled via potentiometer R30 and ac-coupled to modulate the current through Q3. IC3, Q2, and Q3 is a class B amplifier that is made to operate as a constant-current source.

Impedance measurements obtained through the conductance catheter cannot be used directly, since important intrinsic errors are generated by the nonhomogeneous distribution of the current lines within the measured volume as well as because of leakage currents through the heart muscle and extracardiac tissues. The first error affects the linear relationship between conductance and volume, while the latter adds a "phantom" parallel volume. Special calibration and signal processing techniques are usually applied to scale and linearize the impedance measurements against true ventricular volume.

Conductance catheters are often used by researchers and physicians to generate a graph of the left-ventricular pressure–volume relationship. This relationship, known as a *PV loop*, gives important information on the type of cardiomyopathy and possible treatments. Pressure is measured using a miniature sensor that is part of the conductance catheter. PV diagrams clearly distinguish the four phases of the cardiac cycle (isovolumetric contraction,

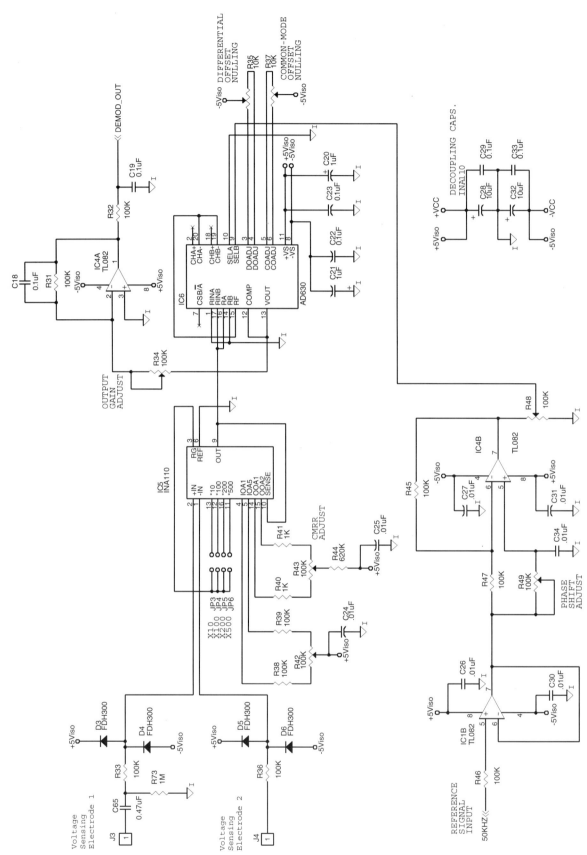

Figure 8.18 A lock-in amplifier is used to detect 50-kHz voltage signals in the general-purpose impedance plethysmograph. The circuit comprises a standard linear amplifier (IC5) followed by a synchronous detector (IC6). The reference signal is buffered through IC1B. Demodulation is maximized by correcting the reference signal phase through the phase-shifting circuit built around IC4B. The lock-in amplifier's output is scaled and low-pass filtered via the circuit built around IC4A.

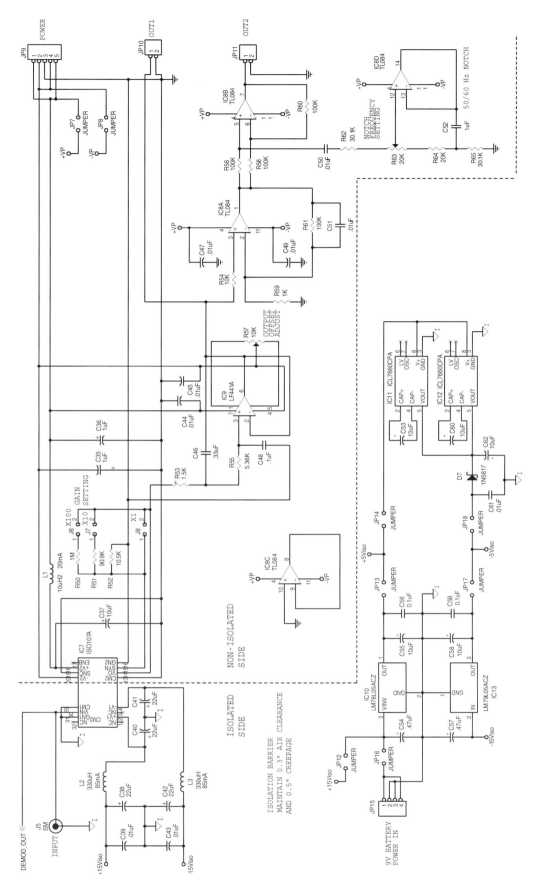

Figure 8.19 The demodulated voltage signal detected by the general-purpose impedance plethysmograph is galvanically isolated from recording instruments via isolation amplifier IC7. A notch filter built around IC8 filters power line interference. IC7 also generates isolated ±15 V, which is regulated to ±5 V to power the 50-kHz oscillator, constant-current source, and synchronous demodulator. Alternatively, a 9-V battery powers these circuits via +5-V linear regulator IC10. If a battery is used, −5 V is generated using voltage pumps IC11 and IC12.

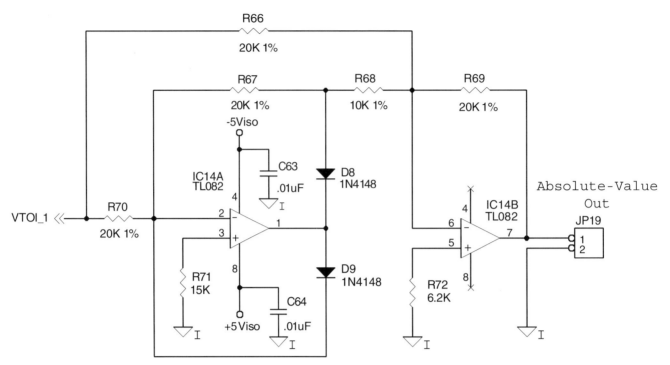

Figure 8.20 A precision full-wave rectifier built around IC14A is available for testing the general-purpose impedance plethysmograph. It provides the absolute value of the voltage generated by the current source across the current-injection electrodes.

ejection, relaxation, and passive filling) and allow estimation of the heart's contractile state (inotropy). Areas under the PV loop are related to cardiac work, and thus heart efficiency can also be calculated. Finally, the shape of the PV loop can be analyzed to assess valvular function as well as the coupling between the ventricles and the arterial load.

INTRACARDIAC IMPEDANCE SENSOR

Impedance sensors for use as hemodynamic sensors in implantable pacemakers don't need to be as accurate as those used for the generation of PV loops. Relative rather than absolute indications of volumes are usually sufficient. As such, impedance sensors used in pacemakers often make certain assumptions that simplify the volume-estimation problem at the expense of precision. The circuit represented in Figure 8.22 is a simple, yet highly effective impedance sensor suitable for implantable cardiac stimulators. This technique is known as *compensated capacitor discharge* (CCD) *impedance* sensing [Prutchi, 1996]. It makes use of very small energy probe pulses (orders of magnitude subthreshold) to estimate the resistive component of the lead impedance. The output of the circuit is an analog voltage proportional to the lead impedance.

In its simplest form, the circuit comprises a first capacitor, C_a (e.g., 0.01 μF), referred to as the *active capacitor*. At the beginning of each impedance measurement cycle, C_a is charged to a preselected voltage level V_{src} (e.g., −1 V). At the same time, a capacitor C_p (of the same value as C_a), referred to as the *passive capacitor* or *presample capacitor*, is discharged to 0 V. After C_a is fully charged to V_{src} and C_p is fully discharged, a switch connects C_p to the body, allowing it to sample the potential across the lead system and a dc-blocking capacitor C_b (e.g., 1 μF) for a brief interval t_{CCD} (e.g., 10 μs). The voltage

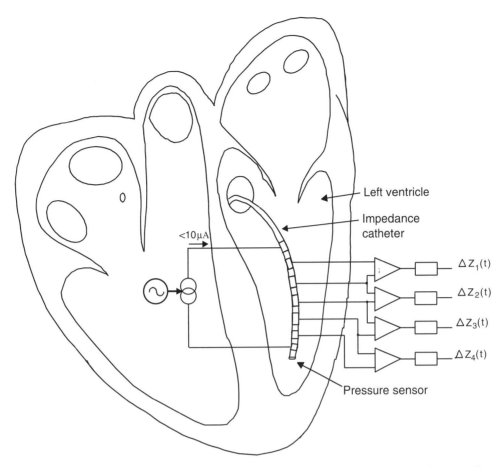

Figure 8.21 Ventricular volume can be measured using the conductance catheter. A small ($< 10\,\mu$A), high-frequency (e.g., 50 kHz) constant current is injected via the two extreme electrodes of a multielectrode catheter. Voltages are measured at several intermediate pairs of electrodes to electrically divide the ventricular cavity into a number of cylindrical segments. The various voltage signals are each proportional to the volume held by each segment. The sum yields the total ventricular volume. Pressure is measured using a miniature sensor that is part of the conductance catheter, since these measurements are used most often to determine the left-ventricular pressure–volume relationship.

resulting across C_p represents the effect of integrating charge transferred from sources in the body (e.g., the intracardiac electrogram signal) or lead system (e.g., electrode polarization potentials) on a capacitor of value $C_a = C_p$.

Immediately thereafter, C_a is discharged across the lead system by closing a switch for the same amount of time t_{CCD} that sampling was done on C_p. With no other sources in the circuit, the voltage on the active capacitor decays exponentially according to

$$V_{C_a} = V_{src} e^{-t/RC_a}$$

where V_{C_a} is the voltage remaining on the active capacitor after a time t, V_{src} the initial voltage of the active capacitor, R the lumped resistance of the circuit, and C_a the capacitance of the active capacitor. The resistance of the circuit to the narrow pulse would then be determined from

$$R = \frac{-t}{C_a \ln[V_{C_a}(t)/V_{src}]}$$

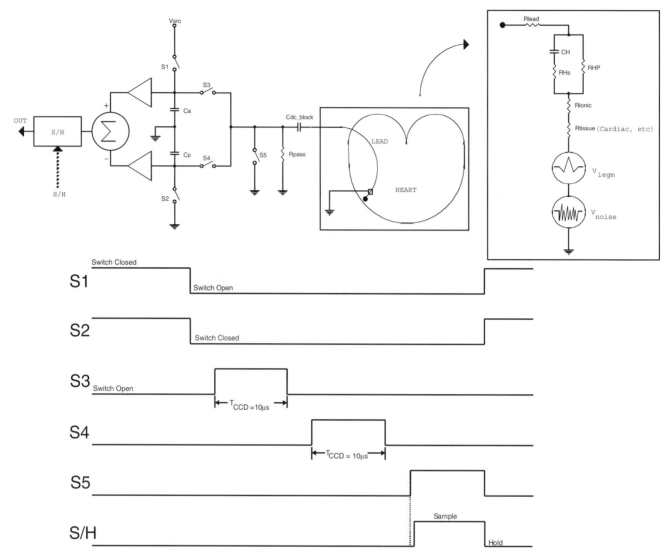

Figure 8.22   The CCD impedance sensor for implantable cardiac stimulators: (*a*) simplified circuit diagram of the sensor; (*b*) simplified timing diagram. At the beginning of each measurement cycle, C_a is charged to V_{src} while C_p is discharged. C_p is then connected to the body, allowing it to sample the potential across the lead system for a brief interval t_{CCD}. Immediately thereafter, C_a is discharged across the lead system for the same amount of time t_{CCD}. The subtraction of V_{C_p} from V_{C_a} is a value proportional to the tissue impedance,

$$R = \frac{-t_{CCD}}{C_a \ln\{[V_{C_a}(t_{CCD}) - V_{C_p}(t_{CCD})]/V_{src}\}}$$

In reality, however, other sources in the circuit (e.g., intrinsic electrical activity of the heart, electrode polarization potentials) have a strong effect on $V_{C_a}(t)$ and make the measurement of R imprecise.

By using the voltage sampled in C_p, the effects of these sources of error can be canceled. This compensation process is carried out by subtracting V_{C_p} from V_{C_a} before determining the resistive component R of the impedance:

$$R = \frac{-t}{C_a \ln\{[V_{C_a}(t) - V_{C_p}(t)]/V_{src}\}}$$

At the end of the measurement cycle, a switch is closed for interval t_{ACTD} to discharge the dc-blocking capacitor and the capacitance at the electrode interface. After this switch is opened, a fairly high value resistor, Rpas (e.g., $100\,k\Omega$), completes the "passive" discharge of these capacitances. The charge movement during both active and passive discharge causes balancing of the injected charge, resulting in a net-zero charge flow through the tissue for each measurement cycle.

Improved performance over the CCD technique is possible by modifying the impedance-measurement circuit configuration as shown in Figure 8.23. Here, a first capacitor C_{a1} (e.g., $0.01\,\mu F$), referred to as the *first active capacitor*, is charged through switch S1 to a preselected

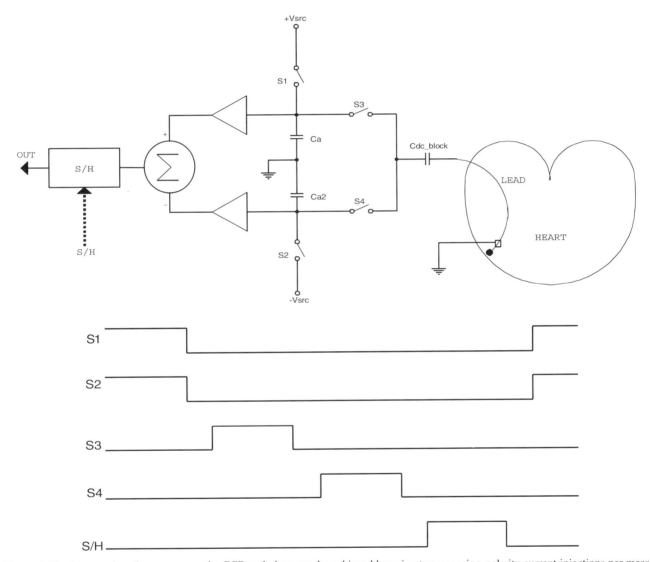

Figure 8.23   Improved performance over the CCD technique can be achieved by using two opposing-polarity current injections per measurement cycle: (*a*) simplified circuit diagram of the sensor; (*b*) simplified timing diagram. C_{a1} is charged to V_{src} and C_{a2} to $-V_{src}$. These capacitors are then discharged in sequence. The discharge of both capacitors in reverse polarity through the tissue cancels sources of error and yields twice the amount of signal as the circuit of Figure 8.22. The impedance is then given by

$$R = \frac{-t_{CCD}}{C_a \ln\{[V_{C_{a1}}(t_{CCD}) - V_{C_{a2}}(t_{CCD})]/2V_{src}\}}$$

voltage level V_{src} (e.g., $+1$ V). At the same time, a capacitor C_{a2} (of the same value as C_{a1}), referred to as the *second active capacitor*, is charged through switch S2 to the negative of the voltage level V_{src} (e.g., -1 V).

After C_{a1} is fully charged to V_{src} and C_{a2} is fully charged to $-V_{src}$, a switch (S3) connects C_{a1} to the body, allowing it to discharge through the lead system and a dc-blocking capacitor $C_{dc\ block}$ (e.g., $1 \mu F$) for a brief interval t_{CCD} (e.g., $10 \mu s$). Immediately thereafter, C_{a2} is discharged across the lead system by closing a switch (S4) for the same amount of time that C_{a1} was discharged. With no other sources in the circuit, the voltage on each active capacitor decays exponentially according to

$$V_{C_{a(i)}} = V_{src} e^{-t/RC_{a(i)}}$$

where $V_{C_{a(i)}}$ is the voltage remaining on active capacitor i after a time t, V_{src} the initial voltage of the active capacitor, R the lumped resistance of the circuit, and $C_{a(i)}$ the capacitance of the active capacitor i. The resistance of the circuit to the narrow pulse would then be determined from

$$R = \frac{-t}{C_{a(i)} \ln[V_{C_{a(i)}}(t)/V_{src}]}$$

As explained above, however, other sources in the circuit (e.g., the intracardiac electrogram, electrode polarization potentials) have a strong effect on $V_{C_{a(i)}}(t)$ and make measurement of R imprecise.

By using the discharge of both capacitors, which happens in reverse polarity through the tissue, the effects of these sources of error are virtually canceled. This compensation process is carried out by subtracting $V_{C_{a1}}$ from $V_{C_{a2}}$ before determining the resistive component R of the impedance:

$$R = \frac{-t}{C_a \ln\{[V_{C_{a1}}(t) - V_{C_{a2}}(t)]/2V_{src}\}}$$

Since the discharge polarity through the body is reversed for each phase, the subtraction of capacitor voltages results in twice the voltage signal while canceling interfering signals:

$V_{C_{a1}}(t) - V_{C_{a2}}(t) =$ voltage decay on a capacitor of size C_a due to discharge through the resistive path + effect of interference sources on a capacitor of size C_a $-$ ($-$voltage decay on a capacitor of size C_a due to discharge through the resistive path + effect of interference sources on a capacitor of size C_a)
$= 2$(voltage decay on a capacitor of size C_a due to discharge through the resistive path)

Active discharge is not needed at the end of the measurement cycle. The charge injected by each phase is substantially similar but in the opposite direction. This results in the desired net-zero charge flow through the tissue for each measurement cycle.

Also at the end of the measurement cycle, the voltage difference between the capacitors is measured and sampled via a sample-and-hold circuit. Figure 8.24 is an actual dual opposing capacitor discharge (DOCD) prototype circuit. The core circuit of the sensor is presented in Figure 8.25. In it, capacitor C_{a1} (C39) is charged to $+1.2$ V through switch IC9D and current-limiting resistor R11. At the same time, C_{a2} (C47) is charged to -1.2 V through IC9A. The ground path during this process is established through IC9C. All other

Figure 8.24   This sensor circuit generates low-energy capacitive-discharge pulses to measure two-terminal impedance across electrodes implanted in the heart. An ISO107 isolation amplifier is used for galvanic isolation of the applied part from external signal acquisition equipment.

switches are open during the charging of C_{a1} and C_{a2}. This is the normal state of the state machine implemented by the microcontroller (IC12) of Figure 8.26.

When an impedance measurement is to be performed, the external data acquisition system generates an interrupt that is received through line IMPED_START on the state-machine microcontroller IC12. Switches IC9D, IC9C, and IC9A are opened. Almost simultaneously, IC10C connects the reference terminal of the capacitors to one of the leads. Then, the other lead is connected to C_{a1} through IC10D and dc-blocking capacitor C48. The discharge of C_{a1} through the body–lead system is sampled for 10 µs. IC10D is then opened, and C_{a2} is allowed to discharge through the lead system for 10 µs by way of IC10A, C48, and IC10C.

After the samples are taken, IC10C is opened to float the lead system in relationship to the system ground. (Optionally, IC10B is closed to discharge the dc-blocking capacitor actively for 768 µs. This was left in the circuit to make it possible to implement CCD measurements.) IC9C is closed and the differential signal corresponding to the compensated impedance measurement is developed by IC8A, IC8D (unity-gain buffers, used only to preserve the charge on the active capacitors), and instrumentation amplifier IC11. At the end of 768 µs, the sample-and-hold implemented through IC9B, C44, and IC8B holds the voltage level corresponding to the impedance measurement. This level is then scaled and filtered through IC8C and its associated components.

The circuit of Figure 8.27 isolates sensor signals from recording instruments connected to the sensor. This circuit also generates isolated power for the applied part of the intracardiac impedance sensor. With the component values shown, the circuit produces an output voltage as a function of impedance (resistance) as shown in the graph of Figure 8.28. It should be noted that other implementations of the same circuit, which are particularly efficient for use in implantable medical devices, are possible, as shown in Figure 8.29.

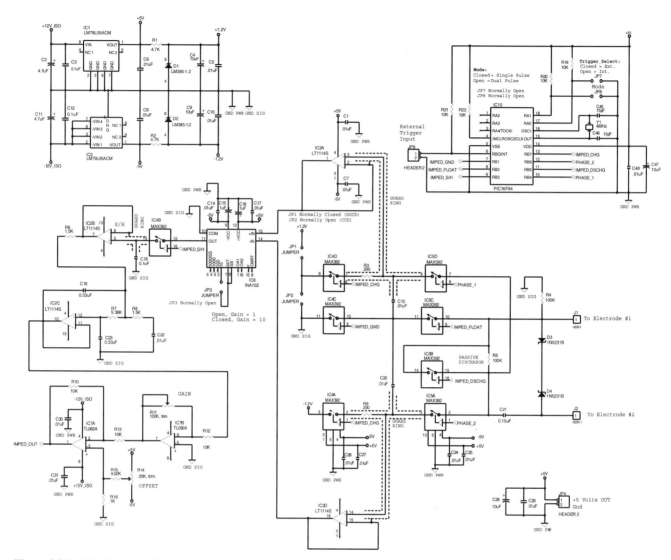

Figure 8.25 Circuit for the dual opposing capacitor discharge intracardiac impedance sensor core. C_{a1} (C39) is charged to +1.2 V through IC9D and C_{a2} (C47) is charged to −1.2 V through IC9A with ground established through IC9C. IC9D, IC9C, and IC9A then open while IC10C connects the reference terminal C39 and C47 to one of the intracardiac electrodes. C_{a1} discharges into the heart through IC10D and C48 for 10 μs. C_{a2} then discharges for 10 μs by way of IC10A, C48, and IC10C. IC10C is then opened and IC9C is closed. The differential compensated impedance measurement is generated by instrumentation amplifier IC11. A sample-and-hold circuit (IC9B, C44, and IC8B) holds the measurement. The impedance signal is scaled and filtered by IC8C.

Here most of the active signal-processing circuitry (buffers and differential amplifier) are replaced by a switched-capacitor differential to a single-ended converter, which also acts as a sample-and-hold circuit.

Figure 8.30 shows some impedance signals obtained through an instrument that uses the sensor circuit of Figure 8.24. These signals were acquired from a human subject using the electrode configuration shown in Figure 8.30*b*. Theoretical and experimental studies [Hoekstein and Inbar, 1994] have shown that the largest contributor to the impedance signal detected through pacing electrodes is the near-field movement of the cardiac walls in the largely inhomogeneous field around the distal electrode.

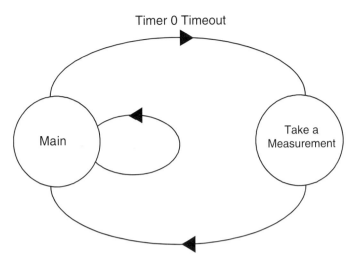

Figure 8.26 Timing of the various events of the intracardiac impedance sensor is performed by a state machine that runs on microcontroller IC12. The interrupt that causes a measurement cycle to start can be received from an external source (JP9) or from an internal clock programmed at 300 Hz.

TRANSCUTANEOUS PACING

Cardiac pacing can be achieved by delivering pacing currents through skin-surface electrodes. Transcutaneous external cardiac pacing (TEP) [Bocka, 1989] is used as a temporary life-support measure in patients with symptomatic bradycardia and a pulse, but has shown little benefit in pulseless situations. Common TEP pulse durations are between 20 and 40 ms, and pacing currents are in the range 50 to 200 mA. TEP is limited in practice to stimulating the ventricles, since higher currents are required to stimulate the atria, making it difficult (impossible, really, under clinical circumstances) to pace the chambers selectively.

Pain and discomfort are the limiting factors to TEP use. A current of 100 mA applied into an average impedance of 50 Ω for 20 ms will deliver 0.1 J of energy. This is well below the 1 to 2 J required to cause severe pain sensation at the skin using electrode pads with a large surface area (e.g., 100 cm^2). However, the force of skeletal muscle contraction is what leads to TEP discomfort. Placing electrodes over areas of least skeletal muscle can minimize discomfort. Placement is generally best in the midline chest and just below the left scapula. Sedation is often used to control discomfort.

You may be thinking about the way in which magnetic stimulation is much less uncomfortable than electrical stimulation of the brain (see Chapter 7). Pulsed magnetic cardiac pacing has been proposed [Irwin et al., 1970]. Unfortunately, however, much more energy is required to pace the heart than to stimulate a few neurons. The magnetic field energy thresholds for cardiac pacing are in the range 30 to 100 kJ. Don't know about you, but we take cover before we fire the <2-kJ magnetic stimulator that we presented in Chapter 7. The design of a pulser that can put out 100-kJ fields once a second is certainly not a trivial matter. All that hassle and the discomfort problem will probably not even go away since skeletal muscle will probably be stimulated along the way.

VENTRICULAR TACHYARRHYTHMIAS

Ventricular arrhythmias are a major cause of cardiovascular mortality. Therapy for serious ventricular arrhythmias has evolved over the past decade, from treatment with

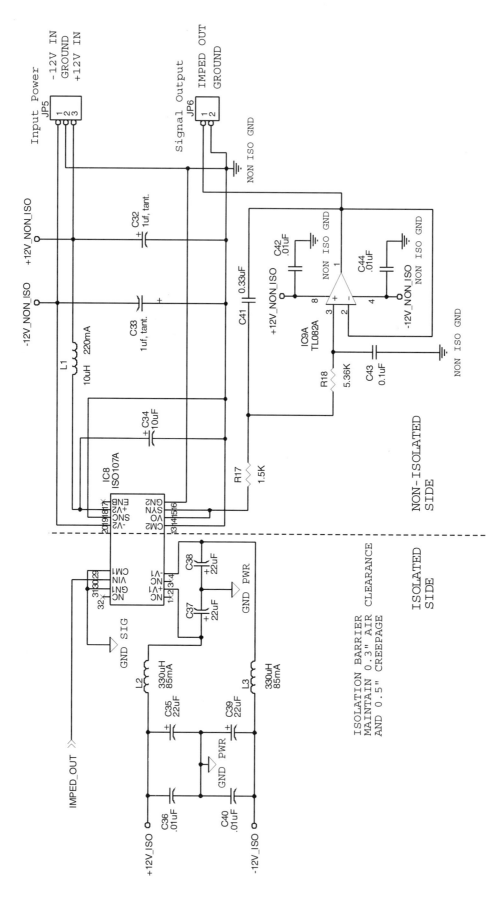

Figure 8.27 Impedance signals detected by the applied part of the intracardiac impedance sensor are galvanically isolated by IC8 from signal acquisition equipment that would be connected to JP6. A medical-grade ±12-V power supply powers the sensor. Isolated power for the applied part is generated by IC8.

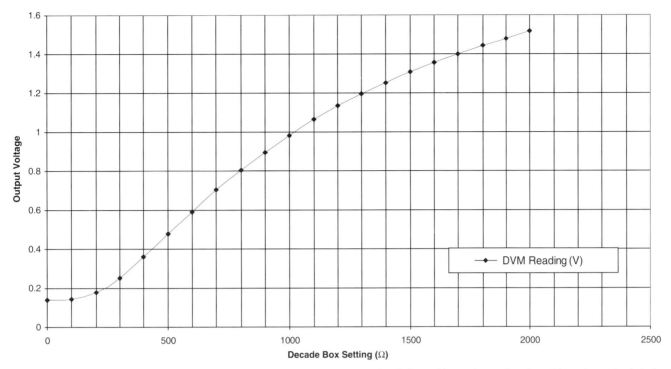

Figure 8.28 The intracardiac impedance sensor produces an output voltage that follows this graph as a function of impedance (resistive).

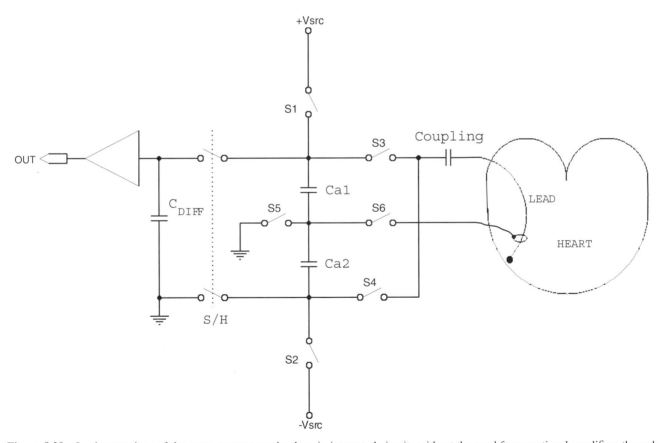

Figure 8.29 Implementations of the same concept can be done in integrated circuits without the need for operational amplifiers through the use of a switched-capacitor differential to single-ended converter, which also acts as a sample-and-hold circuit.

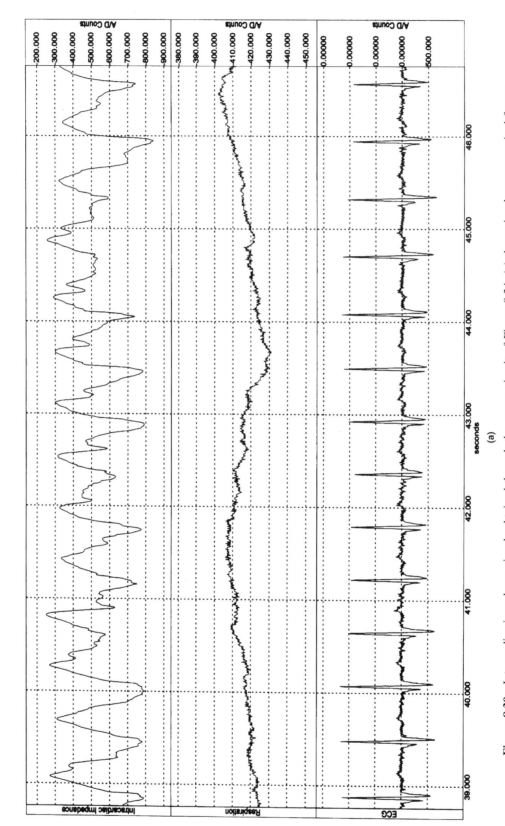

Figure 8.30 Intracardiac impedance signals obtained through the sensor circuit of Figure 8.24: (*a*) these signals were acquired from a human subject, (*b*) impedance signals were collected with a bipolar pacing lead located in the right ventricle. The largest contributor to the impedance signal detected through pacing electrodes is the near-field movement of the cardiac walls in the largely inhomogeneous field around the distal electrode.

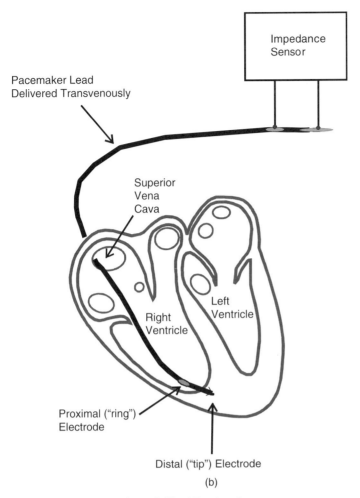

Figure 8.30 (*Continued*)

antiarrhythmic drugs to implanted devices. Ventricular tachycardia, a condition that occurs in approximately 2 out of 10,000 people, is a potentially lethal arrhythmia that often causes the heart to become inefficient at pumping blood through the body. Ventricular rates of 160 to 240 beats/minute are usually considered to be the result of a nonphysiologic tachycardia (i.e., a high rate that is not the result of the body's metabolic demand).

Ventricular tachycardia can occur spontaneously. It can also develop as a complication of a heart attack, cardiomyopathy, mitral valve prolapse, or myocarditis, and after heart surgery. It may be a result of scar tissue formed after an earlier heart attack or as an undesired effect of antiarrhythmic drugs. It may be triggered by disrupted blood chemistries (such as a low potassium level), pH (acid–base) changes, or insufficient oxygenation.

AV nodal reentry tachycardia (AVNRT) is the most common form of *paroxysmal supraventricular tachycardia* (PSVT, a tachycardia not directly of ventricular origin which comes in sudden attacks). Patients with this arrhythmia do not usually have other structural problems with their heart. PSVT originates in tissues near the AV node, which as we discussed earlier, is the electrical structure that transmits impulses between the atria and the ventricles. Susceptible persons have two pathways that can conduct impulses to and from the AV node. Under certain circumstances, usually following a premature beat, these pathways can form an electrical circuit that allows an impulse to reenter the electrical pathway repeatedly. Each reentry leads to an impulse propagation to the ventricles, and thus a rapid heart beat.

Wolff–Parkinson–White syndrome (WPW) is another tachyarrhythmia that does not originate directly in the ventricles. This type of arrhythmia involves an extra electrical pathway from the atria to the ventricles known as the *accessory pathway*. Under certain circumstances, electrical impulses may be transmitted rapidly via this additional pathway, causing palpitations and dizziness. Other times, the pathway allows atrial impulses conducted to the ventricles to reenter and restimulate the atria, leading to a rapid arrhythmia.

A real *ventricular tachycardia* (VT) is an arrhythmia that originates in the ventricular tissues. It is usually seen in patients who have damaged ventricular muscle, possibly as the result of a heart attack or myocardial infarction. Ventricular scar tissue alters many local electrical properties of the myocardium and sets up conditions favorable to the formation of local reentrant electrical circuits. When the reentrant circuit activates, it is capable of delivering rapid stimuli to the ventricles, leading to a rapid arrhythmia.

Because the reentry time of the local circuit delivers stimuli faster than the heart's natural electrical activity, it takes over the heart beat for the duration of the arrhythmia. VT most often results in poor pumping of blood by the heart because the ventricular rate is high during the arrhythmia and because the activation sequence of the chamber does not follow the normal pattern for efficient and effective pumping. VT often causes severe drops in blood pressure, which leads to unconsciousness. In its most extreme form, ventricular tachycardia can be fatal, requiring immediate medical attention.

If a critical number of reentrant loops appear, the arrhythmia degenerates into ventricular fibrillation (VF). Here, the ventricles depolarize repeatedly in an erratic, uncoordinated manner. The ECG in ventricular fibrillation shows random, apparently unrelated waves with no recognizable QRS complex. Ventricular fibrillation is often caused by drug toxicity, electrocution, drowning, and myocardial infarction. Ventricular fibrillation is almost invariably fatal because the uncoordinated contractions of ventricular myocardium result in highly ineffective pumping and little or no blood flow to the body. VF is characterized by a lack of pulse and pulse pressure and patients rapidly lose consciousness.

For a patient to survive, VT and VF require prompt termination which can be accomplished most readily by the administration of an electrical shock passed across the chest. A normal rhythm can sometimes be restored through the defibrillation current because it stimulates each myocardial cell of the ventricles to depolarize simultaneously. Following synchronous repolarization of all ventricular cells, the SA node can once again assume the role of pacemaker, and the ventricular myocardial cells can resume the essentially simultaneous depolarization of normal sinus rhythm.

DEFIBRILLATION

Virtually all modern defibrillators store energy in capacitors, which are charged to a certain voltage, depending on the energy needed. The defibrillation waveform is based on the discharge of this capacitor, either with a wave-shaping inductor (damped sine waveform) or with a switching circuit that truncates the capacitor's exponential decay (truncated exponential waveform). Defibrillation waveforms are described by the number of phases in the defibrillation waveform, the tilt of the defibrillation current, and the duration of the defibrillation waveform.

In a monophasic waveform, the polarity of the electrodes remains the same during the entire pulse, while biphasic waveforms are formed when the polarity of the electrodes reverses at some time during the pulse. *Tilt* is defined as the fractional reduction in the stimulus waveform over the pulse duration. It describes how steeply the capacitor discharges into the patient's body and gives an indication of how much of the stored energy is delivered to the tissues. Tilt is dependent on the *RC* time constant (device capacitance times the impedance of the transthoracic discharge pathway). *Duration* refers to the extent of the defibrillation waveform. Commonly, duration is in the range 8 to 40 ms.

Bare-Bones Defibrillator

Figure 8.31 shows a damped sine waveform defibrillator. A transformer T1 steps up the power line voltage to a high voltage (a few kilovolts). Capacitor C1 is charged through rectifier D1, current-limiting resistor R1, and charge switch S1 to some voltage *V* (measured through voltmeter M1) in order to store energy $E = \frac{1}{2}(C1)V^2$. When the defibrillation switch S2 closes, the defibrillation current I flows through the inductor L1 and the patient, who has a transthoracic impedance of $R_{patient}$. The discharge waveform depends on the values of C1, L1, and total impedance ($R_{inductor} + R_{patient}$). Note that the critical damping resistance of the circuit is $R_{critical} = 2\sqrt{L1/C1}$. Since defibrillators are commonly designed assuming that a patient impedance of 50 Ω, and if we assume an inductor impedance $R_{inductor}$ of 10 Ω, a suitable $R_{critical}$ could be 67 Ω. The actual energy delivered to a patient depends on patient impedance and is given by

$$E_{delivered} = E_{C1} \frac{R_{patient}}{R_{patient} + R_{inductor}}$$

For this example, the discharge is underdamped (biphasic, also referred to as a *Gurvich waveform*) when the patient resistance is less than about 56 Ω because

$$R_{patient} + R_{inductor} = 56\,\Omega + 10\,\Omega = 66\,\Omega < R_{critical} = 67\,\Omega$$

In this case, the waveform is underdamped and produces a biphasic (oscillating) waveform.

If the patient impedance is higher than 67 Ω, the waveform is overdamped (monophasic, also referred to as an *Edmark waveform*). In this case the inductor slows the rate of rise

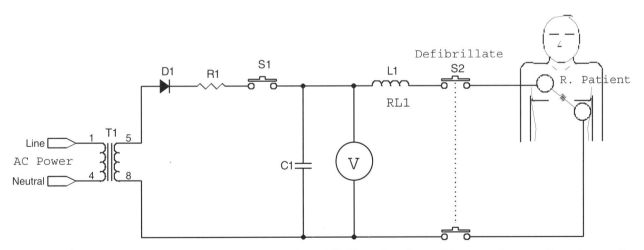

Figure 8.31 Simplified block diagram of a damped sine waveform defibrillator. Transformer T1 steps up the power line voltage to charge C1 through rectifier D1, R1, and charge switch S1 to some voltage *V* (measured through voltmeter M1) in order to store energy $E = \frac{1}{2}(C1)V^2$. When the defibrillation switch S2 closes, the defibrillation current flows through L1 and the patient.

of the discharge current, reduces the maximum voltage applied to the patient, and shapes the waveform to produce a damped sinusoidal waveform. The current delivered to the patient gradually rises to a rounded peak and drops back to zero. The discharge current pulse duration is about $2.5\sqrt{L1/C1}$, about 2.5 to 3.5 ms for most defibrillators.

The circuit would produce an exponential waveform if the inductor L1 is eliminated. However, long-duration exponential-decay waveforms are unreliable for defibrillation because the low-amplitude long-duration currents at the end of the defibrillation waveform could refibrillate the heart. If the exponential decay is truncated, however, defibrillation success is markedly increased, with an efficacy approaching that of a damped sinusoidal waveform defibrillator. Truncation allows larger-value capacitors to be used, which means that the needed energy can be stored at lower voltages. This makes it possible to use solid-state devices in the switching circuitry.

Truncated exponential waveforms are more sensitive to patient impedance changes than to damped sinusoidal waveforms. In a typical truncated exponential waveform defibrillator, current drops drops from about 27 A for a patient impedance of 50 Ω to 10 A for a patient impedance of 150 Ω. For damped sinusoidal waveform defibrillators, the peak current goes from about 60 A for a patient impedance of 50 Ω to 29 A if the patient impedance increases to 150 Ω. Considering that about 30 to 40 A is commonly necessary for successful defibrillation, both types of defibrillators are suitable for low- and medium-value patient impedances. However, damped sinusoidal waveform defibrillators are more effective in defibrillating high-impedance patients.

A simple, practical damped sinusoidal waveform defibrillator circuit is shown in Figure 8.32. This circuit is designed to deliver defibrillation energies of up to 320 J into a 50-Ω load through a 5-ms Edmark (monophasic) waveform. When the power switch SW1 is on, depressing and holding the charge pushbutton SW2 energizes the primary of high-voltage transformer T1, a of 110 V to 3 kV current-limiting transformer rated at 150 mA. The high-voltage rectifier network formed by diodes D1–D4 charges capacitor C1 through current-limiting resistor R1. Meter M1 measures the voltage across the energy storage capacitor. Its scale should be calibrated so that it provides an estimate of energy (in joules) delivered to the patient, assuming a load impedance of 50 Ω.

Once C1 is charged to the desired voltage, defibrillation energy can be delivered to the patient by pressing on pushbuttons SW3 and SW4 simultaneously. In commercial defibrillators, the insulating handles for the paddle electrodes usually house one pushbutton each. This ensures that the physician administering the defibrillation shock is in control of the discharge and that the paddle electrodes do not become energized by accident. The debouncing circuit energized by SW3 and SW4 presents 12 V dc across the coil terminals of relay K1, which is used to transfer the defibrillation charge from capacitor C1 to the patient. Charge from capacitor C1 is delivered to the patient via pulse-shaping inductor L1 and DPDT high-voltage relay K1. A suitable choice for this relay is the Kilovac Products KM-14 DPDT gas-filled "patient" relay (about $600 in low volumes).

A 5-kΩ resistor formed by R4 and R5 is connected across C1 via DPDT relay K2 during the discharge mode. The high value of this resistor has negligible effect when the pulse is being delivered to a patient. However, these resistors discharge the capacitor if the defibrillation buttons are depressed without a suitable load across the paddle electrodes. The capacitor is also discharged if the defibrillator is powered down because SW1 is turned off, the power cord is unplugged, or because safety interlock switch SW5 opens (to protect maintenance personnel from dangerous voltages when the instrument's cabinet is opened). The leakage current through the meter circuit slowly dissipates the stored energy if the defibrillation pushbutton switches SW3 and SW4 are not depressed soon after capacitor C1 is charged.

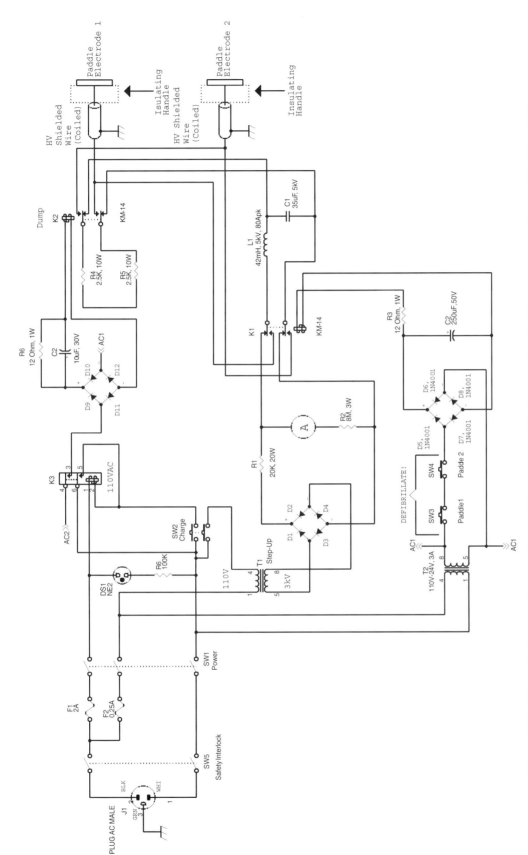

Figure 8.32 Schematic diagram of a damped sinusoidal waveform defibrillator capable of delivering energies of up to 320 J into a 50-Ω patient load through a 5-ms Edmark (monophasic) waveform. Charge pushbutton SW2 energizes high-voltage transformer T1. C1 is charged through the high-voltage rectifier network D1–D4 and R1. Meter M1 is calibrated to yield an estimate of energy (in joules) delivered to the patient, assuming a load impedance of 50 Ω. Defibrillation energy is delivered to the patient by simultaneously pressing on pushbuttons SW3 and SW4, which energize relay K1, which is used to transfer the defibrillation charge from capacitor C1 to the patient via pulse shaping inductor L1. R4 and R5 discharge C1 if the defibrillation buttons are depressed without a suitable load across the paddle electrodes or the defibrillator is powered down.

Implantable Cardioverter Defibrillators

The father of the implantable defibrillator is widely recognized to be Michael Mirowski. His work on an implantable defibrillator began in the late 1960s after his friend and mentor, Harry Heller, died as a result of ventricular arrhythmia. His concept was to develop a compact defibrillator that could provide continuous rhythm monitoring and deliver appropriate electrical shock therapy when necessary.

The first successful human implant of a totally implantable defibrillator occurred in 1980. Five years later the FDA approved the release of an implantable defibrillator for use in the United States. These early defibrillators were simple devices that would only deliver high-energy shocks to interrupt ventricular tachyarrhythmias. Much progress has been made since, and today, the implantable cardioverter–defibrillator (ICD) is the best therapy for patients who have experienced an episode of ventricular fibrillation not accompanied by an acute myocardial infarction or other transient or reversible cause. It is also superior therapy in patients with sustained ventricular tachycardia causing syncope or hemodynamic compromise.

The implantable defibrillator is connected to leads positioned inside the heart or on its surface. These leads are used to deliver the defibrillation shocks as well as to sense the cardiac rhythm and sometimes pace the heart, as needed. The various leads are tunneled to the device, which is implanted in a pouch beneath the skin of the chest or abdomen. They can be installed through blood vessels, eliminating the need for open chest surgery. Modern implantable cardioverter defibrillators have a volume of 30 to 40 cm^3. Microprocessor-based circuitry within the device continuously analyzes the patient's cardiac rhythm. When ventricular tachycardia or fibrillation is detected, the device shocks the heart to restore normal rhythm. Defibrillators are also able to provide "overdrive pacing" to convert a sustained ventricular tachycardia electrically, and pacing if bradycardia occurs. Implantable defibrillators are usually powered by internal lithium–silver vanadium oxide (Li/SVO) batteries capable of delivering ampere-level currents needed when charging the high-voltage capacitors.

Implantable defibrillators usually generate biphasic truncated exponential decay waveforms, where a second phase of opposite polarity follows the first shock phase. Biphasic waveforms are as effective as monophasic waveforms, but at lower energy and with fewer postshock complications. The mechanism by which biphasic pulses reduce the threshold voltage is probably related to the first phase of a biphasic pulse not having to synchronize as many cells, since the second phase removes the residual charge caused by the first phase, which could otherwise reinitiate refibrillation. Besides lower energy requirements, biphasic shocks are not as traumatic to the heart as monophasic shocks.

The optimal durations of the two phases are relatively independent of one another. Theoretically, the optimal first phase is identical to the optimal monophasic waveform—about 2.5 ms for a wide range of *RC* values. The optimal second-phase duration is based on the membrane time constant (about 3 ms). Despite this, several studies have shown that the second phase should be shorter than the first phase. The second phase of some biphasic waveforms is so short that it cannot reverse the transmembrane potential caused by the first phase, yet it greatly increases defibrillation efficacy. In these studies, as the second-phase duration increased, the pulse defibrillated less and less effectively and had higher voltage requirements, which implies that the second phase should be as short as possible, yet long enough to return the transmembrane potential to a level close to the one that existed before the shock.

A major difference between transthoracic (external) defibrillators (which apply currents across the chest from electrodes placed on the skin) and internal defibrillators (which apply the defibrillation currents directly to the myocardium) is the energy needed to cardiovert or defibrillate. Transthoracic defibrillation at 100 J is successful in half the cases, and defibrillation on the order of 200 to 300 J is successful in 85% of the cases. A combination of antiarrhythmic drugs and defibrillation has a 95% success rate. In contrast, internal defibrillation requires far less energy. Depending on electrode configuration, energy requirements for defibrillation can be less than 5 J. Typical implantable defibrillators can deliver a maximum of about 30 J per shock. Of course, the electrodes used for internal defibrillation, especially those used with implantable defibrillators, are different than the gelled paddles common for external defibrillators.

Originally, open-chest surgery was required to implant the large, flat patch electrodes that were sewn to the outer surface of the heart. However, the advent of the first transvenous lead systems in the early 1990s meant that physicians could maneuver the leads through a vein into the heart, eliminating the need to open the chest. Today, most implantable defibrillators use a single defibrillation lead. The defibrillator's titanium can is implanted in the upper chest, and it acts as the return electrode for the defibrillation current.

SHOCK BOX PROTOTYPE

Modern implantable defibrillators are true marvels of microelectronic packaging. Figure 8.33 shows the innards of one such device. This level of miniaturization is achieved using packaging technologies that are outside of the typical hobbyist's budget. In fact, many startup companies developing implantable devices often chose not to pursue the technologies required for miniaturization (e.g., custom ICs, chip-level packaging, ceramic substrates), and instead, use off-the-shelf components and inexpensive manufacturing technologies (e.g., surface-mounted components, low-power commercial ICs, printed circuit boards) so they can invest their efforts into developing the technologies that differentiate them from the rest of the pack. For this reason it is not easy to build an experimental implantable defibrillator. Instead, we chose to present the instrument of Figure 8.34 only as a demonstrator of the internal workings of an implantable defibrillator. It shows the considerations included in implementation of the various modules of a *shock box*, the circuitry responsible for generating high-voltage defibrillation pulses. This instrument does not include simulation of the parts of an automatic defibrillator that are responsible for detecting ventricular arrhythmias.

As shown in Figure 8.35, power for the circuit is obtained through a power line–operated medical-grade power supply. The +15-V line is used to power an isolated dc/dc converter that yields isolated 30 V dc. The 30 V is used to operate a smart gel-cell battery charger which charges two 12-V, 1.2-Ah gel-cell batteries in series. The battery powers the module's microcontroller constantly. Whenever the defibrillation module is enabled, the battery is made to power a high-voltage power supply which charges the energy-storage capacitor bank (165 µF) to a programmable level (up to 50 J).

The level of charge to be stored in the capacitor bank is selected through a digital-to-analog converter controlled by the microcontroller. The actual voltage across the capacitor bank is monitored by the microcontroller through an analog-to-digital converter which samples the voltage divider internal to the high-voltage power supply. Once charged to the desired level, the defibrillation pulse is generated by commuting the capacitor bank onto the defibrillation load through an H-bridge switch matrix. The switches in the H-bridge are under the control of the microcontroller.

Internal discharge of the capacitor bank is possible through a circuit that dumps stored charge into a dummy load. This makes it possible to discharge capacitor banks after a capacitor reform procedure[1] as well as to disarm the defibrillation module after an aborted

---

[1]The capacitance of electrolytic capacitors changes as a function of use and other factors. Whenever they are not used for some time, they require *"reforming"* such that they can be made to store the full desired charge. Reforming is accomplished by periodic charging of the capacitors to their full capacity.

Figure 8.33 Modern implantable defibrillators are true marvels of microelectronic packaging. The level of miniaturization is achieved using advanced circuit integration and packaging technologies, such as custom ICs, chip-level packaging, and ceramic substrates.

defibrillation. Voltage across the capacitor bank as well as the capacitor charge and discharge currents can be monitored using an oscilloscope or other data acquisition system by way of isolation amplifiers. The module is controlled by an onboard microcontroller. It receives parameter information and commands through an isolated RS232 line. Defibrillation commands are entered via a control computer.

In addition to running the charge and defibrillate sequences, the microcontroller also performs housekeeping functions (e.g., verify clocks, verify stored energy, perform capacitor reform). To ensure that the required energy has been stored in the capacitor bank prior to defibrillation, the microcontroller reads the voltage across the capacitor bank (using a suitable voltage divider) through an analog-to-digital converter. To enable emergency manual charging and defibrillation (i.e., not through commands from the computer), isolated pushbutton switches are available on the instrument. Whenever activated in the manual mode, the defibrillation module charges to the full energy selection, displays the charge status, and awaits for the manual command to defibrillate. Defibrillation in the manual mode is done at preset waveform parameters.

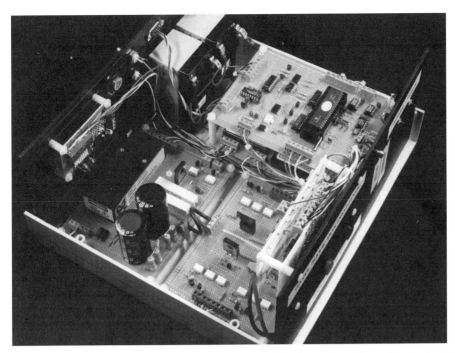

Figure 8.34 This prototype instrument is a demonstrator of the internal workings of an implantable defibrillator. This circuit is capable of delivering defibrillation energies of up to 50 J to the heart. This energy level is suitable when at least one of the electrodes is in direct contact with the ventricular muscle.

Lead impedance is estimated prior to delivery of a shock or under command from the control computer. Inappropriate lead impedances cause the defibrillation command to be aborted since delivering high energies into inappropriately low loads can be dangerous, because the developed currents pose fire or explosion risks. The different modules of the circuit are presented in the following sections. The interconnection between these modules takes place as depicted in Figure 8.36.

Power Supply Section

A Condor medical-grade (low-leakage, redundantly insulated) power supply is used to generate +15 V dc for the defibrillation module. The power line is applied to the power supply input terminals through a medical-grade connector/switch/fuse/filter/voltage-selector module. The power supply has low leakage and a redundant isolation on the power transformer, and its linear regulator is capable of producing 15 V at 0.4 A.

As shown in Figure 8.37, a Power Convertibles HB04U15D15Q (C&D Power Technologies) isolated dc/dc converter powers the battery charger circuitry. The isolation barrier of this converter is rated at 3000 V (continuous) and tested at 8000 V with a maximum 60-Hz leakage of 2 µA. This dc/dc isolation rating is the main power line leakage barrier for the applied part of the circuit. The output of the dc/dc converter is filtered through a Murata BNX002 filter block and a 470-µF capacitor (C22). A TIP42 transistor and a 33-V zener diode are configured as an "amplified zener diode" to clamp the dc/dc output of the dc/dc converter at approximately 34 V.

A single-chip gel-cell charge controller is used to charge two sealed, leakproof gel cells in series. Charging of the two 12-V batteries is controlled through a Unitrode (now owned by Texas Instruments) UC3906 IC, which implements an optimal-charge algorithm with

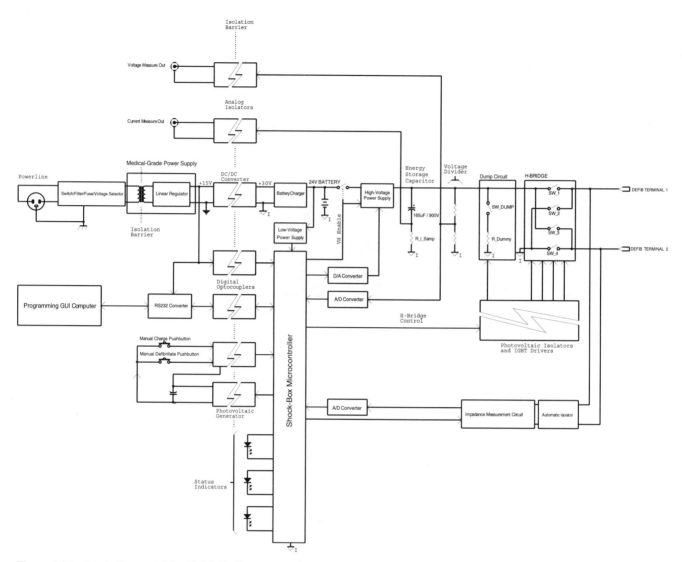

Figure 8.35   Block diagram of the 50-J defibrillator. Power is supplied by a medical-grade power supply. A current of $+15$ V from this supply operates a dc/dc converter to charge two 12-V gel-cell batteries. Battery power is supplied to the microcontroller constantly. The battery is also used to power a high-voltage power supply which charges the energy-storage capacitor bank (165 μF) up to 50 J, as selected through a DAC. A defibrillation pulse is generated by commuting the capacitor bank onto the defibrillation load through an H-bridge switch matrix. Capacitor bank voltage and current can be monitored by way of isolation amplifiers. Parameter information and commands are entered via a control computer through isolated RS232 line. The instrument includes circuitry to measure load impedance without delivering shock currents.

fast, float, and trickle-charge phases to charge the gel cells safely. Two Yuasa NP1.2-12 batteries are connected in series to generate approximately 24 V for the high-voltage power supply. Each battery has a nominal voltage of 12 V and a capacity of $C = 1.2$ Ah.

Lead–acid batteries with a gelled electrolyte are best kept charged and maintained by a charger at a float voltage of 2.25 to 2.3 V per cell. To obtain a full charge, the battery is charged to about 2.4 V per cell. Therefore, the charger IC switches back to the float level when necessary. Terminal voltages that exceed the float level place the battery in what is known as the *overcharge region*. During fast charge, the charger limits the current to a safe level known as the *bulk rate* and then tapers off to the overcharge region at about one-tenth of the bulk rate.

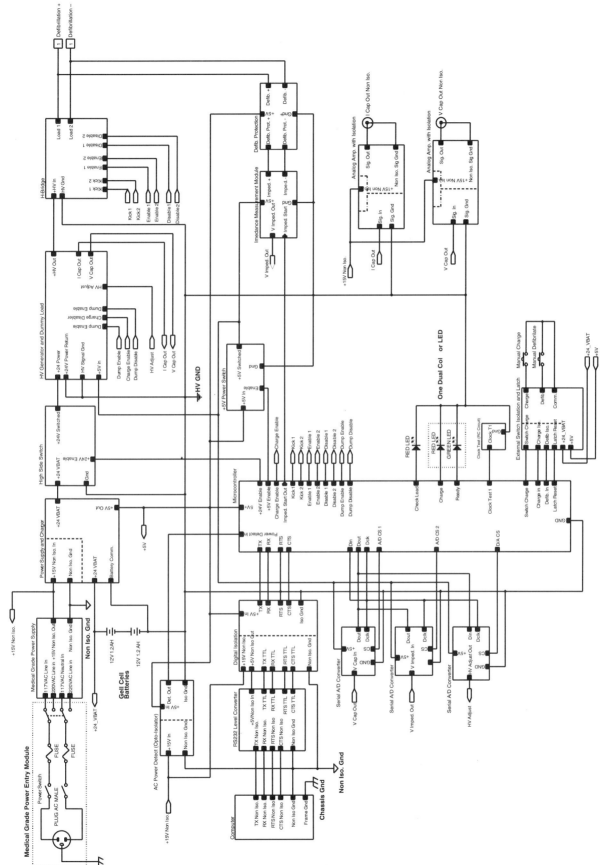

Figure 8.36 Interconnection diagram between the various modules of a shock-box instrument.

421

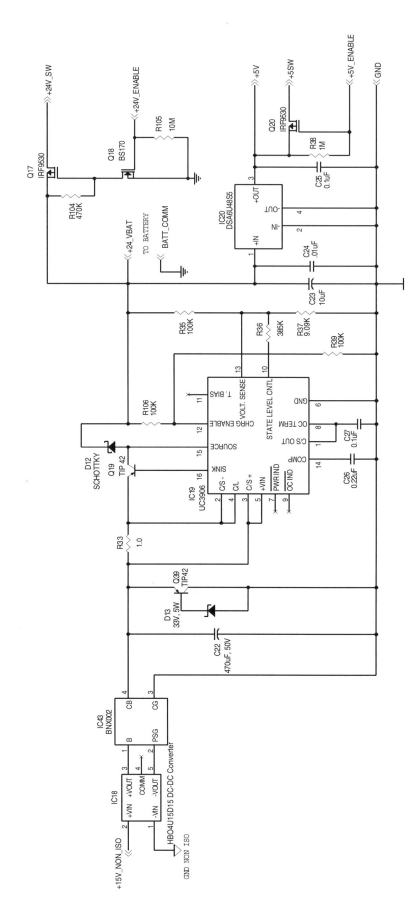

Figure 8.37 An isolated dc/dc rated at 3 kV with a maximum 60-Hz leakage of a 2-μA converter powers the battery charger circuitry. A single-chip gel-cell charge controller IC (UC3906) charges two sealed 12-V batteries. Q4 switches power to the high-voltage power supply from the +24-V battery. This switch is turned-on by Q5 upon receipt of the appropriate command from the shock-box microcontroller. IC9 produces 5 V to run the microcontroller and associated logic.

In operation, the UC3906 ensures that power is available from the dc/dc converter and that the batteries are in a good state. Pin 5 monitors the supply voltage and enables the chip when at least 4.5 V is available. Pin 12 senses the battery terminal voltage. If the voltage is too low (indicative of a dead battery or reverse polarity), the charger is disabled.

Upon detection of charging power and a good battery, the UC3906 puts two watchdogs to work: One regulates the charge current and the other looks at the battery terminal voltage. The current regulator (which uses Q19 as the power transistor) senses the voltage across series resistor R33 and limits it to 0.25 V by controlling the charging current. Thus, the bulk charging rate is determined solely by the value of this resistor. As such, the bulk charging current using a 1-Ω resistor is 250 mA. This current corresponds to a C/4.8 charging rate. A current of 250 mA is also well within the 300-mA output range of the dc/dc converter.

Battery terminal voltage sensed at pin 13 is compared to the IC's internal reference voltage. The actual terminal voltage is prescaled appropriately through R35 and R37. The resistor divider values were selected such that when the critical voltage is reached, the voltage at pin 13 equals 2.3 V. At this point, pin 10 is latched and R36 no longer participates in the circuit. When the terminal voltage rises to a level that is just below float, the voltage regulator takes control away from the bulk-current regulator and goes into the overcharge state. The current then tapers as the voltage continues to rise toward 2.4 V per cell, the point at which the float state is started.

As the current tapers, the voltage across R33 drops. Another watchdog looks at this voltage to determine when it goes below 0.025 V. When 0.025 V is sensed, a latch is toggled and pin 10 ungrounded. Float conditions are established and the battery voltage drifts back to 2.3 V per cell, which is maintained until the battery becomes discharged or the power is switched off and back on. A MOSFET (Q17) switches power to the high-voltage power supply from the +24-V battery. This switch is turned on by Q18 upon receipt of the appropriate command from the shock-box microcontroller.

High-Voltage Capacitor Charger

Implantable defibrillators typically use flyback converters to charge the energy-storage capacitor bank. Crude feedback loops are used in these devices to control the charge level. Instead of designing a custom high-voltage converter, the shock-box prototype uses an OEM module designed specifically for charging capacitor banks: the Ultravolt model IC24-P30 programmable high-voltage power supply. The high-voltage charge section is shown in Figure 8.38. This module utilizes a dual-ended forward topology with a nominal switching frequency of < 100 kHz. A soft-start circuit brings the converter to full power over a 1-ms period. A constant-frequency PWM regulation system controls a MOSFET push-pull power stage and HV transformer. The power stage is protected from output current overloads via a secondary current limit circuit. The current limit is optimized for low-impedance capacitor charging. HV ac is rectified and multiplied internally. The HV developed by this multiplier generates feedback voltage which is sent to the control circuit to maintain regulation. The ac feedback network is configured for no overshoot into capacitive loads.

The module has high efficiency (up to 92%) and requires +23 to +30 V dc to operate. The module will remain operational (derated performance) down to +9 V. The HV power output is not isolated from the input. The module produces an output that is proportional to the level presented to its control input. A voltage of 0 to +5 V at the input results in 0 to +1 kV (at 30 W) at the output. The module also has a TTL-controlled enable function. When disabled, the module remains on standby mode at <30 mA.

The module's dimensions are 3.7 in. $\times$ 1.5 in. $\times$ 0.77 in. Although the specific module used in our prototype is encased in plastic, the same module is available in an RF-tight case with a six-sided mu-metal shield. The module was originally selected by using the following formula, used to calculate the rise time required to charge an external capacitor

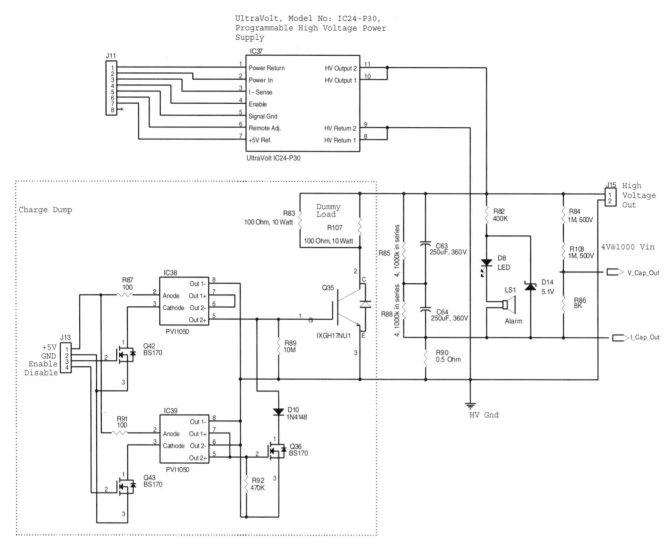

Figure 8.38   Photoflash electrolytic capacitors are the most commonly used capacitors in implantable defibrillators. Two Panasonic TS-HB-series 330-μF capacitors at 450 V are used in the shock box for an equivalent capacitance of 165 μF at 900 V. The capacitors are charged in series directly from a programmable HV power supply. 400-kΩ bleeder resistors equalize the voltage across the capacitors during charging. A blinking LED and a piezo buzzer are powered directly from the HV line to warn the user of the presence of high voltage in the capacitor bank.

load (C_L) to 99% of the desired value:

$$T(\text{ms}) = V\frac{C_L + C_{\text{int}}(\mu F)}{I_{\text{short}}(A)}$$

where $C_L = 165\,\mu F$, $C_{\text{int}} = 0.018\,\mu F$, $V = 820\,V$ (for ~50 J), and $I_{\text{short}} = 30\,\text{mA}$. As such, the worst-case charge time is just over 4 s (as long as the necessary power is available at the supply input).

Energy Storage Capacitor

Photoflash electrolytic capacitors are the most commonly used capacitors in implantable defibrillators. These capacitors have been tested thoroughly in the photoflash and strobe markets and are considered to be very reliable. However, their capacitance changes as a function of use and other factors. For this reason, whenever they are not used for some time, they require reforming such that they can be made to store the full desired charge. Reforming is accomplished by charging the capacitors to their full capacity periodically. Recently, high-voltage tantalum capacitors have been made available for use in defibrillators. These capacitors do not require reforming, and their charge acceptance remains more or less constant despite its use. However, high-voltage tantalum capacitors are very expensive (a few hundred dollars each), and they are not available in the voltage and capacitance ratings necessary for our application.

For the prototype, we used two Panasonic TS-HB-series 330-μF capacitors at 450 V in series for an equivalent capacitance of 165 μF at 900 V. As shown in the Figure 8.38, the capacitors are charged in series directly from the HV power supply. 400-kΩ bleeder resistors are used to equalize the voltage across the capacitors during charging. In addition, a 0.5-Ω resistor was added in the series connection to enable monitoring of the charge–discharge current. A circuit comprising a blinking LED and a piezo buzzer is powered directly from the HV line to warn the user (especially during experimentation with the circuit) that there is energy stored in the capacitor bank.

Switching Devices

External defibrillators (which deliver up to 400 J) use high-voltage high-current relays to connect the defibrillation load to the storage capacitor. Although they are simple to control, these devices are bulky and clearly unsuitable for an implantable device. A common way of delivering stored energy to the load in implantable defibrillators is through insulated gate bipolar transistors (IGBTs). Noteworthy properties of IGBTs are the ease of voltage control and the low losses at high voltages. These characteristics are similar to those of MOSFETs. However, the effective ON resistance of IGBTs is significantly lower than that of MOSFETs.

We selected the IXYS IXGH17N100U1 IGBT for the prototype circuit. A very similar device in bare-die is used in commercially available implantable defibrillators. This device features a second-generation HDMOS process with very low $V\_CE(SAT) = 3.5$ V for minimum ON-state conduction losses. The IXGH17N100U1 is rated for a V_CES of 1,000 V at 34 A (or 64 A for 1 ms).

The common practice in implantable devices is to drive the IGBTs (or MOSFETs) using pulse transformers and a driving circuit. For the sake of simplicity, however, this circuit uses new photovoltaic optocouplers such as the International Rectifier PVI1050. These devices generate their own dc current at the output, and as such, can be used to implement much simpler control circuits. Each PVI1050 has two photovoltaic cells driven from a single LED source. Each photovoltaic cell produces a maximum of 8 V at 10 μA.

As shown in Figure 8.39, the two photovoltaic cells inside each PVI1050 are wired in series. Two PVI1050s (IC_Enable_1 and IC_Enable_2) are wired in parallel to charge the IGBT gate capacitance to the saturation voltage. R_Gate_Bleed_IGBT is used to passively bleed charge accumulated at the gate (e.g., through the Miller effect) of the IGBT. This ensures that the IGBT remains off until turned on by activating the photovoltaic isolators.

Current for the LEDs of the photovoltaic isolators is switched under the control of the defibrillation module microcontroller through a small switching FET. Since charge buildup with constant light output takes some time, this instrument implements a technique to improve the response time of the photovoltaic isolators [Prutchi and Norris, 2002]. The current is switched on through the LED through the Q_ENABLE line and is limited to a safe continuous level through the 39- and 4.7-Ω resistors. For a brief period (100 μs), however, a very strong current is sent through the LED by way of the Q_KICK line to boost the output and yield an IGBT turn-on time of under 25 μs at full energy (full saturation).

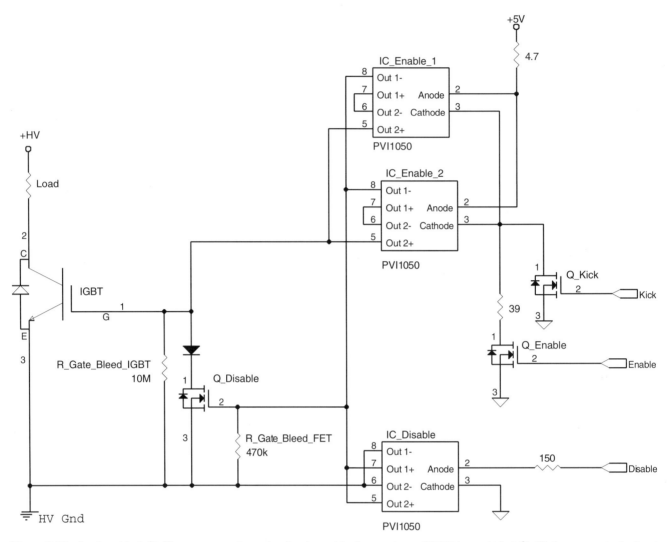

Figure 8.39 Implantable defibrillators commonly use insulated-gate bipolar transistors (IGBTs) to switch defibrillation energy to the heart. In a shock-box circuit, photovoltaic optocouplers are used to control the IGBTs. The two photovoltaic cells inside each PVI1050 are wired in series to produce up to 16 V. Two PVI1050s (U_Enable_1 and U_Enable_2) are wired in parallel to charge the IGBT gate capacitance to the saturation voltage. R_Gate_Bleed_IGBT bleeds charge accumulated at the IGBT gate to ensure that it remains off until turned on by activating the photovoltaic isolators. Fast IGBT turn-off is accomplished by discharging the IGBT gate through a FET (Q_Disable).

There is no need for further protection of the gate since the 10 MΩ loaded output of the photovoltaic cells in series/parallel is approximately 24 V, and the allowed transient VGEM of the IGBT is ±30 V. Fast IGBT turn-off is accomplished by discharging the IGBT gate through a FET (Q_DISABLE). This FET is actuated via a third photovoltaic isolator.

As shown in Figure 8.39, a switching stage is available on the prototype to enable fast dump of stored energy into a 50-Ω load. This feature may be used to deliver sequential shocks with decreasing energy (since the remaining energy of the previous shock may be larger than the energy desired for the succeeding shock) or to return the module to a safe state after aborting the delivery of a shock.

As shown in Figure 8.40, an H-bridge switch configuration was used in the prototype to generate biphasic waveforms onto the defibrillation load. Each switch acts as

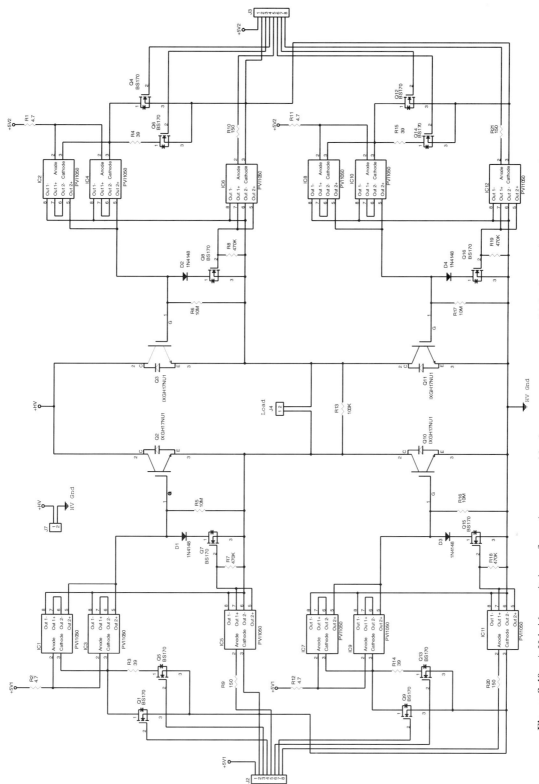

Figure 8.40 A H-bridge switch configuration was used in the prototype to generate biphasic waveforms onto the defibrillation load. Each switch acts as an independent, floating solid-state relay. A 100-kΩ resistor across the bridge output is used to maintain stability of the bridge's switches by establishing a common-mode horizon whenever there is no defibrillation load connected to the output.

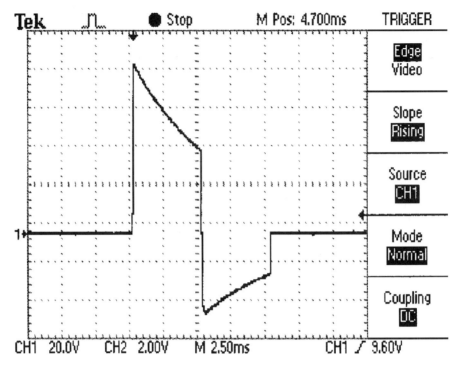

Figure 8.41 Oscilloscope display of the voltage across a 50-Ω resistive load connected to the output of the H-bridge. The capacitor bank was charged to approximately 90 V, and the circuit was made to generate a biphasic pulse with 5-ms-duration phases. The interphase interval was chosen arbitrarily to be 50 µs.

an independent, floating solid-state relay. A 100 kΩ resistor across the bridge output is used to maintain stability of the bridge's switches by establishing a common-mode horizon whenever there is no defibrillation load connected to the output. It must be noted that the photovoltaic isolators are not used as part of the applied-part isolation (safety) barrier. Rather, they are used to simplify the driving circuitry for the H-bridge, since they act as floating sources that establish a potential difference only across the gate–emitter of their related IGBT.

Figure 8.41 presents the oscilloscope display of the voltage across a 50-Ω resistive load connected to the output of the H-bridge. For this figure the capacitor bank was charged to approximately 90 V, and the circuit was made to generate a biphasic pulse with 5-ms-duration phases. The interphase interval was arbitrarily chosen to be 50 µs.

Isolation Amplifiers

As shown in Figure 8.42, an AD210AN isolation amplifier (IC29) is used to sample the voltage across the capacitor bank (using the voltage divider formed by resistors R84 and R86 on the high-voltage capacitor charging circuit). A second AD210AN (IC35) samples the instantaneous capacitor charge–discharge current by sampling the voltage developed across the 0.5-Ω current-measurement resistor (R90 on the high-voltage capacitor charging circuit). The output of the AD210s is made available through BNC connectors to make it possible to sample and analyze the charge–discharge waveforms with an oscilloscope or other data acquisition system.

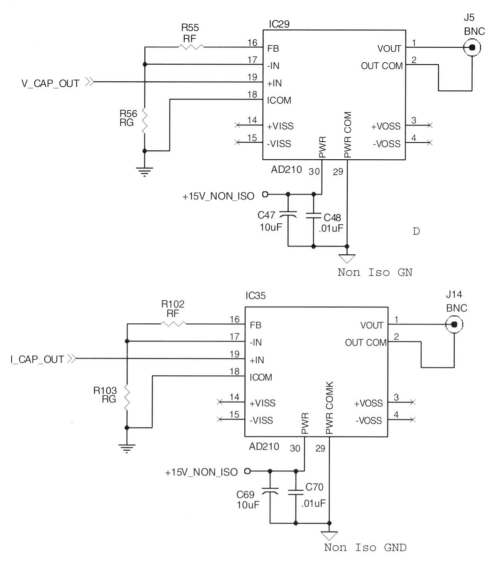

Figure 8.42   An AD210AN isolation amplifier is used to sample the voltage across the capacitor bank (using the voltage divider formed by resistors R84 and R86 on the high-voltage capacitor charging circuit). A second AD210AN samples the instantaneous capacitor charge–discharge current by sampling the voltage developed across the 0.5-Ω current-measurement resistor (R90 on the high-voltage capacitor charging circuit).

Control of the Shock Box

As shown in Figure 8.43, a Microchip PIC 16C77 microcontroller (IC24) is the main shock-box controller. A 32-kHz crystal-controlled clock operates the timers, and a 4-MHz crystal-controlled clock runs processing upon wake-up. Correct clock operation is verified through an independent *RC* circuit (R42 and C31). This microcontroller is powered from one of the module's batteries through a 5-V regulator (IC20 on the battery-charging circuit) and will thus have constant supply of energy available for housekeeping of the module during system power-downs.

The microcontroller communicates with the programming GUI computer through a serial interface. A four-line RS232 protocol (TX/RX,RTS/CTS) protocol is implemented

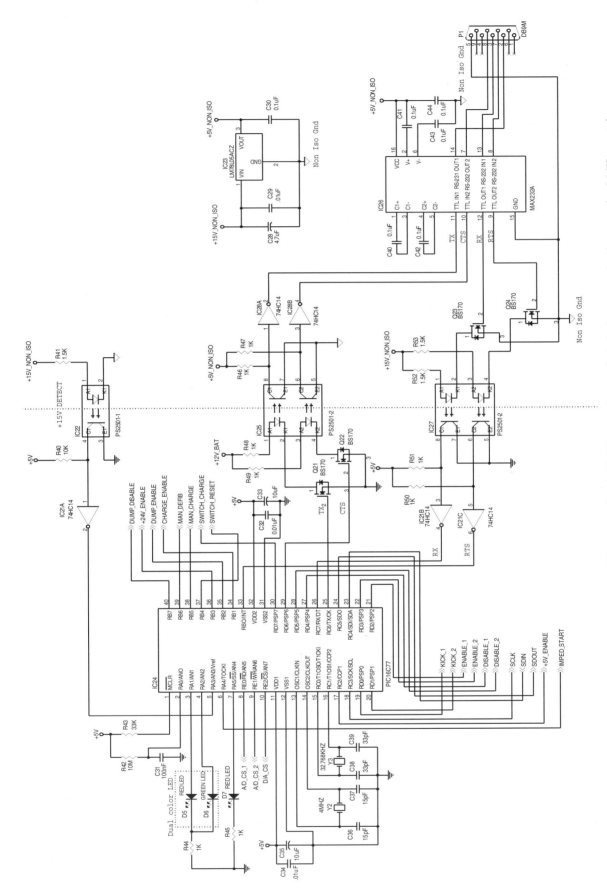

Figure 8.43 A PIC 16C77 is the main shock-box controller. A 32-kHz crystal-controlled clock operates the timers, and a 4-MHz crystal-controlled clock runs processing upon wake-up. Communications with the computer is done through an isolated four-line RS232 protocol. Proper RS232 levels are generated through a TTL-to-RS232 interface chip. A dual-color LED (D5/D6) indicates the state of the high-voltage capacitor charge. LED D7 warns of lead-impedance overrange conditions.

onboard the PIC. The lines are optically isolated by way of IC25 and IC27. The LEDs in the optoisolators are driven by FETs Q21, Q22, Q23, and Q24. The optoisolator outputs are buffered and inverted through IC21B, IC21C, IC28A, and IC28B. Proper RS232 levels are generated through a TTL-to-RS232 interface chip (IC26). The circuit connected to the programming GUI computer is powered by $+5$ V derived from the $+15$-V Condor power supply through linear regulator IC23.

The shock-box controller knows the state of the power line through optoisolator IC22. The LED in this optoisolator is powered from the $+15$-V output of the Condor power supply through resistor R41. The output of the optoisolator is conditioned and inverted by IC21A and fed to the PIC through pin 5. A dual-color LED (D5/D6) controlled from pins 3 and 4 of the PIC is used as an indicator of the high-voltage capacitor charge state (red, charging; green, ready to defibrillate; red blinking, charge fault). A red LED (D7) interfaced to pin 7 of the PIC is used to indicate a lead-impedance overrange condition.

D/A and A/D Converters

The PIC communicates with two D/A converters (one to measure impedance, the other to measure the level of high-voltage charge) and a D/A converter (used to determine the set point for the high-voltage power supply) through a SPI serial interface (pins 18, 23, and 24). Chip enables to address these converters are generated through pins 8, 9, and 10. As shown in Figure 8.44, a LTC1451 D/A converter (IC33) is interfaced to the shock-box microcontroller via a three-line serial (SPI) bus ($+$ enable). IC33 is operated with its internal reference. The maximum of 4.096 V output from this D/A limits the voltage output of the HV power supply to approximately 860 V, preventing accidental overcharge of the capacitor bank.

IC33 is powered through the switched $+5$-V (isolated) line. For this reason, the enable line is controlled through a FET used as an isolation mechanism between the converter IC and the microcontroller. Two MAX187 A/D converters (IC30, IC31) are interfaced to the shock-box microcontroller via serial (SPI) bus ($+$ separate enables) to measure the voltage across the capacitor bank as well as the analog output of the impedance measurement circuit. The converters use their internal references and are powered through the switched $+5$-V (isolated) line. For this reason, the enable lines are controlled through FETs (Q25 and Q26), which are used as isolation mechanisms between the converter ICs and the microcontroller.

Isolated Manual-Mode Pushbuttons

A photovoltaic isolator powered from the module's battery in a pulsed regime is controlled by the microcontroller to generate an isolated supply to power the LEDs of optoisolators controlled by the manual defibrillation control switches. This is done to maintain full isolation of the manual-control switches from the applied part, yet still be able to operate the module in manual mode whenever the main power line fails. As shown in the circuit of Figure 8.45, photovoltaic isolator IC40 charges two 10-μF capacitors in parallel. When power is first applied to the shock-box circuit, charging occurs for 4 s. From that point, the capacitors are topped off every 10 s through a 1-s pulse. The series diode prevents the capacitors from being discharged through the relatively high dark leakage current of the photovoltaic isolator during the no-charge periods.

Whenever one of the switches (SW1 or SW2) is pressed, the energy stored in the capacitors is used to pulse the LED of the photocoupler (in IC42) associated with the switch depressed. Activation of the photocoupler is conditioned through the associated inverter/buffer (IC21D or IC21E) and latched by the associated flip-flop (IC41A or IC41B). Upon wake-up, the shock-box microcontroller reads the status of the latches. If a switch closure is found, the microcontroller begins the behavior requested (i.e., charge to 50 J or defibrillate), recharges the isolated 10-μF capacitors for 4 s, and then resets the latches (IC41A and IC41B).

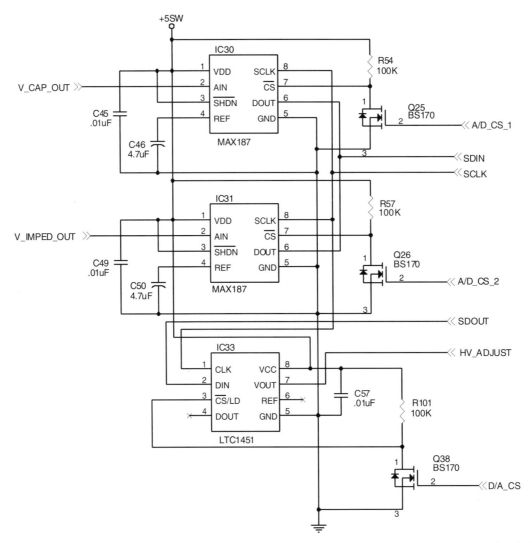

Figure 8.44 The PIC communicates with two A/D converters (one to measure impedance, the other to measure the level of high-voltage charge) and a D/A converter (used to determine the set point for the high-voltage power supply) through a SPI serial interface. The maximum of 4.096 V output from this D/A limits the voltage output of the HV power supply to approximately 860 V. Two MAX187 A/D converters measure the voltage across the capacitor bank as well as the analog output of the impedance measurement circuit.

Lead Impedance Measurement Circuit

Lead impedance measurements are made with the compensated capacitor discharge technique described earlier, which uses very small energy probe pulses to estimate the resistive component of the lead system impedance. As explained before, the output of the circuit is an analog voltage proportional to the lead impedance. Figure 8.46 presents the actual impedance measurement circuit implemented in the shock box. In it, the "active" capacitor C_a (C15) is charged to 1.2 V through switch IC15A and current-limiting resistor R29. At the same time, the "passive" capacitor C_p (C7) is discharged through IC15D. The ground path during this process is established through IC15C. All other switches are open during the charging of C_a and discharging of C_p. This is the normal state of the state machine.

When an impedance measurement is to be performed, the shock-box microcontroller generates an interrupt (through line IMPED_START) to the impedance measurement

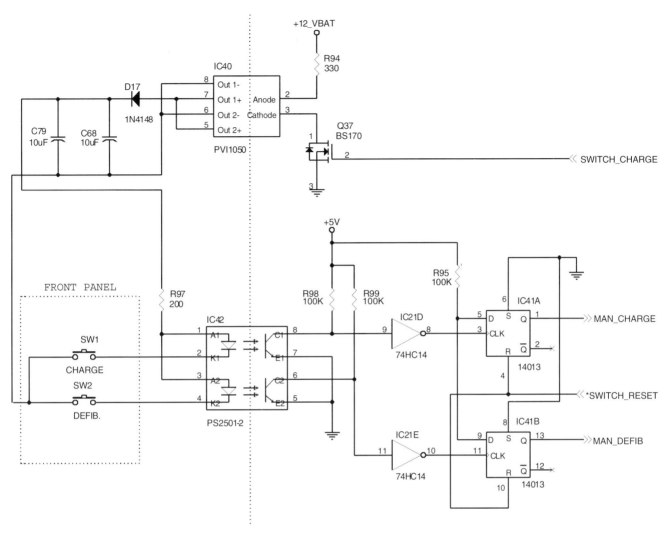

Figure 8.45   Manual control of the defibrillation is made possible through isolated front-panel pushbuttons. A photovoltaic isolator charges two 10-μF capacitors in parallel every 10 s through a 1-s pulse. Whenever one of the switches (SW1 or SW2) is pressed, the energy stored in the capacitors is used to pulse the LED of the photocoupler associated with the depressed switch, starting the behavior requested (i.e., charge to 50 J or defibrillate).

circuit's state-machine microcontroller IC13. IC15D, IC15C, and IC15A are opened. Almost simultaneously, IC16C connects the reference terminal of the capacitors to one of the leads. Then the other lead is connected to C_p through IC16D and dc-blocking capacitor C16. The effect of the body–lead system on a 0.1-μF capacitor is sampled for 10 μs. IC16D is then opened, and C_a is allowed to discharge through the lead system for 10 μs by way of IC16A, C16, and IC16C.

After the passive and active samples are taken, IC16C is opened to float the lead system in relationship to the system ground. IC16B is closed to actively discharge the dc-blocking capacitor for 768 μs. While active discharge takes place, IC15C is closed and the differential signal corresponding to the compensated impedance measurement is developed by IC14A, IC14D, and instrumentation amplifier IC17. At the end of the 768 μs, the sample-and-hold implemented through IC15B, C12, and IC14B holds the voltage level corresponding to the impedance measurement. This level is then scaled and filtered through IC14C and its associated components. Finally, the analog level is converted to a digital

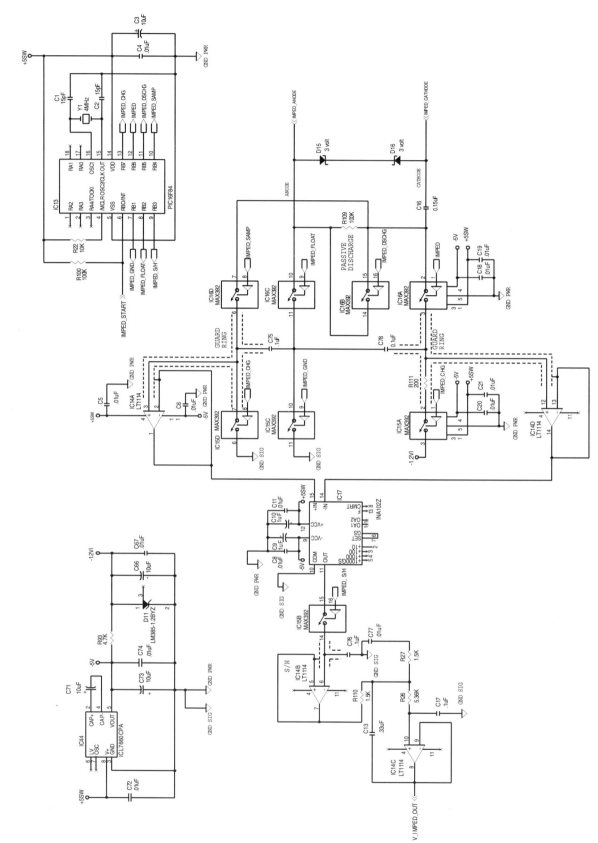

Figure 8.46 A lead impedance sensor measures the load impedance without using a high-voltage shock. This circuit used the "compensated capacitor discharge" sensing principle of Figure 8.22.

value by the corresponding A/D associated to the shock-box microcontroller. Impedance limit values are determined within the shock-box microcontroller through a preloaded lookup table.

To conserve power, the impedance-measurement circuit is powered only when necessary. This is done by supplying the circuit through a switch (MOSFET Q20 in the battery-charging circuit) whenever deemed necessary by the shock-box controller (through the line +5V_ENABLE). The input of the impedance measurement circuit is connected to the defibrillation electrode terminals by way of a self-activated isolation circuit. A solid-state switching circuit that reacts to current flow protects both the impedance measurement circuit from defibrillation pulses as well as the patient from faults in the impedance measurement circuit.

Referring to the upper half of the circuit of Figure 8.47, the gates of FETs Q29 and Q30 are held at a positive voltage in excess of the conduction threshold by the +5-V source through resistor R63. This allows impedance measurement to take place via these FETs and resistor R70. However, when a defibrillation pulse is applied, current increases through R70 (shunted at the impedance measurement circuit by the back-to-back zener diodes), and depending on the polarity of the pulse, one of the bipolar transistors (Q7 or Q28) starts to conduct, placing its associated FET in the high-impedance state.

The bipolar transistor remains conductive until the voltage drops below a safe level for the impedance measurement circuit, at which time the low-impedance path is reestablished through the FET. The lower portion of the circuit in the figure is essentially the same as that described above. This portion is used to protect the second terminal of the impedance-measurement circuit from the high-voltage defibrillation pulses.

Firmware for the Shock Box

The embedded software source code listing includes comments, to make it easy to understand its operation. It is supplied in the book's ftp site as SHOCKBOX.ZIP. The following are short descriptions of the modules implemented by this software.

GUI Command Processor   This module processes commands sent by the programming GUI computer to the shock-box microcontroller. There are 15 commands that can be sent by the programming GUI computer to the shock box via RS232. These commands are sent in frames with the following format:

Command Number | Data Count | Data | LSB Checksum | MSB Checksum

where Command Number is one of 15 possible commands (see the code for a list of commands) Data Count is the number of data bytes in the parameter data message to follow, Data the Parameter data, LSB Checksum the least-significant byte of the 16-bit checksum, and MSB Checksum the most-significant byte of the 16-bit checksum. Every time the module executes an accepted command, it either echoes back the GUI command (whenever data are not expected back as a response for the command), or sends the command number and the data expected with the same format as for GUI to shock-box communication (where the data are not a command parameter but rather, the response).

Main Wake-up Module   This module wakes the processor up every 250 ms. Upon wake-up, it does the following:

1. Checks if front-panel switches (manual charge or manual defibrillate) have been pressed.

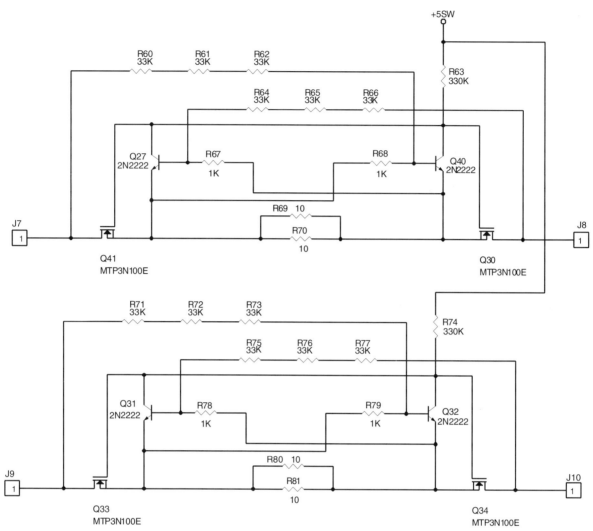

Figure 8.47   The input of the impedance measurement circuit of Figure 8.46 is connected to the defibrillation electrode terminals by way of a self-activated isolation circuit. A solid-state switching circuit that reacts to current flow protects both the impedance measurement circuit from defibrillation pulses as well as the patient from faults in the impedance measurement circuit. The gates of Q29 and Q30 are held at a positive voltage in excess of the conduction threshold by the +5-V source through resistor R63. This allows impedance measurement to take place via these FETs and resistor R70. However, when a defibrillation pulse is applied, current increases through R70, causing one of the bipolar transistors (Q7 or Q28) to conduct and place its associated FET in the high-impedance state.

   2. If the switches are pressed, takes appropriate action, including:
   a. Initiates charge.
   b. Checks for completion of charge cycle.
   c. Awaits front-panel command to shock.
   d. Dumps remaining stored energy after shock.
   e. Informs GUI of charge due to manual request, and defibrillates due to manual request.
   f. Checks for GUI information requests.

3. Every 1 s (every fourth 250-ms wake-up) this module performs the following actions:

   a. If a charge command has been issued, checks that once the desired charge has been reached, charge is not maintained for more than 30 s, and if past, initiates charge dump.

   b. Tests the lead impedance and sets the "check lead impedance" LED appropriately.

   c. Checks if ac power is available and turns peripheral power on constantly (if ac power is not available, peripherals such as the A/Ds, D/A, and impedance measurement circuit are powered only when needed).

4. Every minute, this module increments the reform counter, which keeps track of the time elapsed since last capacitor reform.

Lead Impedance Verification   This module initiates lead impedance measurement and runs the A/D that reads the impedance sensor output. This function is initiated either manually (through a GUI request for lead impedance) or before defibrillation is delivered to the load. A submodule drives the impedance-measurement A/D and takes eight readings (2.048 ms apart) and averages the measurements prior to using them to enable or disable defibrillation as well as to communicate the impedance measurement to the GUI.

Charge Control   This software module enables the high-voltage power supply, establishes DAC output proportional to the desired stored energy level, and verifies successful completion of the charge cycle.

Deliver   This software module runs a sequencer to turn the H-bridge switches on and off according to the defibrillation waveform desired. This module also initiates a lead impedance check prior to delivery.

Assessment of Charge Delivered   This module measures the remaining voltage on the capacitor bank after shock delivery and communicates the measured value to the control computer. Estimation of charge delivery is done by the control computer. A submodule takes care of controlling the A/D that reads the voltage across the capacitor bank and averages eight readings.

Reform   This module keep tabs on the time elapsed since the last full-charge MANUAL charge (through front-panel control) or reform. If the time elapsed runs over 30 days, this module enables a full charge of the capacitor bank in order to reform. The module then dumps the charge into the dummy load after the reform operation.

Verification of Correct Clock Operation   This module verifies correct operation of the timing mechanisms of the microcontroller. If a fault is found, the module deactivates automatically.

Read Manual Switches   This module takes care of charging the isolated capacitor used to power the optocouplers that respond to the manual charge and manual defibrillation pushbutton commands.

CARDIAC FIBRILLATOR

No, it's not a typo—a cardiac fibrillator is an instrument for electrical arrest of the action of the heart. Deliberate cardiac arrest is frequently required during cardiac surgery after arrangements are made for cardiopulmonary bypass. Cardiac paralysis allows a surgeon to perform delicate procedures on a quiet, motionless heart. A common way of causing elective cardiac arrest used to be by applying medium-frequency sine-wave stimulation through electrodes on the surface of the heart to bring about ventricular fibrillation, but this method is seldom used today. Instead, a potassium-salt solution is usually injected to stop the heart intentionally (cardoplegia). When a patient is placed on the cardiopulmonary bypass machine, the heart and body are cooled to reduce the heart's metabolic requirements and oxygen consumption. The cardiplegia solution is then injected to cause cardiac standstill and further reduce the heart's oxygen consumption.

When the surgery has been completed, the cardioplegic agents are flushed out of the heart with warm blood. The heart resumes contraction, and use of the heart–lung machine is discontinued. If normal contractions do not begin at once, a defibrillation shock is used. Temporary pacing leads ("heart wires") are sometimes left stitched on the surface of the heart for connection to a temporary pacemaker, since it is not uncommon that the heart remains bradycardic until all pharmacologic agents are flushed out and normal metabolism is resumed. Chemical cardioplegia is preferred to fibrillation because the heart's quivering during fibrillation wastes the heart muscle's energy resources that are needed once beating is resumed. Despite this, fibrillators are still used in animal research connected to the specific mechanisms of fibrillation as well as in the assessment of defibrillation systems.

Figure 8.48 shows a battery-powered fibrillator. The circuit uses a 555 timer IC to generate a 50-Hz square wave. This signal drives two sets of six paired CMOS buffers each. This output stage is actually a full bridge which causes doubling of the effective voltage across the heart. Capacitors C2 and C4 block dc components on the output signal. Resistors R2 and R3 are used to protect the circuit against defibrillation currents. Since a certain current *density* threshold has to be exceeded to cause fibrillation, the fibrillating current and electrode surface contact area are proportional. Using pediatric-use defibrillation paddle electrodes to fibrillate a pig's heart typically requires 10 to 12 V at 50 Hz. Using the outside surface of alligator clips as electrodes (which have a much lower contact surface area) may take only 3 to 4 V.

As simple as this circuit is, an even simpler battery-powered fibrillator is the most common method for inducing fibrillation in the animal lab. Briefly touching the terminals of a standard 9-V battery to the ventricle (e.g., to the right-ventricular outflow tract) almost always fibrillates the heart. Moral of the story: Don't lick-test 9-V batteries laying around an electrophysiology lab.

CONCLUDING REMARKS

There is much more to the design of active implantable medical devices (implants that rely for their functioning on a source of electrical energy or any source of power other than that generated directly by the human body or by gravity) than the short explanations and demonstrations presented in this chapter. The European Union has published a directive that regulates the level of performance and testing applicable to active implantable medical devices. This standard is:

- EN-45502-1, *Active Implantable Medical Devices—Part 1: General Requirements for Safety, Marking and Information to Be Provided by the Manufacturer*, 1998

As a specialized subset of the implantable devices field, cardiac pacemakers and defibrillators have been around for quite a few years now. Their clinical success have turned them almost into commodity items with standardized features. Regulators recognize this fact and are trying to establish specific standards for the performance of these devices. Although at the time of this writing the relevant European standards are still in

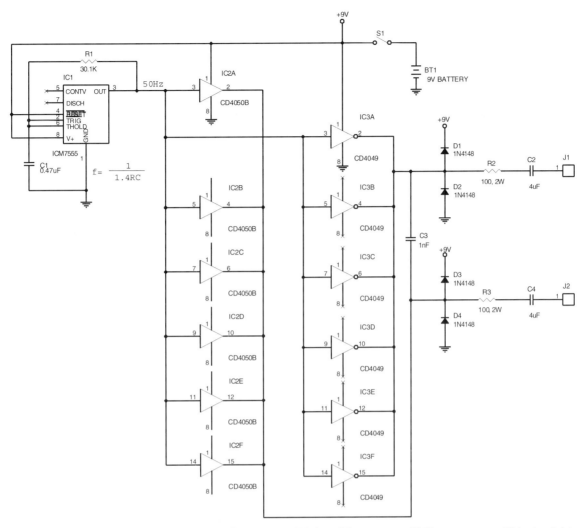

Figure 8.48 Schematic diagram of a battery-powered fibrillator. A 555 timer IC generates a 50-Hz square wave. This signal drives paired CMOS buffers. This output stage is actually a full bridge which causes doubling of the effective voltage across the heart.

draft form, they are very instructive as to the specific methods applied to the design and testing of implantable pacemakers and defibrillators:

- prEN-45502-2-1, *Active Implantable Medical Devices—Part 2-1: Particular Requirements for Active Implantable Medical Devices Intended to Treat Bradyarrhythmia (Cardiac Pacemakers)*, Final Draft, January 2001
- prEN-45502-2-2, *Active Implantable Medical Devices—Part 2-2: Particular Requirements for Active Implantable Medical Devices Intended to Treat Tachyarrhythmia (Includes Implantable Defibrillators)*, 1998

On the U.S. side, the FDA has not yet adopted a set of standards harmonized with EN-45502. Officially, the FDA still refers manufacturers to a guidance document written in the 1970s:

- Donald F. Dahms, Implantable Pacemaker Testing Guidance, FDA FOD#383, posted January 12, 1990, available at www.fda.gov/cdrh/ode/doc383.pdf

However, this document simply outlines the areas in which implantable pacemakers should be tested (e.g., electrical characterization, interference, biocompatibility, animal testing) so that they can be considered safe for human implant. This document does not provide the specific performance requirements or testing methods available in the EN-45502 series of standards. This is not to say that the FDA doesn't expect to see much, *much* more than what the guidance document sketches. Until a standard is adopted by FDA, however, it is up to the manufacturers to find ways of justifying the performance of their devices and to find ways of testing the devices for safety and effectiveness to satisfy the FDA.

Nonimplantable pacemakers and defibrillators are also governed by particular standards:

- IEC-601-2-31, *Particular Requirements for the Safety of External Cardiac Pacemakers with Internal Power Source*, 1994, specifies the particular safety requirements for external pacemakers powered by an internal electrical power source. Applies also to patient cables but does not apply to equipment that can be connected directly or indirectly to a power line.

- IEC-601-2-4, *Particular Requirements for the Safety of Cardiac Defibrillators and Cardiac Defibrillator Monitors*, 1983, specifies requirements for the safety of cardiac defibrillators and cardiac defibrillator monitors incorporating a capacitive energy storage device.

- ANSI/AAMI DF2, *Cardiac Defibrillator Devices*, 1989, provides minimum labeling, performance, and safety requirements for cardiac defibrillator devices. Requirements for performance cover energy range, limited energy output, energy accuracy, pulse shape and duration, controls and indicators, energy-level indicator, reduce-charge capability, paddle electrode contact area, and other characteristics.

- AAMI DF39, *Automatic External and Remote Control Defibrillators*, 1993, covers testing and reporting performance results of automatic external defibrillators, primarily through description of the performance for the arrhythmia detection component.

REFERENCES

AAMI, *AAMI/ANSI Standard HF18: Electrosurgical Devices, Annex B*, 1994.

Bocka, J. J., External Transcutaneous Pacemakers, *Annals of Emergency Medicine*, 18(12), 1280–1286, 1989.

Hoekstein, K. N., and G. F. Inbar, *Cardiac Stroke Volume Estimation from Two Electrode Electrical Impedance Measurements*, EE Publication 911, Technion Institute of Technology, Haifa, Israel, 1994.

Irwin, D. D., S. Rush, R. Evering, E. Lepeschkin, D. B. Montgomery, and R. J. Weggel, Stimulation of Cardiac Muscle by a Time-Varying Magnetic Field, *IEEE Transactions on Magnetics*, 6(2), 321–322, 1970.

Prutchi, D., Rate Responsive Cardiac Pacemaker with Filtered Impedance Sensing and Method, U.S. patent 5,431,772, July 2, 1996.

Prutchi, D., and M. Norris, Electro-optically Driven Solid State Relay System, U.S. patent 6,335,538, January 1, 2002.

EPILOGUE

Out of clutter, find simplicity
from discord, find harmony.
In the middle of difficulty lies opportunity.

—Albert Einstein

Think you have a great idea for a new medical device? You're not alone. Every year, many physicians and engineers try to develop their ideas and market them commercially. With a prototype in hand, you may be thinking about how to turn it into an actual product that can be commercialized. As a first step, you may proceed to conduct some preliminary safety and functional bench testing on your prototype device. Concurrently, you may invest a little time visiting the library to research for possible patent infringement. Then you may show your prototype to a physician, who may help evaluate your idea by testing the device in animals and human subjects. If all goes well, and testing shows that your idea would truly constitute an advancement to medicine, you may be tempted to start investing in the infrastructure, tools, and materials required to mass-produce and commercialize your device. Well, don't, at least not yet.

REGULATORY PATH

Unlike many other high-tech businesses, the medical devices industry is highly regulated. Especially in the United States the level of regulation is so extreme that compliance concerns often outweigh all other technical and financial considerations. In fact, even if you could build your device out of your own garage at a cost that you could afford to cover from your own pocket, the extensive clinical trials and lengthy submission process that are needed to satisfy the Food and Drug Administration's (FDA) requirements would probably make the home-based venture completely nonfeasible. To give you an idea of the

Design and Development of Medical Electronic Instrumentation   By David Prutchi and Michael Norris
ISBN 0-471-67623-3   Copyright © 2005 John Wiley & Sons, Inc.

magnitude of the issue, just consider that it is not uncommon for the costs connected to a bare-bones clinical trial and submission for even a simple device to reach several million dollars.

The FDA reasons that these regulatory burdens are absolutely necessary to ensure the safety and efficacy of a new medical device before it is used by the general population. We agree. Although you may have a different opinion regarding the validity of the justifications behind FDA's strict policies, one thing is clear: The extensive clinical trials, complex submission requirements, and lengthy approval process require resources that are beyond the budget of the average midnight engineer.

We don't mean to discourage you from pursuing your idea, but we want you to be aware of the unique characteristics of the biomedical startup. First, you will probably need to recruit resources from a specialized venture capital firm. There, investors will not only be interested in the size of the market opportunity or the gross margin potential but will be looking for a business plan that includes detailed information regarding the following topics:

1. Preliminary animal and clinical data which objectively demonstrate efficacy and cost-effectiveness of the device
2. A clearly defined regulatory pathway that will secure regulatory approval ahead of competitive technologies
3. A comprehensive assessment of the risks posed by the proposed technology
4. A strategic analysis of the market that will identify potential alliances with established companies that can help bring the product more quickly to market
5. A financial plan for sustaining research, development, and day-to-day operations of the company for the duration of the clinical trials and submission process

FDA CLASSIFICATIONS

Prior to 1976, the FDA was limited to reacting to hazardous devices already on the market and, then, only after patient injuries occurred. However, in 1976, Congress chartered the FDA with enforcing the Federal Food, Drug, and Cosmetic (FD&C) Act, which requires that reasonable assurance of safety and effectiveness be demonstrated for medical devices intended for human use before being marketed to the public. Important amendments expanding on the scope of this act were introduced in 1992 and 1997.

The users of a device, usually physicians and patients, don't necessarily have the expertise and time required to evaluate independently the safety and effectiveness of each device they use. Because of this, the FDA requires the manufacturer of the device to demonstrate safety and effectiveness scientifically. At the FDA, "safe" means that the expected benefits to the patient's health for the intended use of the device outweigh probable risks of harm or injury by the device. *Effective* means that the device has the clinical effect claimed by the manufacturer in a reliable manner.

This last definition is especially important for the budding inventor, since the claim of effectiveness relates to clinical usefulness. It is not sufficient to demonstrate that the device accomplishes some physiological effect, but it must provide some medical benefit. Once a manufacturer submits technical and scientific data that demonstrate safety and effectiveness, the FDA conducts an independent review of all the information and makes decisions about marketing and labeling approvals on the basis of these evaluations.

Medical devices are categorized by the FDA depending on the degree of risk they present to the patient and/or user, and the level of regulatory control that the FDA deems necessary to provide sufficient assurance of their safety and effectiveness. Device classification depends on the intended use of the device and also upon indications for use. For example, a scalpel's intended use is to cut tissue. A subset of intended use arises when a more specialized indication is added in the device's labeling, such as "for making incisions in the cornea."

Class I devices are those for which the FDA requires the lowest level of regulatory control. This classification is given to devices for which there is sufficient information to conclude that safety and effectiveness can be ensured by "general controls" alone. General controls give FDA authority to enforce against misbranding and adulteration. In addition, the FDA requires manufacturers to register and list their devices, notify consumers of problems associated with devices, and follow good manufacturing practices (GMPs) in their manufacturing operations.

Class II devices are those for which the FDA does not believe that general controls suffice, but for which there is sufficient information that the FDA can establish special controls to ensure their safety and effectiveness. As such, class II devices are not only subject to general controls but are also subject to compliance with specific standards, guidelines, and patient or special follow-up once in the market.

Class III devices are those for which the FDA requires the strictest regulatory control. This class is reserved for devices for which general and special controls cannot provide reasonable assurance of safety and effectiveness or those for which there is insufficient information to make that assessment. Typically, a class III device is "represented to be for a use in supporting or sustaining human life or for a use which is of substantial importance in preventing impairment of human health, or presents a potential unreasonable risk of illness or injury."

PATH THROUGH THE FDA

The requirements for medical devices to be used in human patients vary according to the regulatory classification of the device. Regardless of classification, all medical devices need to be cleared by the FDA before commercial distribution. Most devices are cleared for marketing either through approval of a premarket approval (PMA) application or through clearance of a premarket notification submission on the basis of section 510(k) of the FD&C Act.

Class I and class II devices are often subject to section 510(k) of the act, which is used to determine whether a new device is substantially equivalent to a device in use prior to May 28, 1976 (a preamendment device), or should be classified as a postamendments device. Substantial equivalence is based on the device's intended use, its technical characteristics, as well as safety and effectiveness aspects. A manufacturer submitting a 510(k) application must supply information identifying a medical device that was marketed legally prior to the amendment. If the device is found to be substantially equivalent, it can be marketed immediately, whereas a new device that is not substantially equivalent requires to go through the PMA process. The advantage of the 510(k) route is that it is substantially faster and less costly than the PMA application.

Preparing a PMA application is definitely not a trivial pursuit. A PMA application must include:

- Full reports of all information known to the manufacturer (regardless of whether it is good or bad) concerning all research done to demonstrate safety and effectiveness
- Detailed description of all components, properties, and principles of operation of the device
- Detailed manufacturing data
- Proposed labeling (which includes any text or informational figures on the device, packaging, and literature)

All data supplied by the manufacturer is subjected to thorough, independent review by FDA employees and medical advisors, including specialized medical devices advisory panels composed of scientists and practicing physicians. If the reviewing team makes a positive recommendation for approval, a manufacturing site inspection is conducted by the FDA. A final notification of the FDA decision is mailed to the manufacturer following concluding discussions.

Exceptions

There are some exceptions to these rules. Medical devices that have not been approved by the FDA for marketing can be used with human subjects under a Humanitarian Device Exemption (HDE), an Investigational Device Exemption (IDE), or a Product Development Protocol.

Humanitarian Use Devices (HUDs) are devices intended to diagnose or treat a condition that affects fewer than 4000 U.S. patients per year. The HDE application is similar to a PMA but does not require the manufacturer to demonstrate effectiveness, only safety. This path was provided through the 1990 Safe Medical Devices Act as a way to reduce the financial burden of R&D in the development of devices with limited market potential, but which may provide benefit to the patients who need them.

A clinical trial for a device may be conducted through an investigational exemption to the restraint on commercial distribution of unapproved medical devices. An IDE application requires sufficient evidence of the device's safety and a reasonable expectation of effectiveness to warrant its testing in humans. The FDA recognizes the investigational nature of a clinical trial done under an IDE, allowing fewer rigors than in a PMA application. Nevertheless, important safety issues must be addressed through thorough bench and animal testing before using the device on human subjects. The use of a medical device under an IDE is very restricted:

- The device can be made available only to designated investigators.
- The device is to be used only under the terms of the approved IDE application.
- Designated investigators are responsible for ensuring that the investigation adheres to the investigational plan as well as any and all conditions imposed by the host hospital's Institutional Review Board (IRB) and the FDA.
- The rights, safety, and welfare of subjects is maintained according to ethical guidelines of the World Medical Association Declaration of Helsinki.
- Strict control is maintained over the devices under investigation.

Finally, section 515(f) of the FD&C Act provides an alternative to the PMA process for class III devices. A Product Development Protocol (PDP) should be submitted by the manufacturer to the FDA very early in a device's development cycle. The PDP defines the types of data and specific safety and performance levels that the device must achieve to receive market clearance. The product can be introduced to the market as soon as the FDA verifies the manufacturer's data, showing that it meets the preestablished safety and performance levels.

MARKETING AN IDEA

Despite the strong skepticism that is commonly encountered by new biomedical technologies, market statistics show that investors are eager to back medical device ventures. In fact, a recent survey by *Medical Device and Diagnostic Industry* magazine indicated that the industry is undergoing profound changes which may make it easy for small entrepreneurs to enter the health-care marketplace. Supporting this conclusion are data such as a decrease in the percentage of executives who reported that FDA's policies were having harmful effects on their business (47% in 1994, 46% in 1995, and 31% in 1996), a similar decrease in the percentage of respondents reporting delays in the premarket approval of their products (53% in 1994, 54% in 1995 and 45% in 1996), as well as an increase in the median sales volume change (10% in 1995 to 20% in 1996). Now, these changes are certainly not a quantum leap toward a most favorable business climate. However, operating with cautious optimism, you can certainly adapt and prosper in this friendlier (or at least less hostile) environment.

A different strategy for medical device startups that has become very popular is to try to sell the idea or invention directly to a company that would market it and pay royalties. This is a business model that has made at least a few inventors very rich. However, attracting the interest of a serious company takes much more than a good idea. A common misconception is that companies purchase raw ideas for development. Many inventors make the mistake of believing that a manufacturing company will agree to sign a Non-Disclosure Agreement (NDA) which would allow the inventor to submit an unpatented idea in confidence. The argument is that the inventor will not have to go through the expensive and time-consuming processes of proper R&D, patenting and marketing research before the idea can be evaluated by a big-league player.

Inventors often try to use NDAs to shortcut the process of patenting. But no NDA confers property rights on an idea; only a patent does. In fact, most reputable manufacturing companies will not sign NDAs just to listen to unsolicited ideas. Companies are legitimately concerned of external contamination. Companies worry that someone with an undeveloped idea will come to them when they are already working in the same area. If they sign an NDA, there is the possibility that the "inventor" will later sue the company for misappropriation, claiming that they stole the idea for a successful product. Large companies have learned the lesson, and most will refuse to accept anything in confidence, requiring inventors to sign a waiver before they look at unsolicited ideas. The waiver requires the person or company making the disclosure of the idea to give up all rights except those protected under the patent laws.

NDAs have a legitimate use. They may be needed with potential licensees of a technology to protect trade secrets. However, don't live under the illusion that someone will pay you a million dollars for an idea that you disclosed to them with or without an NDA in place. There is no replacement for thorough and secret R&D followed by your own market research to prove the workability, patentability, and marketability of the device. Armed with this information, you can openly court licensees or distributors.

Other inventors don't find easy access to companies or are concerned that a large corporation will steal their ideas upon disclosure. These people are often lured to use the services of an invention or patent promotion firm. These companies promise to evaluate, develop, patent, and market inventions. Beware, however, that in many cases these firms do little or nothing for the thousands of dollars that inventors are required to pay in exchange for services. The stories abound of dishonest promoters who take advantage of an inventor's enthusiasm for a new product. They not only urge inventors to patent their ideas, but also make false and exaggerated claims about the market potential of the invention. Our advice is that you proceed with caution and in assessing the true market potential of your product. At all costs, avoid falling for the sweet-sounding promises of a fraudulent promotion firm.

Finally, a note on the issue of infringing on someone else's intellectual property. Patents protect many circuits and algorithms used in medical devices. The responsibility for avoiding infringement is yours. However, the responsibility for policing patent rights belongs to the patent holder. It is your call as to whether you'd rather learn about a "problem" patent by doing research or by waiting for a cease and desist letter. Your choice can be influenced by many factors: What does it cost you to set up your business? How quickly can you recover your investment? What are your costs of closing down quickly likely to be? How well do you know the technical field that your product is in? What do you know about how aggressive existing companies in the industry are at policing their patents?

We recommend running a thorough search on one of the various Web-based patent search engines (e.g., www.delphion.com, www.uspto.gov) for patents that may be related to the device or technique that you are developing. In addition, the U.S. Patent and Trademark Office offers free information about patents, trademarks, and copyrights, and every state has a Patent and Trademark Depository Library that maintains collections of current and previously issued patents and patent and trademark reference materials.

More than the description of the invention, *read the patent's claims*. If your product or method is described by a claim, you have a problem. Avoiding a claim is generally a matter of not having one of the features claimed. One way of "getting around a patent" is to study the patent in detail and engineer around the patent. In other words, determine how to modify your device to operate outside the scope of the patent claims. For example, in a medical device patent for a specific biopotential signal application, operating outside the input voltage ranges claimed might be sufficient to avoid infringement. However, an improvement that has all the features claimed and a few additional ones is likely to be covered by the patent. If you do this, it is recommended that you consult a patent attorney for an infringement opinion that presents the risk factors of whether your modifications would be sufficient to avoid liability.

When there's a question as to whether or not your product infringes, you may want to ask a patent attorney for an opinion of noninfringement. Getting that opinion does not mean that you don't infringe, but it can protect you from having damages trebled in the event a court finds that you did infringe. Another thing that you can do is get a validity opinion about the risks that a court of law would find the patent valid.

Neither the infringement opinion nor the validity opinion come with any guarantees, they merely explain the risks of doing a particular activity. The decision to operate within or near the scope of a claimed invention is a business decision based on the risks of being sued and the risks of losing that lawsuit. If you make the wrong business decision, you may have to cease production and pay hefty damages. If the court determines that your infringement was "willful," your damages may be tripled and you may end up paying the other side's attorney fees.

One of our favorite methods of avoiding patent infringement is to copy the product described in a patent that expired at the end of its full term. Patents issued before June 5, 1995, had a 17-year term measured from the date of issue. Patents filed after that date have a 20-year term measured from the filing date. Patents filed before that date, and pending or issued on that date, have the longer of the two terms. Once a patent has expired, the invention is in the public domain, and you are free to use it.

Note that the safest way is to look for patents that have expired at the end of their "full term," not just "expired," because it is possible that a product described in a patent which expired early for lack of payment of maintenance fees might infringe the claims of an older "dominating" patent that is still in effect. If a patent expires after its full term, it is likely that any older patent would have expired first. However, various factors, including term extensions, or perhaps very prolonged periods pending resolution due to appeals, make this less than 100% certain, but highly likely.

So what are you waiting for? Get your soldering pen, help humanity, and make a decent living along the way.

APPENDIX A

SOURCES FOR MATERIALS AND COMPONENTS

Alfa Aesar (Johnson Matthey)
www.alfa.com
30 Bond Street
Ward Hill, MA 01835
(978) 521-6300
(978) 521-6350 (fax)

Analog Devices, Inc.
www.analog.com
One Technology Way
Norwood, MA 02062
(617) 329-4700
(617) 326-8703 (fax)

Apex Microtechnology Corporation
www.apexmicrotech.com
5980 North Shannon Road
Tucson, AZ 85741
(520) 690-8600
(520) 888-3329 (fax)

Axon Instruments, Inc.
www.axon.com
3280 Whipple Road
Union City, CA 94587
(510) 675-6200
(510) 675-6300 (fax)

Burr-Brown Corporation
www.burr-brown.com
6730 South Tucson Boulevard
Tucson, AZ 85706
(520) 746-1111
(502) 889-1510 (fax)

Capteur Sensors & Analysers Ltd.
www.capteur.co.uk
Walton Road
Portsmouth, Hants, PO61SZ, UK
+44 (0) 1235-750300
+44 (0) 2392-386 611 (fax)

CCC del Uruguay
www.ccc.com.uy
General Paz 1371
11400 Montevideo, Uruguay
598-2-600-7629
598-2-601-6286 (fax)

C&D Technologies' Power Electronics
www.cdpowerelectronics.com
3400 East Britannia Drive
Tucson, AZ 85706
(800) 547-2537
(520) 770-9369 (fax)

Design and Development of Medical Electronic Instrumentation   By David Prutchi and Michael Norris
ISBN 0-471-67623-3   Copyright © 2005 John Wiley & Sons, Inc.

Chomerics
www.chomerics.com
77 Dragon Court
Woburn, MA 01888-4014
(781) 935-4850
(781) 933-4318 (fax)

Cornell Dubilier Electronics, Inc.
www.cornell-dubilier.com
140 Technology Place
Liberty, SC 29657
(864) 843-2626
(864) 843-2402 (fax)

Dallas Semiconductor Corporation
(part of Maxim Integrated Products, Inc.)
www.maxim-ic.com
4401 Beltwood Parkway South
Dallas, TX 75244-3292
(214) 450-0448
(214) 450-0470 (fax)

Electronic Design & Research, Inc.
www.vsholding.com
7331 Intermodal Drive
Louisville KY 40258
(502) 933-8660
(502) 933-3422 (fax)

Elpac Electronics, Inc.
www.elpac.com
1562 Reynolds Avenue
Irvine, CA 92614-5612
(949) 476-6070
(949) 476-6080 (fax)

Fair Radio Sales
www.fairradio.com
1016 East Eureka Street
Lima, OH 45802
(419) 227-6573
(419) 227-1313 (fax)

Fair-Rite
www.fair-rite.com
1 Commercial Row
Wallkill, NY 12589
(845) 895-2055
(845) 895-2629 (fax)

Fischer Scientific
www.fishersci.com
One Liberty Lane
Hampton, NH 03842
(603) 926-5911
(603) 929-2215 (fax)

Hammond Manufacturing Co., Inc.
(Electronics Group)
www.hammondmfg.com
256 Sonwil Drive
Cheektowaga, NY 14225
(716) 651-0086
(716) 651-0726 (fax)

Harris Semiconductor
(part of Intersil Corporation)
www.intersil.com
1301 Woody Burke Road
Melbourne, FL 32902
(407) 724-3000
(407) 724-3937 (fax)

Harvard Apparatus, Inc.
www.harvardapparatus.com
84 October Hill Road
Holliston, MA 01746
(508) 893-8999
(508) 429-5732 (fax)

Heraeus W.C. GmbH & Co.
www.wc-heraeus.com/medical-technology
KG Heraeusstrasse 12-14 D-63450
Hanau, Germany
+49 (6181) 353830
+49 (6181) 359448 (fax)

Honeywell Solid State Electronics Center
www.ssec.honeywell.com
12001 State Highway 55
Plymouth, MN 55441
(763) 954-2539
(763) 954-2720 (fax)

International Rectifier
www.irf.com
233 Kansas Street
El Segundo, CA 90245
(310) 322-3331
(310) 322-3332 (fax)

Intersil Corporation
www.intersil.com
7585 Irvine Center Drive
Suite 100
Irvine, CA 92618
(949) 341-7000
(949) 341-7123 (fax)

In Vivo Metric
www.invivometric.com
910 Waugh Lane
Ukiah, CA 95482
(707) 462-4121
(707) 462-4011 (fax)

Kilovac Corporation
www.kilovac.com
P.O. Box 4422
Santa Barbara, CA 93140
(805) 684-4560
(805) 684-9679 (fax)

LEM U.S.A., Inc.
www.lem.com
6643 West Mill Road
Milwaukee, WI 53218
(414) 353-0711
(414) 353-0733 (fax)

MacDermid Incorporated
www.macprintedcircuits.com
245 Freight Street
Waterbury, CT 06702
(203) 575-5700
(203) 575-7916 (fax)

Magnetic Shield Corporation
www.magnetic-shield.com
740 North Thomas Drive
Bensenville, IL 60106
(888) 766-7800
(630) 766-2813 (fax)

Maxim Integrated Products, Inc.
www.maxim-ic.com
120 San Gabriel Drive
Sunnyvale, CA 94086
(408) 737-7600
(408) 737-7194 (fax)

Microchip Technology, Inc.
www.microchip.com
2355 West Chandler Boulevard
Chandler, AZ 85224
(480) 792-7966
(480) 792-4338 (fax)

Mini-Circuits
www.minicircuits.com
P.O. Box 350166
Brooklyn, NY 11235-0003
(718) 934-4500
(718) 332-4661 (fax)

MSI Scantech Limited
www.msi-scantech.co.uk
The Centre, Reading Road
Eversley, Hampshire, RG27 0NB, UK
+44 (0) 118 973 7926
+44 (0) 118 973 7927 (fax)

muRata Electronics (USA)
www.murata.com
2200 Lake Park Drive
Smyrna, GA 30080-7604
(770) 436-1300
(770) 436-3030 (fax)

Noble-Met Ltd.
http://www.uticorporation.com/
global/noblemet/noblemet.html
200 South Yorkshire Street
Salem, VA 24153
(540) 389-7860
(540) 389-7857 (fax)

Ohmic Instruments Co.
www.ohmicinstruments.com
508 August Street
Easton, MD 21601
(410) 820-5111
(410) 822-9633 (fax)

Panasonic Industrial Co.
www.panasonic.com
Two Panasonic Way, 7H-4
Secaucus, NJ 07094
(201) 348-5232
(201) 392-4441

PASCO Scientific
www.pasco.com
P.O. Box 619011
10101 Foothills Boulevard
Roseville, CA 95747-9011
(916) 786-3800
(916) 786-8905 (fax)

Philips Semiconductors
www.phillips.semiconductors.com
811 East Arques Avenue
Sunnyvale, CA 94088-3409
(800) 234-7381
(708) 296-8556 (fax)

Pragmatic Instruments, Inc.
www.pragmatic.com
7313 Carroll Road
San Diego, CA 92121
(800) 772-4628
(619) 271-9567 (fax)

Radio Shack
www.radioshack.com
200 Taylor Street, Suite 600
Fort Worth, TX 76102
(800) 843-7422
(817) 415-6880 (fax)

Sensortechnics, Inc.
www.sensortechnics.com
1420 Providence Highway, Unit 267
Norwood, MA 02062
(781) 762-1674
(781) 762-2564 (fax)

ST Microelectronics
www.st.com
1310 Electronics Drive
Carrollton, TX 75006
(972) 466-6000
(972) 466-8130 (fax)

Surplus Sales of Nebraska
www.surplussales.com
1502 Jones Street
Omaha, NE 68102
(402) 346-4750
(402) 346-2939 (fax)

TDK Corporation of America
www.component.tdk.com
1600 Feehanville Drive
Mount Prospect, IL 60056
(847) 803-6100
(847) 803-6296 (fax)

Texas Instruments Incorporated
www.ti.com
12500 TI Boulevard
Dallas, TX 75243
(800) 336-5236

The MathWorks, Inc.
www.mathworks.com
3 Apple Hill Drive
Natick, MA 01760-2098
(508) 647-7000
(508) 647-7001 (fax)

Thermometrics New Jersey
www.thermometrics.com
808 U.S. Highway 1
Edison, NJ 08817
(732) 287-2870
(732) 287-8847 (fax)

UFI
www.ufiservingscience.com
545 Main C-2
Morro Bay, CA 93442
(805) 772-1203
(805) 772-5056 (fax)

UltraVolt, Inc.
www.ultravolt.com
CS 9002
Ronkonkoma, NY 11779
(631) 471-4444
(631) 471-4696 (fax)

VWR Scientific Products
www.vwrsp.com
3000 Hadley Road
South Plainfield, NJ 07080
(908) 757-4045
(908) 757-0313 (fax)

Wilson Greatbatch Technologies, Inc.
www.greatbatch.com
10,000 Wehrle Drive
Clarence, NY 14031
(716) 759-6901
(716) 759-8579 (fax)

WireTronic, Inc.
www.wiretron.com
19698 State Highway, 88, C
Pine Grove, CA 95665
(209) 296-8460
(209) 296-8462 (fax)

Xylem
www.xylemcompany.com
1480 Lake Drive West
Chanhassen, MN 55317
(952) 368-9040
(952) 368-9041 (fax)

Yuasa Battery, Inc.
www.yuasabatteries.com
2366 Bernville Road
Reading, PA 19612
(866) 431-4784

APPENDIX B

FTP SITE CONTENT
ftp://ftp.wiley.com/public/sci_tech_med/medical_electronic/

DISCLAIMER

All of the files in the book's ftp site (collectively, the *software*) are provided to you free of charge for your convenience. This software is presented only as examples of engineering building blocks used in the design of experimental biopotential signal acquisition and processing systems. The authors do not suggest that the software presented herein can or should be used by the reader or anyone else to acquire or process signals from human subjects or experimental animals. Neither do the authors suggest that the software can or should be used in place of or as an adjunct to professional medical treatment or advice. Sole responsibility for the use of the software or of systems incorporating the software lies with the reader, who must apply for any and all approvals and certifications that the law may require for its use.

The authors do not make any representations as to the completeness or the accuracy of the software, and disclaim any liability for damages or injuries, whether caused by or arising from the lack of completeness, inaccuracies of the software, misinterpretations of the directions, misapplication of the software, or otherwise. References to other manufacturers' products made in this software do not constitute an endorsement of these products but are included for the purpose of illustration and clarification. Since some of the software may relate to or be covered by U.S. or other patents, the authors and the publisher disclaim any liability for the infringement of such patents by the making, use, or selling of such equipment or software, and suggest that anyone interested in such projects seek proper legal counsel.

The authors and the publisher are not responsible to the reader or third parties for any claim of special or consequential damages. The software in the book's ftp site is provided "as-is" without any warranty whatsoever. No technical support or application assistance are available. In no event will the authors or the publisher be liable to you, under any legal theory, for indirect or consequential damages resulting from loss of use, profits, downtime, goodwill, damage to, or replacement of equipment or property arising out of or in connection with the use or performance of the software. In addition, no liability is assumed by

Design and Development of Medical Electronic Instrumentation   By David Prutchi and Michael Norris
ISBN 0-471-67623-3   Copyright © 2005 John Wiley & Sons, Inc.

the authors or the publisher for loss of data or any costs for recovering, reprogramming, or reproducing any data stored in any machine used in connection with the software.

REDISTRIBUTED FREEWARE

For your convenience, the book's ftp site also contains some of the programs listed in the book as *freeware* that can be downloaded from the Web. These programs are included here free of charge with the understanding that "freeware" refers to software that can be downloaded without having to pay money for them. These files may be distributed freely without paying anything in the way of royalties or distribution fees as long as they are not modified prior to distribution. Note that this also applies to you. If you intend to distribute further the files listed as freeware, you must make sure that you hand them out in their original form. You must also distribute things that come in multiple parts as a complete set (e.g., many of the computer programs have extra files; you must distribute them all together, the way you retrieved them from the ftp site). The reason is quite simple: The hard work that went into these products should be distributed in the way that its author intended, not the way that someone else thinks it should be.

CONTENTS OF MEDICAL_ELECTRONIC.ZIP

Software for Introduction

Folder: Redistributed Freeware

- Little Stimulus Maker v2.0 (LSM2_SFX.EXE freeware for Windows 9x) by John Kelly.

Software for Chapter 5: Signal Conditioning, Data Acquisition, and Spectral Analysis

Folder: Universal Sensor Interface

VB5.0

- LPT8_DVM.VBP is a VisualBasic (v5.0) application project that shows how to develop a virtual instrument to acquire analog and digital data as well as to control the D/A and digital outputs of the Universal Sensor Interface. INPOUT32.DLL is used to allow input and output operations to be performed on the printer port. If 16-bit operation is required, modify the programs to make use of the 16-bit CUSER2.DLL file.
- LPT8_LOGGER.VBP is basically the same as LPT8_DVM.VBP, but a file dialog has been added to make it possible to log acquired data directly to disk.
- LPT8_THERM is used to read temperature in °C and °F using a thermistor.

QuickBASIC

- LPTAN8.BAS is a simple program for driving the Universal Sensor Interface A/D.
- LPTAN8.EXE is the compiled (DOS) version of this program.
- ACQUIRE8.BAS implements an eight-channel oscilloscope/four-channel logic analyzer.

- LPT8FAST.BAS is the same as ACQUIRE8.BAS, but provisions have been made to allow acquired data to be recorded on disk. The compiled version (LPT8FAST.EXE) of this program is intended to be run from a bootable diskette in which CONFIG.SYS first initializes a RAM drive.

- ATOD_SL8.BAS is similar to LPTAN8.BAS, but acquisition is regulated through the TIMER command. Data frames are acquired at desired intervals in the range 1 to 86,400 seconds. ATOD_SL8.EXE is the compiled (DOS) version of this program.

- DTOA.BAS is a simple program that implements the serial protocol to write values to the D/A converters. DTOA.EXE is the compiled (DOS) version of this program.

- THERMOM.BAS is used to read temperature in °C and °F using a thermistor. THERMOM.EXE is the compiled (DOS) version of this program.

- SCALE.BAS implements an auto-zeroing digital scale by controlling the sensor's output offset through the Universal Sensor Interface's D/A. SCALE.EXE is the compiled (DOS) version of this program.

Folder: Spectral Analysis

- SPECTRUM.BAS is a QuickBASIC 4.5 program that estimates and displays the spectrum of a signal via the zero-padded FFT, an averaged-periodogram method (Welch's estimator), and a parametric estimator (Marple's autoregressive method). SPECTRUM.EXE is the compiled (DOS) version of this program.

- LPT8SPEC8.BAS is essentially the same as SPECTRUM.BAS but drives the Universal Sensor Interface to acquire evenly sampled data. LPT8SPEC.EXE is the compiled (DOS) version of this program.

- MARPLE.DAT is a short 64-point complex test data set used to evaluate different spectral analysis methods.

- NOISE.BAS is a QuickBASIC program to generate data useful for evaluating the resolving power of spectral estimators. NOISE.DAT is an ASCII data file generated through this program.

Folder: Redistributed Freeware

Signal Generators for the PC Sound Card

- BIP Electronics Labs Sine Wave Generator v3.0 (SINE30.ZIP, freeware for Windows 3.1, but works well in most cases under Windows 9x) by Marcel Veldhuijzen.

- Sweep Sine Wave Generator v2.0 (SWPGEN20.ZIP, freeware for Windows 9x) by David Taylor.

- PC function generator (PLAY.EXE, freeware for Windows 9x, 2000, NT) by the Physics Lab of Rutgers University.

- Sound interference (INTERFERENCE.EXE, freeware for Windows 9x, 2000, NT) by the Physics Lab of Rutgers University. This program generates sine waves through both channels of the sound card. The phase difference between the right and left channels can be set through software, allowing demonstrations of active noise control and classical wave interference.

- Beat-tone generation (BEATINBRAIN.EXE, freeware for Windows 9x, 2000, NT) by the Physics Lab of Rutgers University. This program generates two sine waves of different frequencies through the left and right sound card, allowing experiments in the generation and detection of beat tones.

Audio Oscilloscopes for the PC Sound Card

- BIP Electronics Labs Digital Scope v3.0 (SCOPE30.ZIP, freeware for Windows 3.1, but works well in most cases under Windows 9x) by Marcel Veldhuijzen.
- Oscilloscope for Windows v2.51 (OSC2511.ZIP, freeware for Windows 9x) by Konstantin Zeldovich.

Audio Spectrum Analyzers for the PC Sound Card

- Spectrogram v5.0.5, Dual Channel Audio Spectrum Analyzer (GRAM501.ZIP freeware for Windows 9x) by Richard Horne.
- Audio Wavelet Analyzer v1.0 (AUDIOWAVELETANALYZE.ZIP freeware for Windows 9x) by Christoph Lauer.
- 16- or 24-bit Sound Card Oscilloscope and FFT Analyzer (16BITFFTSCOPE.EXE and 24BITFFTSCOPE.EXE, freeware for Windows 9x, 2000, NT) by the Physics Lab of Rutgers University.

Transfer Function Analyzer for the PC Sound Card

- RightMark Audio Analyzer v2.5 (RMAA25.ZIP freeware, open-source code for Windows 9x, 2000 and NT) by Alexey Lukin and Max Liadov.
- WaveTools v1.0 Audio Analysis software (Signal Generator, Spectrum Analyzer, Oscilloscope and Audio Meter) (WAVETOOL.ZIP freeware for Windows 9x) by Paul Kellett.

Audio Frequency Counter for the PC Sound Card

- BIP Electronics Labs Digital Frequency Counter (COUNTER.ZIP, freeware for Windows 3.1, but works well in most cases under Windows 9x) by Marcel Veldhuijzen.

Software for Chapter 6: Signal Sources for Simulation, Testing, and Calibration

- REALECG.MAT is a Matlab file that contains 60 s of real ECG signal digitized at a rate of 5 kHz. The digitized data vector is named "ecg." The time vector associated with this ECG signal is named "time."

Folder: Arbitrary Waveform Generator

- MAT2ARB.M is a Matlab function that saves two vectors from the Matlab environment as a file that can be loaded into the two-channel Arbitrary Waveform Generator.
- Arb Loader subfolder contains the installation files for the program used to load waveforms into the two-channel Arbitrary Waveform Generator through the PC's parallel printer port.
- Arb Calibration Files subfolder contains signal/marker files that can be used to test the two-channel Arbitrary Waveform Generator.

Folder: Responsive Simulator Firmware

- This file contains the PIC firmware for the Responsive Cardiac Simulator (QUICK-VIEW.ZIP).

Folder: Redistributed Freeware

- SoundArb v1.02 (SASETUP.EXE freeware for Windows 9x, NT) by David Sherman Engineering Co.

Software for Chapter 7: Stimulation of Excitable Tissue

Folder: Hodgkin–Huxley

- HODGKINHUXLEY.M is a Matlab program that simulates the response of a giant squid axon to an electrical stimulus. ALPHA_H.M, ALPHA_M.M, ALPHA_ N.M, BETA_H.M, BETA_M.M and BETA_N.M are Matlab functions that calculate the rate constants for the Hodgkin–Huxley model.

Folder: Redistributed Freeware

- Finite Element Methods for Magnetics v3.1 (FEEM_SETUP.EXE freeware for Windows 9x) by David Meeker.
- WireTronic wire information software (WTSETUP.EXE freeware for Windows 9x) by WireTronic, Inc.
- VirtualCV v1.0 cyclic voltammetry simulator (VTLCV10.ZIP freeware for Windows 9x) by Andre Laouenan.
- Beat-tone generation (BEATINBRAIN.EXE, freeware for Windows 9x, 2000, NT) by the Physics Lab of Rutgers University. This program generates two sine waves of different frequencies through the left and right sound card. Can be used with isolated high-voltage output stages for interferential muscle stimulation.

Software for Chapter 8: Cardiac Pacemakers and Defibrillators

Folder: VVI

- VVI.ZIP implements a simple VVI state machine that disregards the VOO mode. This is the program that was discussed above.
- VVI WITH EMBEDDED VOO.ZIP implements a VVI state machine that can be placed in VOO mode by disregarding sensed events. The state machine is identical to the one implemented in VVI.ZIP, but the EnableSense routine enables or disables sensing depending on the desired state.
- VVI VOO WITH VOO STATE.ZIP is a VVI state machine in which the VOO state is implemented through a new state during which there is no sensing.

Folder: DOCD

- This file contains the PIC firmware for the DOCD Impedance Sensor (DOCD.ASM).

Folder: Shock-Box Firmware and Control Software

- This file contains the firmware for the defibrillation shock-box PICs and the VB Control software for the defibrillation shock box (SHOCKBOX.ZIP).

INDEX

AAMI, *see* Association for the Advancement of Medical Instrumentation
Ablation, *see* Electrosurgery
Action current, 305
Action potential, 305
   transmembrane recording of, 42
Aliasing, 76, 224, 236, 242
American National Standards Institute (ANSI), 98, 440
Amplifier, 1
   AC-coupled, 46, 49
   array, 8, 15, 22, 38, 123, 249
   blanking, 88, 93
   bootstrapped, 49
   buffer, 7, 38
   CMR of, 1, 2, 24, 25, 29, 33, 46
   CMRR of, 2, 26, 29, 49
   DC-coupled, 41, 44
   differential, 23
   drift of, 3
   follower, 7
   frequency response of, 1, 29, 32, 41, 108
   gain of, 1, 6, 17, 23, 25, 28, 39, 108
   ICIA, 29, 49
   input impedance of, 3, 8
   instrumentation, 27, 29, 33
   inverting, 6
   isolation, 87, 104, 106, 108, 428
   lock-in, 395
   noise of, 3, 18, 32
   operational, 6, 23, 27, 325
   recovery of, 3
   saturation, 3
   sense, *see* Pacemaker, sense amplifier
   sensor, 213

single-ended, 6
switched-capacitor, 33
synchronous, 395
wideband, 41
Analog-to-digital converter, 118, 206, 226, 417, 431
Aneurysm, *see* Embolic therapy
ANSI, *see* American National Standards Institute
Antenna
   applicator, 361
   biconical, 177
   dipole, 156
   loop, 156
   measurement, 154
   ridged-horn, 177
   TEM horn, 179
Association for the Advancement of Medical Instrumentation (AAMI), 5, 81
   load, 129, 131, 341
   standards, 5, 98, 290, 340, 394, 440
Atrio-ventricular node, 370, 371
Atrium, 369
Autoregressive model (AR), 239
Autoregressive moving average model (ARMA), 239
AV block, 371, 376

Bandwidth
   FM, 230, 274, 276
   telemetry, 230
   telephone, 230, 276
Barrier
   against fire, 98, 145
   dielectric, 99, 109, 139
   testing of, 139

Battery
   charger, 419
   implantable, 317, 359, 372, 386, 416
   powered equipment, 125, 141, 386, 417
Biocompatibility, 145, 328
Bioimpedance, *see* Sensor, impedance
Biopotential signals, 1
   frequency range of, 41
   model of, 3, 4
Bipole source, 293, 297
Blumlein generator, 179
Body potential driving, 18, 85
Bovie, *see* Electrosurgery
Bradycardia, 312, 371
Bundle of His, 370

Cancer treatment, 362, 363
Cardiac
   ablation, *see* Electrosurgery
   arrhythmia, 357, 369, 407
   conduction system, 369
   contractility modulation, 359
   defibrillation, *see* Defibrillator
   defibrillator, *see* Defibrillator
   electrophysiology, 42
   fibrillation, 102, 412, 437
   fibrillator, 437
   output, 369, 393
   pacemaker, *see* Pacemaker
   pacing, *see* Pacemaker
   stroke volume, 392
Cardiomyoplasty, 312
CENELEC, *see* Comité Européen de Normalisation Electrotechnique

Design and Development of Medical Electronic Instrumentation By David Prutchi and Michael Norris
ISBN 0-471-67623-3 Copyright © 2004 John Wiley & Sons, Inc.